Arterial Chemoreceptors

ADVANCES IN EXPERIMENTAL MEDICINE AND BIOLOGY

Editorial Board:

NATHAN BACK, *State University of New York at Buffalo*
IRUN R. COHEN, *The Weizmann Institute of Science*
ABEL LAJTHA, *N.S. Kline Institute for Psychiatric Research*
JOHN D. LAMBRIS, *University of Pennsylvania*
RODOLFO PAOLETTI, *University of Milan*

Recent Volumes in this Series

Volume 640
MULTICHAIN IMMUNE RECOGNITION RECEPTOR SIGNALING
Edited by Alexander Sigalov

Volume 641
CELLULAR OSCILLATORY MECHANISMS
Edited by Miguel Maroto and Nick Monk

Volume 642
THE SARCOMERE AND SKELETAL MUSCLE DISEASE
Edited by Nigel G. Laing

Volume 643
TAURINE 7
Edited by Junichi Azuma

Volume 644
TROPOMYOSIN
Edited by Peter Gunning

Volume 645
OXYGEN TRANSPORT TO TISSUE XXX
Edited by Per Liss, Peter Hansell, Duane F. Bruley, and David K. Harrison

Volume 646
EARLY NUTRITION PROGRAMMING AND HEALTH OUTCOMES IN LATER LIFE
Edited by Berthold Koletzko, Tamás Desci, Denes Molnár, and Anne De la Hunty

Volume 647
THERAPEUTIC TARGETS OF THE TNF SUPERFAMILY
Edited by Iqbal Grewal

A Continuation Order Plan is available for this series. A continuation order will bring delivery of each new volume immediately upon publication. Volumes are billed only upon actual shipment. For further information please contact the publisher.

Constancio Gonzalez · Colin A. Nurse · Chris Peers
Editors

Arterial Chemoreceptors

Springer

Editors

Prof. Constancio Gonzalez
Depto. Bioquímica y Biología
Molecular y Fisiología
Facultad de Medicina
c/Ramón y Cajal, n°7
47005 Valladolid
Spain

Prof. Colin A. Nurse
Department of Biology
McMaster University
1280 Main St. W
Hamilton ON L8S 4K1
Canada

Prof. Chris Peers
Division of Cardiovascular and Neuronal
Remodelling LIGHT
Faculty of Medicine and Health
Clarendon Way
Leeds LS29JT
United Kingdom

ISBN 978-90-481-2258-5   e-ISBN 978-90-481-2259-2

DOI 10.1007/978-90-481-2259-2

Library of Congress Control Number: 2009920977

© Springer Science+Business Media B.V. 2009
No part of this work may be reproduced, stored in a retrieval system, or transmitted
in any form or by any means, electronic, mechanical, photocopying, microfilming, recording
or otherwise, without written permission from the Publisher, with the exception
of any material supplied specifically for the purpose of being entered
and executed on a computer system, for exclusive use by the purchaser of the work.

Printed on acid-free paper

9 8 7 6 5 4 3 2 1

springer.com

Preface

This volume contains the Proceedings of the XVIIth ISAC Meeting held in Valladolid, Spain, July 1–5, 2008. As such, it contains the most permanent records of the combined efforts of all attendants. The meeting was held at the School of Medicine of Valladolid, that had the privilege of a recent celebration. The celebration was none other than its 600th anniversary, implying that all participants were surrounded by historical landmarks, from the historical building of the University, to the Museum of polychrome sculpture of Valladolid, to the Monastery of Clarisas in Tordesillas, to the beauty and charm of Salamanca. In this ambience we had three days of intense work, distributed in several oral sessions, preceded by plenary lectures given by our invited speakers who were kind enough to provide us with the latest progress in their specific fields. We also had time allotted to poster viewing. As regular attendants to the XVIIth ISAC Meeting, we want to express our appreciation for the valuable discussions surrounding each poster, the enthusiastic presentation of data, the comments of colleagues with suggestions for improvement, and the plans for collaborations that emerged from these discussions.

Needless to say that XVIIth ISAC Meeting was the fruit of many collaborative efforts. The Local Organizing Committee profited from the advice of several colleagues from around the world, namely, Prof. Chris Peers from Leeds, UK, Prof. Prem Kumar from Birmingham, UK, Prof. Nanduri Prabhakar from Chicago, USA, Prof. Colin Nurse from Hamilton, Canada, and Prof. Rodrigo Iturriaga from Santiago, Chile. They helped the local committee in selecting the most convenient dates for the meeting as well as in contributing to the initial program outline. We want to thank the invited speakers and authors of reviews because their contributions will no doubt enhance the impact of this volume.

Among the local organizing committee, there is one person to whom we, all ISAC members, are really indebted. We are referring to Dra. Asunción Rocher, Secretary and soul of XVIIth ISAC Meeting. She was prompt, efficient, and patient, and with the help of Ms. Josefina Revuelta, our Departmental Secretary, answered the hundreds of mails received before and after the meeting.

In the XVIIth ISAC Meeting all areas on Arterial Chemoreceptors were covered in the presentations and compiled in this volume. There were presentations on the structure and developmental aspects of the carotid body chemoreceptors, on the molecular biology and biophysical aspects of the ion channels expressed in chemoreceptor cells, on the neurotransmitters and their receptors expressed in the carotid body, on the central integration of the carotid body generated activity, and on systemic effects of chemoreceptor reflexes. Some important studies on central chemoreceptors, on neuroepithelial bodies and other lung receptors, on hypoxic pulmonary vasoconstriction and on oxygen sensing in endothelial cells widened the scope and enriched the meeting. Perhaps the areas generating the most enthusiastic discussions dealt with the mechanisms of chemoreception.

The Heymans-De Castro-Neil Awards for young investigators were selected by Profs. D. Riccardi, A. Obeso and P. Kumar, and the awardees were Nikol Piskuric (Hamilton, Ontario), Mark L. Dallas (Leeds, UK) and Olaia Colinas (Valladolid, Spain). ISAC wishes the recipients a fruitful development of their current research projects and successful scientific careers.

The XVIIth ISAC Meeting was supported by funds of the Spanish Ministery of Science and Innovation, by the Council of Education of the Government of Castile and León and by the University of Valladolid. We are grateful to them all. We are also grateful to Mr. Max Haring of Springer for his expert management of the production of this volume.

At the business meeting, it was decided that the next Symposium will be held at McMaster University, Hamilton, Ontario, Canada in 2011 with Prof. Colin Nurse as ISAC and Meeting President.

Valladolid, Spain	Constancio Gonzalez
Hamilton, Canada	Colin A. Nurse
Leeds, UK	Chris Peers

Contents

The Discovery of Sensory Nature of the Carotid Bodies – *Invited Article* .. 1
F. De Castro

Fifty Years of Progress in Carotid Body Physiology – *Invited Article* 19
R.S. Fitzgerald, C. Eyzaguirre and P. Zapata

Carotid Body: New Stimuli and New Preparations – *Invited Article* 29
Colin A. Nurse

Enzyme-Linked Acute Oxygen Sensing in Airway and Arterial
Chemoreceptors – *Invited Article* 39
Paul J. Kemp and C. Peers

Cysteine Residues in the C-terminal Tail of the Human $BK_{Ca}\alpha$ Subunit
Are Important for Channel Sensitivity to Carbon Monoxide 49
S.P. Brazier, V. Telezhkin, R. Mears, C.T. Müller, D. Riccardi and P.J. Kemp

Modulation of O_2 Sensitive K^+ Channels by AMP-activated Protein
Kinase 57
M.L. Dallas, J.L. Scragg, C.N. Wyatt, F. Ross, D.G. Hardie,
A.M. Evans and C. Peers

Hydrogen Sulfide Inhibits Human BK_{Ca} Channels 65
V. Telezhkin, S.P. Brazier, S. Cayzac, C.T. Müller, D. Riccardi and P.J. Kemp

DPPX Modifies TEA Sensitivity of the Kv4 Channels in Rabbit Carotid
Body Chemoreceptor Cells 73
O. Colinas, F.D. Pérez-Carretero, E. Alonso, J.R. López-López
and M.T. Pérez-García

Sustained Hypoxia Enhances TASK-like Current Inhibition by Acute Hypoxia in Rat Carotid Body Type-I Cells 83
F. Ortiz, R. Iturriaga and R. Varas

Inhibition of L-Type Ca^{2+} Channels by Carbon Monoxide 89
M.L. Dallas, J.L. Scragg and C. Peers

Effects of the Polyamine Spermine on Arterial Chemoreception 97
S. Cayzac, A. Rocher, A. Obeso, C. Gonzalez, P.J. Kemp and D. Riccardi

RT-PCR and Pharmacological Analysis of L-and T-Type Calcium Channels in Rat Carotid Body 105
A.I. Cáceres, E. Gonzalez-Obeso, C. Gonzalez and A. Rocher

Functional Characterization of Phosphodiesterases 4 in the Rat Carotid Body: Effect of Oxygen Concentrations 113
A.R. Nunes, J.R. Batuca and E.C. Monteiro

Calcium Sensitivity for Hypoxia in PGNs with PC-12 Cells in Co-Culture 121
G.P. Patel, S.M. Baby, A. Roy and S. Lahiri

Modification of Relative Gene Expression Ratio Obtained from Real Time qPCR with Whole Carotid Body by Using Mathematical Equations 125
J.H. Kim, I. Kim and J.L. Carroll

Neurotransmitters in Carotid Body Function: The Case of Dopamine – *Invited Article* 137
R. Iturriaga, J. Alcayaga and C. Gonzalez

Adenosine in Peripheral Chemoreception: New Insights into a Historically Overlooked Molecule – *Invited Article* 145
S.V. Conde, E.C. Monteiro, A. Obeso and C. Gonzalez

The A_{2B}-D_2 Receptor Interaction that Controls Carotid Body Catecholamines Release Locates Between the Last Two Steps of Hypoxic Transduction Cascade 161
S.V. Conde, A. Obeso, E.C. Monteiro and C. Gonzalez

Benzodiazepines and GABA-$GABA_A$ Receptor System in the Cat Carotid Body 169
A. Igarashi, N. Zadzilka and M. Shirahata

Contents

Evidence for Histamine as a New Modulator of Carotid Body Chemoreception 177
R. Del Rio, E.A. Moya, J. Alcayaga and R. Iturriaga

Fluoresceinated Peanut Agglutinin (PNA) is a Marker for Live O_2 Sensing Glomus Cells in Rat Carotid Body 185
I. Kim, D.J. Yang, D.F. Donnelly and J.L. Carroll

Neuroglobin in Aging Carotid Bodies 191
V. Verratti, C. Di Giulio, G. Bianchi, M. Cacchio, G. Petruccelli, C. Di Giulio, L. Artese, S. Lahiri and R. Iturriaga

Oxygen Sensing and the Activation of the Hypoxia Inducible Factor 1 (HIF-1) – *Invited Article* 197
Joachim Fandrey and Max Gassmann

Upregulation of Erythropoietin and its Receptor Expression in the Rat Carotid Body During Chronic and Intermittent Hypoxia 207
S.Y. Lam, G.L. Tipoe and M.L. Fung

Iron Chelation and the Ventilatory Response to Hypoxia 215
Mieczyslaw Pokorski, Justyna Antosiewicz, Camillo Di Giulio and Sukhamay Lahiri

Systemic Effects Resulting from Carotid Body Stimulation – *Invited Article* 223
Prem Kumar

Bicarbonate-Regulated Soluble Adenylyl Cyclase (sAC) mRNA Expression and Activity in Peripheral Chemoreceptors 235
A.R. Nunes, E.C. Monteiro, S.M. Johnson and E.B. Gauda

Developmental Maturation of Chemosensitivity to Hypoxia of Peripheral Arterial Chemoreceptors – *Invited Article* 243
Estelle B. Gauda, John L. Carroll and David F. Donnelly

Physiological Carotid Body Denervation During Aging 257
C. Di Giulio, J. Antosiewicz, M. Walski, G. Petruccelli, V. Verratti, G. Bianchi and M. Pokorski

Does Ageing Modify Ventilatory Responses to Dopamine in Anaesthetised Rats Breathing Spontaneously? 265
T.C. Monteiro, A. Obeso, C. Gonzalez and E.C. Monteiro

The Role of the Carotid Bodies in the Counter-Regulatory Response to Hypoglycemia 273
Denham S. Ward, William A. Voter and Suzanne Karan

The Respiratory Responses to the Combined Activation of the Muscle Metaboreflex and the Ventilatory Chemoreflex 281
C.K. Lykidis, P. Kumar and G.M. Balanos

Cardiovascular Responses to Hyperoxic Withdrawal of Arterial Chemosensory Drive 289
Patricio Zapata, Carolina Larrain, Marco-Antonio Rivera and Christian Calderon

Time-Dependence of Hyperoxia-Induced Impairment in Peripheral Chemoreceptor Activity and Glomus Cell Calcium Response 299
J.L. Carroll, I. Kim, H. Dbouk, D.J. Yang, R.W. Bavis and D.F. Donnelly

Long-Term Regulation of Carotid Body Function: Acclimatization and Adaptation – *Invited Article* 307
N.R. Prabhakar, Y.-J. Peng, G.K. Kumar, J. Nanduri, C. Di Giulio and Sukhamay Lahiri

Effects of Intermittent Hypoxia on Blood Gases Plasma Catecholamine and Blood Pressure 319
M.C. Gonzalez-Martin, V. Vega-Agapito, J. Prieto-Lloret, M.T. Agapito, J. Castañeda and C. Gonzalez

Cardioventilatory Acclimatization Induced by Chronic Intermittent Hypoxia 329
R. Iturriaga, S. Rey, R. Del Rio, E.A. Moya and J. Alcayaga

Ventilatory Drive Is Enhanced in Male and Female Rats Following Chronic Intermittent Hypoxia 337
D. Edge, J.R. Skelly, A. Bradford and K.D. O'Halloran

Contrasting Effects of Intermittent and Continuous Hypoxia on Low O_2 Evoked Catecholamine Secretion from Neonatal Rat Chromaffin Cells ... 345
Dangjai Souvannakitti, Ganesh K. Kumar, Aaron Fox and Nanduri R. Prabhakar

Hypoxic Pulmonary Vasoconstriction – *Invited Article* 351
A. Mark Evans and Jeremy P.T. Ward

Contents

Impact of Modulators of Mitochondrial ATP-Sensitive Potassium Channel (mitoK$_{ATP}$) on Hypoxic Pulmonary Vasoconstriction 361
R. Paddenberg, P. Faulhammer, A. Goldenberg, B. Gries, J. Heinl and W. Kummer

Oxygen Sensing in the Brain – *Invited Article* 369
Frank L. Powell, Cindy B. Kim, Randall S. Johnson and Zhenxing Fu

The Central Respiratory Chemoreceptor: Where Is It Located? – *Invited Article* 377
Y. Okada, S. Kuwana, Z. Chen, M. Ishiguro and Y. Oku

Anatomical Architecture and Responses to Acidosis of a Novel Respiratory Neuron Group in the High Cervical Spinal Cord (HCRG) of the Neonatal Rat 387
Y. Okada, S. Yokota, Y. Shinozaki, R. Aoyama, Y. Yasui, M. Ishiguro and Y. Oku

Systemic Inhibition of the Na$^+$/H$^+$ Exchanger Type 3 in Intact Rats Activates Brainstem Respiratory Regions 395
R. Pasaro, J.L. Ribas-Salgueiro, E.R. Matarredona, M. Sarmiento and J. Ribas

Nitric Oxide in the Solitary Tract Nucleus (STn) Modulates Glucose Homeostasis and FOS-ir Expression After Carotid Chemoreceptor Stimulation 403
M. Lemus, S. Montero, S. Luquín, J. García and E. Roces De Álvarez-Buylla

Airway Receptors and Their Reflex Function – *Invited Article* 411
J. Yu

Airway Chemosensitive Receptors in Vagus Nerve Perform Neuro-Immune Interaction for Lung-Brain Communication 421
H.F. Li and J. Yu

The Role of NOX2 and "Novel Oxidases" in Airway Chemoreceptor O$_2$ Sensing 427
Ernest Cutz, Jie Pan and Herman Yeger

Recruitment of GABA$_A$ Receptors in Chemoreceptor Pulmonary Neuroepithelial Bodies by Prenatal Nicotine Exposure in Monkey Lung .. 439
X.W. Fu and E.R. Spindel

Concluding Remarks 447
Colin A. Nurse

Index 451

Carlos Eyzaguirre Edwards † (1923–2009)

Lecturing at the reception of his title of *Doctor Honoris Causa* by the University of Barcelona (Spain) 22-02-2001

Prof. Carlos Eyzaguirre[†]

Born in 1923, and passed away on February 2, 2009, in Santiago, Chile.

His medical studies were initiated at the Universidad Católica de Chile and continued at the Universidad de Chile, where he was granted the MD degree in 1947. His post-doctoral training was performed in the United States of America, first as research resident in the Department of Medicine, Johns Hopkins University, and then as fellow of the Guggenheim Foundation at the Wilmer Institute of the same institution, where he collaborated with the famous Professor Stephen Kuffler, performing the first intracellular recordings on sensory mechanoreceptors, a research mentioned up today as classic in most Physiology textbooks.

He is mostly known by his prolonged and successful research career at the Department of Physiology, University of Utah Medical School (USA), initiated as Assistant Professor in 1957, then Associate Professor, full Professor (1962), Chairman of the Department (1965), and finally Emeritus Professor (2005). Working there, in 1961, he excised the carotid body of the cat for its superfusion in vitro with oxygenated saline solution, recording the electrical responses of the carotid (sinus) nerve to isolated and combined chemical stimuli applied to the receptor organ. This was an extraordinary achievement, since the carotid body is a tangle of blood vessels, with extremely high oxygen demand. This preparation, by allowing the study of chemoreceptor activity for many hours in vitro, was pivotal for modern studies on respiratory regulation.

His scientific work attracted a large number of young researchers to pursue their postdoctoral training under his guidance (5 from North America, 6 from South America, 6 from Europe and 16 from Asia), as well as many visiting professors working as his research associates. All his disciples and associates will remember him always as an intelligent mentor, a tireless and meticulous researcher, a precise, concise and elegant writer of scientific papers, a fair and cheerful gentleman, and a great friend.

He was granted three doctorates *honoris causa*, by the Pontificia Universidad Católica de Chile (which he considered as *his alma mater*), the Universidad Complutense de Madrid and the Universitat de Barcelona. He received the medals Claude Bernard from France and Ramón y Cajal from Spain.

Prof. Eyzaguirre was part of the group of physiologists that decided the organization of the International Society for Arterial Chemoreceptors (ISAC), of which he was his first President, in charge of organizing the ISAC Symposium held in Park City (1990). He was an enthusiastic participant in all symposia on arterial chemoreceptors, except the last two ISAC meetings (2005 and 2008).

He was author of nearly 150 full papers published in international scientific journals, including some extensive reviews that are widely cited. His scientific career was initiated with the publication of a paper when he was 24 years old, and is prolonged after his death, since he is one of the authors of a paper to be published within this volume.

Patricio Zapata
Santiago, Chile

Contributors

M.T. Agapito Departmento de Bioquímica y Biología Molecular y Fisiología. Facultad de Medicina. Universidad de Valladolid, IBGM/CSIC. C/. Ramon y Cajal, 47005 Valladolid, Spain, teresaa@bio.uva.es

J. Alcayaga Laboratoria de Fisiología Celular, Facultad de Ciencias, Universidad de Chile, Santiago, Chile, jalcayag@uchile.cl

E. Alonso Instituto de Biología y Genética Molecular (IBGM), Universidad de Valladolid, C/Sanz y Forés s/n, 47003 Valladolid, espe@ibgm.uva.es

J. Antosiewicz Medical Research Centre, Polish Academy of Sciences, 02-106 Warsaw, Poland, jantosiewicz@cmdik.pan.pl

R. Aoyama Department of Orthopaedic Surgery, Keio University School of Medicine, Tokyo, Japan, ryoma@asahi-net.email.ne.jp

L. Artese Medical and Dental Schools, University of Chieti-Pescara, 66100 Chieti, Italy, artese@unich.it

S.M. Baby Department of Physiology, University of Pennsylvania Medical Center, Philadelphia, PA, 19104, USA

G.M. Balanos School of Sport and Exercise Sciences, University of Birmingham, Edgbaston, Birmingham, B15 2TT, UK

J.R. Batuca Department of Pharmacology, Faculty of Medical Sciences, New University of Lisbon, Campo Mártires da Pátria, 130, Lisbon 1169-056, Portugal, joana.batuca@fcm.unl.pt

R.W. Bavis Department of Biology, Bates College, Lewiston, Maine, USA, rbavis@bates.edu

G. Bianchi Department of Basic and Applied Medical Sciences. University of Chieti-Pescara, 66100 Chieti, Italy, gbianchi@unich.it

A. Bradford Department of Physiology, Royal College of Surgeons in Ireland, St Stephen's Green, Dublin 2, Ireland, abradfor@rcsi.ie

S.P. Brazier School of Bioscience, Museum Avenue, Cardiff University, Cardiff CF11 9BX, UK, braziersp@cardiff.ac.uk

M. Cacchio Department of Basic and Applied Medical Sciences. University of Chieti-Pescara, 66100 Chieti, Italy, cacchio@unich.it

A.I. Cáceres Yale University School of Medicine Department of Pharmacology, New Haven, CT 06520-8066, USA, ana.caceres@yale.edu

Christian Calderón Laboratorio de Neurobiología, Pontificia Universidad Católica de Chile, Santiago, Chile

John L. Carroll Department of Pediatrics, University of Arkansas for Medical Sciences. Little Rock, Arkansas, USA, carrolljohnl@uams.edu

J. Castañeda Departmento de Bioquímica y Biología Molecular y Fisiología, Facultad de Medicina, Universidad de Valladolid, IBGM/CSIC. C/. Ramon y Cajal, 47005 Valladolid, Spain, fcastaneda@hcuv.sacyl.es

S. Cayzac Cardiff University, Museum Avenue, Cardiff CF10 3US, UK, cayzacs@cardiff.ac.uk

Z. Chen Department of Biochemical and Analytical Pharmacology, GlaxoSmithKline, Research Triangle Park, NC USA, zibin8@gmail.com

O. Colinas Instituto de Biología y Genética Molecular (IBGM), Universidad de Valladolid, C/ Sanz y Forés s/n, 47003 Valladolid, Spain, olaiacm@ibgm.uva.es

Silvia V. Conde Department of Pharmacology, Faculty of Medical Sciences, New University of LIsbon, Campo Mártires da Pátria, 130, 1169-056 Lisboa, Portugal, silvia.conde@fcm.unl.pt

Ernest Cutz The Hospital for Sick Children, 555 University Avenue, Toronto, ON, Canada, M5G1X8 ernest.cutz@sickkids.ca

Mark L Dallas Division of Cardiovascular and Neuronal Remodelling, Leeds Institute of Genetics, Health & Therapeutics University of Leeds, Leeds LS2 9JT, UK, m.l.dallas@leeds.ac.uk

H. Dbouk Department of Pediatrics, University of Arkansas for Medical Sciences, Little Rock, Arkansas, USA, hassan_dbouk@ouhsc.edu

F. De Castro Grupo de Neurobiología del Desarrollo-GNDe, Hospital Nacional de Parapléjicos, Finca "La Perelada" s/n, 45071 Toledo, Spain, fdec@sescam.jccm.es

R. Del Rio Laboratoria de Neurobiología, Facultad de Ciencias Biológicas, P. Universidad Católica de Chile, Santiago, Chile, radelrio@uc.cl

Camilo Di Giulio Department of Basic and Applied Medical Sciences, University of Chieti-Pescara, 66100 Italy, concettadigiulio@yahoo.it, digiulio@unich.it

David F. Donnelly Division of Respiratory Medicine, Department of Pediatrics, Yale University School of Medicine, New Haven, Connecticut 06520, USA, David.Donnelly@Yale.edu

D. Edge UCD School of Medicine and Medical Science, University College Dublin, Dublin 4, Ireland

A. Mark Evans Centre for Integrative Physiology, College of Medicine and Veterinary Medicine, Hugh Robson Building University of Edinburgh, Edinburgh EH8 9XD, UK, Mark.Evans@ed.ac.uk

Carlos Eyzaguirre† Facultad de Medicina, Clínica Alemana – Universidad del Desarrollo, Av. Las Condes 12438, Lo Barnechea, Santiago, Chile

Joachim Fandrey Institut fuer Physiologie, Universitaet Duisburg-Essen, Hufelandstrasse 55, D-45147 Essen, Germany, joachim.fandrey@uni-due.de

P. Faulhammer Institute of Anatomy and Cell Biology, Justus-Liebig-University, ECCPS, Giessen, Germany

Robert S. Fitzgerald Department of Environmental Health Sciences, Division of Physiology, Baltimore, Maryland 21205, USA, rfitzger@jhsph.edu

Aaron Fox The University of Chicago, Department of Neurobiology, Pharmacology and Physiology, Chicago, IL 60637, USA, aaronfox@uchicago.edu

X.W. Fu Health & Science University, Division of Neuroscience, Oregon National Primate Research Center, Oregon, Beaverton, OR, USA, fuxiao@ohsu.edu

Zhenxing Fu Department of Medicine, University of California, San Diego, CA 92093-0623. USA, zfu@ucsd.edu

M.L. Fung Department of Physiology, The University of Hong Kong, Pokfulam, Hong Kong, China, fuxiao@ohsu.edu

J. García Facultad de Medicina, Universidad de Guadalajara, Guadalajara, México

Max Gassmann Institute of Veterinary Physiology, Vetsuisse Faculty and Zurich Center for Integrative Human Physiology (ZIHP), University of Zürich, CH-8057 Zürich, Switzerland, maxg@access.uzh.ch

Estelle B. Gauda Johns Hopkins Medical Institutions, Department of Pediatrics, Division of Neonatology, Baltimore, MD 21287, USA, egauda@mail.jhmi.edu

A. Goldenberg Institute of Anatomy and Cell Biology, Justus-Liebig-University, ECCPS, Giessen, Germany

Constancio Gonzalez Departmento de Bioquímica y Biología Molecular y Fisiología. Facultad de Medicina. Universidad de Valladolid, IBGM/CSIC. C/. Ramon y Cajal, 47005 Valladolid, Spain, constanc@ibgm.uva.es

M.C. González-Martín Departmento de Bioquímica y Biología Molecular y Fisiología. Facultad de Medicina. Universidad de Valladolid, IBGM/CSIC. C/. Ramon y Cajal, 47005 Valladolid, Spain, mameng13@ibgm.uva.es

E. González-Obeso Departmento de Bioquímica y Biología Molecular y Fisiología. Facultad de Medicina. Universidad de Valladolid, IBGM/CSIC. C/. Ramon y Cajal, 47005 Valladolid, Spain, elvirago_2000@yahoo.com

B. Gries Institute of Anatomy and Cell Biology, Justus-Liebig-University, ECCPS, Giessen, Germany

D.G. Hardie Division of Molecular Physiology, College of Life Science, University of Dundee, Dundee DD1 5EH, UK, d.g.hardie@dundee.ac.uk

J. Heinl Institute of Anatomy and Cell Biology, Justus-Liebig-University, ECCPS, Giessen, Germany

A. Igarashi Department of Environmental Health Sciences, Division of Physiology, The Johns Hopkins University, Baltimore, MD, USA; Department of Anesthesiology, Yamagata Prefectural Shinjo Hospital, Shinjo, Yamagata, Japan, igarashi-ayu@umin.ac.jp

Makio Ishiguro The Institute of Statistical Mathematics, Research Organization of Information and Systems, Tokyo, Japan, ishiguro@ism.ac.jp

Rodrigo Iturriaga Laboratory of Neurobiology, Faculty of Biological Sciences, Pontificia Universidad Católica de Chile, Casilla 114-D, Portugal 49, Santiago, Chile, riturriaga@bio.puc.cl

Randall S. Johnson Division of Biological Sciences, University of California, San Diego, CA 92093-0377, USA, rjohnson@biomail.ucsd.edu

S.M. Johnson Physiology and Biophysics, Howard University College of Medicine, Washington, DC, USA, smjohnson@howard.edu

Suzanne Karan Department of Anesthesiology, University of Rochester School of Medicine and Dentistry, Rochester, NY, 14642 USA, suzanne_karan@urmc.rochester.edu

Paul J. Kemp School of Biosciences, Cardiff University, Museum Avenue, Cardiff CF10 3US, UK, kemp@cardiff.ac.uk

Cindy B. Kim Department of Medicine, University of California, San Diego, CA 92093-0623, USA, um06ck@leeds.ac.uk

I. Kim Department of Pediatrics, University of Arkansas for Medical Sciences, Little Rock, AR, USA, kiminsook@uams.ed

J.H. Kim University of Arkansas at Little Rock, Little Rock, AR 72204 USA, jhkim@ualr.edu

Ganesh Kumar Department of Medicine, Center for System Biology of Oxygen Sensing, Chicago, IL 60637, USA, gkumar@medicine.bsd.uchicago.edu

Contributors

Prem Kumar School of Experimental Medicine, College of Medical and Dental Sciences, University of Birmingham, B15 2TT, UK, p.kumar@bham.ac.uk

W. Kummer Institute for Anatomy and Cell Biology, Justus-Liebig-University, Aulweg 123, 35385 Giessen, Germany, wolfgang.kummer@anatomic.med.uni-giessen.de

Shun-ichi Kuwana Faculty of Health Sciences, Uekusa Gakuen University, Chiba, Japan, s-kuwana@uekusa.ac.jp

S. Lahiri Department of Physiology, University of Pennsylvania, Philadelphia, PA 19104, USA, lahiri@mail.med.upenn.edu

S.Y. Lam Department of Physiology, The University of Hong Kong, Pokfulam, Hong Kong, Japan, ssylam@hkusua.hku.hk

Carolina Larrain Facultad de Medicina, Clínica Alemana, Universidad del Desarrollo, Santiago, Chile, clarrain@udd.cl

M. Lemus Centro Universitario de Investigaciones Biomédicas, Universidad de Colima, Colima, Col. 28045, Mexico

Huafeng Li Department of Pulmonary Medicine, University of Louisville, Louisville, KY 40292 USA, h0li0013@louisville.edu

José Ramón López-López Instituto de Biología y Genética Molecular (IBGM), C/Sanz y Forés s/n, 47003 Valladolid, jrlopez@ibgm.uva.es

S. Luquín Centro de Investigaciones de Occidente, IMSS Guadalajara, Mexico

C.K. Lykidis School of Sport and Exercise Sciences, University of Birmingham, Edgbaston, Birmingham, B15 2TT, UK, CXL584@bham.ac.uk

E.R. Matarredona Departamento de Fisiología y Zoología, Facultad de Biología, Avda. Reina Mercedes 6, 41012-Sevilla, Spain, matarredona@us.es

R. Mears School of Bioscience, Museum Avenue, Cardiff University, Cardiff CF11 9BX, UK

E.C. Monteiro Department of Pharmacology, Faculty of Medical Sciences, New University of LIsbon, Campo Mártires da Pátria, 130, 1169-056 Lisboa, Portugal, emilia.monteiro@fcm.unl.pt

T.C. Monteiro Departamento de Farmacologia, Faculdade de Ciências Médicas, Universidade Nova de Lisboa, Campo Mártires da Pátria, 130, 1169-056 Lisboa, Portugal, teresa.monteiro@fcm.unl.pt

S. Montero Centro Universitario de Investigaciones Biomédicas, Universidad de Colima, Colima, Col. 28045, Mexico

Esteban A. Moya Laboratoria de Neurobiología, Facultad de Ciencias Biológicas, P. Universidad Católica de Chile, Santiago, Chile, elmoya@uc.cl

C.T. Müller School of Bioscience, Museum Avenue, Cardiff University, Cardiff CF11 9BX, UK

J. Nanduri Department of Medicine, Center for Systems Biology of O_2 sensing, University of Chicago, IL, USA, nanduri@uchicago.edu

Rita Nunes Department of Pharmacology, Faculty of Medical Sciences, New University of Lisbon, Campo Mártires da Pátria, 130, Lisbon 1169-056, Portugal, rita.nunes@fcm.unl.pt

Colin A. Nurse Department of Biology. McMaster University, ON, Canada L8S 4K1, nursec@mcmaster.ca

Ana Obeso Department of Biochemistry, Molecular biology and Physiology, Faculty of Medicine, University of Valladolid, IBGM/CSIC, Calle Ramón y Cajal, 47005 Valladolid, Spain, aobeso@ibgm.uva.es

K. ÒHalloran UCD School of Medicine and Medical Science, University College Dublin, Dublin 4, Ireland, ken.ohalloran@ucd.ie

Y. Okada Department of Medicine, Keio University Tsukigase Rehabilitation Center, Izu, Japan, yasumasaokada@1979.jukuin.keio.ac.jp

Y. Oku Department of Physiology, Hyogo College of Medicine, Nishinomiya, Japan, yoku@hyo-med.ac.jp

F. Ortiz Laboratory of Neurobiology, Faculty of Biological Sciences, Pontificia Universidad Católica de Chile, Casilla 114-D, Portugal 49, Santiago, Chile, induchuna@yahoo.es

R. Paddenberg Institute of Anatomy and Cell Biology, Justus-Liebig-University, ECCPS, Giessen, Germany, Renate.Paddenberg@anatomie.med.uni.giessen.de

Jie Pan The Hospital for Sick Children, 555 University Avenue, Toronto, ON, Canada, M5G1X8. jiepan627@gmail.com

Rosario Pásaro Departamento de Fisiología y Zoología, Facultad de Biología, Avda. Reina Mercedes 6, 41012-Sevilla, Spain, mrpasaro@us.es

G.P. Patel Department of Physiology, University of Pennsylvania Medical Center, Philadelphia, PA, 19104, USA

Chris Peers Division of Cardiovascular and Neuronal Remodelling, Leeds Institute of Genetics, Health & Therapeutics University of Leeds, Leeds LS2 9JT, UK, c.s.peers@leeds.ac.uk

Y.-J. Peng Department of Medicine, Center for Systems Biology of O_2 sensing, University of Chicago, IL, USA

F. D. Pérez-Carretero Instituto de Biología y Genética Molecular (IBGM), C/Sanz y Forés s/n, 47003 Valladolid, Spain, fperezca@cofis.es

Contributors

M.T. Pérez-García Instituto de Biología y Genética Molecular (IBGM), C/Sanz y Forés s/n, 47003 Valladolid, Spain, tperez@ibgm.uva.es

G. Petruccelli Department of Basic and Applied Medical Sciences, University of Chieti-Pescara, 66100 Italy, g.petruccelli@unich.it

M. Pokorski Polish Academy of Sciences, 02 106 Warsaw, Poland, mpokorski@cmdik.pan.pl

Frank L. Powell Department of Medicine and White Mountain Research Station, University of California, San Diego, La Jolla, CA 92093, USA, fpowell@ucsd.edu

Nanduri R Prabhakar Department of Medicine, Center for System Biology of O_2 Sensing, Chicago, IL 60637 USA, nprabhak@medicine.bsd.uchicago.edu

J. Prieto-Lloret King's College London, Division of Asthma, Allergy and Lung Biology, London SE1 9RT, UK, jplloret@hotmail.com

S. Rey Laboratoria de Neurobiología, Facultad de Ciencias Biológicas, P. Universidad Católica de Chile, Santiago, Chile, srkeim@med.puc.cl

J. Ribas Departamento de Fisiología Médica y Biofísica, Facultad de Medicina, c/ Dr. Fedriani s.n., 41009-Sevilla, Spain, jribas@us.es

D. Riccardi School of Bioscience, Museum Avenue, Cardiff University, Cardiff CF11 9BX, UK, riccardi@cardiff.ac.uk

Marco-Antonio Rivera Laboratorio de Neurobiología, Pontificia Universidad Católica de Chile, Santiago, Chile

E. Roces De Álvarez-Buylla Universidad de Colima Av. 25 de Julio S/N, Col. villas de San Sebastián, Colima, Col. 28045 Mexico, rab@ucol.mx

Asun Rocher Departmento de Bioquímica y Biología Molecular y Fisiología, Facultad de Medicina, Universidad de Valladolid, IBGM/CSIC. C/. Ramon y Cajal, 47005 Valladolid, Spain, rocher@ibgm.uva.es

F. Ross Division of Molecular Physiology, College of Life Science, University of Dundee, Dundee DD1 5EH, UK

A. Roy Department of Physiology & Biophysics, University of Calgary, Calgary, Alberta, T2N4L3, Canada

Juan Luis Ribas Salgueiro Departamento de Fisiología y Zoología, Facultad de Biología, Avda. Reina Mercedes 6, 41012-Sevilla, Spain, jlribas@us.es

M. Sarmiento Departamento de Bioquímica y Medicina Legal, Facultad de Farmacia, Sor Gregoria de Santa Teresa, s/n., 41012-Sevilla, Spain, msarmiento@us.es

Jason L. Scragg Division of Cardiovascular and Neuronal Remodelling, Leeds Institute of Genetics, Health & Therapeutics University of Leeds, Leeds LS2 9JT, UK, J.scragg@leeds.ac.uk

Yoshio Shinozaki Department of Orthopaedic Surgery, Keio University School of Medicine, Tokyo, Japan, y-shino@themis.ocn.ne.jp

M. Shirahata Bloomberg School of Public Health, Johns Hopkins University, Baltimore, MD 21218 USA, mshiraha@jhsph.edu

J.R. Skelly UCD School of Medicine and Medical Science, University College, Dublin, Dublin 4, Ireland

Dangjai Souvannakitti Department of Medicine, Center for System Biology of O_2 Sensing, Chicago, IL 60637, USA, dangjai@uchicago.edu

E.R. Spindel Division of Neuroscience, Oregon National Primate Research Center, Oregon Health & Science University, Beaverton, OR, USA, Spindele@ohsu.edu

V. Telezhkin School of Bioscience, Museum Avenue, Cardiff University, Cardiff CF10 3AX, UK, telezhkinv@cardiff.ac.uk

G. L. Tipoe Department of Anatomy, The University of Hong Kong, Pokfulam, Hong Kong, UK, tgeorge@hkucc.hku.hk

R. Varas Laboratory of Neurobiology, Faculty of Biological Sciences, Pontificia Universidad Católica de Chile, Casilla 114-D, Portugal 49, Santiago, Chile, rvaras@bio.puc.cl

V. Vega-Agapito Departmento de Bioquímica y Biología Molecular y Fisiología, Facultad de Medicina, Universidad de Valladolid, IBGM/CSIC. C/. Ramon y Cajal, 47005 Valladolid, Spain, toya@ibgm.uva.es

V. Verratti Department of Basic and Applied Medical Sciences, University of Chieti-Pescara, 66100 Italy, vittorelibero@hotmail.it

William A. Voter Department of Anesthesiology, University of Rochester School of Medicine and Dentistry, Rochester, NY, 14642 USA

M. Walski Medical Research Centre, Polish Academy of Sciences, 02 106 Warsaw, Poland, walski@cmdik.pan.pl

Denham S. Ward Department of Anesthesiology, University of Rochester School of Medicine and Dentistry, Rochester, NY, 14642 USA, Denham_Ward@URMC.Rochester.edu

Jeremy P.T. Ward Division of Asthma, Allergy and Lung Biology, King's College London, London SE1 9RT, UK, Jeremy.ward@kcl.ac.uk

C.N. Wyatt Department of Neuroscience, Cell Biology and Physiology, Wright State University, Dayton, OH 45435, USA, christopher.wyatt@wright.edu

D.J. Yang Department of Pediatrics, University of Arkansas for Medical Sciences. Little Rock, Arkansas, USA, yangdong@uams.edu

Yukihiko Yasui Department of Anatomy and Morphological Neuroscience, Shimane University School of Medicine, Izumo, Japan, yyasui@shimane-med.ac.jp

Contributors

Herman Yeger The Hospital for Sick Children, 555 University Avenue, Toronto, ON, Canada, M5G1X8, herman.yeger@sickkids.ca

J. Yu Department of Pulmonary Medicine, University of Louisville, Louisville, KY 40292 USA; Department of Physiology and Pathophysiology, Shanghai Medical College, Fudan University, Shanghai, China 200032, j0yu0001@louisville.edu

N. Zadzilka Division of Physiology, Department of Environmental Health Sciences, The Johns Hopkins University, Baltimore MD, USA, nicole.zadzilka@utoledo.edu

Patricio Zapata Facultad de Medicina, Clínica Alemana – Univrsidad del Desarrollo, Av. Las Condes 12438, Lo Barnechea, Santiago, Chile, pzapata@udd.cl

The Discovery of Sensory Nature of the Carotid Bodies – *Invited Article*

F. De Castro

Abstract Although the carotid body (or *glomus caroticum*) was a structure familiar to anatomists in the XVIIIth century, it was not until the beginning of the XXth century that its role was revealed. It was then that the German physiologist Heinrich Hering described the respiratory reflex and he began to study the anatomical basis of this reflex focusing on the carotid region, and the carotid sinus in particular. At this time, the physiologists and pharmacologists associated with Jean-François Heymans and his son (Corneille) in Ghent (Belgium) adopted a different approach to resolve this issue, and they centred their efforts on the cardio-aortic reflexogenic region. However, at the *Laboratorio de Investigaciones Biológicas* (Madrid, Spain), one of the youngest and more brilliant disciples of Santiago Ramón y Cajal, Fernando De Castro, took advantage of certain technical advances to study the fine structure of the carotid body (De Castro, 1925). In successive papers (1926, 1928, 1929), De Castro unravelled most of the histological secrets of this small structure and described the exact localisation of the "chemoreceptors" within the *glomus*. Indeed, his was the first description of cells specifically devoted to detect changes in the chemical composition of blood. Heymans was deeply interested in the work of De Castro, and he extended two invitations to the Spanish neurologist to visit Ghent (1929 and 1932) so that they could share their experiences. From 1932–1933, Corneille Heymans focused his attention on the carotid body and his physiological demonstration of De Castro's hypothesis regarding chemoreceptors led to him obtaining the *Nobel Prize in Physiology or Medicine* in 1938, while Spain was immersed in its catastrophic Civil War.

Keywords Carotid body · Glomus caroticum · De Castro · Heymans · Sensory organ · Intracarotid gland · Microganglia

---

F. De Castro (✉)
Grupo de Neurobiología del Desarrollo-GNDe, Hospital Nacional de Parapléjicos,
Finca "La Peraleda" s/n, 45071 Toledo, Spain
e-mail: fdec@sescam.jccm.es

1 Introduction

The carotid body (also known as the carotid corpuscule, carotid ganglion, carotid gland and *glomus caroticum*[1]) is an anatomical structure situated at the bifurcation of the internal and external carotid from the primitive carotid artery. This structure became known in the XVIIIth century, although its function remained elusive to scientists for centuries. In the mid XIXth century, the experiments performed by the Weber brothers and Henle to modulate cardiac frequency by electrical stimulation of the vagus nerve (even stopping the heart) can be considered to have set the first stone in a new line of research in the cardio-aortic-carotid region. The studies that were carried out in the following decades explored the cardio-respiratory reflexes and the physiology of this system in general. The work published by Marey in 1859 was also particularly relevant, demonstrating the direct and opposing relationship between arterial pressure and cardiac frequency: if one increases, the other falls, and vice versa. Although Cyon and Ludwig (1866) demonstrated the implication of the vagal nerves in the origin of bradycardia and hypotension, the studies of François-Franck (1876–1878) situated the origin of this reflex in the brainstem (that is, in the central nervous system -CNS-).

However, Pagano and Siciliano proposed in 1900 that the cardio-respiratory reactions were reflexes originated in the carotid region independent of the CNS. Indeed, for more than fifty years, this contradiction with François-Franck's postulates was ignored by the scientists studying this problem. However, finally it was Heinrich Hering and his collaborators who elegantly demonstrated that either the electrical or mechanical stimulus of the carotid sinus (a dilatation of the bifurcation of both carotid arteries from the primitive one) triggers a reflex mechanism that results in bradycardia and arterial hypotension, baptised as the "sinus reflex". Indeed, this physiologist from Köln (Germany) also discovered that this region was innervated by a branch of the glossopharyngeal nerve, named the "sinus nerve" or "Hering's nerve" (1923–1927).

In parallel, Jean-François Heymans and his son Corneille, and their disciples,[2] applied the famous technique of two living dogs in parabiosis to reach the conclusion that hypertensive bradycardia is a reflex mechanism, independent of the central structures (Heymans and Ladon, 1925). They stated that the "main regulation of the respiration depends on the cardio-aortic region, and it is conditioned by the pressure and composition of circulating blood"[3] (Heymans and Heymans, 1927). Once this had been established, the study of cardio-respiratory reflexes and physiology began to move its centre of gravity towards Madrid, the capital of Spain. There, one of the youngest and the last direct disciple of the reputed Santiago Ramón y

---

[1] Throughout this text, we will preferentially use two terms: carotid body and *glomus caroticum*, which represent the more modern concepts of this anatomical structure.

[2] All the physio-pharmacologists who worked with J.-F. and C. Heymans are collectively known as the "Ghent school", since the Institute for Pharmacology was situated in this Belgian city.

[3] Translated into English by the author of this article.

Cajal,[4] Fernando de Castro, began to publish his anatomo-histological observations that were to transform this research field.

2 De Castro, 1926

In the middle of the 1920's, the rich blood supply and sympathetic innervation of the *glomus caroticum* had become evident, but the fine details behind the organization of this anatomical structure, as well as its physiological relevance, remained completely unknown. After his remarkable studies on the structure and organization of the sensory and sympathetic ganglia (the main articles related to this research are: De Castro, 1921; De Castro 1922), Fernando de Castro began to study the aorto-carotid region. He found that the entire heads of animals could be fixed by adding nitric acid (3–4%) to the classic fixatives (urethane, chloral hydrate, formol or especially somnifene), perfectly preserving the nervous structures within their skeletal casing, as well as the peripheral innervations to which Cajal's famous reduced silver impregnation method could be applied (De Castro, 1925). The technique De Castro developed is especially effective in small animals, such as rodents, and it reduces the staining of the connective tissue, thereby increasing the contrast of peripheral nerve structures like the *glomus caroticum*. In his own words, this technique was crucial to be able to study the detailed innervation of the carotid sinus and the carotid body (De Castro, 1926). In fact, the first microphotographs of the rich vascular supply and the innervation of the *glomus caroticum* were published by the De Castro in his methodological paper (De Castro, 1925).

The first detailed study of the carotid region was published by De Castro in 1926, using tissue from both adults and embryos of different mammalian species (small rodents, cats, dogs, rabbits and humans). With the aid of microphotography (a technique introduced into the Cajal Institute after 1922), he could perform an exhaustive and intensive study of this area enabling him to make the first crucial clarification regarding this structure. Accordingly, De Castro showed that the nerve fibres do not form a closed plexus around the carotid body but rather, they project into this structure from the superior cervical ganglion (sympathetic system) and to a lesser extent, from the glossopharyngeal nerve (the intercarotid branch of the glossopharyngeal nerve that exclusively innervates the carotid body). Years previously, Hering had identified this intercarotid branch as the "sinus nerve", which innervates the carotid sinus and is responsible for the "sinus reflex" (see above). However, it was De Castro who clearly stated that the sinus nerve is a branch of the vagus nerve, which while sometimes lying close to the intercarotid branch of the glossopharyngeal nerve, is not the same at all (De Castro, 1926).

In this paper, De Castro described the complicated structure of the *glomus caroticum* in detail: a real tangle of small blood vessels, sympathetic axons and

---

[4] Santiago Ramón y Cajal (1852–1934) and Camillo Golgi (1843–1926) shared the Nobel Prize in Physiology or Medicine in 1906.

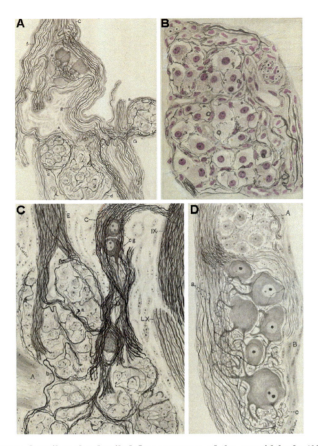

Fig. 1 De Castro describes the detailed fine structure of the carotid body (1926). Original drawings from Fernando De Castro, all of them included in his first publication on the innervation of the carotid region (De Castro, 1926). (**A**) Arrival of the inter-carotid nerve (a branch of the glossopharyngeal nerve; *c*) to the carotid body (*d*) of a young human. In the carotid body, different glomeruli with fine innervation can be identified. A sympathetic microganlgion is also illustrated (*e*). (**B**) Illustration of the *glomus caroticum* of a young human. Glomic cells are illustrated with coloured nuclei. The sensitive innervation surrounds these cells, but they do not comprise a closed plexus around the carotid body. (**C**) Illustration of the carotid body, forming different glomeruli close to the carotid artery (*A*). The sympathetic nerve (E) arrives from the superior cervical ganglion, but its contribution to the innervation of the carotid body is not the most relevant. The same can be said of the vagus nerve (*LX*) in the vicinity of the carotid body. By contrast, the most relevant contribution comes from the intercarotid nerve (*C*), a branch of the glossopharyngeal nerve (*IX*). Within this nerve, a sympathetic microganglion can be seen (*cg*). (**D**) Detailed illustration of one of the sympathetic microganglia that can be observed within the intercarotid nerve (see **A, C**)

glandular cells, which may form small glomeruli within the carotid body as well as a minuscule and complicated plexuses of glossopharyngeal fibres surrounding these glomeruli (Fig. 1A, C–D). The complexity of these nervous plexuses within the carotid body is even greater in humans than in the other mammals studied, although by contrast, there is less complexity around the carotid body (Fig. 1A, B).

With this paper (De Castro, 1926), Fernando de Castro began a detailed study of the innervation of the *glomus caroticum* and its physiological implications, which he continued in his second paper on this subject (De Castro, 1928) and intermittently, in successive papers until the end of his life. At that time, the controversy among the physiologists interested in the function of the carotid body and the anatomical basis of the Hering's reflex was being strongly pursued. In 1924, Hering attributed the profuse innervation of the carotid sinus to the fact that the mechanical or electrical stimulation of this region results in bradychardia and a decrease in arterial pressure (Hering, 1924). One year later, through Drüner, merely due to his intuition and without the support of any experimental data, Drüner launched the hypothesis after which that the intercarotid gland (the carotid body) was responsible for Hering's reflex (Drüner, 1925). In the same year, Hering responded that the mechanical pressure of the intercarotid territory, as well as that of the carotid body, does not fire the reflex, confirming that the "Sinus reflex" must be due to the excitation of the arterial wall in the sinus (Hering, 1925).

3 De Castro, 1928

Although his first paper directly affected the controversy between Hering and Drüner, De Castro decisively resolved this scientific joust with his second article on the subject, published in 1928. Using material from humans, monkeys, cows, cats, rabbits and rodents, De Castro fell back on the methylene blue reaction of Erlich-Dogiel and Bielschowsky's method (the protocol modified by Boecke) to complement the observations obtained with Cajal's technique. This paper can be considered in two main blocks: the confirmation of the physiological existence of the carotid sinus and the study of its innervation (baro-receptors); the detailed study of the innervation of the carotid body and the description of a different kind of receptors (chemoreceptors). Although the present work focuses on the chemoreceptors, we must briefly review the results obtained by Fernando de Castro on the innervation of the carotid sinus.

In agreement with the prior description by Hering (1927), De Castro confirmed the existence of the carotid sinus in all the species and at all ages studied, except in the human foetus in which the sinus is not macroscopically evident. Similarly, in the cow where there is no internal carotid artery, the carotid sinus is present in the bifurcation of the primitive carotid and the occipital arteries. This study ruled out other hypotheses about the nature of the carotid sinus, such as that of a pathological malformation (Binswanger and other anatomists in the last part of the XIXth century), or the existence of enlargements in each vascular bifurcation, as suggested by other prestigious scientists (Henle, Luschka, etc.).

In this second paper, Fernando de Castro described the sensory innervation of the carotid region in detail. He showed that the fibres concentrate in the part of the carotid sinus immediately prior to the origin of the internal carotid artery. This is where the vessel's wall is thinner and it is bordered by two concentric bands, one with little sensory innervation and the second with almost no such terminals,

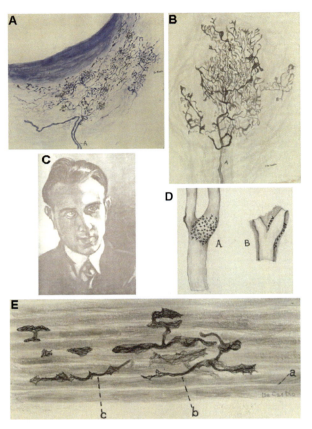

Fig. 2 De Castro describes the baroreceptors in the carotid sinus in detail (1928). Fernando De Castro's original drawings from his second publication on the innervation of the carotid region (De Castro, 1928). (**A**) Illustration of a type I (diffuse) baroreceptor close to the adventitia of the artery from a young human, stained with methylene *blue*. (**B**) Detailed illustration of a type II (circumscribed) baroreceptor of the human carotid sinus stained by silver impregnation. The myelin trunk is marked by *A*. (**C**) Portrait of Fernando De Castro by the Hungarian painter Férenc Miskolzy, brother of the famous neurologist Deszö Miskolzy, the founder of modern Hungarian Neurology. The portrait reflects the aspect of the Spanish neurohistologist at the time of his studies on the innervation of the carotid body and sinus. (**D**) Schematic distribution of the baro-receptors in the human carotid sinus. The symbols identify dense terminals from (*o*) to (>), passing through (+). In the section of the artery (*B*) the situation of the terminals in the thinner part of the arterial wall is illustrated. (**E**) Fine terminals of baroreceptors (*b*, *c*) intermingled with the collagen fibres (in diffuse *grey*, *a*), a distribution that allows the changes in volume of the blood vessel to be better detected. In some cases, the nerve terminals form a meniscus

a kind of "sensory penumbra" (Fig. 2D). This region shows virtually the densest innervation of the entire arterial tree and the fact that this distribution exists in all the species studied, including the cow, coupled with the specific location of the carotid sinus (see above), confirms that it is not merely anecdotic. Rather, in all animals the sensory fibres are specifically situated at the origin of the artery that irrigates the brain. Depending on their morphology and distribution, De Castro identified the different types of sensory receptors (basically either "disperse" or "circumscribed" receptors; Fig. 2A, B). However, it is even more interesting that he described that the fibres are terminals that extend and ramify through all the different planes of the adventitial layer of the artery, even the deepest one. Moreover, he described them as nude terminals, devoid of any kind of cover, so that they can sense the changes in the volume of the vessel derived from the changes in arterial pressure (Fig. 2E). Unfortunately, the silver method used by De Castro for this study does not stain the elastic fibres of the arterial vessels, so he could not study the intimate relationship between the sensory receptors and the elastic fibres that determine the movement of the artery. In none of the thousands of sensory terminals studied by De Castro did he observe any kind of cellular anastomosis. Indeed, the author highlights that this observation was not due to the *esprit d'ecole*, in clear reference to Cajal's "neuronal theory", even though he and the Master himself were the most ardent defenders of this theory against the continued attacks of the supporters of the "reticularist theory". However, these observations confirmed those published earlier, in 1924 and 1926, where De Castro showed that the arborization innervating the carotid sinus is more profuse in the big mammals and simpler in the small rodents (De Castro, 1928).

De Castro ingeniously studied the nature of carotid sinus innervation. To disprove the hypothesis of Boecke about the origin of these nerves in a sympathetic ganglion (approx., 1918), he ablated all the ganglia of the cervical chain and he failed to detect any sign of degeneration in the terminals innervating the carotid sinus. The same results were obtained when the glossopharyngeal, vagus and spinal nerves were sectioned. From the results of such trophic deprivation, De Castro concluded that the fibres innervating the carotid sinus must be sensory fibres, neurons projecting towards the central nervous system, which provides them with sufficient trophic support to survive. At that point, he proposed that these observations reflect the morphological basis of the physiological responses described by Hering (1927), the "sinus reflex". The hypothesis launched by Drüner was definitively defeated by the combination of the physiological studies of Heinrich Hering (1927) and the anatomo-histological ones of Fernando De Castro (1926 and mainly, 1928).

Having reviewed the studies of the carotid sinus and the baroreceptors, I will now focus on the second big block of results published in 1928, regarding the *glomus caroticum* and the chemoreceptors. Just when his first observations on this subject were published in 1926, De Castro postulated that the cells of the carotid body were probably cells with a paracrine or autocrine function, in accordance with the suggestion of many scientists at that time that the *glomus caroticum* was a real gland (the intercarotid gland). However, when he performed the elegant ablation experiments described above, De Castro observed that a few motile sympathetic fibres within the *glomus* did not degenerate once the sympathetic ganglia of the cervical chain were extirpated, indicating that they originated from the sympathetic neurons of the microganglia present within the *glomus* (see above: De Castro, 1926). By contrast, sectioning of the glossopharyngeal nerve where it exits the skull resulted in the fast (a mere 5–6 days after lesion) and almost total degeneration of the fibres forming the nervous plexus in the carotid body and the terminals connecting with glomic cells. This degeneration affected the cell cytoplasm and mitochondria, but the feochrome reaction of Henle and Vulpian was not observed, ruling out the possibility that the *glomus* might be a paraganglion, as suggested by Kohn (1900–1903). Moreover, De Castro thought that a organ as exquisitely innervated as the carotid body would not be a residual or involutive organ.

Accordingly, he decided to paint the surface of the carotid artery of adult cats with phenol to kill the terminals innervating the carotid sinus, and to study the possibility that the *glomus caroticum* might be involved in regulating arterial pressure. However, he detected only minimal changes in the latter, which suggested that the carotid body contributed little to the sinus reflex. This minimal contribution to the control of blood pressure might be due to motile fibres from the sympathetic microganglia within the carotid body (or even to a minimal branch coming from the superior cervical ganglion). However, this would not seem to be the main function of the carotid body because electrical stimulation of the intercarotid nerve (which remains intact when the glossopharyngeal nerve is sectioned) did not result in any relevant vasoconstriction of the artery. Also in 1928, a group of Rumanian scientists insisted that the *glomus caroticum* participated in the regulation of arterial pressure (Jacoborici et al., 1928), as suggested by Drüner three years ago, although De Castro's experiments completely ruled out this possibility.

De Castro performed a second series of experiments to clarify the function of the *glomus* in which he sectioned the glossopharyngeal and vagus nerves in cats and dogs, and studied the effects of this procedure on the carotid body. The degeneration of the fibres was fast and almost total, from which he deduced that the terminals innervating the *glomus* belong to sensory neurons from the nuclei of both nerves (glossopharyngeal and vagus).[5] Given that this sensory nature is different from that which determines the arterial pressure through Hering's "sinus reflex", De Castro hypothesised that these nerve receptors within the *glomus* would detect changes in the chemical composition of the blood. Indeed, blood pressure needs more urgent control exerted by the baroreceptors in the carotid sinus, while qualitative changes in the composition of the blood should be detected by a second system, not far from the first, such as the chemoreceptors of the *glomus caroticum*. De Castro even declared in his article that the changes in composition would not be detected directly by the nerve terminals, because they were not in direct contact with the blood. Rather, he stated, the glomic epithelial cells should perform this task via an "active protoplasmic process" extending from these cells, centripetally transmitted to the nerve terminals.

---

[5] The sensitive ganglia of the IXth craneal nerve are the Ehrenritter's and Andersch's ganglia, those from the Xth nerve are the jugular, the nodose and the petrose ganglia.

4 Heymans in the 1930's and the Nobel Prize in Physiology or Medicine

As stated previously, in the decade from 1920 to 1930 the two most important figures in cardiac and respiratory physiology research were Heinrich Hering (Köln University, Germany), and Jean-François Heymans along with his son Corneille (both working at the University of Ghent, Belgium). Fernando de Castro communicated his studies in different meetings of the *Association d'Anatomistes* held at Liege in 1926 (Belgium), at London in 1927 (Great Britain), and at Bourdeaux in 1929 (France), and together with the aforementioned publications, the work from Spain raised the attention of the entire scientific community interested in the field, including some physiologists. As a result, Professor Goormaghtigh, the chair of Pathology at the University of Ghent, transmitted an invitation from Corneille Heymans to Fernando de Castro to visit Ghent and to discuss his work with him. Just after the meeting, De Castro visited his sister who lived in La Roche-Chalais, in the French Périgord (not far from Bourdeaux), where he received a letter from Heymans again inviting him to visit Ghent (Fig. 3A). De Castro accepted and his own words are better than any other description:

> I went to Ghent and I was with him [*Heymans*] for two or three days, during which we performed some experiments on the carotid sinus of dogs. At that time, he was not interested in the carotid body. Rather, he had studied the vasoconstriction phenomena on the cardio-aortic region and he had just started to study the carotid region. His work on the carotid body came afterwards, and he said to me. *I'm deeply interested in your idea about this. It would be great if I could demonstrate it!*[6] (Gómez-Santos, 1968).

This was Fernando de Castro's first visit to the laboratory of Corneille Hyemans in Ghent (April, 1929) and there was to be a second one. However, the Belgian physiologists did not directly address the study of the physiological role of the carotid body and they took years to abandon their hypothesis that the respiratory reflex originated in the carotid sinus. For example, although in his first work about the physiological reflexes originated in the carotid sinus (Heymans, 1929), Heymans recognised that the anatomy and the histology of the sinus region had been described "recently, in a precise and detailed manner by De Castro",[7] he did not pay any attention to the *glomus caroticum*, and even less so to De Castro's hypothesis on the function of this small structure in chemoreception (Heymans, 1929). In their detailed study the following year, Heymans and his collaborators described the primacy of the carotid sinus (even ahead than the respiratory centre in the brainstem) to detect changes in arterial pressure and in the concentration of H^+, CO_2 and O_2 in the blood. However, although they cited the work of De Castro (1928) in the

---

[6] Translated into English by the author of this article.
[7] Translated into English by the author of this article.

Fig. 3 Corneille Heymans and Fernando De Castro, a long friendship full of mutual admiration. (**A**) Letter from Heymans, dated the 28th March 1929, inviting Fernando De Castro to his lab at Ghent University. As can be seen on the envelope (upper right), the letter was sent to "Chez Lagoubie, La Roche-Chalais, Dordogne", the family name of De Castro's French brother-in-law where the Spanish histologist was staying for a few days after the meeting of the Association d'Anatomistes in Bourdeaux (France). (**B**) De Castro's drawing showing two dogs in parabiosis (the famous physiological technique in which the Heymans -father and son- were consumed masters). The notes in Spanish complete the information for the experiments. (**C**) Letter from Heymans, dated 29th December 1939, in answer to a previous letter from Fernando De Castro (dated 15th December 1939) where he congratulated the Belgian physio-pharmacologist for the Nobel Prize award in Physiology or Medicine. (**D**) Picture of the portrait of Corneille Heymans, dedicated to De Castro (see hand-writing in blue at the lower-right corner of the image) and dated 1939 by the painter. Fernando De Castro kept this dedicated picture on his bureau until he died in 1967. (**E**) De Castro prepared to perform one of his famous and complicated nerve anastomosis in a cat, in the precarious conditions of his laboratory in Madrid (circa 1941)

bibliography, they did not discuss his studies in the text (Heymans and Bouckaert, 1930; Heymans et al., 1930). This is likely to be the origin, at least in part, of the persistent error in failing to differentiate between the site at which blood pressure and it composition are detected (Gallego, 1967).

It was not until 1931, when these researchers published a study (focused on the study of bradycardia induced by different drugs) in which they explicitly accepted the hypothesis of De Castro:

> "...our experiments indicate that the starting point of the reflexes triggered by the injection of chemical substances may be localised in the region of the carotid bifurcation and, more exactly, in the glomus caroticum, as proposed by De Castro" (Heymans et al., 1931).

And even more explicit, perhaps the reference published in a second paper about the same subject one year later:

> ...the substances injected in the common carotid abandon this slowly through the small arterial branches which have not been sutured, and then they can act for a relatively long period of time, on the region of the carotid ganglion which is, as De Castro and we have demonstrated, the starting point of the reflexes triggered by the chemical excitants[8] (Heymans et al., 1932).

Coinciding with the second visit of Fernando de Castro to the *Pharmacological Institute* at the University of Ghent (where he repeated experiments on dogs together with Heymans; Fig. 3B), the Belgian physio-pharmacologists once and for all adopted the hypothesis of Fernando de Castro. At that time, they began to study more closely the region of the *glomus caroticum* and the physiological importance of the chemoreceptors:

> The reflexogenic hypothesis of vascular sensitivity in the regions of the carotid sinus was already postulated by De Castro in 1928 based on experimental morphological observations[9] (Heymans et al., 1932).

In 1933, Heymans grouped all the discoveries published by his group up to that date in a book (Heymans et al., 1933). In this text, the different types of reflexes are profiled and it is reflected that changes in arterial pressure (detected by the baroreceptors from the carotid sinus) or in the chemical composition of blood (detected by the chemoreceptors of the *glomus caroticum* or carotid body) control respiratory frequency, the heart rate and the arterial pressure. From this year on, Heymans and his collaborators will cite this book and less frequently their initial publications on the subject (1929–1932).

5 The Scientific Path of Fernando De Castro Between 1929 and 1936

During this period, Fernando de Castro was following difficult and complex approaches in Madrid using experimental surgery to demonstrate that the neurons of the *glomus* respond to changes in the chemical composition of blood. For this purpose, he developed different anastomosis between the glossopharyngeal, vagus and hypoglossal nerves to create artificial reflex arches. This approach required the

---

[8] Translated into English by the author of this article.
[9] Translated into English by the author of this article.

prior and detailed study of the regeneration of the sectioned preganglionic branches. These studies (De Castro, 1930, 1933, 1934, 1937), necessary to the ulterior demonstration of the physiological role of arterial chemoreceptors, distracted Fernando de Castro from his original goal by raising other scientific questions, which led him to work with the world famous and reputed biologist Giusseppe Levi at Turin (Italy), both in 1932 and particularly in 1934. One of Fernando de Castro's last collaborators, Prof. Jaime Merchán, could not understand why de Castro "did not try to put two living animals in parabiosis, a classical physiological technique and certainly easy enough for someone with the surgical skills of your grand-father [*Fernando de Castro*]".[10] It is possible that these parabiosis experiments, after all similar to those performed by the Heymans, and by Fernando de Castro and Heymans, in Ghent (see above), were those he referred to at the end of his work published in 1928 (see above).

In these years, a scientist close to the octogenarian Santiago Ramón y Cajal was worried about the poor international repercussion of the studies carried out by many members of his school (mostly published in Spanish; Penfield, 1977) and in particular, of the works of De Castro on the carotid sinus and carotid body (see for example the letters from Cajal to Fernando de Castro, dated February 18th, 1932 and March 15th, 1933 -both conserved in the Fernando de Castro Archive-). This led Cajal to take two significant steps: (i) from 1923 onwards, the journal published by the Cajal Institute ("Trabajos del Laboratorio de Investigaciones Biológicas") was published in French as "Travaux du Laboratoire de Recherches Biologiques"; (ii) together with Fernando de Castro, Cajal decided to compile all the different Histological protocols and experimental procedures that made up the technical corpus of the Spanish School of Neurologists (or Cajal's School), and to publish them in a book (Cajal and De Castro, 1933).

However, it was Rafael Lorente de Nó, another young disciple of Cajal and a close friend of Fernando de Castro, who tried to draw the attention of their Maestro and of his colleague and friend to the error in their ways. In his opinion, they should not forsake the study of the carotid region and of its physiological implications in order to start working on the regeneration of the nervous system in vitro, which seems clear after the letter written by Cajal on May 19th 1934, to Fernando de Castro, ill in Turin (Italy) at that time:

> I received a letter from Lorente regretting, that after many years of work in a difficult histological speciality, with an excellent orientation and dominion of scientific bibliography, you have changed your bearing to work in a field that, if not exhausted, does not initially offer to the researcher unexpected fruits. When I answer him, I will inform him that both paths converge and not only will no harm be done, but it is favourable to air one's intelligence in other scientific domains.

This episode suggests that the change of course in Fernando de Castro's scientific career was adopted with Cajal's agreement. Although there are additional events that may help understand the stance adopted by Cajal, as indicated in the letter above,

---

[10] Professor Jaime Merchán (UMH, Spain), personal communication (March, 2008).

including events associated with the research into arterial chemoreceptors, it seems clear that Lorente de Nó was not entirely wrong.

In Febraury1934, De Castro went to Turin to work with Levi, and there his life ran a serious risk due to a rare gastric hemorragia. Santiago Ramón y Cajal died in October 1934 (with the ensuing organizational changes at the Cajal Institute), and the political and social events in Spain (and Europe) became complicated, leading up to the Spanish Civil War in July 1936.

6 Corneille Heymans, Nobel Prize in Physiology or Medicine in 1938

In 1938, the Nobel Committee evaluated the proposal of Cornelius Heymans for the Nobel Prize in Physiology or Medicine. It was a year in which the number of scientists nominated was quite numerous.[11] Professor Liljestrand, member of the committee, together with Ulf von Euler[12] (both collaborators of Cornelius Heymans) evaluated the candidature of the Belgian physio-pharmacologist,[13] and they left this testimony in the records of the Nobel Foundation regarding the prize for Physiology or Medicine:

> Hering's work was twice submitted to special investigation. In 1932, the reviewer exposed some doubts about his qualification for a prize like this, taken into account the obvious analogy with previous research performed on the aortic nerves and those from their predecessors mentioned above [*Pagano, Siciliano, Sollmann y Brown*]; afterwards, it was considered the division of the prize between Hering and Heymans, but the Committee was still sceptic about the merits of the first one. Pagano was also nominated by his compatriots, but only in 1943 (Liljestrand, 1962).

In fact, when consulting the database of the Nobel Foundation, it can be seen that the German physiologist Heinrich Hering was proposed for the prize in 1932, 1933, and 1937, as well as sharing a nomination with the Louis Lapique and Corneille Heymans in 1934. But in 1938, no one nominated the physiologist from Köln for the Nobel Prize.

The case of Heymans merits certain attention because he was nominated for the Nobel Prize in 1934, 1936, 1938 (when it was awarded to him) and also in 1939. Curiously, only in the proposals from 1938 and 1939 is there any explicit mention of his studies on the baro- and chemoreceptors of the blood, while on the previous occasions he was nominated for his studies on blood circulation in general. It also seems interesting that while in 1938 he was nominated by the Hungarian professor Mansfeld, on all the other occasions he was supported by Belgian

---

[11] Consulting the database of the Nobel Foundation, there were a total of 96 scientists nominated for the Nobel Prize in Physiology or Medicine in 1938. Among them, Hess, Houssay, Stanley, Sasaki, Lapique, Erlanger, Gasser, Cushing... to cite only some of the most relevant names.

[12] Future Nobel Prize in Physiology or Medicine in 1970.

[13] Database of the Nobel Foundation: http://nobelprize.org/nomination/medicine/nomination.php?action=simplesearch&string=Heymans&start=11

professors and medical doctors, including the deluge of proposals supporting his candidature in 1939 (either alone or in conjunction with Louis Lapique and Walter Hess).

So what about Fernando De Castro? A legend circulates that during the deliberation of the Nobel Committee over lunch, someone argued in favour of Fernando De Castro's nomination and that another committee member, remembering the war that was desolating Spain at that time, asked: "But, does someone know if De Castro is still alive?".[14] Once again, the database of the *Nobel Foundation* is clear (which opens all its documents to the public 50 years after the concession of each Prize): nobody either in Spain or in any other country nominated Fernando De Castro for the Nobel Prize in those years.[15] By contrast, the Spaniard Pío del Río-Hortega was nominated in 1929 by a professor at the *University of Valladolid* and in 1937 (in the middle of the Spanish Civil War) by two professors from the *University of Valencia*.[16] After carefully consulting the database of the Nobel Foundation, it can be seen that Santiago Ramón y Cajal supported only one nomination during his life (together with another eight Spanish scientists and medical doctors, and he did not lead the proposal): that of the French immunologist Richet in 1912.[17] It remains a mystery that during the 28 years after he was awarded the Nobel prize and as active as they were until close to his death in 1934, this Spanish genius remained at the margin of these scientific jousts.

The draft written by Fernando De Castro to felicitate Heymans for the award of the prize, dated December 15th 1939, is conserved in his archives, as is the hand-written answer from the Nobel Prize dated December 29th of the same year. Following on from the studies of the Ghent group, Comroe discovered the chemoreceptors in the so called *glomus aorticum* (innervated by the depressor nerves) had a minor role in the respiratory reflexes given that they merely respond to extreme cases of hypoxemia (Comroe, 1939). And Fernando De Castro continued working in the fine structure and physiology of the carotid body just from the end of the Spanish Civil War (April, 1939) in penury of technical support and in difficult personal and economical circumstances (Fig. 3E), something that did not really change until the decade of 1950.[18]

In the presentation of the Nobel Prize in Physiology or Medicine to professor Corneille Hyemans, at Ghent on January 1940,[19] the Swedish professor

---

[14] Professor Gunnar Grant's personal communication (meeting of the *Cajal Club*; Stockholm, May 2006).

[15] This information was corroborated by a letter dated April 3rd, 2007 written by the Administrator of the Nobel Committee in answer to the author of the present paper.

[16] While his 1929 nomination was evaluated, that in 1937 was not.

[17] Charles Richet was awarded the Nobel Prize in Physiology or Medicine in 1913 "in recognition of his work on anaphylaxis".

[18] This was his main research labour till he died in April, 1967.

[19] Although the IInd World War commenced in September 1939, it was not until May 10th 1940 that the German troops invaded Belgium, The Netherlands and Luxembourg in their race towards France. Thus, life in these countries would have been relatively normal in January 1940, which

The Discovery of Sensory Nature of the Carotid Bodies

G. Liljestrand recognised the work of Fernando De Castro as fundamental in the path that Heymans followed towards his final success:

> Since the end of the 18th century we know of the existence of a curious structure in the region of the sinus, the glomus caroticum or carotid body which, in man, extends over only a few millimetres. The glomus consists of a small mass of very fine intertwining vessels arising from the internal carotid and enclosing various different types of cells. It has been considered by some as being a sort of endocrine gland similar to the medulla of the suprarenal glands. De Castro, however, in 1927 demonstrated that the anatomy of the glomus could in no way be compared to that of the suprarenal medulla. De Castro suggested rather that the glomus was an organ whose function was to react to variations in the composition of the blood, in other words an internal gustatory organ with special ≪chemo-receptors≫. In 1931, Bouckaert, Dautrebande, and Heymans undertook to find out whether these supposed chemo-receptors were responsible for the respiratory reflexes produced by modifications in the composition of the blood. By localized destruction in the sinus area they had been able to stop reflexes initiated by pressure changes, but respiratory reflexes could still continue to occur in answer to changes in the composition of the blood. Other experiments showed that Heymans's concepts on the important role played by the glomus in the reflex control of respiration by the chemical composition of the blood were undoubtedly correct".[20]

Due to the IInd World War, Heymans did not give his Nobel lecture until December 12th, 1945 (when he received the prize, *de facto*). It was then surprising that he scarcely cited the earlier work of Fernando De Castro:

> Histological research carried out by de Castro, Meyling and Gosses, and our own experimental findings, obtained with J. J. Bouckaert and L. Dautrebande in particular, has led to the locating of the carotid sinus chemo-receptors in the glomus caroticum and of the presso-receptors in the walls of the large arteries arising from the carotid artery.[21]

When asked both indirectly as well as directly by the interviewer to explain why Cornelius Heymans did not share the Nobel Prize he received in 1938 with Fernando De Castro, he explained:

> He performed the physiological demonstration, not the anatomical one. Obviously, there was a tremendous loss of time from 1936 to 1938 due to the war and because at that time I was performing the experiments on nervous anastomosis. This was to automatically register the phenomenon of the carotid reflexes on the eye, with the nervous anastomosis detecting chemical changes in the blood. Years after, I presented this work in a symposium held at Stockholm, in which the opening conference was entrusted to Heymans and the second to me. There, I presented the work that I couldn't finish during the war. I had no more cats during that difficult time, they died of hunger or I was forced to sacrifice them. For this reason, I could not complete my experiments at that time (Gómez-Santos, 1968).

In this respect, the comments on this chapter of the life of Fernando De Castro published by the Chilean scientist Juan de Dios Vial, who worked with De Castro in the fifties, are very interesting:

---

allowed this ceremony to be celebrated. These circumstances delayed holding the Nobel ceremony in Stockholm for six years.

[20] Prof. G. Liljestrand (Karolinska Institutet); presentation of the 1938 Nobel Prize in Physiology or Medicine to Prof. C. Hyemans (Ghent, Belgium; January 16th, 1940).

[21] Nobel Lecture, Prof. C. Heymans (Stockholm, Sweden; December 12th, 1945).

The period of intense activity around 1930, which had ended by the proposal of the idea that the glomus was a chemoreceptor, was followed by a lull. This may be due to the fact that his contribution was widely ignored as coming from Spain. He opened the way to Heymans' discoveries but did not receive due credit, even at the moment when the latter was awarded the Nobel Prize. I never heard De Castro himself refer to that circumstance, but his disciples and friends often did, and were somewhat bitter about it. These were also the years of Cajal's death, and of organizational changes in his Institute (Vial, 1996).

7 Heymans and De Castro: A History of Mutual Admiration

In Fernando De Castro's archives, dozens of letters remitted by Corneille Heymans and other close collaborators are conserved, as well as several drafts of the correspondence maintained by Fernando De Castro with all the members of the Ghent School. Although most of the correspondence conserved corresponds to the period between 1930 and 1960, the first letters sent by Heymans date from the early twenties. Some of these documents are especially significant, like the draft of the felicitation for the concession of the Nobel Prize sent by De Castro as mentioned above, the dedicated photograph of Heymans' portrait that was all the time in De Castro's office (Fig. 3D), two invitations sent in 1952 by Bouckaert and Heymans himself inviting Fernando De Castro to take part in the symposium held in the honour of the Belgian Nobel prize winner and even a letter from Heymans' wife (written on February 17th, 1953),[22] in which she personally welcomes Fernando De Castro to the cited hommage or inviting him to the wedding of Heymans' daughter in 1957. All these are examples of the friendship between Corneille Heymans and Fernando de Castro, as published by the son of the latter: "full of mutual admiration, their friendship continued over the years until April 15th 1967, the date of my father's death".[23]

Acknowledgments In deeply in debt with my PhD student Diego García-González, who helped me with different aspects of the final edition of this manuscript.

The work is supported with grants of Spanish institutions: Ministerio de Ciencia e Innovación (SAF2007-65845), Instituto de Salud Carlos III-FIS (PI04-2591), Consejerías de Sanidad (ICS06024-00) y de Educación (PAI08-0242-3822) de la Junta de Comunidades de Castilla-La Mancha y FISCAM (GCS2006-C24 y PI2007-66).

References

Comroe, J.H. (1939) The location and function of the chemoreceptors of the aorta. *Am. J. Physiol.* 127, 176–191.

Cyon, E., and Ludwig, C. (1866) Die Reflexe eines der sensiblen Nerven des Herzens auf die Motorischen der Blutgefässe. *Ber. Sächs. Ges. Wiss. Leipzig.* 18, 307–329.

---

[22] Fernando De Castro invited Heymans to the symposium in honour of Santiago Ramón y Cajal, celebrated to commemorate the centenary of his birth, and held in Madrid in 1952.

[23] F.-G. de Castro in letter to the Director of the journal ABC (Madrid) published on November 20th, 1974.

De Castro, F. (1921) Estudio sobre los ganglios sensitivos del hombre adulto en estado normal y patológico. *Trab. Lab. Invest. Biol.* 19, 241–340.

De Castro, F. (1922) Evolución de los ganglios simpáticos vertebrales y prevertebrales. Conexiones y citoarquitectonia en algunos grupos de ganglio en el niño y en el hombre adulto. *Trab. Lab. Invest. Biol.* 20, 113–208.

De Castro, F. (1925) Technique pour la coloration du système nerveux quand il est pourvu de ses étuis osseux. *Trav. Lab. Rech. Biol.* 23, 427–446.

De Castro, F. (1926) Sur la structure et l'innervation de la glande intercarotidienne (glomus caroticum) de l'homme et des mammifères, et sur un nouveau système d'innervation autonome du nerf glosopharyngien. *Trav. Lab. Rech. Biol.* 24, 365–432.

De Castro, F. (1928) Sur la structure et l'innervation du sinus carotidien de l'homme et des mammifères. Nouveaux faits sur l'innervation et la fonction du glomus caroticum. *Trav. Lab. Rech. Biol.* 25, 331–380.

De Castro, F. (1929) Ueber die Struktur und innervation des glomus caroticum beim Menschen aund bei den Säugetieren. Anatomisch-experimentelle Untersuchungen. *Zeist. Anat. Entwicklungs.* 89, 250–265.

De Castro, F. (1930) Recherches sur la dégénération et la régénération du système nerveux sympathique. Quelques observations sur la constitution des synapses dans les ganglions. *Trav. Lab. Rech. Biol.* 26, 357–456.

De Castro, F. (1933) Quelques recherches sur la transplantation de ganglions nerveux (cérebrospinaux et sympathiques) chez les mammifères. Etudes comparatives sur la capacité réactionnelle et la résistance vitale des neurones survivants dans les greffes. *Trav. Lab. Rech. Biol.* 27, 237–302.

De Castro, F. (1934) Note sur la régénération fonctionnelle hétérogène dans les anastomoses des nerfs pneumogastrique et hypoglose avec le sympathique cervical. *Trav. Lab. Rech. Biol.* 29, 307–316.

De Castro, F. (1937) Sur la régénération fonctionnelle dans le sympathique (anastomoses croisées avec des nefs de type iso et hétéromorphe). Une référence spéciale sur la constitution des synapses. *Trav. Lab. Rech. Biol.* 31, 271–345.

Drüner, L. (1925) Ueber die anatomischen Unterlagen der Sinusreflexe Herings. *Deutsche Mediz. Wochenschr.* 51, 559.

François-Franck, C.E. (1876) Physiologie experimentale. *Trav. Lab. Marey* 2, 221–288.

François-Franck, C.E. (1877) *Trav. Lab. Marey* 3, 273–288.

François-Franck, C.E. (1878) Sur les effets cardiaques et respiratoires des irritations de certains nerfs sensibles du coeur, et sur les effets cardiaques produits par l'irritation des nerfs sensibles de l'appareil respiratoire. *Trav. Lab. Marey* 6, 73.

Gallego, A. (1967) La contribución de Fernando De Castro al descubrimiento y estudio de los preso-receptores vasculares y quimio-receptores sanguíneos. *Arch. Fac. Med. Madrid* 11, 406–439.

Gómez-Santos, M. (1968) *Cinco grandes de la Ciencia española.* Ed. Biblioteca Nueva, Madrid, Spain.

Hering, H.E. (1923) Der Karotidsdruckversuch. *Münch. Med. Wschr.* 70, 1287–1290.

Hering, H.E. (1924) Der Sinus caroticus an der Ursprungsstelle der Carotis interna als Ausgangsort eines hemmenden Herzreflexes und depressorischen Gefässreflexes. *Münch. Med. Wschr.* 71, 701–705.

Hering, H.E. (1927) *Die Karotissinusreflexe auf Herz und Gefässe.* Ed. T. Steinkopff, Dresden-Leipzig, Germany.

Heymans, C. (1929) *Arch. Int. Pharmacodyn.* 35, 269–313.

Heymans, C., and Bouckaert, J.J. (1930) Sinus Caroticus and Respiratory Reflexes. *J. Physiol.* 69, 254–266.

Heymans, C., Bouckaert, J.J., and Dautrebande, L. (1930) Sinus carotidien et réflexes respiratoires. II. Influences respiratoires réflexes de l'acidôse de l'alcalose, de l'anhydride carbonique, de l'ion hydrogéne et de l'anoxémie: Sinus carotidiens et échanges respiratoires dans le poumons et au delá des poumons. *Arch. Int. Pharmacodyn.* 39, 400–448.

Heymans, C., Bouckaert, J.J., and Dautrebande, L. (1931) Au sujet du mécanisme de la bradycardie provoquée par la nicotine, la lobéline, le cyanure, le sulfure de sodium, les nitrites et la morphine, et de la bradycardie asphyxique. *Arch. Int. Pharmacodyn.* 41, 261–289.

Heymans, C., Bouckaert, J.J., Von Euler, U.S., and Dautrebande, L. (1932) Sinus carotidiens et reflexes vasomoteurs. *Arch. Int. Pharmacodyn.* 43, 86–110.

Heymans, C., Bouckaert, J.J. and Regniers, P. (1933) *Le Sinus Carotidien.* Ed. G. Doin, Paris.

Heymans, J.F., and Heymans, C. (1927) Sur les modifications directes et sur la régulation réflexe de l'activité du centre respiratoire de la tête isolée du chien. *Arch. Int. Pharmacodyn.* 33, 273–370.

Heymans, C., and Ladon, A. (1925) Recherches physiologiques et pharmacologiques sur la tête isolée et le centre vague du chien. I: Anémie, asphyxie, hypertension, adrénaline, tonus pneumogastrique. *Arch. Int. Pharmacodyn.* 30, 145.

Jacoborici, J., Mitzescu, I.I., and Pop, A. (1928) Sur la fonction de la glande (paraganglion) carotidienne. La glande et le réflexe du sinus carotidien. *Compt. Rend. Soc. Biol.* 98(9).

Kohn, A. (1900) Ueber den bau und die entwicklung der sogen carotidsdrüse. *Arch. F. Mikroscop. Anat. Bd.* 56, 81–148.

Kohn, A. (1903) Die Paraganglien. *Arch. F. Mikroscop. Anat. Bd.* 62, 263–365.

Liljestrand, G. (1962) *Nobel the Man and His Prices.* Ed. Elsevier, Amsterdam, The Netherlands.

Marey, E.-J. (1859) Des causes d'erreur dans l'emploi des instruments pour mesurer la pression sanguine, et des moyens de les éviter. *Compt. Rend. Soc. Biol.* 1, 55–58.

Pagano, G. (1900) Sur la sensibilité du Coeur et des vaisseaux sanguins. *Arch. Ital. Biol.* 33.

Penfield, W. (1977) *No Man Alone: A Neurosurgeon'S Life.* Ed. Little Brown & Company, Boston-Toronto, USA-Canada.

Ramón y Cajal, S. and De Castro, F. (1933) *Técnica micrográfica del sistema nervioso.* Ed. Tipografía Artística, Madrid, Spain.

Siciliano (1900) Les effets de la compression des carotides sur la pression, sur le coeur et sur la respiration. *Arch. Ital. Biol.* 33, 338–344.

Vial, J.D. (1996) A tribute to Fernando De Castro on the centennial of his birth. In: *Arterial Chemoreception.* Eds. Zapata et al., Plenum Press, New York, USA.

Fifty Years of Progress in Carotid Body Physiology – *Invited Article*

R.S. Fitzgerald, C. Eyzaguirre and P. Zapata

Abstract Research on arterial chemoreceptors, particularly on the carotid body, has been fruitful in the last fifty years, to which this review is addressed. The functional anatomy of the organ appears to be well established. The biophysical bases by which glomus cells transduce chemical changes in the *milieu intérieur* (hypoxia, hypercapnia, acidosis) into electrical and biochemical changes in glomus cells have received much attention. Physical changes (in temperature, flow and osmolarity) are also detected by the carotid body. Electrical coupling between glomus cells themselves appears as very extensive. Sustentacular cells classically considered as ensheathing glia for glomus cells and nerve endings now appear to behave as stem cells precursors for glomus cells under chronic hypoxic conditions. Many papers have been devoted to transmitters released from glomus cells (acetylcholine, dopamine, ATP) and well as to their effects upon chemosensory nerve activity. Chemosensory neurons have been explored from generation of action potentials at peripheral nerve endings, passing to properties of perikarya at petrosal ganglia and finally at characterization of synaptic transmission at solitary tract nuclei. There is abundant literature on ventilatory and cardiovascular reflexes elicited from arterial chemoreceptors. The transient effects of sudden and brief withdrawal of chemosensory discharges by hyperoxia also provide clues on the role played by carotid bodies in the homeostasis of full organisms.

Keywords Carotid body physiology · Progress in research · Research last 50 years

1 Prolegomena

January 31, 1743 appears to be the date of the first historically available document about the anatomy of the carotid body. The thesis of Louis Taube was presented in the lab of the great German physiologist, Albrecht von Haller. Over 200 years

P. Zapata (✉)
Facultad de Medicina, Clínica Alemana – Universidad del Desarrollo, Santiago, Chile; Facultad de Medicina, P. Universidad Católica de Chile, Santiago, Chile
e-mail: pzapata@udd.cl

followed without much significant work. In 1926 and especially in 1928, the eminent Spanish histologist Fernando De Castro presented his studies and his thoughts on the function of the carotid body. Though Corneille Heymans was awarded the 1938 Nobel Prize in Physiology or Medicine, his work was based on both the anatomy presented in De Castro's work and on his hypothesis as to its physiological function (for refs. see Eyzaguirre et al., 1983).

2 Organization of the Carotid Body

Several electron-microscopic studies in the 1960's showed that glomus cells (De Castro's epithelioid cells) were rich in dense-core granules and abundantly supplied with apposing nerve endings containing synaptic-like vesicles, suggesting from morphological grounds that glomus cells were some sort of endocrine cells controlled by efferent nerve fibers. Then, Arthur Hess performed several electron-microscopic studies in normally innervated, fully denervated and solely deafferented carotid bodies, demonstrating that glomus cells had reciprocal synapses with apposing nerve endings, which were part of the peripheral processes of sensory neurons with perikarya located in the petrosal ganglion, thus confirming De Castro's earlier hypothesis (see refs. in Eyzaguirre et al., 1983).

The electron microscopic studies performed by Donald McDonald showed that the carotid body was vascularly supplied by convoluted capillaries in the center of glomoids (glomus cells islets) and straight capillaries and arterio-venous anastomoses at the periphery of the organ (see McDonald, 1981), thus providing a vascular system capable of regulating blood flow through the parenchyma of the carotid body.

The autonomic innervation of the carotid body was also a matter of much debate, but the elegant microscopic studies of Alain Verna led to the conclusion that most sympathetic fibers arriving from the superior cervical ganglion were innervating carotid body blood vessels and that a parasympathetic innervation of the organ was scarce or absent (see Verna, 1997). However, autonomic neurons located in microganglia along the glossopharyngeal nerve may provide efferent inhibition to the carotid body (Campanucci and Nurse, 2007).

Sustentacular cells (expressing the glial marker GFAP) were for long time considered only as enveloping bags for glomus cells and nerve endings appositions. But, clustered glomus cells in culture (retaining their sustentacular envelope) behave differently from isolated cells. Moreover, recent observations by López-Barneo and coworkers suggest that -when exposed to prolonged hypoxia- sustentacular cells may also behave as stem cells precursors for glomus cells (see Pardal et al., 2007).

3 Testing Carotid Body Physiological Responses

The initial information on carotid body physiology was obtained by the elicitation of ventilatory chemoreflexes in preparations in toto, a procedure that was followed by the recording of carotid nerve chemosensory discharges in situ. Thus, Mulligan and Lahiri (1981) showed that every agent acting as inhibitor of cytochrome oxidase (such as cyanide) or uncoupler of electron transport (such as 2,4-dinitrophenol) was an effective stimulant for carotid body chemosensory discharges. Attempts to perfuse the carotid body in situ with saline resulted in short-lived preparations, but perfusing the organ with blood was more successful (O'Regan, 1979). Nevertheless, it was not possible to establish whether carotid body responses to a single chemical stimulus (e.g., hypoxia) or to a given pharmacological agent (e.g., nicotine) resulted from their direct effects on chemoreceptors themselves or were secondary effects associated with changes in other chemoreceptor natural stimuli or to changes in blood flow through carotid body tissue.

The above problems were solved by Eyzaguirre and coworkers through the use of an isolated preparation of the carotid body in vitro superfused with saline solutions flowing at a controlled rate, and in which one controlled change in chemical constituents or physical conditions would be introduced at a time without affecting other variables of the experiment. This type of preparation revealed that the carotid body was a "multimodal" receptor, responding directly to various chemical changes (oxygen and carbon dioxide tensions; hydrogen ion and potassium concentrations) as well as to physical changes (temperature, osmolarity and fluid flow) that may occur under physiological conditions (see Eyzaguirre et al., 1983).

Further advances have been made by studies on carotid body tissue slices in vitro, cells acutely dissociated from excised carotid bodies, carotid body cells in tissue culture, petrosal ganglion cells in tissue culture, and co-cultures of carotid body and sensory ganglion cells. However, cultured glomus cells, and for that matter preparations in vitro, do not necessarily behave the same as glomus cells in the animal, and studies on carotid bodies in situ or in whole animals -in which the natural environment of the carotid body is better preserved- are still required to validate current observations made on isolated preparations in vitro.

4 Glomus Cell Responses to Chemical Stimuli

Since another review is addressed to the effects of "natural" chemical stimuli on membrane potential and ion channels of glomus cells, we will briefly mention that different stimuli (hypoxia, hypercapnia, acidity) depolarize about half of glomus cells while the others undergo hyperpolarization, an effect conditioned by the presence or absence of sustentacular cells (see Eyzaguirre et al., 1989).

The demonstration by José López-Barneo and Constancio González and their co-workers that hypoxia produces a reversible inhibition of a transient K^+ current in glomus cells (López-Barneo et al., 1988) initiated a prolific area of research on oxygen-dependent ionic channels. The studies performed by Chris Peers and Keith Buckler had been particularly clarifying on this issue (see Peers and Buckler, 1995).

It has been found that glomus cells are dye and electrically coupled between themselves, through gap junctions revealed by high resolution electron microscopy and biochemical characterization of connexins. Chemoreceptor stimulants (acute and chronic hypoxia, hypercapnia, acidity, cholinergic agents and dopamine) uncouple most glomus cells, a change accompanied by cell depolarization and decreased amplitude of junction channels activity. Coupling is mostly resistive from glomus cells to nerve endings, but it is mostly capacitive from nerve endings to glomus cells (see Eyzaguirre, 2007). Thus, slow electric events originating in the glomus cells can be transferred to the nerve endings.

We know from several studies that hypoxia and acidity increase the intracellular concentration of calcium ($[Ca^{2+}]_i$). This may be due to calcium inflow through the glomus cell membrane (González et al., 1994), but it may also come from intracellular stores (Biscoe and Duchen, 1990). An acute increase in calcium concentration is required for excitation-secretion coupling to occur, as shown in every place in which it has been studied, and therefore it will be necessary for release of transmitters to convey excitation from glomus cells to sensory nerve endings. Nevertheless, carotid body preparations bathed in zero $[Ca^{2+}]_o$ -that should eliminate transmitter release- show a reduced chemosensory nerve discharge, but it is still there.

5 Chemical Transmission Between Glomus Cells and Chemosensory Nerve Endings

Several observations showing that destruction of glomus cells by criocoagulation or ischemia suppresses the chemoreceptive properties of the carotid body, and that regenerating nerve fibers acquire chemosensory activity when they come into contact with glomus cells (see Eyzaguirre et al., 1983) indicated that glomus cells were indeed required for hypoxic chemoreception. Furthermore, intracellular recordings from carotid nerve terminals showed small depolarizing potentials in unstimulated preparations, which upon stimulation with NaCN or ACh increased their frequency to the point of fusion, resulting in a larger total depolarization accompanying the increased sensory discharge (see Eyzaguirre et al., 1983). At a time when chemical transmission between nervous structures appeared to have defeated electrical transmission, a frantic search for chemical transmitters between glomus cells and nerve endings was initiated.

The **cholinergic hypothesis** was based on several observations accumulated during the first half of the 20th century (for refs. see Eyzaguirre et al., 1983). The idea of acetylcholine (ACh) playing the role of transmitter between glomus cells and chemosensory nerve endings gained acceptance by the bioassays revealing the persistence of ACh within the carotid body after denervation of the organ (Eyzaguirre and Koyano, 1965), suggesting that ACh was probably stored within glomus cells. More recently, Shirahata and Fitzgerald (1996) -using high performance liquid chromatography-electrochemical detection- confirmed the release of ACh from both cat and pig cultured glomus cells in vitro incubated in hypoxic solutions.

The possibility that ACh effects on the carotid body were vascularly mediated was discarded by the persistence of such effects on cat carotid bodies superfused in vitro (Eyzaguirre and Koyano, 1965). The sensitivity to ACh and nicotine of petrosal ganglion cells provided further support to the idea that chemosensory neurons were indeed cholinoceptive (Fitzgerald et al., 2000; Alcayaga et al., 1998). Furthermore, Fitzgerald and Shirahata (1994) reported that selective perfusion of cat's carotid bifurcation with a mixture of cholinergic antagonists (α-bungarotoxin, mecamylamine and atropine) reduces the chemosensory response of the carotid body to hypoxia, while carotid sinus barosensory discharges were unaffected.

The **dopaminergic hypothesis** was based on the detection of high levels of dopamine (DA) within the carotid bodies of all mammalian species studied (see Eyzaguirre et al., 1983). The observation by Fidone et al. (1982) that ^{3}H-DA outflow was highly correlated with the degree of hypoxia to which rabbit carotid bodies superfused in vitro had been exposed revealed that hypoxia releases DA from glomus cells and gave strong support to the idea that DA was the chemical transmitter between such cells and sensory nerve endings. The later availability of voltammetric-amperometric techniques employing carbon fibre electrodes made possible the fast temporal resolution of DA release from the carotid body, confirming that hypoxia was a strong stimulus for DA release from the organ, but revealing that such DA release was slower than chemosensory excitation (Donnelly, 1993; Buerk et al., 1995; Iturriaga et al., 1996).

The main problem with the dopaminergic hypothesis is that DA inhibits carotid body chemosensory discharges in most preparations and depresses ventilation in most mammalian species, with the possible exception of the rabbit. The transient increases in carotid body chemosensory discharges or ventilation observed after administration of dopaminergic blockers suggest that endogenous DA released from glomus cells may serve as an inhibitory modulating agent (see Zapata, 1997).

The **purinergic hypothesis**. The changing levels of ATP within carotid body tissues and the chemosensory excitatory effects observed upon its administration were initially considered as part of ATP metabolic role, but the demonstration of membrane receptors for ATP in other tissues made necessary to consider its possible role as transmitter between glomus cells and sensory nerve endings. Thus, $P2X_2$ and $P2X_3$ receptors subunits were detected in afferent nerve terminals surrounding clusters of glomus cells, as well as in the perikarya of many petrosal ganglion neurons (Prasad et al., 2001). Recently, Conde and Monteiro (2006) report that incubated carotid bodies release larger amounts of ATP when exposed to hypoxic media than when exposed to normoxic and hyperoxic media.

McQueen et al. (1998) reported that ATP evokes cardiorespiratory effects in rats, but that its antagonists suramin and PPADS did not affect carotid body chemosensory responses to cyanide-induced hypoxia or to asphyxia, while Rong et al. (2003) reported that PPADS reduces hypoxia-induced carotid nerve discharge in mice.

The **cholinergic-purinergic hypothesis**. Colin Nurse and co-workers reported that synapses re-established in co-cultures of rat's glomus cells with sensory ganglion neurons had their hypoxic-induced discharges only partially blocked by nicotinic cholinergic antagonists (hexamethonium or mecamylamine), or by P2X purinergic blocker suramin, but that such evoked activity was suppressed by a mixture of mecamylamine and suramin, leading to the proposal that ACh and ATP were co-released from glomus cells during hypoxic stimulation, and thus responsible for the ensuing excitation of the juxtaposed sensory neurons (Zhang et al., 2000; Nurse and Zhang, 2001). While ACh and ATP were confirmed as strong stimuli for chemosensory discharges recorded from cat's carotid bodies in situ and in vitro, and these responses were respectively blocked by mecamylamine and suramin, the combination of such antagonists did not suppress chemosensory excitation evoked by hypoxic stimulation (see Zapata, 2007).

6 Chemoreflexes Originated from the Carotid Bodies

The **respiratory response** to carotid body stimulation seems to be the most obvious and the most powerful of the chemoreflex responses. Increases in tidal volume and frequency, end expiratory volume, airways secretions, and airways resistance are the most prominent respiratory responses to hypoxic, hypercapnic or acidotic stimulation of the carotid bodies. The Wisconsin group led by Bisgard, Dempsey, and Forster and more recently by Mitchell has contributed further to these studies with unanesthetized goats and other species. References to the abundant literature on this issue can be found in the reviews by Fitzgerald and Lahiri (1986), and Fitzgerald and Shirahata (1997).

A problem to be solved was the role played by the carotid bodies under resting (normoxic, eucapnic) conditions. Pierre Dejours (1957) proposed to test such "chemosensory drive" by applying an abrupt and brief ventilation of pure oxygen to silence arterial chemoreceptors and therefore withdraw chemoreflex influences. This Dejours' maneuver results in an almost immediate but brief decrease in tidal volume and/or respiratory rate. This hyperoxia-evoked transient hypoventilation was absent in animals after bilateral sectioning of carotid sinus nerves, as well as in humans previously subjected to bilateral glomectomy (Honda et al., 1979). It must be mentioned that awake dogs remain hypercapnic for at least 19 days after removal of their carotid bodies (Rodman et al., 2001).

The **cardiovascular responses** to carotid body stimulation have been a subject of controversy. They are less intense and somehow slower than respiratory responses, they are modified by respiratory responses, and some of them may be secondary to respiratory responses. It must be mentioned that Andrzej Trzebski and associates in Warsaw reported a transient reduction in arterial pressure upon breathing 100% O_2 in hypertensive patients. In view of recent studies by Schultz and coworkers (2007) showing the prominent role played by the carotid bodies in chronic heart failure, it might be helpful to review some of the findings from the mid-20th century. Whereas stimulation of the carotid sinus as, for example, in high blood pressure acts as a brake on sympathetic activity, carotid body stimulation increases sympathetic output. Carotid body stimulation initially produces bradycardia which upon continued stimulation is followed by tachycardia. The impact of carotid body stimulation on cardiac contractility is still under debate. Experimental evidence supports both an increase and a decrease, differences arising from methods of measurement, species under study, state of wakefulness, sleep, or anesthesia, and control of ventilatory conditions (see reviews by Eyzaguirre et al., 1983; Fitzgerald and Shirahata, 1997).

Carotid body-stimulated increases in sympathetic output constrict the vasculature in most, but not all vascular beds. Carotid body stimulation blunts the classical hypoxic pulmonary vasoconstrictor response. It also provokes a dilation in the bronchial vasculature. Coronary circulation is certainly impacted by carotid body stimulation, but this is a classical example of the difficulty of predicting the outcome of the stimulation due to the presence of other reflex responses impacting on the vascular resistance. The increases in sympathetic output also produce a decrease in venous capacitance, which will in turn increase venous return and therefore cardiac output. The impact of carotid body stimulation on the cerebral vasculature remains perhaps the most controversial area. However, it appears that chemoreceptor stimulation does reduce vascular resistance in the neurohypophysis and in the eye. (See Fitzgerald and Shirahata, 1997).

7 Concluding Remarks

In summary, we can say that the last 50 years have produced a *dramatic advance* (see Figs. 1, 2 and 3) in our understanding of:

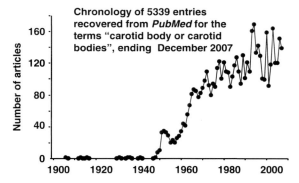

Fig. 1 Chronology of 5339 entries recovered from *PubMed* (service of the National Library of Medicine) for the terms "carotid body or carotid bodies", for the period ending at December 2007. For the same period, only 269 papers appeared for the terms "aortic body or aortic bodies", and 2604 for the term "arterial chemoreceptors". Number of papers on the ordinate; year of publication on the abscissa. The first 5 papers (from 1906 to 1935) are reports on carotid body tumors

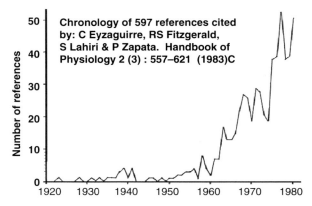

Fig. 2 Chronology of 597 papers cited as references for the chapter on "Arterial chemoreceptors" of the "Handbook of Physiology" (Section 2, Volume 3), printed in 1983. Number of papers on the ordinate; year of publication on the abscissa. The first reference corresponds to a paper by MW Gerard and PR Billingsley (1923) on the innervation of the carotid body. The second paper (1928) is De Castro's foundational contribution to this topic

Levels:	Studies:
Molecules	O_2-dependent ion channels
Organelles	DA uptake in isolated granules from glomus cells
Cells	Membrane potential changes in glomus cells
	Intracellular Ca^{++} changes in glomus cells
Tissue	Gap junctions between glomus cells
	Glomus cells to nerve endings transmission
Organ	Functional organization of the CB
	O_2 consumption of CB in vitro
System	Ventilatory reflexes originated from CBs
	Cardiovascular adjustments evoked from CB
Organism	Arousal evoked from CBs excitation
	'Defense' reaction evoked from CBs
Environmental	CB adjustment at high altitude
	CB function after chronic hypoxic challenges
Populations	Genetic traits in CB tumours
	Geographical distribution of CB tumors

Fig. 3 Levels of studies performed on the carotid body. Most studies on molecules, organelles, organ and tissue had been performed along the last 50 years, as well as those on environmental and populations levels

- what carotid body's impact is on the organism's need to maintain or reestablish homeostasis;
- what mechanisms are involved in generating the neural traffic from the carotid body to the nucleus tractus solitarius;
- which activity, passed on to various other nuclei in the brainstem and down the cord, initiates and sustains homeostasis-benefitting reflex responses;
- Not all controversies, conflicts of data, contradictory interpretations have been resolved, but at least an overall general agreement regarding the basic events required to increase CB outward neural traffic to the nucleus tractus solitarius seems to exist;
- Up to the middle of the 20th century most of today's remaining questions weren't even being asked.

References

Alcayaga J, Iturriaga R, Varas R, Arroyo J, Zapata P (1998) Selective activation of carotid nerve fibres by acetylcholine applied to the cat petrosal ganglion in vitro. *Brain Res* 786: 47–54

Biscoe TJ, Duchen MR (1990) Responses of type I cells dissociated from the rabbit carotid body to hypoxia. *J Physiol* 428: 39–59

Buerk DG, Lahiri S, Chugh D, Mokashi A (1995) Electrochemical detection of rapid DA release kinetics during hypoxia in perfused-superfused cat CB. *J Appl Physiol* 78: 830–7

Campanucci VA, Nurse CA (2007) Autonomic innervation of the carotid body: Role in efferent inhibition. *Respir Physiol Neurobiol* 157: 83–92

Conde SV, Monteiro EC (2006) Profiles for ATP and adenosine release at the carotid body in response to O_2 concentrations. *Adv Exp Med Biol* 580: 179–84

Dejours P (1957) Methodological importance of the study of a living organism at the initial phase of interruption of a physiological equilibrium. *C R Hebd Seances Acad Sci* 245: 1946–8

Donnelly DF (1993) Electrochemical detection of catecholamine release from rat carotid body in vitro. *J Appl Physiol* 74: 2330–7

Eyzaguirre C (2007) Electric synapses in the carotid body-nerve complex. *Respir Physiol Neurobiol* 157: 116–122

Eyzaguirre C, Fitzgerald RS, Lahiri S, Zapata P (1983) Arterial chemoreceptors. In: *American Physiological Society: Handbook of Physiology*, sect. 2: The Cardiovascular System, vol. 3: Peripheral Circulation and Organ Blood Flow. Baltimore, MD: Williams & Wilkins, pp. 557–621

Eyzaguirre C, Koyano H (1965) Effects of hypoxia, hypercapnia, and pH on the chemoreceptor activity of the carotid body in vitro. *J Physiol* 178: 385–409

Eyzaguirre C, Koyano H, Taylor JR (1965) Presence of acetylcholine and transmitter release from carotid body chemoreceptors. *J Physiol* 178: 463–76

Eyzaguirre C, Monti-Bloch L, Baron M, Hayashida Y, Woodbury JW (1989) Changes in glomus cell membrane properties in response to stimulants and depressants of carotid nerve discharge. *Brain Res* 477: 265–279

Fidone S, González C, Yoshizaki K (1982) Effects of low oxygen on the release of dopamine from the rabbit carotid body in vitro. *J Physiol* 333: 93–110

Fitzgerald RS, Lahiri S (1986) Reflex responses to chemoreceptor stimulation. In: *American Physiological Society (eds) Handbook of Physiology*, sect. 3, vol. 2, pp. 313–362

Fitzgerald RS, Shirahata M (1994) Acetylcholine and carotid body excitation during hypoxia in the cat. *J Appl Physiol* 76: 1566–1574

Fitzgerald RS, Shirahata M (1997) Systemic responses elicited by stimulating the carotid body: primary and secondary mechanisms. In: González C (ed) *The Carotid Body Chemoreceptors.* Berlin: Springer-Verlag, pp. 171–191

Fitzgerald RS, Shirahata M, Wang HY (2000) Acetylcholine is released from in vitro cat carotid bodies during hypoxic stimulation. *Adv Exp Med Biol* 475: 485–94

González C, Almaraz L, Obeso A, Rigual R (1994) Carotid body chemoreceptors: From natural stimuli to sensory discharges. *Physiol Rev* 74: 829–898

Honda Y, Watanabe S, Hashizume I, Satomura Y, Hata N, Sakakibara Y, Severinghaus JW (1979) Hypoxic chemosensitivity in asthmatic patients two decades after carotid body resection. *J Appl Physiol* 46: 632–8

Iturriaga R, Alcayaga J, Zapata P (1996) Dissociation of hypoxia-induced chemosensory responses and catecholamine efflux in cat carotid body superfused in vitro. *J Physiol* 497: 551–64

López-Barneo J, López-López JR, Ureña J, González C (1988) Chemotransduction in the carotid body: K^+ current modulated by PO_2 in type I chemoreceptor cells. *Science* 241: 580–582

McDonald DM (1981) Peripheral chemoreceptors: structure-function relationships of the carotid body. In: Hornbein TF (ed) *Regulation of Breathing. Lung Biology in Health and Disease*, vol 17. NY: Marcel Dekker. pp. 105–319

McQueen DS, Bond SM, Moores C, Chessell I, Humphrey PP, Dowd E (1998) Activation of P_2X receptors for adenosine triphosphate evokes cardiorespiratory reflexes in anaesthetized rats. *J Physiol* 507: 84–55

Mulligan E, Lahiri S (1981) Dependence of carotid chemoreceptor stimulation by metabolic agents on P_aO_2 and P_aCO_2. *J Appl Physiol* 50: 884–891

Nurse CA, Zhang M (2001) Synaptic mechanisms during re-innervation of rat arterial chemoreceptors in co-culture. *Comp Biochem Physiol A - Mol Integr Physiol* 130: 241–251

O'Regan RG (1979) Responses of the chemoreceptors of the cat carotid body perfused with cell-free solutions. *Ir J Med Sci* 148: 78–85

Pardal R, Ortega-Sáenz P, Durán R, López-Barneo J (2007) Glia-like stem cells sustain physiologic neurogenesis in the adult mammalian carotid body. *Cell* 131: 364–377

Peers C, Buckler KJ (1995) Transduction of chemostimuli by the type I carotid body cell. *J Memb Biol* 144: 1–9

Prasad M, Fearon IM, Zhang M, Laing M, Vollmer C, Nurse CA (2001) Expression of P_2X_2 and P_2X_3 receptor subunits in rat carotid body afferent neurones: role in chemosensory signalling. *J Physiol* 537: 667–77

Rodman JR, Curran AK, Henderson KS, Dempsey JA, Smith CA (2001) Carotid body denervation in dogs: eupnea and the ventilatory response to hyperoxic hypercapnia. *J Appl Physiol* 91: 328–35. Erratum in: *J Appl Physiol* 91(5): following table of contents

Rong W, Gourine AV, Cockayne DA, Xiang Z, Ford APDW, Spyer KM, Burnstock G (2003) Pivotal role of nucleotide $P2X_2$ receptor subunit of the ATP-gated ion channel mediating ventilatory responses to hypoxia. *J Neurosci* 23: 11315–11321.

Schultz HD, Li YL, Ding Y (2007) Arterial chemoreceptors and sympathetic nerve activity: implications for hypertension and heart failure. *Hypertension* 50: 6–13

Shirahata M, Fitzgerald RS (1996) Release of acetylcholine from cultured cat and pig glomus cells. *Adv Exp Med Biol* 410: 233–237

Verna A (1997) The mammalian carotid body: morphological data. In: González C (ed) *The Carotid Body Chemoreceptors.* Berlin: Springer; Austin, TX: Landes, pp. 1–29

Zapata P (1997) Chemosensory activity in the carotid nerve: Effects of pharmacological agents. In: González C (ed) *The Carotid Body Chemoreceptors.* Berlin: Springer-Verlag, pp. 119–146

Zapata P (2007) Is ATP a suitable co-transmitter in carotid body arterial chemoreceptors? *Respir Physiol Neurobiol* 157: 106–115

Zhang M, Zhong HJ, Vollmer C, Nurse CA (2000) Co-release of ATP and ACh mediates hypoxic signalling at rat carotid body chemoreceptors. *J Physiol, Lond* 525: 143–158

Carotid Body: New Stimuli and New Preparations – *Invited Article*

Colin A. Nurse

Abstract Beginning with the pioneering work of Heymans and collaborators in the 1930's, investigations into the role of the mammalian carotid body (CB) in the control of ventilation have attracted much attention. Progress for many years was restricted to the whole animal and organ level, resulting in characterization of the stimulus-response characteristics of the CB with its afferent nerve supply during exposure to chemostimuli such as low PO_2 (hypoxia), elevated PCO_2 (hypercapnia), and low pH (acidity). Major advances on the cellular and molecular mechanisms of chemotransduction occurred ~20 years ago with the use of freshly-dissociated CB preparations and single cell studies using patch clamp and spectrofluorimetric techniques. This review will focus on more recent advances based on novel preparations including co-cultures of isolated CB receptor clusters and dispersed sensory or autonomic neurons, thin CB tissue slice preparations, and transgenic models. These preparations have contributed significantly, not only to our understanding of the transduction and neurotransmitter mechanisms that operate in the CB during sensory processing, but also to the identification and characterization of novel CB stimuli such as hypoglycemia. Though the complexity of this remarkable organ still belies its tiny size, these recent advances are slowly unraveling the intricacies surrounding its ability to act as a polymodal detector of blood-borne chemicals and to alter its sensitivity to patterned stimuli.

Keywords Type I cells · Petrosal neurons · Carotid body co-cultures · Carotid body slice · Chemostimuli · Hypoglycemia

1 Introduction

The mammalian carotid body (CB) is a small chemoreceptor organ that senses chemicals in arterial blood and initiates compensatory reflex responses via increased afferent nerve activity, resulting in corrective changes in ventilation (Gonzalez

C.A. Nurse (✉)
Department of Biology, McMaster University, 1280 Main St. West, Hamilton, Ontario, Canada L8S 4K1
e-mail: nursec@univmail.cis.mcmaster.ca

et al. 1994). Aided by a rich vascularization, blood-borne chemical signals such as low PO_2 (hypoxia), elevated PCO_2 (hypercapnia), and low pH (acidosis) are transduced by endogenous chemoreceptor (type I) cells, which in turn release excitatory neurotransmitters onto terminals of the carotid sinus nerve (CSN). Our early understanding of the role of the CB in the control of ventilation was based on whole animal studies, aided subsequently by extracellular recordings of sensory discharge from the intact CSN. However, major breakthroughs on cellular and molecular mechanisms of chemotransduction occurred in the late 1980's and early 1990's with the application of patch clamp electrophysiological techniques and the monitoring of intracellular calcium signals in single isolated type I cells, following enzymatic dissociation of the CB. The reader is referred to other reviews for a comprehensive summary of these studies, as well as more recent advances using these 'reduced' single cell preparations (Gonzalez et al. 1994; Peers and Buckler 1995; Lopez-Barneo et al. 2001). A major focus of the present review is on more recent studies following the advent of other reduced preparations that came to fruition in the late 1990's and early 2000's. The following preparations have aided significantly in our understanding of sensory processing in the CB: (i) monolayer co-culture preparations of rat type I cell clusters and dissociated afferent (petrosal) and efferent (glossopharyngeal) neurons; (ii) thin living CB tissue slice preparations; and (iii) transgenic mice with knockout of specific protein targets. The main contributions from these preparations will be discussed in the following sections, and in some cases unifying results from the different preparations will be integrated for ease of presentation.

2 Co-Culture Preparations of Rat Carotid Body

In order to study synaptic transmission and afferent signaling in the CB, a co-culture model was developed and formally introduced in the late 1990's (Zhong et al. 1997). A central feature of this model is the development de novo of functional chemosensory units comprising type I receptor clusters and sensory (petrosal) neurons in monolayer cultures. A phase contrast photomicrograph from a typical co-culture of rat type I cells and petrosal neurons is shown in Fig. 1A. As is evident from this figure, in some cases both the neuronal soma and type I cell cluster may be fortuitously juxtaposed, thereby increasing the chances for successful contacts. In preparing these co-cultures, the usual first step is to plate dissociated CB cells such that islands or clusters of type I cells, arising from incomplete dissociation of the tissue, settle on the culture substrate. During the first few days in culture background cells including fibroblasts and other non-neuronal cells proliferate, while the type I cell islands (and contiguous glial-like type II cells) remain relatively stable even though type I cells do have the capacity for division (Nurse, 2005; Pardal et al. 2007).

The co-culture is then initiated a few days later by adding an overlay of enzymatically-dissociated neurons from the petrosal ganglia. Consequently, the co-culture contains a random distribution of type I clusters (2–20 cells) and individual

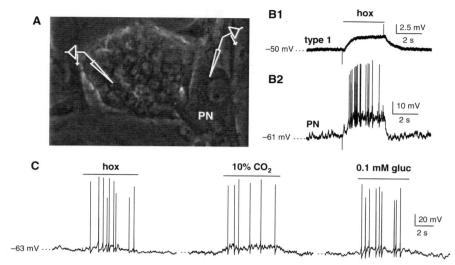

Fig. 1 Sensory transmission in co-cultures of rat type I cells and petrosal neurons. (A) Phase contrast photomicrograph of a typical co-culture containing a type I cluster and juxtaposed petrosal neuron (PN). Simultaneous dual perforated patch recordings can be obtained from both a presynaptic type I cell (within the cluster) and nearby postsynaptic PN as illustrated by the electrode placements in A. (B) Fast application of a hypoxic stimulus (PO$_2$ ~ 5 mmHg) caused a depolarizing receptor potential in the type I cell that was coincident with a suprathreshold chemoexcitatory response in the adjacent PN. (C) Recording from a functional PN in co-culture showed that the same chemosensory unit could be excited by three stimuli, i.e. hypoxia (*left*), isohydric hypercapnia (10% CO$_2$, pH=7.4; *middle*), and low glucose (0.1 mM; *right*)

neurons within a monolayer (Fig. 1A). Though not easily visible under phase contrast microscopy, neurofilament-positive processes extend from the petrosal cell bodies and ramify over the background cell monolayer, sometimes contacting type I clusters (Zhong et al. 1997). The key feature is that functional synaptic interactions often develop between outgrowing petrosal neurites and neighboring type I clusters, or between petrosal somas and contiguous type I cells. The main advantage of the preparation is that both presynaptic (i.e. type I cells) and postsynaptic (i.e. petrosal neuron) elements are clearly visible under phase or differential interference contrast optics and can therefore be conveniently studied by electrophysiological and/or imaging techniques (Fig. 1A). This contrasts with another commonly-used in vitro model consisting of the *intact* CB-attached sinus nerve preparation, where access and visibility at the synaptic sites are severely hampered, and recording of afferent activity is limited to propagating action potentials rather than synaptic potentials (Nurse 2005; Kumar and Bin-Jaliah 2007).

It should also be noted that as a result of incomplete dissociation, many type I cells in culture will have retained contacts with their normal type I cell neighbors as initially present in situ. The preservation of these contacts is advantageous because, unlike the situation for completely dissociated single type I cells, autocrine and paracrine interactions as well as electrical coupling can contribute

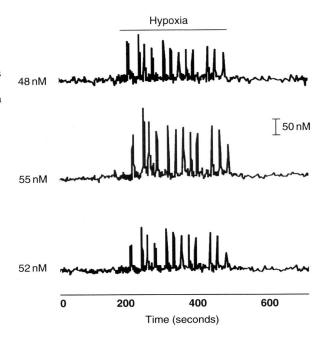

Fig. 2 Simultaneous ratiometric fura-2 intracellular Ca^{2+} measurements from type I cells within a cluster. Traces show responses from 3 non-contiguous cells within a cluster of ~12 cells before, during, and after exposure to hypoxia (PO_2 ~15 mmHg). Note intracellular calcium spikes or oscillations in each cell during hypoxia, and periods of apparent synchronous spike activity. Cell 1 (*upper trace*) was approximately 4 cell diameters away from cells 2 (*middle trace*) and 3 (*lower trace*); data from Dookhoo and Nurse, unpublished observations

normally to sensory processing under these monolayer conditions (Nurse 2005; Eyzaguirre 2007). Though not studied in detail, it is also conceivable that coordinated Ca^{2+}-signaling mechanisms that lead to synchronized activity may occur within type I cell clusters, as has been described in pancreatic ß-cells. For example, within larger clusters spontaneous Ca^{2+}-dependent action potentials have sometimes been recorded from a type I cell (Nurse 2005). More recently, our preliminary fura-2 imaging experiments have revealed intracellular Ca^{2+} spikes or oscillations in several cells within a cluster (Fig. 2), though it remains to be determined whether cell-cell interactions are required for this phenomenon.

A more recent and novel CB co-culture preparation was developed to permit studies of the role of autonomic efferent innervation on type I cell function (Campanucci et al. 2006). In this case, dissociated cholinergic and nitric oxide (NO)-positive autonomic neurons that are embedded in the glossopharyngeal and carotid sinus nerves were co-cultured with type I cell clusters. It is unclear whether *bona fide* synaptic contacts develop under these conditions, but these glossopharyngeal (GPN) neurons appear to have an 'attraction' for type I cells and form intimate, functional associations in culture (Campanucci et al. 2006).

In spite of the success of these models, the preparation of these co-cultures is both tedious and labor intensive, and therefore not amenable for routine use in most laboratories. Clearly, a readily available source of type I cells, for example an appropriate propagating cell line, would greatly facilitate studies on CB function. A particularly attractive recent development offers much promise in this area, with the discovery of glial-like stem cells in the adult rat CB (Pardal et al. 2007).

Under certain culture conditions, these stem cells form multipotent, self-renewing colonies and, dramatically, could differentiate into clones of type I cells with the same complex chemoreceptor properties of native, mature type I cells in situ.

3 Electrophysiological Experiments on Carotid Body Co-Cultures

Since its introduction the co-culture preparation of rat type I cells and petrosal neurons has been employed to uncover several neurotransmitter and neuromodulatory mechanisms that operate in the rat carotid body (Nurse 2005). In the majority of experiments reported to date, a depolarizing response obtained during perforated-patch recordings from petrosal neurons during application of the sensory stimulus has been used as a 'reporter' of functional connectivity with juxtaposed type I clusters (Fig. 1A,B). In some cases, simultaneous dual pre- and post-synaptic recordings have been obtained from a functional chemosensory unit (Zhang et al. 2007). An example is illustrated in Fig. 1B, where exposure to hypoxia caused type I cell depolarization associated with a depolarization and burst in spike activity in the adjacent neuron. Interestingly, as illustrated in Fig. 1C, the *same* chemosensory unit in co-culture could process several sensory modalities including hypoxia, hypercapnia, and the novel CB stimulus, low glucose (Pardal and López-Barneo 2002a). The fact that the neuronal afferent response to these chemostimuli could usually be blocked by a combination of purinergic (e.g. suramin) and nicotinic (e.g. mecamylamine) receptor blockers, and by low Ca^{2+}/high Mg^{2+}-containing extracellular solutions, provided compelling evidence for chemical transmission involving co-release of ATP and ACh from type I cells (Zhong et al. 1997; Zhang et al. 2000; Nurse 2005). Because in a few cases, blockade was not complete, other mechanisms including the additional co-release of 5-HT and/or electrical transmission, could not be ruled out (Zhang et al. 2000). Pharmacological characterization of the purinergic receptors expressed in identified chemosensory neurons in co-culture suggested contributions from both $P2X_2$ and $P2X_3$ subunits (Prasad et al. 2001). Importantly, $P2X_2$ and $P2X_3$ subunits were located at petrosal afferent terminals apposed to type I cells in the rat CB in situ using confocal immunofluorescence (Prasad et al. 2001). Confirmation of this prediction, together with complementary pharmacological studies in the isolated CB-sinus nerve preparation in vitro (Zhang et al. 2000), laid the cornerstone for the validity of the co-culture method as a reliable technique for studying CB function. Moreover, the subsequent demonstration that the hypoxic ventilatory response and hypoxia-induced sensory discharge in the sinus nerve were severely impaired in P2X2- (but not P2X3-) knockout mice provided compelling evidence for a central role of the purinergic $P2X_2$ subunit in sensory transmission (Rong et al. 2003). The co-culture model has also helped in uncovering autocrine/paracrine roles of CB neuromodulators. For example, endogenous release of 5-HT and GABA was found to enhance and inhibit CB chemosensory function respectively, via presynaptic G-protein coupled $5-HT_{2A}$ and $GABA_B$ receptors on type I cells (Nurse 2005).

In a more recent co-culture model of type I cells and efferent nNOS-positive glossopharyngeal (GPN) neurons, another potential role of ATP in mediating feedback inhibition was uncovered (Campanucci et al. 2006). In this study, application of ATP activated P2X receptors on juxtaposed GPN neurons, leading to type I cell hyperpolarization that was prevented by the NO scavenger, carboxy-PTIO. This raises the possibility of a dual excitatory/ inhibitory role of ATP during CB chemoexcitation. In the inhibitory pathway, released ATP may activate P2X receptors on nearby GPN terminals, causing an increase in intra-terminal Ca^{2+} that leads to activation of nNOS and NO synthesis/release.

4 Carotid Body Slice Preparation

Another key development in the late 1990's was the CB slice preparation that was elegantly exploited in several studies carried out in the laboratory of Professor José López-Barneo (Pardal et al. 2000; Pardal and López-Barneo 2002a,b). The earliest version of the CB slice preparation provided useful information on cholinergic receptive sites in the CB, but the small size of the type I cells combined with the use of intracellular electrodes resulted in low resting potentials and unstable recordings (Eyzaguirre 2007). In the more recent versions, the whole CBs are first collected in ice-cooled O_2-saturated Tyrode's solution and then embedded in a 3% (wt/vol) low-melting point agarose gel. After rapid cooling, the agarose block is glued to the stage of a vibratome or tissue vibroslicer, and 100–150 μm thick slices are cut with standard razor blades (Pardal et al. 2000; Pardal and López-Barneo 2002a,b). The slices may be used shortly after preparation or, more commonly, after culture for 24–48 h in DMEM medium supplemented with 10% fetal bovine serum (Pardal et al. 2000). The morphological appearance of a typical thin rat CB slice, viewed within a few hours after tissue section, is illustrated in Fig. 3A. After 24-48 h in culture, cell aggregates representing type I glomeruli become more easily discernible on the superficial aspects of the slice as discrete entities from the surrounding tissue, and are more accessible for cellular studies (Pardal et al. 2000; Fig. 3B). These glomeruli are immunopositive for tyrosine hydroxylase (TH), the rate-limiting enzyme in catecholamine biosynthesis, confirming their identity as type I cells (Fig. 3C). The major advantages of the slice preparation are: (i) avoidance of enzymatic and mechanical dissociation of the tissue (as in dissociated cell culture preparations), resulting is better preservation of the cellular physiology and morphology; (ii) minimal disruption of cell-to-cell contacts among neighboring type I and/or type II cells, allowing intercellular interactions to be studied; (iii) the ability to perform high-resolution patch clamp techniques on single cells within the slice using infra-red DIC optics; and (iv) the ability to apply non-invasive techniques such as carbon fiber amperometry to monitor secretory activity from single, 'intact' type I cells. For example, such tissue slice preparations have been used to confirm the role of K^+ channel inhibition, membrane depolarization, and voltage-gated Ca^{2+} entry in hypoxia-induced catecholamine (CA) secretion from intact type I cells present

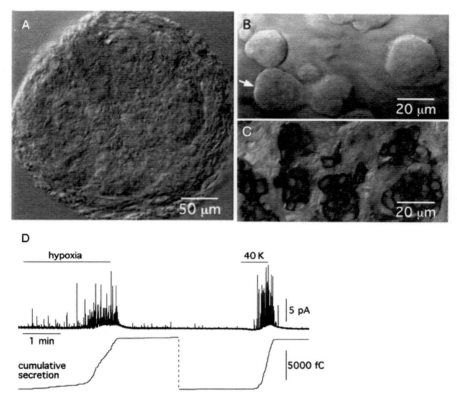

Fig. 3 Appearance and secretory function of thin slices of rat carotid body. Low power micrograph of a thin carotid body slice soon after sectioning (**A**). Note prominent clusters or glomeruli containing type I cells after culture of the slice for ~72 h (**B**). In fixed slices, these clusters are immunopositive for the catecholaminergic biosynthetic enzyme, tyrosine hydroxylase (**C**). In (**D**) carbon fiber amperometry was used to detect quantal catecholamine (CA) secretion from a type I cell in a slice preparation, following exposure to acute hypoxia (PO$_2$ ~ 20 mmHg; *upper trace*). Secretion was also evoked by the depolarizing stimulus high extracellular K$^+$. By integrating the area (i.e. charge) under the amperometric spikes, an estimate of total cumulative CA secretion during stimulus application was obtained (*lower trace*). This figure was modified from Pardal and López-Barneo (2002a; Figs. 1,5), and reproduced here with permission

within glomeruli (Pardal and López-Barneo 2002a; Fig. 3D). This preparation has also been used to shed light on a controversial issue arising from the use of dispersed cells as to whether hypoxia-induced closure of Ca^{2+}-dependent K$^+$ (BK) channels contribute to the PO$_2$ sensitivity of resting rat type I cells. While blockers of BK channel (e.g. TEA and iberiotoxin or IBTX) failed to affect membrane potential and intracellular Ca^{2+} levels in single isolated rat type I cells (Buckler 2007), these agents did indeed stimulate CA (Pardal et al. 2000) and ATP (Buttigieg and Nurse 2004) secretion from rat CB tissue slices. Because IBTX also stimulated CA secretion as determined by HPLC from clusters of cultured rat type I cells (see

Nurse 2005), it appears that at least some BK channels are open under resting (normoxic) conditions when rat type I cells are present in clusters or glomeruli, but may not be, when present as single dissociated cells. While the importance of isolated single-cell studies has been amply demonstrated, the above studies emphasize the additional need for studying type I cells within their native clustered arrangement for a complete understanding of CB function.

5 The Carotid Body as a Glucosensor: Contributions from Slice and Co-Culture Preparations

The slice preparation played a pioneering role in expanding our appreciation of the CB as a polymodal chemosensor and, particularly, in providing the initial evidence that type I cells may act as direct glucosensors (Pardal and López-Barneo 2002b). Though the CB has long been proposed to participate in glucose homeostasis, carbon fiber amperometric studies using the slice preparation provided the first compelling evidence that low glucose stimulated CA release from type I cells, an effect that was abolished by blockers of voltage-gated Ca^{2+} channels (Pardal and López-Barneo 2002b). Similarly, in fresh CB tissue slices low glucose was found to stimulate Ca^{2+}-dependent secretion of a key excitatory CB neurotransmitter, ATP (Zhang et al. 2007). In more recent electrophysiological studies on CB slices and cultured type I cells, hypoglycemia induced a depolarizing receptor potential due to activation of background cationic Na^+-permeable channels, possibly from the transient receptor potential C family (Garcia-Fernández et al. 2007; Zhang et al. 2007). These studies provided evidence that low glucose also caused inhibition of a voltage dependent K^+ conductance and interacted additively with low PO_2, consistent with the involvement of separate transduction pathways.

The main conclusions derived from the CB slice model, concerning the mechanisms of glucosensing and the interaction between PO_2 and low glucose, have received strong support from the co-culture model. For example, in simultaneous recordings from a type I cell and juxtaposed petrosal neuron in co-culture, physiological hypoglycemia caused type I cell depolarization and an increase in petrosal sensory discharge (Zhang et al. 2007). An important common feature in physiological studies on the thin CB slice and co-culture preparations is that all chemostimuli, including low glucose, have *direct* access to the type I cells when applied to the bathing solution. As is evident from the above discussion, these two experimental approaches have led to the simplest conclusion that the action of low glucose on type I cells is direct. This contrasts with the conclusions from other studies based on the superfused, intact CB-sinus nerve preparation in vitro (Kumar and Bin-Jaliah 2007). In these studies, low glucose solutions had no effect on sensory discharge in the isolated whole CB-sinus nerve preparation in vitro, even though insulin-induced mild hypoglycemia caused hyperventilation in the whole animal that was prevented by CB denervation (Kumar and Bin-Jaliah 2007). Though these studies supported a role of the CB in the counter-regulatory response to hypoglycemia, they argue for an *indirect*, rather than a direct, action of low glucose on CB receptors. The reasons for these discrepancies still await final resolution. However, in view of the negative findings, it is important that there be independent verification that glucose concentrations at the chemoreceptive sites are comparable to those in the bathing solution in these superfused (or perfused) in vitro whole CB preparations.

6 Transgenic Mouse Models

Another recent and important development in the carotid body field is the use of transgenic mouse models. One key contribution, i.e. the use of $P2X_2$ and $P2X_3$ single and double knockout mice to demonstrate the central role of ATP and the $P2X_2$ subunit in CB sensory transmission (Rong et al. 2003), has already been mentioned above. The reader is referred to a recent comprehensive review on the contributions of transgenic models to our understanding of CB function (Ortega-Sáenz et al. 2007). These models have included transgenic mice deficient in putative O_2-sensor proteins, e.g. NADPH oxidase, hemeoxygenase-2, succinate dehydrogenase (SDHD), and HIF-1α. While these models will become increasingly important in our quest to understand CB function, interpretation of the results will need caution because of the possibility of compensatory developmental changes, and the fact that the transduction pathways (e.g. O_2-sensitive K^+ channel subtypes) may vary among different species.

Acknowledgments The work attributed to the author's laboratory was supported by grants from the Canadian Institutes of Health Research (MOP 12037). I am indebted to Cathy Vollmer for expert technical assistance in developing the co-culture preparations, and to several colleagues especially Min Zhang, Huijun Zhong, and Verónica Campanucci.

References

Buckler, K.J. (2007). TASK-like potassium channels and oxygen sensing in the carotid body. *Respir. Physiol. Neurobiol.* 157, 55–64.

Buttigieg, J., Nurse, C.A. (2004). Detection of hypoxia-evoked ATP release from chemoreceptor cells of the rat carotid body. *Biochem. Biophys. Res. Comm.* 322, 82–87.

Campanucci, V.A., Zhang, M., Vollmer, C., Nurse, C.A. (2006). Expression of multiple P2X receptors by glossopharyngeal neurons projecting to rat carotid body O_2-chemoreceptors: role in nitric oxide-mediated efferent inhibition. *J. Neurosci.* 26, 9482–9493.

Eyzaguirre, C. (2007). Electrical synapses in the carotid body-nerve complex. *Respir. Physiol. Neurobiol.* 157, 116–122.

Garcia-Fernández, M., Ortega-Sáenz, P., Castellano, A., López-Barneo, J. (2007). Mechanisms of low-glucose sensitivity in carotid body glomus cells. *Diabetes* 56, 2893–2900.

Gonzalez, C., Almaraz, L., Obeso, A., Rigual, R. (1994). Carotid body chemoreceptors: from nature stimuli to sensory discharges. *Physiol. Rev.* 74, 829–898.

Kumar, P., Bin-Jaliah. (2007). Adequate stimuli of the carotid body: more than an oxygen sensor? *Respir. Physiol. Neurobiol.* 157, 12–21.

Lopez-Barneo, J., Pardal, R., Ortega-Saenz, P. (2001). Cellular mechanisms of oxygen sensing. *Ann. Rev. Physiol.* 63, 259–287.

Nurse, C.A. (2005). Neurotransmission and neuromodulation in the chemosensory carotid body. *Auton. Neurosci.* 120(1–2), 1–9.

Ortega-Sáenz, P., Pascual, A., Piruat, J.L., López-Barneo, J. (2007). Mechanisms of acute oxygen sensing by the carotid body: lessons from genetically modified animals. *Respir. Physiol. Neurobiol.* 157, 140–147.

Pardal, R., López-Barneo, J. (2002a). Carotid body thin slices: responses of glomus cells to hypoxia and K^+ channel blockers. *Respir. Physiol. Neurobiol.* 132, 69–79.

Pardal, R., López-Barneo, J. (2002b). Low glucose-sensing cells in the carotid body. *Nature Neurosci.* 5, 197–198.

Pardal, R., Ludewig, U., Garcia-Hirschfeld, J., Lopez-Barneo, J. (2000) Secretory responses of intact glomus cells in thin slices of rat carotid body to hypoxia and tetraethylammonium. *Proc. Natl. Acad. Sci. USA* 97(5), 2361–2366

Pardal, R., Ortega-Sáenz, P., Durán, R., López-Barneo, J. (2007). Glia-like stem cells sustain physiologic neurogenesis in the adult mammalian carotid body. *Cell* 131, 364–367.

Peers, C., Buckler K.J. (1995). Transduction of chemostimuli by the type I carotid body cell. *J. Membr. Biol.* 144, 1–9.

Prasad, M., Fearon, I.M., Zhang, M., Laing, M., Vollmer, C., Nurse, C.A., 2001. Expression of $P2X_2$ and $P2X_3$ receptor subunits in rat carotid body afferent neurones: role in chemosensory signalling. *J. Physiol.* 537, 667–677.

Rong, W., Gourine, A.V., Cockayne, D.A., Xiang, Z., Ford, A.P.D.W., Syyer, K.M. (2003). Pivotal role of nucleotide $P2X_2$ receptor subunit of the ATP-gated ion channel mediating ventilatory responses to hypoxia. *J. Neurosci.* 23, 11315–11321.

Zhang, M., Zhong, H., Vollmer, C., Nurse, C.A. (2000). Co-release of ATP and ACh mediates hypoxic signalling at rat carotid body chemoreceptors. *J. Physiol.* 525, 143–158.

Zhang, M., Buttigieg J., Nurse, C.A. (2007). Neurotransmitter mechanisms mediating low-glucose signaling in co-cultures and fresh tissue slices of rat carotid body. *J. Physiol.* 578, 735–750.

Zhong, H., Zhang, M., Nurse, C.A. (1997). Synapse formation and hypoxic signalling in co-cultures of rat petrosal neurons and carotid body type I cells. *J. Physiol.* 503, 599–612.

Enzyme-Linked Acute Oxygen Sensing in Airway and Arterial Chemoreceptors – *Invited Article*

Paul J. Kemp and C. Peers

Abstract Researchers have speculated as to the molecular basis of O_2 sensing for decades. In more recent years, since the discovery of ion channels as identified effectors for O_2 sensing pathways, research has focussed on possible pathways coupling a reduction in hypoxia to altered ion channel activity. The most extensively studied systems are the K^+ channels which are inhibited by hypoxia in chemoreceptor tissues (carotid and neuroepithelial bodies). In this review, we consider the evidence supporting the involvement of well defined enzymes in mediating the regulation of K^+ channels by hypoxia. Specifically, we focus on the roles proposed for three enzyme systems; NADPH oxidase, heme oxygenase and AMP activated protein kinase. These systems differ in that the former two utilise O_2 directly (to form superoxide in the case of NADPH oxidase, and as a co-factor in the degradation of heme to carbon monoxide, bilirubin and ferrous iron in the case of heme oxygenase), but the third responds to shifts in the AMP:ATP ratio, so responds to changes in O_2 levels more indirectly. We consider the evidence in favour of each of these systems, and highlight their differential importance in different systems and species. Whilst the evidence for each playing an important role in different tissues is strong, there is a clear need for further study, and current awareness indicates that no one specific cell type may rely on a single mechanism for O_2 sensing.

Keywords Hypoxia · NADPH-oxidase · Hemeoxygenase · AMP-activated kinase · CO · Superoxide anion · Reactive oxygen species

1 Introduction

Formation of the majority of cellular ATP requires the reduction of O_2 as the terminal step in mitochondrial electron transport. For this reason alone, most eukaryotic cells require O_2 and so can be termed "O_2 sensitive", since their behaviour will

---

P.J. Kemp (✉)
School of Biosciences, Cardiff University, Museum Avenue, Cardiff CF10 3US, UK
e-mail: kemp@cardiff.ac.uk

change if O_2 levels decline below a level critical for production of adequate amounts of ATP. However, "O_2 sensing" has come to mean something quite distinct from "O_2 sensitive" and is commonly referred to as a property of specialised tissues which act to optimise the uptake of O_2 from the atmosphere and its delivery throughout the body. These tissues include airway chemoreceptors, arterial chemoreceptors and, although not discussed in this chapter, specialized parts of the circulation. They also act as an early warning system when O_2 availability is compromised, initiating reflex responses which are designed to ensure that cells throughout the body are provided with sufficient O_2 for their varying needs.

The mechanisms by which these specialized tissues sense and respond to acute hypoxia are still topics of much research and debate. It is clear that there are several means by which a cell might act as an O_2 sensor and this diversity, together with the number of different types of O_2 sensing cells that exist, could not be discussed adequately within the confines of this article. Instead, we focus on the concept that O_2 sensing is an enzyme-based process. Several types of enzyme have been proposed as central to O_2 sensing mechanisms. These include; nicotinamide adenine dinucleotide phosphate oxidases (NADPH oxidases), hemeoxygenases (HOs), adenosine monophosphate-activated protein kinase (AMPK), cytochrome P450 monooxygenases and proline/asparagine hydroxylases. Since the proline and asparagine hydroxylases are associated with the chronic regulation of transcriptional events under the control of hypoxia inducible factor (HIF), and cytochrome P450 monooxygenases have only been implicated in vascular responses to hypoxia, these systems are not discussed here. For further information on these two enzyme systems, we refer the reader to other chapters in this volume and to several recent reviews (e.g. (Fandrey et al., 2006; Ward, 2008). Here, we discuss the potential involvement of NADPH oxidases, HOs and AMPK specifically with respect to acute O_2 sensing by airway and arterial chemoreceptors – neuroepithelial bodies (NEBs) of the lung and type I (glomus) cells of the carotid body, respectively. Central to O_2 sensing in both cell types is the hypoxic inhibition of K^+ channels, which causes membrane depolarization, Ca^{2+} influx and hence neurotransmitter release. Our proposal that O_2 sensing is an enzyme-linked process is discussed in relation to this membrane hypothesis for chemoreception.

2 NADPH Oxidase

In the field of immunology, a member of the NADPH oxidase family is well known as the source of the rapid and dramatic generation by neutrophils of reactive oxygen species (ROS) that are required to kill invading micro-organisms (Lambeth et al., 2007). A schematic of the general composition of NADPH oxidase enzymes is shown in Fig. 1A and highlights the fact that they are complex heteromultimers. Several homologues of the catalytic subunit are now known to exist, which are termed Nox1 to Nox5 as well as Duox1 and Duox2. In neutrophils, the catalytic subunit is Nox2 (also known as gp91^{phox}) and, in response to protein kinase C

Enzyme-Linked Acute Oxygen Sensing

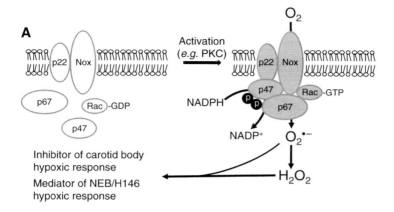

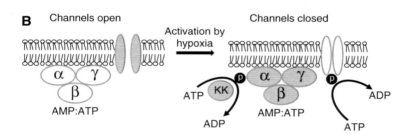

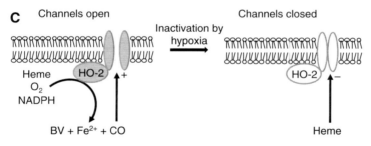

Fig. 1 Enzyme-linked mechanisms for O_2 sensing. (**A**) Activation of NADPH oxidase requires phosphorylation of p67^{phox} (p67) and p47^{phox} (p47) which leads to their recruitment into the active complex, minimally consisting of the subunits shown. Once active, this enzyme complex generates superoxide ($O_2 \cdot^-$) from molecular O_2 using NADPH as an electron donor. In the carotid body, this leads to inhibition of the hypoxic response (He et al., 2005). In NEBs (Fu et al., 2000) and H146 cells (O'Kelly et al., 2000), NADPH products are the mediators of the hypoxic response. (**B**) In the normoxic carotid body, the trimeric AMP-activated protein kinase is present but inactive. During hypoxia, the rise in the cellular AMP:ATP ratio results in phosphorylation of the α subunit by an upstream kinase (KK). This results in BK_{Ca} channel phosphorylation leading to its inhibition (Wyatt et al., 2007). (**C**) Hemeoxygenase-2 (HO-2) is constitutively active and bound to the BK_{Ca} α subunit. In the presence of molecular O_2 it produces biliverdin (BV), Fe^{2+} and CO. CO and BV are activators of BK_{Ca}. During hypoxia, production of CO BV are reduced whilst heme levels begin to rise; all these factors contribute to channel inhibition (Williams et al., 2004; Tang et al., 2003)

activation, auxiliary subunits (which include p47^{phox} and p67^{phox}) co-localise with it at the plasma membrane to form the active enzyme complex. However, it is now known that various forms of the oxidase can be constitutively active, and can also reside in differing intracellular locations.

Activity of the enzyme results in formation of superoxide from molecular O_2 using NADPH as an electron donor (Fig. 1A). Superoxide may itself act as a signalling entity, but is usually rapidly dismutated to H_2O_2 to exert biological activity. NADPH oxidase was initially proposed as an O_2 sensor in carotid body by Acker and colleagues, based in part on the ability of diphenylene iodonium (DPI), a rather non-specific inhibitor of the oxidase, to suppress hypoxic excitation of carotid body afferent sensory nerves (Cross et al., 1990). Much evidence has since accumulated to oppose this proposition. Perhaps the best evidence is genetic; hypoxic inhibition of K^+ channels in type I cells of mice lacking the gp91^{phox} subunit was similar to that seen in wild type mice (Roy et al., 2000). Furthermore, carotid bodies of mice lacking the p47^{phox} subunit (or rats exposed to the more selective Nox inhibitor AEBSF) showed greater sensitivity to hypoxia, as determined by measurements of $[Ca^{2+}]_i$ in type I cells, and afferent chemosensory discharge (He et al., 2005). Thus, inhibition of NADPH oxidase led to an augmentation of hypoxia-evoked rises in intracellular Ca^{2+} concentration ($[Ca^{2+}]_i$) in type I cells. Furthermore, the ability of hypoxia to elevate ROS was completely lacking in the p47^{phox} knockout mice, suggesting that hypoxia may actually activate the oxidase to produce ROS, which then suppress chemosensitivity.

In marked contrast to its proposed actions in the carotid body, NADPH oxidase appears to be of central importance to O_2 sensing by NEBs. This idea was initially proposed in the first report of O_2 sensitive K^+ channels in NEBS (Youngson et al., 1993). The notion gained momentum with the finding that the NEB whole-cell K^+ current was augmented by H_2O_2 (Wang et al., 1996) and the mechanistic link was further strengthened by the demonstration that K^+ currents were insensitive to hypoxia in mice engineered to express a non-functional form of gp91^{phox} (Fu et al., 2000). Furthermore, although currents were augmented by H_2O_2 in both groups, DPI inhibited K^+ currents only in wild type mice and not mutants and this effect could be reversed by H_2O_2. These compelling findings were further complemented by concurrent studies in an immortalized NEB counterpart, the H146 cell. In this model system, activation of protein kinase C increased ROS production by NADPH oxidase and counteracted hypoxic inhibition of the whole cell K^+ current (O'Kelly et al., 2000).

Thus, NAPDH oxidase can be considered an important O_2 sensory mechanism in O_2 sensing cells. However, whilst its role in NEB cells is fundamental, it has a more subtle involvement in carotid body chemoreception (Fig. 1A). Furthermore, even in specialised sensory cells, other means of O_2 sensing can co-exist with the oxidase, as observed in H146 cells (O'Kelly et al., 2001) and type I carotid body cells (Williams et al., 2004). Finally, constitutively active NADPH oxidase can also act as an O_2 sensor to regulate numerous functions in cells not considered primarily as O_2 sensing cells (e.g. vascular endothelial cells (Aley et al., 2005)).

3 Hemeoxygenase

Hemeoxygenases (HOs) are a triumvirate of enzymes which are fundamental to the breakdown of cellular heme (see (Maines and Gibbs, 2005) for a comprehensive review of the HO system). Although this family of enzymes contains two proteins – hemeoxygenase 1 (HO-1) and hemeoxygenase (HO-2) – whose enzyme activities have been well known for decades, the characterisation of the third member – Hemeoxygenase 3 (HO-3) – is still in its infancy, and substrate specificity/enzyme activity for HO-3 has yet to be firmly established. HO-1 and HO-2 catalyse the oxidation of heme, in the presence of NADPH and molecular O_2, to generate carbon monoxide (CO), iron and biliverdin (Fig. 1B); biliverdin is then rapidly converted to bilirubin by the action of biliverdin reductase. HO-2 is constitutively expressed/active in all cells thus far studied, whereas HO-1 is typically induced by a large gamut of canonical cellular stressors, such as heat shock, serum withdrawal, perturbation in redox state and chronic hypoxia/hyperoxia. Due to their ability to generate products which are antioxidant in nature, HOs have been classically considered as an important first-line of defence in combating oxidative stress in a variety of tissues. However, more recently it has been shown that HOs contribute to many other cellular processes including the immune system, the cardiovascular system, central nervous system and, most importantly for this review, acute O_2 sensing (see (Kemp, 2005; Ryter et al., 2006)). The first observation that HOs may be involved in O_2 sensing came from the work of Prabhakar and colleagues who demonstrated that HO-2 was expressed in the carotid body and that one of the products of its activity, CO, produced strong inhibition of carotid body O_2-dependent function (Prabhakar et al., 1995). Some time later, it became clear that such an inhibitory influence of CO on chemosensitivity was via specific activation of one of the K^+ channels strongly implicated in the rat glomus cell hypoxic response (Peers, 1990; Riesco-Fagundo et al., 2001; Lewis et al., 2002); this large conductance, calcium-activated K^+ channel is variably known as BK_{Ca}, maxiK and slo1.

In 2004, Williams and colleagues (Williams et al., 2004) demonstrated that recombinant BK_{Ca} was a protein partner of HO-2 and that it was activated by CO. Thus, following immunoprecipitation with an antibody directed against the α-subunit of BK_{Ca} (KCNMA1), they employed MALDI-TOF mass spectrometry to identify HO-2 as a component of the BK_{Ca} channel complex. Furthermore, using siRNA-evoked protein knock-down, they went on to demonstrate that this constitutively active enzyme was integral to the full-blown O_2-sensitivity of the channel, both in recombinant expression systems and in isolated rat glomus cells. These data led to the hypothesis that co-localization of BK_{Ca} with HO-2 resulted in tonic, CO (and possibly BV)-dependent channel activity in normoxia and that decreased O_2 might result in their reduced production and consequent channel closure (Fig. 1B). For this proposed mechanism to begin to explain hypoxia-evoked cell depolarisation, BK_{Ca} channels must contribute to resting membrane potential in normoxia, a suggestion fully borne out by the recent observation that recombinant BK_{Ca} activity is significant around resting membrane potential only in the presence of CO (Hou et al., 2008).

The molecular mechanism underlying activation of BK_{Ca} channels by CO is still controversial, although it is clear that β-subunits are not required (Williams et al., 2008; Hou et al., 2008). However, an earlier proposal that CO affects channel activity via interacting with a heme group bound to the α-subunit at histidine 616 (Jaggar et al., 2005) seems unlikely now that it is clear that neither mutation of histidine 616 to arginine (Williams et al., 2008) nor redox modulation (Williams et al., 2008; Hou et al., 2008) affects the ability of CO to activate the channel. Indeed, the binding of CO appears to be dependent either upon residues in the vicinity of the "calcium bowl" (Williams et al., 2008) or within the so-called "RCK domain" (Hou et al., 2008), or both. Such CO binding is likely dependent upon transition metal centres, although, for now, this remains conjecture.

Since the proposal of the HO-2 hypothesis, one study has probed genetically the role of this enzyme in rat carotid body oxygen-sensing. However, of genetic necessity, the model was tested using mouse carotid body (Ortega-Saenz et al., 2006). These data, employing HO-2 knock-out mice, showed very clearly that hypoxia-evoked catecholamine release from carotid body slices was essentially no different in knock-out and wild type mice. This apparent lack of involvement of HO-2 in O_2-sensitivity of the carotid body may reflect the fact that it is widely believed, based on evidence from Perez-Garcia and colleagues, that mouse carotid body does not utilise BK_{Ca} in its response to hypoxia (Perez-Garcia et al., 2004); indeed, where an involvement of BK_{Ca} has been suggested, there are even clear disparities between different mouse strains (Yamaguchi et al., 2003). Of course, based on our current appreciation that several different mechanisms exist to sense a reduction in O_2 within a single cell type, it may also be the case that role of HO-2 in glomus cell O_2-sensing might be easily accommodated by alternate O_2 sensors, especially if one of them has been ablated genetically.

4 AMP-Activated Kinase

The adenosine monophosphate activated protein kinase (AMPK) has until recent years been considered almost exclusively as a regulator or "guardian" of cellular energy status. It is widely distributed and has been implicated in numerous metabolic disorders including obesity and type II diabetes. The kinase is a trimeric protein consisting of α, β, and γ subunits (of which up to three isoforms of each are known) and, as indicated in Fig. 1C, is typically activated by a shift in the delicately balanced and dynamic AMP:ATP ratio (Hardie et al., 2006). The ubiquitous nature of this kinase would argue against a specialized role in O_2 sensing tissues, as would the widely-held idea that the O_2 sensing process does not involve a compromise of metabolism that would significantly alter ATP levels. However, studies to date suggest that it may indeed serve a fundamental role in the process of O_2 sensing. The role of AMPK in O_2 sensing by NEBs is entirely unexplored, and its role in hypoxic pulmonary vasoconstriction is considered elsewhere in this volume (see Chapter by Ward and Evans). We shall, therefore, only consider the evidence for its potential role in O_2 sensing by the carotid body. Studies thus far have relied to a large extent on the use of AICAR, a drug which is taken up into cells and metabolised to an AMP mimetic. It is now known that this compound can exert cellular effects which do not involve AMPK activation. Nevertheless, AICAR has been shown to mimic hypoxic excitation of the rat carotid body at various levels (Wyatt et al., 2007). In brief, AICAR inhibits both BK_{Ca} (Fig. 1C) and TASK-like leak K^+ channels, depolarizes type I cells, causes Ca^{2+} influx and hence a rise of glomus cell $[Ca^{2+}]_i$ and an increase in afferent chemosensory fibre discharge in a manner which is dependent on the presence of extracellular Ca^{2+}. All of these effects are similar to the actions of acute hypoxia, and can be reversed, or prevented, by the AMPK antagonist, compound C. Importantly, compound C also reverses the effects of hypoxia. These data, together with the observation that the catalytic AMPKα1 co-localizes at the plasma membrane with BK_{Ca} channels, strongly implicate AMPK as an important mediator of hypoxic chemotransduction in the carotid body.

Although the involvement of AMPK in O_2 sensing by the carotid body appears attractive, several issues remain to be determined, not least of which is whether there is a suitably sensitive shift in the AMP:ATP ratio to permit kinase activation. This is unknown in the carotid body but such a shift has been observed in the pulmonary vasculature (Evans et al., 2005). Furthermore, it remains to be established whether AMPK plays an important role in carotid body chemoreception in species other than rat. Clearly, and as discussed briefly above, mouse, rat (and rabbit) type I cells display marked differences in terms of their electrophysiological properties. Therefore, it will be particularly important to determine whether the O_2 sensitive K^+ channels of the mouse (most likely members of the Kv3.x family) or rabbit (Kv4.x family) carotid bodies are modulated by AMPK. If true, this would open up the opportunity to exploit transgenic models (e.g. targeted AMPK knockout models) and avoid potentially confounding pharmacological approaches to determining the role of this specific kinase as a mediator of hypoxic chemotransduction in the carotid body.

5 Conclusions

Since the proposal of the membrane hypothesis for cellular O_2 sensing, numerous molecular mechanisms have been suggested. Several have been subsequently discounted, but the evidence for others remains strong. However, definitive progress has been hampered by species and tissue differences, imprecise pharmacology and potential compensation following genetic manipulation. Nevertheless, it is abundantly clear that multiple systems are available for O_2 sensing, suggesting that this process is so important to cellular function that cells express a large repertoire of, possibly overlapping, systems in order to ensure rapid homeostatic responses are maintained in the face of reduced O_2. Many of these systems are enzyme-linked.

Acknowledgments The authors would like to thank the British Heart Foundation and the Wellcome Trust for continuous support for their work.

References

Aley, P. K., Porter, K. E., Boyle, J. P., Kemp, P. J., & Peers, C. (2005). Hypoxic modulation of Ca^{2+} signaling in human venous endothelial cells. Multiple roles for reactive oxygen species. *J Biol. Chem.*, 280, 13349–13354.

Cross, A. R., Herderson, L., Jones, O. T. G., Delpiano, M. A., Hentschel, J., & Acker, H. (1990). Involvement of an NAD(P)H oxidase as a pO_2 sensor protein in the rat carotid body. *Biochem. J.*, 272, 743–747.

Evans, A. M., Mustard, K. J., Wyatt, C. N., Peers, C., Dipp, M., Kumar, P., Kinnear, N. P., & Hardie, D. G. (2005). Does AMP-activated protein kinase couple inhibition of mitochondrial oxidative phosphorylation by hypoxia to calcium signaling in O_2-sensing cells? *J Biol. Chem.*, 280, 41504–41511.

Fandrey, J., Gorr, T. A., & Gassmann, M. (2006). Regulating cellular oxygen sensing by hydroxylation. *Cardiovasc. Res.*, 71, 642–651.

Fu, X. W., Wang, D., Nurse, C. A., Dinauer, M. C., & Cutz, E. (2000). NADPH oxidase is an O_2 sensor in airway chemoreceptors: evidence from K^+ current modulation in wild type and oxidase-deficient mice. *Proc. Natl. Acad. Sci. USA*, 97, 4374–4379.

Hardie, D. G., Hawley, S. A., & Scott, J. W. (2006). AMP-activated protein kinase–development of the energy sensor concept. *J. Physiol.*, 574, 7–15.

He, L., Dinger, B., Sanders, K., Hoidal, J., Obeso, A., Stensaas, L., Fidone, S., & Gonzalez, C. (2005). Effect of p47phox gene deletion on ROS production and oxygen sensing in mouse carotid body chemoreceptor cells. *Am. J. Physiol. Lung Cell. Mol. Physiol.*, 289, L916–L924.

Hou, S., Xu, R., Heinemann, S. H., & Hoshi, T. (2008). The RCK1 high-affinity Ca^{2+} sensor confers carbon monoxide sensitivity to Slo1 BK channels. *Proc. Natl. Acad. Sci. USA*, 105, 4039–4043.

Jaggar, J. H., Li, A., Parfenova, H., Liu, J., Umstot, E. S., Dopico, A. M., & Leffler, C. W. (2005). Heme is a carbon monoxide receptor for large-conductance Ca^{2+}-activated K^+ channels. *Circ. Res.*, 97, 805–812.

Kemp, P. J. (2005). Hemeoxygenase-2 as an O_2 sensor in K^+ channel-dependent chemotransduction. *Biochem. Biophys. Res. Commun.*, 338, 648–652.

Lambeth, J. D., Kawahara, T., & Diebold, B. (2007). Regulation of Nox and Duox enzymatic activity and expression. *Free Radic. Biol. Med.*, 43, 319–331.

Lewis, A., Peers, C., Ashford, M. L. J., & Kemp, P. J. (2002). Hypoxia inhibits human recombinant maxi K^+ channels by a mechanism which is membrane delimited and Ca^{2+}-sensitive. *J. Physiol.*, 540, 771–780.

Maines, M. D. & Gibbs, P. E. (2005). 30 some years of heme oxygenase: from a "molecular wrecking ball" to a "mesmerizing" trigger of cellular events. *Biochem. Biophys. Res. Commun.*, 338, 568–577.

O'Kelly, I., Lewis, A., Peers, C., & Kemp, P. J. (2000). O_2 sensing by airway chemoreceptor-derived cells: protein kinase C activation reveals functional evidence for involvement of NADPH oxidase. *J. Biol. Chem.*, 275, 7684–7692.

O'Kelly, I., Peers, C., & Kemp, P. J. (2001). NADPH oxidase does not account fully for O_2 sensing in model airway chemoreceptor cells. *Biochem. Biophys. Res. Comm.*, 283, 1131–1134.

Ortega-Saenz, P., Pascual, A., Gomez-Diaz, R., & Lopez-Barneo, J. (2006). Acute oxygen sensing in heme oxygenase-2 null mice. *J. Gen. Physiol.*, 128, 405–411.

Peers, C. (1990). Hypoxic suppression of K^+ currents in type-I carotid-body cells – selective effect on the Ca^{2+}-activated K^+ current. *Neurosci. Lett.*, 119, 253–256.

Perez-Garcia, M. T., Colinas, O., Miguel-Velado, E., Moreno-Dominguez, A., & Lopez-Lopez, J. R. (2004). Characterization of the Kv channels of mouse carotid body chemoreceptor cells and their role in oxygen sensing. *J. Physiol.*, 557, 457–471.

Prabhakar, N. R., Dinerman, J. L., Agani, F. H., & Snyder, S. H. (1995). Carbon monoxide: a role in carotid body chemoreception. *Proc. Natl. Acac. Sci. USA*, 92, 1994–1997.

Riesco-Fagundo, A. M., Perez-Garcia, M. T., Gonzalez, C., & Lopez-Lopez, J. R. (2001). O_2 modulates large-conductance Ca^{2+}-dependent K^+ channels of rat chemoreceptor cells by a membrane-restricted and CO-sensitive mechanism. *Circ. Res.*, 89, 430–436.

Roy, A., Rozanov, C., Mokashi, A., Daudu, P., Al-mehdi, A. B., Shams, H., & Lahiri, S. (2000). Mice lacking in gp91 phox subunit of NAD(P)H oxidase showed glomus cell $[Ca^{2+}]_i$ and respiratory responses to hypoxia. *Brain Res.*, 872, 188–193.

Ryter, S. W., Alam, J., & Choi, A. M. (2006). Heme oxygenase-1/carbon monoxide: from basic science to therapeutic applications. *Physiol Rev.*, 86, 583–650.

Tang, X. D., Xu, R., Reynolds, M. F., Garcia, M. L., Heinemann, S. H., & Hoshi, T. (2003). Haem can bind to and inhibit mammalian calcium-dependent Slo1 BK channels. *Nature*, 425, 531–535.

Wang, D., Youngson, C., Wong, V., Yeger, H., Dinauer, M. C., Vega-Saenz de Miera, E., Rudy, B., & Cutz, E. (1996). NADPH-oxidase and hydrogen peroxide sensitive K^+ channel may function as an oxygen sensor complex in airway chemoreceptors and small cell lung carcinoma cell lines. *Proc. Natl. Acad. Sci. USA*, 93, 13182–13187.

Ward, J. P. (2008). Oxygen sensors in context. *Biochim. Biophys. Acta*, 1777, 1–14.

Williams, S. E., Brazier, S. P., Baban, N., Telezhkin, V., Muller, C. T., Riccardi, D., & Kemp, P. J. (2008). A structural motif in the C-terminal tail of slo1 confers carbon monoxide sensitivity to human BK_{Ca} channels. *Pflugers Arch.*, 456, 561–572.

Williams, S. E., Wootton, P., Mason, H. S., Bould, J., Iles, D. E., Riccardi, D., Peers, C., & Kemp, P. J. (2004). Hemoxygenase-2 is an oxygen sensor for a calcium-sensitive potassium channel. *Science*, 306, 2093–2097.

Wyatt, C. N., Mustard, K. J., Pearson, S. A., Dallas, M. L., Atkinson, L., Kumar, P., Peers, C., Hardie, D. G., & Evans, A. M. (2007). AMP-activated protein kinase mediates carotid body excitation by hypoxia. *J Biol. Chem.*, 282, 8092–8098.

Yamaguchi, S., Balbir, A., Schofield, B., Coram, J., Tankersley, C. G., Fitzgerald, R. S., O'Donnell, C. P., & Shirahata, M. (2003). Structural and functional differences of the carotid body between DBA/2J and A/J strains of mice. *J. Appl. Physiol.*, 94, 1536–1542.

Youngson, C., Nurse, C., Yeger, H., & Cutz, E. (1993). Oxygen sensing in airway chemoreceptors. *Nature*, 365, 153–155.

Cysteine Residues in the C-terminal Tail of the Human BK$_{Ca}\alpha$ Subunit Are Important for Channel Sensitivity to Carbon Monoxide

S.P. Brazier, V. Telezhkin, R. Mears, C.T. Müller, D. Riccardi and P.J. Kemp

Abstract In the presence of oxygen (O_2), carbon monoxide (CO) is synthesised from heme by endogenous hemeoxygenases, and is a powerful activator of BK$_{Ca}$ channels. This transduction pathway has been proposed to contribute to cellular O_2 sensing in rat carotid body. In the present study we have explored the role that four cysteine residues (C820, C911, C995 and C1028), located in the vicinity of the "calcium bowl" of C-terminal of human BK$_{Ca}\alpha$-subunit, have on channel CO sensitivity. Mutant BK$_{Ca}\alpha$-subunits were generated by site-directed mutagenesis (single, double and triple cysteine residue substitutions with glycine residues) and were transiently transfected into HEK 293 cells before subsequent analysis in inside-out membrane patches. Potassium cyanide (KCN) completely abolished activation of wild type BK$_{Ca}$ channels by the CO donor, tricarbonyldichlororuthenium (II) dimer, at 100 μM. In the absence of KCN the CO donor increased wild-type channel activity in a concentration-dependent manner, with an EC$_{50}$ of ca. 50 μM. Single cysteine point mutations of residues C820, C995 and C1028 affected neither channel characteristics nor CO EC$_{50}$ values. In contrast, the CO sensitivity of the C911G mutation was significantly decreased (EC$_{50}$ ca. 100 μM). Furthermore, all double and triple mutants which contained the C911G substitution exhibited reduced CO sensitivity, whilst those which did not contain this mutation displayed essentially unaltered CO EC$_{50}$ values. These data highlight that a single cysteine residue is crucial to the activation of BK$_{Ca}$ by CO. We suggest that CO may bind to this channel subunit in a manner similar to the transition metal-dependent co-ordination which is characteristic of several enzymes, such as CO dehydrogenase.

Keywords BK · Maxi K · Carbon monoxide · CO · Cysteine · Potassium channel · Mutagenesis · KCN

S.P. Brazier (✉)
School of Bioscience, Museum Avenue, Cardiff University, Cardiff CF11 9BX, UK
e-mail: braziersp@cardiff.ac.uk

1 Introduction

Large conductance, calcium-dependent potassium channels (BK_{Ca}), expressed either natively in carotid body glomus cells (Riesco-Fagundo et al., 2001) and arterial smooth muscle (Wang et al., 1997), or heterologously in HEK 293 cells (Williams et al., 2004), are robustly activated by carbon monoxide (CO). Generation of cellular CO is largely due to the enzymatic catalysis of heme by endogenous hemeoxygenases (Tenhunen et al., 1968; Maines et al., 1986; McCoubrey et al., 1997). This process has a strict requirement for the presence of nicotinamide adenine dinucleotide phosphate (NADPH) and molecular oxygen (O_2), suggesting that this transduction pathway may contribute to cellular O_2 sensing in the carotid body. Indeed this hypothesis is strengthened by the observation that hemeoxygenase-2 is a protein partner closely associated with α-subunit of BK_{Ca} ($BK_{Ca}α$) and that channel activation evoked by the co-application of NADPH, O_2 and heme is rapidly reversed upon removal of O_2 (Williams et al., 2004). Although the role that BK_{Ca} channels play in sensing acute changes in cellular O_2 levels has been extensively studied (see (Kemp and Peers, 2007) for recent review), the mechanism by which CO can modulate $BK_{Ca}α$ activity is less clear. To date, two main theories have been proposed: (i) that CO activates $BK_{Ca}α$ directly by binding to histidine residues within the channel protein (Wang and Wu, 1997; Hou et al., 2008) and; (ii) that CO activates the $BK_{Ca}α$ indirectly via heme (Jaggar et al., 2005). However, our recent studies, and data presented herein, suggest that neither of these mechanism holds completely true and that cysteine residues within the C-terminal tail of $BK_{Ca}α$ are central to the CO binding/channel modulation (Williams et al., 2008).

2 Materials and Methods

2.1 Cell Culture

HEK293 cells were maintained in Earle's minimal essential medium (containing L-glutamine) supplemented with 10% fetal calf serum, 1% antibiotic/antimycotic and 1% non-essential amino acids (Gibco BRL, Strathclyde, U.K.) in a humidified incubator gassed with 5% CO_2/95% air. Cells were transfected with $BK_{Ca}α$ constructs using the Amaxa Nucleofector kit V, following the manufacturer's protocols (Amaxa Biosystems, Germany) and incubated for 24 hours before electrophysiological experiments were performed.

2.2 *Site-Directed Mutagenesis*

The Quikchange site-directed mutagenesis kit (Stratagene, La Jolla, CA, USA) was used to mutate four cysteine residues within the tail region of $BK_{Ca}α$ (C820, C911, C995 and C1028). Double and triple mutant constructs were also generated by performing sequential reactions with the same primer sets. The following sample reactions were set up in a final volume of 50 μl: 50ng of template DNA, 125 ng of each primer, 1 μl of dNTP mix, 5 μl of 10x reaction buffer and 2.5 Units of *Pfu-Turbo* DNA polymerase. Tubes were transferred to a thermal cycler and subjected to the following cycling parameters: 1 cycle at 95°C for 30 seconds followed by 18 cycles of 95°C 30 seconds, 55°C 1 minute and 68°C 7 minutes. Reactions were chilled on ice for 2 minutes before the addition of 10 units of *Dpn* I restriction enzyme, followed by incubation for 1 hour at 37°C. *Dpn* I treated DNA was used to transform XL-1 blue competent cells, single colonies were selected and plasmid DNA was isolated for sequencing to verify that individual cysteine residues were correctly modified. The chimeric human BK$_{Ca}$α/mSlo3 channel was generated as previously reported (Williams et al., 2008).

2.3 Electrophysiological Recordings

BK$_{Ca}$ channels were recorded from inside-out patches pulled from transiently transfected HEK293 cells. Pipette and bath solutions contained (in mM): 10 NaCl, 117 KCl, 2 MgCl$_2$, 11 N-2-hydroxyethylpiperazine-N'-2-ethanesulfonic acid (HEPES; pH 7.2) with [Ca^{2+}] adjusted to quasi-physiological 300 nM using ethylene-glycol-tetra-acetic acid (EGTA). The CO donor molecule, tricarbonyldichlororuthenium (II) dimer ([Ru(CO$_3$)Cl$_2$]$_2$), was used to generate free CO in solution (Mann and Motterlini, 2007).

Current recordings were made at room temperature using an Axopatch 200A amplifier and Digidata 1320 A/D interface (Axon Instruments, Forster City, CA, USA). BK$_{Ca}$α single channel currents were studied using two protocols: (1) continuous gap-free recording at various holding potentials, and; (2) a 1s voltage ramp from −30 mV to +120 mV, repeated at 0.1 Hz. All voltages are reported with respect to the inner membrane leaflet. All recordings were filtered with a 8-pole Bessel filter at 5 kHz and digitized at 10 kHz.

3 Results and Discussion

Previous experiments have suggested that CO interacts with the BK$_{Ca}$α subunit either via interacting with histidine residues within the RCK1 domain (Hou et al., 2008) or via heme bound to BK$_{Ca}$α (Jaggar et al., 2005). However, we found that swapping the C-terminal domain of BK$_{Ca}$α with that of mSlo3, a mammalian homologue which shows no calcium sensitivity (Schreiber et al., 1998; Moss and Magleby, 2001), produced a construct which was insensitive to CO. Figure 1A shows that BK$_{Ca}$α channel activity was robustly and reversibly activated by 30 μM CO donor. 30 μM of the breakdown product of this compound, which does not release CO (RuCl$_2$(DMSO)$_4$), did not affect channel activity (data not shown). This suggests that CO was solely responsible for activating BK$_{Ca}$α channels, a notion

Fig. 1 Effect of CO donor on wild type BK$_{Ca}$α and BK$_{Ca}$α/mSlo3 chimera channels. (**A**) Exemplar current recording from an inside-out patch excised from a HEK 293 cell transiently expressing wild type BK$_{Ca}$α channels. (**B**) Exemplar current recording from an inside-out patch excised from a HEK 293 cell transiently expressing BK$_{Ca}$α/mSlo3 chimera channels. Periods of application of 30 μM CO donor (CO donor) and 10 μM of NS1619 (BK channel opener) are shown above each recording

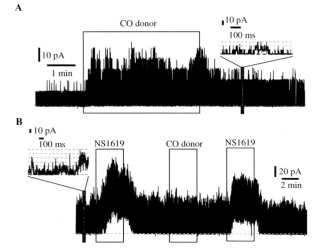

supported by the observation that application of CO gas, dissolved in intracellular solution, was also able to activate the channel (data not shown). In contrast, application of the CO donor to patches containing the BK$_{Ca}$α/mSlo3 chimera had no affect on channel activity, even though these channels were still functional as demonstrated by the application of the specific BK$_{Ca}$ channel opener NS1619 (Fig. 1B). This suggests that modulation of the BK$_{Ca}$α channel is either due to CO binding directly to amino acid residues within this tail region or by a protein partner which associates with the BK$_{Ca}$α channel via the tail domain. As both the RCK1 and the heme binding domains are still present within the BK$_{Ca}$α/mSlo3 chimera, our findings indicate that CO must be acting by a mechanism not yet fully defined.

Carbon monoxide is known to bind to a number of enzymes via metallocluster sites, which contain various metal ions (Iron, Nickel, Copper – (Lemon and Peters, 1999; Bennett et al., 2000; Drennan et al., 2001)). X-ray crystal structures of a number of these enzymes, such as carbon monoxide dehydogenase (Jeoung and Dobbek, 2007), suggest that both thiol groups of cysteine residues and imidazole side chains of histidine residues are essential for metal binding. CO can then interact with the enzymes via these bound transition metal ions.

To determine whether CO modulates channel activity by binding to a transition metal cluster within the tail domain, we have employed potassium cyanide as a substituent of the metal ion, as has been previously achieved at the active centre of carbon monoxide dehydogenase (Ha et al., 2007). In the absence of KCN, the CO donor dramatically increased patch activity and caused a shift of the activation curve to the left (Va$_{50}$ decreased from 75.3 ± 0.1 to 51.3 ± 0.1 mV; $n = 5$ – Fig. 2A). In a concentration-dependent manner, KCN attenuated the ability of 30μM CO donor either to activate the current (KCN IC$_{50}$ = 32.0 ± 3.2 μM; $n = 5$) or shift the Va$_{50}$ value (KCN IC$_{50}$ = 20.1 ± 4.2 μM; n = 5 – Fig. 2A). 100 μM KCN completely

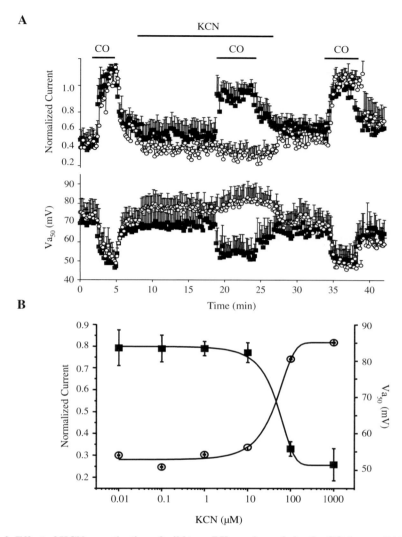

Fig. 2 Effect of KCN on activation of wild type BK$_{Ca}\alpha$ channels by the CO donor. (A) Mean (± S.E.M) normalized current time course showing the inhibitory actions of KCN on current activation (upper −Vp = +60 mV) and Va$_{50}$ (lower) evoked by 30 μM CO donor. Time courses for 10 μM KCN (■) and 1mM KCN (○) are shown. Addition of CO donor and KCN are indicated by the solid lines and labels thereon. (B) Mean (± S.E.M) concentration–response plot showing the effects of KCN on normalized current at a holding voltage of +60 mV (■) and Va$_{50}$ (○) in the presence of 30 μM CO donor

abolished activation by 30 μM CO donor (Fig. 2A). Although CN – can cleave disulphide bridges, such a reaction is not freely reversible in the absence of an oxidizing agent. Therefore, our data suggest that the mode of action for KCN is to compete for CO at the potential metal cluster of the channel.

Alignment of the amino acid sequences of the C-terminal region of human $BK_{Ca}\alpha$ and mSlo3 indicated that cysteine residues are largely conserved between these two homologues. However, four cysteine residues were identified as not being conserved between these two channels, namely residues at C820, C911, C995 and C1028. Thus, we engineered site-specific mutations, with replacement of cysteine residues for glycine residues at these amino acid positions. In a concentration-dependent manner, the CO donor increased open state probability (NPo) of wild type $BK_{Ca}\alpha$ channels with an EC_{50} of $49.5 \pm 8.4\,\mu M$ ($n = 17$; Fig. 3A and B). The single point mutations, C820G, C995G and C1028G, did not significantly change EC_{50}

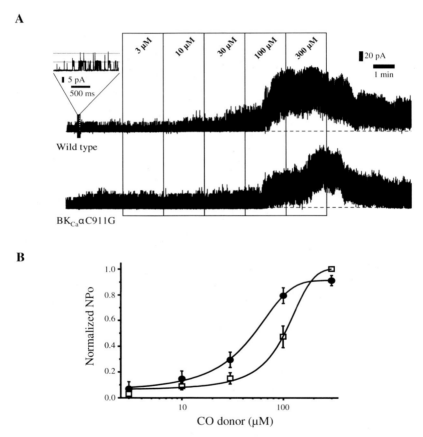

Fig. 3 Effect of a single cysteine mutation (C911G) upon channel activation by the CO donor. (**A**) Exemplar current recordings showing effect of sequential addition of increasing concentration of CO donor (3–300 μM) on wild type $BK_{Ca}\alpha$ channels (*upper*) and $BK_{Ca}\alpha$ C911G mutant channels (*lower*) at a holding potential of +40 mV; black dashed line marks the zero current level. (**B**) Mean (± S.E.M) concentration–response plots displaying the relationship between normalized NPo of wild type $BK_{Ca}\alpha$ channels (●) and $BK_{Ca}\alpha$ C911G mutant channels (□) as a function of CO donor concentration

values for activation of $BK_{Ca}\alpha$ channels by the CO donor (data not shown, $n > 10$ for each). In contrast, the CO sensitivity of the C911G mutant was significantly decreased ($EC_{50} = 103.8 \pm 8.1\,\mu M$; $n = 12$, $P < 0.05$ – Fig. 3A and B). The increase in NPo of wild type $BK_{Ca}\alpha$ channels occurred at low concentrations of CO donor, became significant at $10\,\mu M$ and plateaued at $100\,\mu M$ and above (Fig. 3A). In contrast, the C911G mutant was not significantly activated until the CO donor was raised to $30\,\mu M$ and did not reach a plateau at CO donor concentrations up to $300\,\mu M$ (Fig. 3A and B), above which the CO donor was no longer soluble. All double and triple cysteine $BK_{Ca}\alpha$ mutant constructs which contained the mutation C911G exhibited reduced CO sensitivity, with EC_{50} values shifted towards $100\,\mu M$, whilst those of $BK_{Ca}\alpha$ channel mutants which did not contain mutation at C911G displayed essentially unaltered CO donor EC_{50} values (data not shown).

These data highlight that a single cysteine residue (C911) located in the C-terminal domain of human $BK_{Ca}\alpha$ is crucial to the activation of these channels by CO. We suggest that CO may bind to this channel subunit in a manner similar to the transition metal-dependent co-ordination which is characteristic of several enzymes, such as CO dehydrogenase.

Acknowledgments This work was supported by a grant from the British Heart Foundation (RG/03/001).

References

Bennett B, Lemon BJ, & Peters JW (2000). Reversible carbon monoxide binding and inhibition at the active site of the Fe-only hydrogenase. *Biochemistry* **39**, 7455–7460.

Drennan CL, Heo J, Sintchak MD, Schreiter E, & Ludden PW (2001). Life on carbon monoxide: X-ray structure of Rhodospirillum rubrum Ni-Fe-S carbon monoxide dehydrogenase. *Proc Natl Acad Sci USA* **98**, 11973–11978.

Ha SW, Korbas M, Klepsch M, Meyer-Klaucke W, Meyer O, & Svetlitchnyi V (2007). Interaction of potassium cyanide with the [Ni-4Fe-5S] active site cluster of CO dehydrogenase from Carboxydothermus hydrogenoformans. *J Biol Chem* **282**, 10639–10646.

Hou S, Xu R, Heinemann SH, & Hoshi T (2008). The RCK1 high-affinity Ca^{2+} sensor confers carbon monoxide sensitivity to Slo1 BK channels. *Proc Natl Acad Sci USA* **105**, 4039–4043.

Jaggar JH, Li A, Parfenova H, Liu J, Umstot ES, Dopico AM, & Leffler CW (2005). Heme is a carbon monoxide receptor for large-conductance Ca^{2+}-activated K^+ channels. *Circ Res* **97**, 805–812.

Jeoung JH & Dobbek H (2007). Carbon dioxide activation at the Ni,Fe-cluster of anaerobic carbon monoxide dehydrogenase. *Science* **318**, 1461–1464.

Kemp PJ & Peers C (2007). Oxygen sensing by ion channels. *Essays Biochem* **43**, 77–90.

Lemon BJ & Peters JW (1999). Binding of exogenously added carbon monoxide at the active site of the iron-only hydrogenase (CpI) from Clostridium pasteurianum. *Biochemistry* **38**, 12969–12973.

Maines MD, Trakshel GM, & Kutty RK (1986). Characterization of two constitutive forms of rat liver microsomal heme oxygenase. Only one molecular species of the enzyme is inducible. *J Biol Chem* **261**, 411–419.

Mann BE & Motterlini R (2007). CO and NO in medicine. *Chem Commun* **41**, 4197–4208.

McCoubrey WK, Jr., Huang TJ, & Maines MD (1997). Isolation and characterization of a cDNA from the rat brain that encodes hemoprotein heme oxygenase-3. *Eur J Biochem* **247**, 725–732.

Moss BL & Magleby KL (2001). Gating and conductance properties of BK channels are modulated by the S9-S10 tail domain of the alpha subunit. A study of mSlo1 and mSlo3 wild-type and chimeric channels. *J Gen Physiol* **118**, 711–734.

Riesco-Fagundo AM, Perez-Garcia MT, Gonzalez C, & Lopez-Lopez JR (2001). O_2 modulates large-conductance Ca^{2+}-dependent K^+ channels of rat chemoreceptor cells by a membrane-restricted and CO-sensitive mechanism. *Circ Res* **89**, 430–436.

Schreiber M, Wei A, Yuan A, Gaut J, Saito M, & Salkoff L (1998). Slo3, a novel pH-sensitive K^+ channel from mammalian spermatocytes. *J Biol Chem* **273**, 3509–3516.

Tenhunen R, Marver HS, & Schmid R (1968). The enzymatic conversion of heme to bilirubin by microsomal heme oxygenase. *Proc Natl Acad Sci USA* **61**, 748–755.

Wang R & Wu L (1997). The chemical modification of KCa channels by carbon monoxide in vascular smooth muscle cells. *J Biol Chem* **272**, 8222–8226.

Wang R, Wu L, & Wang Z (1997). The direct effect of carbon monoxide on KCa channels in vascular smooth muscle cells. *Pflugers Arch* **434**, 285–291.

Williams SE, Brazier SP, Baban N, Telezhkin V, Muller CT, Riccardi D, & Kemp PJ (2008). A structural motif in the C-terminal tail of slo1 confers carbon monoxide sensitivity to human BK_{Ca} channels. *Pflugers Arch* **456**, 561–572.

Williams SE, Wootton P, Mason HS, Bould J, Iles DE, Riccardi D, Peers C, & Kemp PJ (2004). Hemoxygenase-2 is an oxygen sensor for a calcium-sensitive potassium channel. *Science* **306**, 2093–2097.

Modulation of O_2 Sensitive K^+ Channels by AMP-activated Protein Kinase

M.L. Dallas, J.L. Scragg, C.N. Wyatt, F. Ross, D.G. Hardie, A.M. Evans and C. Peers

Abstract Hypoxic inhibition of K^+ channels in type I cells is believed to be of central importance in carotid body chemotransduction. We have recently suggested that hypoxic channel inhibition is mediated by AMP-activated protein kinase (AMPK). Here, we have further explored the modulation by AMPK of recombinant K^+ channels (expressed in HEK293 cells) whose native counterparts are considered O_2-sensitive in the rat carotid body. Inhibition of maxiK channels by AMPK activation with AICAR was found to be independent of $[Ca^{2+}]_i$ and occurred regardless of whether the α subunit was co-expressed with an auxiliary β subunit. All effects of AICAR were fully reversed by the AMPK inhibitor compound C. MaxiK channels were also inhibited by the novel AMPK activator A-769662 and by intracellular dialysis with the constitutively active, truncated AMPK mutant, T172D. The molecular identity of the O_2-sensitive leak K^+ conductance in rat type I cells remains unclear, but shares similarities with TASK-1 and TASK-3. Recombinant TASK-1 was insensitive to AICAR. However, TASK-3 was inhibited by either AICAR or A-769662 in a manner which was reversed by compound C. These data highlight a role for AMPK in the modulation of two proposed O_2 sensitive K^+ channels found in the carotid body.

Keywords K^+ channel · Hypoxia · AMP kinase · maxiK channel · Leak K^+ channel · TASK channel · Patch clamp

1 Introduction

O_2-sensitive ion channels are key mediators in mechanisms that monitor O_2 supply to specific tissues and influence changes designed to restore optimal O_2 delivery under pathological situations. Hypoxic inhibition of K^+ channels localised on type

---

C. Peers (✉)
Division of Cardiovascular and Neuronal Remodelling, Leeds Institute of Genetics, Health and Therapeutics University of Leeds, Leeds LS2 9JT, UK
e-mail: c.s.peers@leeds.ac.uk

I cells of the carotid body leads to membrane depolarisation, voltage-gated Ca^{2+} influx and, in turn, neurotransmitter release (Kumar 2007). Various mechanisms have been proposed to account for this inhibition of K^+ channel activity. One such proposal involves AMP-activated protein kinase (AMPK), which acts as a metabolic sensor that is universally expressed in eukaryotes (Hardie et al. 2006). When cells are exposed to metabolic stresses, a rise in the cellular ADP/ATP ratio is partly compensated by the adenylate kinase reaction, which precipitates a much larger increase in the AMP/ATP ratio. Thus, although the fall in ATP may be negligible, the AMP:ATP ratio is increased and consequently AMPK is activated (Hardie et al. 1998). It has therefore been suggested that hypoxia may mediate carotid body excitation by promoting an increase in the AMP:ATP ratio, and consequent AMPK activation, with negligible change in ATP levels (Evans 2006). Consistent with this hypothesis, in type I cells of the rat carotid body, pharmacological activation of AMPK causes membrane depolarisation, inhibition of both maxiK channels and "leak" K^+ channels, and hence increased afferent discharge (Wyatt et al. 2007).

The identity of the O_2 sensing K^+ channels within the carotid body appears species dependent (Kemp and Peers 2007), but within the rat, two distinct O_2 sensitive K^+ channels have been identified; the maxiK channel (high conductance, Ca^{2+} sensitive; (Peers 1990)) and a "leak" K^+ channel which has yet to be fully characterised but is most likely a member of the TASK-like tandem P domain K^+ channel family (Buckler 2007). The relative importance of each of these channels to the chemotransduction mechanism is still unresolved, so we investigated the likely molecular counterparts for these channels to determine a role for AMPK-dependent regulation on their activity.

2 Methods

All experiments were conducted using HEK293 cells stably transfected with maxiK or TASK K^+ channels. For the maxiK cell lines, the co-expressed α- and β-subunits were KCNMA1 and KCNMB1 respectively (Ahring et al. 1997). Cells expressing the same α-subunit but in the absence of the β-subunit were also used for some experiments. For the TASK cell lines, full length cDNAs encoding human TASK-1 (hTASK-1, KCNK3) and TASK-3 (hTASK-3, KCNK9) were used to generate stable lines (cDNAs were a kind gift from Steve A. N. Goldstein, University of Chicago). Lines were maintained in Earle's minimal essential medium (containing L-glutamine) supplemented with 10% fetal calf serum, 1% antibiotic antimycotic, 1% non-essential amino acids and 0.2% gentamicin (all from Gibco BRL, Paisley, UK) and cultured in a humidified atmosphere of air/CO_2 (19:1) at 37°C. Cells were harvested by trypsinization and plated onto coverslips 24–48 h before use in electrophysiological studies.

Coverslip fragments with attached cells were transferred to a continuously perfused recording chamber (perfusion rate 3–5 ml/min, volume ca. 200 μl) mounted on the stage of an inverted microscope. Cells were perfused with a solution containing (in mM): 135 NaCl, 5 KCl, 1.2 $MgCl_2$, 5 HEPES, 2.5 $CaCl_2$, 10 D-glucose (pH 7.4 with KOH). Patch electrodes used for whole cell recordings (resistance 4–7 MΩ) were filled with intracellular solution consisting of (in mM): for the TASK channels, 10 NaCl, 117 KCl, 2 $MgCl_2$, 11 HEPES, 11 EGTA, 1 $CaCl_2$, 2 Na_2ATP or for the maxiK channels: 140 KSCN, 0.5 $CaCl_2$, 1 $MgCl_2$, 5 EGTA, 10 HEPES, 3 MgATP, 0.3 NaGTP (both pH 7.2 with KOH). An addition intracellular solution was used for the maxiK experiments that contained a free $[Ca^{2+}]$ ~140 nM: 140 KSCN, 2.3 $CaCl_2$, 1 $MgCl_2$, 4.5 EGTA, 10 HEPES, 3 MgATP, 0.3 NaGTP (pH 7.2 with KOH). All chemicals were from Sigma-Aldrich (Poole, UK). All test compounds (AICAR, compound C and A-769662) were applied to the cells at the stated concentrations and then incubated for 20mins in a humidified atmosphere of air/CO_2 (19:1) at 37°C.

Outward K^+ currents were recorded at 22 ± 1°C using ramp protocols; briefly, cells were clamped at a holding potential of −70 mV, and a voltage ramp from −100 mV to +100 mV (1s) applied. Data were acquired and digitalised through Digidata 1322A in combination with an Axopatch 200B amplifier and Clampex 9 software (all Molecular Devices, Foster City, CA), sampled at 2 kHz and filtered at 1kHz. Offline data analysis was carried out using Clampfit 9 software (Molecular Devices, Foster City, CA). The effects of drugs were determined using unpaired Student's t-tests. Differences were considered significant when $P < 0.05$. All values stated are as the mean ± SEM.

3 Results

3.1 AICAR Inhibition of maxiK Channel Currents

To examine whether AMPK activation modulated the maxiK channel expressed in HEK293 cells we utilised AICAR, which is metabolised intracellularly to generate an AMP-like molecule, thereby activating AMPK (Corton et al. 1995; Owen et al. 2000). Incubation of maxiK (α and β subunits) expressing HEK293 with AICAR (1 mM, 20 mins), led to an inhibition of channel activity (62.4 ± 5.7% at +30 mV, Fig. 1A), similar to that previously reported (Wyatt et al. 2007). This effect was prevented when AICAR was applied in the presence of compound C (20 μM). Given that maxiK channels are activated by $[Ca^{2+}]_i$ as well as voltage we sought to determine if changing the pipette Ca^{2+} concentration altered the AICAR effects on the channel. Figure 1B shows that altering the free $[Ca^{2+}]_i$ from 15 to 140 nM had no marked effect on the ability of AICAR to inhibit the channel (59.1 ± 4.8% inhibition at +30 mV).

The α subunit of the maxiK channel forms a fully functional channel when expressed alone, but its activity is modified by auxiliary β subunits. We therefore tested the effect of AICAR on the channel, when expressing the α subunit alone. As illustrated in Fig. 1C, AMPK activation still suppressed channel activity in a manner that was reversed by compound C, irrespective of the presence of β subunits.

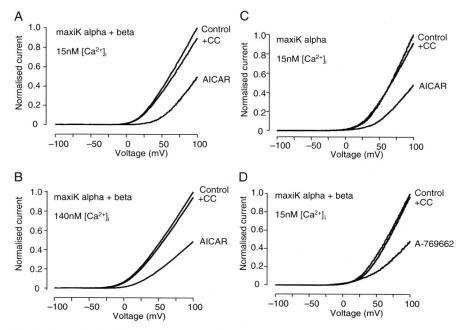

Fig. 1 **Normalised whole-cell currents evoked by ramp depolarizations applied to HEK293 cells expressing maxiK α and β subunits (A,B,D)** or α subunits alone (**C**). Free $[Ca^{2+}]_i$ is indicated for each panel. Cells were pre-incubated at 37°C with AICAR (1 mM, 20 min; **A,B,C**) or A-769662 (**D**; 100 μM), in the absence or presence of compound C (+CC; 20μM) as indicated in each panel

Recent studies have suggested that AICAR targets other AMP sensitive enzymes (e.g. glycogen phosphorylase) as well as activating AMPK (Vincent et al. 1996; Longnus et al. 2003). However, a novel compound (A-769662) that emerged from a high-throughput screen for AMPK activators has been shown to mimic both effects of AMP, i.e. allosteric activation of AMPK and inhibition of dephosphorylation of the AMPKα subunit at the activating Th172 site (Goransson et al. 2007). We tested this compound against the maxiK channel (α and β subunits) and, as with AICAR, we observed an inhibition of channel activity (56.8 ± 3.9% at +30 mV) that was reversed by co-incubation with compound C (Fig. 1D).

3.2 *Intracellular Dialysis of the AMPK Alpha Subunit Mutant (T172D) Inhibits maxiK Channel Currents*

To provide further evidence of an AMPK modulation of the maxiK channel we utilised a molecular approach. The T172D mutant is a constitutively active, truncated form of the α subunit of AMPK that does not require prior phosphorylation for activity (Scott et al. 2002). When the T172D mutant was added to the intracellular

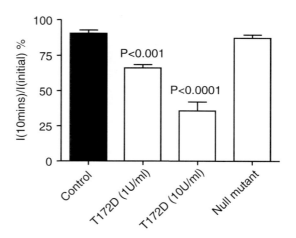

Fig. 2 Mean degree of current inhibition. Current was determined at +50 mV after 10 min dialysis of HEK293 cells expressing maxiK α and β subunits with pipette solution lacking proteins (control) or containing the truncated, constitutively active mutant T172D or the inactive (null) form at 10U/ml. Data are mean ± s.e.m. from at least 6 recordings

(pipette) solution, we observed a time- and concentration-dependent inhibition of the channel (Fig. 2). No such effect was observed over a similar time period when a kinase-inactive mutant (T172D/D139A) was introduced in the same way (Fig. 2). These experiments were conducted at 30°C, since at room temperature the T172D mutant was found to be without effect (not shown).

3.3 AMPK Activation Inhibits TASK-3 Channel Currents

In addition to inhibiting maxiK channels in rat type I cells, hypoxia also targets a leak K^+ conductance (Buckler 1997). The molecular identity of this conductance has yet to be determined, but its pharmacological and biophysical properties suggest it is a TASK-like member of the tandem P domain K^+ channels, sharing similarities with both TASK-1 and TASK-3 (Buckler 2007). Given our previous observations that AICAR inhibited the leak conductance of type I cells (Wyatt et al. 2007), we examined the potential modulation of recombinant TASK-1 and TASK-3 by AMPK. As shown in Fig. 3, AICAR was without effect on TASK-1 currents in HEK293 cells (Fig. 3A), but caused a significant inhibition of TASK-3 channels (45.6 ± 6.3% at +30 mV; Fig. 3B).

4 Discussion

The present study has employed recombinant K^+ channels to probe their potential modulation by AMPK. The work has been prompted by our initial investigations in native type I cells of the rat carotid body. In these cells, AMPK activation inhibited both maxiK channels and the leak K^+ conductance, whilst the delayed rectifier-like component of the macroscopic K^+ current was unaltered (Wyatt et al. 2007). These observations are of potential importance to carotid body chemoreception

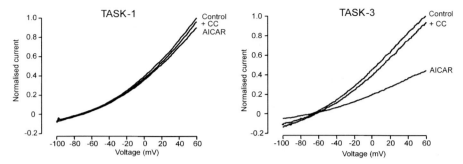

Fig. 3 Normalised whole-cell currents evoked by ramp depolarizations applied to HEK293 cells expressing TASK-1 *(left)* or TASK-3 *(right)* channels. Cells were pre-incubated at 37°C with AICAR (1 mM, 20 min) in the absence or presence of compound C (+CC; 20 μM) as indicated in each panel

since AMPK activation also caused type I cell membrane depolarization and raised $[Ca^{2+}]_i$ and this led to increased afferent chemosensory discharge. Furthermore, hypoxic excitation of the carotid body (determined by type I cell $[Ca^{2+}]_i$ measurements) was inhibited by compound C (Wyatt et al. 2007). These findings implicate AMPK as a crucial mediator of hypoxic chemosensory activity within the carotid body. They also implicate ion channels – particularly certain K^+ channels – as target substrates for AMPK, and indeed we provided evidence that maxiK channels can be directly phosphorylated by this enzyme (Wyatt et al. 2007).

The present work extends these findings, by demonstrating that AMPK activation not only by AICAR, but also by A-769662, inhibits maxiK channels (Fig. 1). This effect was also seen when the constitutively active, truncated form of the kinase α subunit, T172D, was included in the patch pipette (Fig. 2). Interestingly, AMPK-mediated inhibition of maxiK appeared independent of $[Ca^{2+}]_i$ an observation that is of physiological significance since native channel inhibition leads to a rise of $[Ca^{2+}]_i$ (via voltage-gated Ca^{2+} entry) which might otherwise re-activate the channels and inhibit the chemosensory response of these cells.

We also demonstrate that TASK-3, but not TASK-1, is sensitive to AMPK activation in HEK293 cells. This result is intriguing since the native leak conductance in type I cells has yet to be identified at the molecular level, and its pharmacological profile has features reminiscent of (yet also distinct from) both TASK-1 and TASK-3 (Buckler 2007). Our finding in the recombinant expression system argues in favour of the idea that the native channel is unlikely to be TASK-1, and may be a heterodimer of TASK-1 and TASK-3; such heterodimers have been shown to be functional (Czirjak and Enyedi 2003).

In summary, the present study validates known O_2 sensitive K^+ channels as novel targets for AMPK modulation, and further support the concept that this enzyme may be of central importance in hypoxic chemotransduction by the carotid body.

Acknowledgments This work was supported by the Wellcome Trust.

References

Ahring, P. K., Strobaek, D., Christophersen, P., Olesen, S. P., & Johansen, T. E. 1997, Stable expression of the human large-conductance Ca^{2+}-activated K^+ channel alpha- and beta-subunits in HEK293 cells. *FEBS Lett* 415: 67–70.

Buckler, K. J. 1997, A novel oxygen-sensitive potassium channel in rat carotid body type I cells. *J Physiol* 498: 649–662.

Buckler, K. J. 2007, TASK-like potassium channels and oxygen sensing in the carotid body. *Respir Physiol Neurobiol* 157: 55–64.

Corton, J. M., Gillespie, J. G., Hawley, S. A., & Hardie, D. G. 1995, 5-aminoimidazole-4-carboxamide ribonucleoside. A specific method for activating AMP-activated protein kinase in intact cells? *Eur J Biochem* 229: 558–565.

Czirjak, G. & Enyedi, P. 2003, Ruthenium red inhibits TASK-3 potassium channel by interconnecting glutamate 70 of the two subunits. *Mol Pharmacol* 63: 646–652.

Evans, A. M. 2006, AMP-activated protein kinase underpins hypoxic pulmonary vasoconstriction and carotid body excitation by hypoxia in mammals. *Exp Physiol* 91: 821–827.

Goransson, O., McBride, A., Hawley, S. A., Ross, F. A., Shpiro, N., Foretz, M., Viollet, B., Hardie, D. G., & Sakamoto, K. 2007, Mechanism of action of A-769662, a valuable tool for activation of AMP-activated protein kinase. *J Biol Chem* 282: 32549–32560.

Hardie, D. G., Carling, D., & Carlson, M. 1998, The AMP-activated/SNF1 protein kinase subfamily: metabolic sensors of the eukaryotic cell? *Annu Rev Biochem* 67: 821–855.

Hardie, D. G., Hawley, S. A., & Scott, J. W. 2006, AMP-activated protein kinase–development of the energy sensor concept. *J Physiol* 574: 7–15.

Kemp, P. J. & Peers, C. 2007, Oxygen sensing by ion channels. *Essays Biochem* 43: 77–90.

Kumar, P. 2007, Sensing hypoxia in the carotid body: from stimulus to response. *Essays Biochem* 43: 43–60.

Longnus, S. L., Wambolt, R. B., Parsons, H. L., Brownsey, R. W., & Allard, M. F. 2003, 5-Aminoimidazole-4-carboxamide 1-beta -D-ribofuranoside (AICAR) stimulates myocardial glycogenolysis by allosteric mechanisms. *Am J Physiol Regul Integr Comp Physiol* 284: R936–R944.

Owen, M. R., Doran, E., & Halestrap, A. P. 2000, Evidence that metformin exerts its anti-diabetic effects through inhibition of complex 1 of the mitochondrial respiratory chain. *Biochem J* 348 Pt 3: 607–614.

Peers, C. 1990, Hypoxic suppression of K^+ currents in type-I carotid-body cells - selective effect on the Ca^{2+}-activated K^+ current. *Neurosci Lett* 119: 253–256.

Scott, J. W., Norman, D. G., Hawley, S. A., Kontogiannis, L., & Hardie, D. G. 2002, Protein kinase substrate recognition studied using the recombinant catalytic domain of AMP-activated protein kinase and a model substrate. *J Mol Biol* 317: 309–323.

Vincent, M. F., Erion, M. D., Gruber, H. E., & Van den, B. G. 1996, Hypoglycaemic effect of AICAriboside in mice. *Diabetologia* 39: 1148–1155.

Wyatt, C. N., Mustard, K. J., Pearson, S. A., Dallas, M. L., Atkinson, L., Kumar, P., Peers, C., Hardie, D. G., & Evans, A. M. 2007, AMP-activated protein kinase mediates carotid body excitation by hypoxia. *J Biol Chem* 282: 8092–8098.

Hydrogen Sulfide Inhibits Human BK$_{Ca}$ Channels

V. Telezhkin, S.P. Brazier, S. Cayzac, C.T. Müller, D. Riccardi and P.J. Kemp

Abstract Hydrogen sulfide (H$_2$S) is produced endogenously in many types of mammalian cells. Evidence is now accumulating to suggest that H$_2$S is an endogenous signalling molecule, with a variety of molecular targets, including ion channels. Here, we describe the effects of H$_2$S on the large conductance, calcium-sensitive potassium channel (BK$_{Ca}$). This channel contributes to carotid body glomus cell excitability and oxygen-sensitivity. The experiments were performed on HEK 293 cells, stably expressing the human BK$_{Ca}$ channel α subunit, using patch-clamp in the inside-out configuration. The H$_2$S donor, NaSH (100 μM–10 mM), inhibited BK$_{Ca}$ channels in a concentration-dependent manner with an IC$_{50}$ of ca. 670 μM. In contrast to the known effects of CO donors, the H$_2$S donor maximally decreased the open state probability by over 50% and shifted the half activation voltage by more than +16 mV. In addition, although 1 mM KCN completely suppressed CO-evoked channel activation, it was without effect on the H$_2$S-induced channel inhibition, suggesting that the effects of CO and H$_2$S were non-competitive. RT-PCR showed that mRNA for both of the H$_2$S-producing enzymes, cystathionine-beta-synthase and cystathionine-gamma-lyase, were expressed in HEK 293 cells and in rat carotid body. Furthermore, immunohistochemistry was able to localise cystathionine-gamma-lyase to glomus cells, indicating that the carotid body has the endogenous capacity to produce H$_2$S. In conclusion, we have shown that H$_2$S and CO have opposing effects on BK$_{Ca}$ channels, suggesting that these gases have separate modes of action and that they modulate carotid body activity by binding at different motifs in the BK$_{Ca}$α subunit.

Keywords Hydrogen sulfide · BK · maxi-K · Stable transfection · KCN · HEK 293 · Carbon monoxide · Cystathionine-gamma-lyase

---

V. Telezhkin (✉)
School of Bioscience, Museum Avenue, Cardiff University, Cardiff CF10 3AX, UK
e-mail: telezhkinv@Cardiff.ac.uk

1 Introduction

Hydrogen sulfide (H_2S) is generated endogenously from the breakdown of cellular L-cysteine by the actions of the two enzymes, cystathionine-beta-synthase (CBS) and cystathionine-gamma-lyase (CGL) (see (Li and Moore, 2007) for recent review). Together with nitric oxide (NO) and carbon monoxide (CO), H_2S is now considered to be the third member of the so-called "gasotransmitter" family (Wang, 2002). One well characterised molecular target of H_2S is the ATP-sensitive potassium channel (K_{ATP}), and in the peripheral vasculature, H_2S open K_{ATP} to cause smooth muscle cell hyperpolarization and vasodilatation (Cheng et al., 2004). However, K_{ATP} channels may not be a unique target for H_2S in smooth muscle, or elsewhere. Indeed, since NO and CO are well known to modulate large conductance, calcium-sensitive potassium channels (BK_{Ca}) in a variety of tissues and recombinant systems (see, for example, (Wang and Wu, 1997; Riesco-Fagundo et al., 2001; Bolotina et al., 1994; Williams et al., 2004, 2008; Jaggar et al., 2005), it may be that H_2S might have similar actions on this channel. The BK_{Ca} channel is known not only as an essential contributor to regulation of the vascular tone but, importantly, as a key determinant of carotid body excitability ((Edwards et al., 1994; Peers, 1990), and see (Ghatta et al., 2006; Kemp and Peers, 2007) for recent reviews). The current study was designed in order to explore the effects of H_2S on the biophysical characteristics of potassium channels formed by the α-subunit of the human recombinant BK_{Ca} channel ($BK_{Ca}\alpha$).

2 Materials and Methods

2.1 Cell Culture

HEK 293 cells that stably express human BK_{Ca} α subunit alone were grown in Earle's minimal essential medium containing 10% fetal calf serum, 100 U/ml penicillin G, 100 μg/ml streptomycin sulfate, 1% L-glutamine, 50 mg/ml gentamacin, and 1 mg/ml G418 sulfate (Gibco BRL, Paisley, Strathclyde, UK) in a humidified incubator gassed with 5% CO_2/95% air. HEK 293 cells were passaged every 5–7 days in a ratio 1:10 using Ca^{2+}- and Mg^{2+}-free phosphate-buffered saline.

2.2 Electrophysiological Recordings

Cells were grown for several hours on glass coverslips before being transferred to a continuously perfused (5 ml/min) recording chamber (volume ca. 200 μl) mounted on the stage of an inverted microscope equipped with phase-contrast optics. For inside-out, excised patch-clamp recordings, the bath and pipette solutions were composed (in mM) of: 10 NaCl, 117 KCl, 2 $MgCl_2$, 11 N-2-hydroxyethylpiperazine-N'-2-ethanesulfonic acid, free [Ca^{2+}] adjusted to 336 nM using ethylene-glycol-tetra-acetic acid; pH 7.2. NaSH was prepared in the bath solution by isoosmotic replacement of NaCl with NaSH (1 mM–10 mM) whilst 1 mM potassium cyanide (KCN) was dissolved directly in the bath solution. CO was supplied by 30 μM of the CO donor molecule, tricarbonyldichlororuthenium (II) dimer ([Ru(CO$_3$)Cl$_2$]$_2$ – Sigma-Aldrich, Poole, Dorset, U.K.) as previously described (Williams et al., 2004). Single BK$_{Ca}$α currents were recorded at 22 ± 0.5°C. Current recordings were made using an Axopatch 200A amplifier and Digidata 1320 A/D interface (Axon Instruments, Forster City, CA, USA). Currents in inside-out membrane patches were studied using two protocols, with all voltages reported with respect to the inner membrane leaflet (-Vp): (1) continuous gap-free at the indicated holding potentials, and; (2) a 1s voltage-ramp from −30 mV to +120 mV, repeated at 0.1 Hz. All recordings were digitized at 10 kHz and Bessel low-pass Filtered at 5 kHz.

2.3 RT-PCR and Immunohistochemistry

To identify the presence of the H$_2$S-producing enzymes, CBS and CGL, reverse transcriptase-polymerase chain reaction (RT-PCR) and immunochemistry were performed on HEK 293 cell line stably expressing BK$_{Ca}$α and in rat carotid body. Total RNA was extracted using TRIzol reagent (Invitrogen, Paisley, Strathlyde, U.K.) and subjected to first strand cDNA synthesis using Oligo(dT) primers and AffinityScript (Stratagene, La Jolla, CA, U.S.A.). The final mixture (1 μl) was used for PCR amplification using the following primer sequences: human CGL (forward: 5′-agt-act-gtt-tgg-gcc-ttt-gct-tca-3′; reverse: 5′-gtg-gcc-att-cat-gta-ttt-tgt-tgc-3′); human CBS (forward: 5′-cgg-gca-ctg-ggg-ggc-tga-gat-t-3′; reverse: 5′-agc-atg-cgg-gca-aag-gtg-aac-g-3′); rat CGL (forward: 5′-ggc-cag-tcc-tcg-ggt-ttt-gta-3′; reverse: 5′-cac-tgt-ggc-cgt-tca-tgt-att-ttg-3′); rat CBS (forward: 5′-gac-agg-gcg-gtg-gtg-gat-agg-tg-3′; reverse: 5′-tag-gat-ggc-ccc-aga-ctc-gtt-gac-3′). All DNA fragments were amplified using DyNAzyme EXT DNA polymerase (Finnzymes, Espoo, Finland) using the following cycling parameters: 1 cycle at 94°C for 2 minutes followed by 35 cycles of 94°C for 30 seconds, 55°C for 30 seconds, 72°C for 2 minutes and a final extension of 72°C for 30 minutes. PCR products were separated on a 2 % agarose gel and bands were excised and cleaned using QIAquick PCR purification columns (Qiagen, Crawley, West Sussex, U.K.) before being ligated into pGEM-T Easy cloning vector (Promega, Southhampton, Hampshire, U.K.). Constructs were sequenced to determine whether the correct inserts were incorporated into pGEM-T Easy.

Rats were perfusion-fixed with 4% paraformaldehyde, the carotid bifurcations were dissected, embedded in Tissue Tek (Sakura Fintek, Torrance, CA, U.S.A.) and 4 μm-thick cryosections were cut onto Superfrost slides. For immunohistochemistry in HEK 293 cells, cells were plated onto coverslips and fixed with ice-cold methanol for 10 minutes, followed by 3 washes in PBS prior to application of the primary antibodies. Primary antibodies directed against cystathionine-β-synthase (CBS) and cystathionine-γ-lyase (CGL) (1/200, Santa Cruz Biotechnology, Santa Cruz, CA, U.S.A.) were incubated over night at 4°C. Alexa 488- and 594-conjugated secondary antibodies (1/500, Molecular Probes, Eugene, OR, U.S.A.) were incubated for 1 hour at room temperature.

3 Results and Discussion

We have previously shown that the CO donor, $[Ru(CO_3)Cl_2]_2$, is a potent activator of human $BK_{Ca}\alpha$ channels (Williams et al., 2004, 2008). In the current study, 30 μM of the CO donor was used as a reference agonist of $BK_{Ca}\alpha$ channels. In contrast to the effect of the CO donor, the H_2S donor inhibited $BK_{Ca}\alpha$ channels. In experiments performed at holding voltage of +40 mV, NaSH (100 μM–10 mM) decreased open state probability (NPo) of $BK_{Ca}\alpha$ channels in a concentration-dependent manner, with a mean IC_{50} value of 668.0 ± 286.1 μM ($n = 8$) (Fig. 1).

The H_2S donor evoked a concentration-dependent rightward shift of the $BK_{Ca}\alpha$ current activation curve (Fig. 2A). Thus, $BK_{Ca}\alpha$ current half activation voltage (Va_{50}) was significantly increased from $+72.7 \pm 3.2$ mV to $+84.6 \pm 1.5$ mV by the addition of 10 mM NaSH ($n = 11$; $p < 0.01$). This NaSH effect was, again, in contrast to the leftward shift of the activation curve caused by 30 μM CO donor where Va_{50} was significantly decreased to $+58.2 \pm 2.8$ mV ($n = 11$; $p < 0.01$).

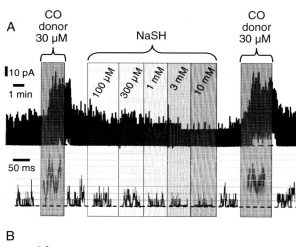

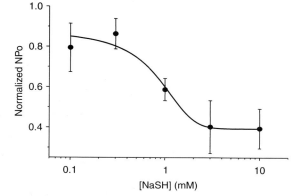

Fig. 1 Inhibitory effect of H_2S donor on human recombinant $BK_{Ca}\alpha$ channels stably expressed in HEK 293 cells. (**A**) Exemplar current recording from an inside-out patch during the addition of 30 μM CO donor, followed by sequential additions of 100 μM to 10 mM NaSH. The CO donor was reintroduced at the end of the NaSH wash-out period (-Vp = +40 mV). (**B**) Mean (± SEM) concentration-response data showing effect of NaSH on normalized NPo; $n = 8$

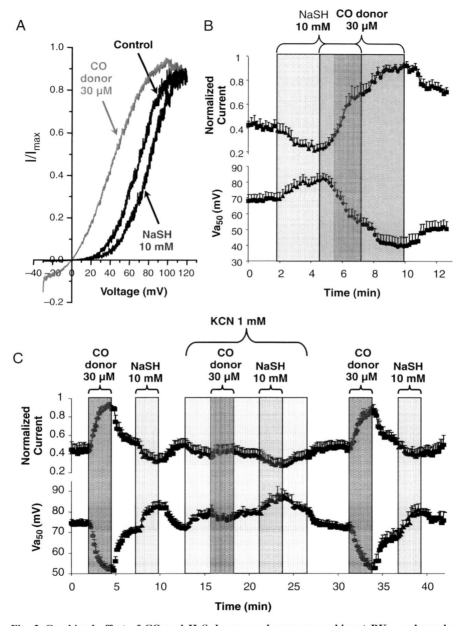

Fig. 2 Combined effect of CO and H₂S donors on human recombinant BK_Caα channels.
(**A**) Exemplar mean I/I_max vs. voltage plots for recombinant BK_Caα currents in absence of donors (Control) and presence of 30 μM CO donor or 10 mM H₂S donor; $n = 7$. (**B**) Mean ($\pm$ SEM) normalized current at $-V_p = +60$ mV (*upper panel*) and Va_{50} (*lower panel*) time-courses for the sequential addition of H₂S and CO donors, as shown; $n = 7$. (**C**) Mean ($\pm$ SEM) normalized current at $-V_p = +60$ mV (*upper panel*) and Va_{50} (*lower panel*) time-courses for sequential addition of H₂S and CO donors in presence and absence of 1 mM KCN, as shown; $n = 9$

10 mM NaSH did not alter BK$_{Ca}\alpha$ single channel conductance (data not shown). All of the effects of both NaSH and of the CO donor on BK$_{Ca}\alpha$ channel activity were fully reversible.

In our previous study, we have shown that the S9–S10 module of the C-terminal tail of BK$_{Ca}\alpha$ is required for CO activation (Williams et al., 2008). Therefore, our next series of experiments was designed to test whether the inhibitory action of H$_2$S on BK$_{Ca}\alpha$ channel targets the same site.

Introduction of 30 μM of the CO donor, in presence of 10 mM NaSH, reversed the NaSH-dependent current inhibition and rightward shift of Va$_{50}$ to produce pronounced current activation and a dramatic shift of Va$_{50}$ towards lower voltages (Fig. 2B). These data may suggest that these two gases are competing for the same regulatory centre of BK$_{Ca}\alpha$ channel in the S9–S10 domain.

Although the CO regulatory domain is located in C-terminal tail (Williams et al., 2008), we (Brazier et al., 2008 in this volume) and others (Hou et al., 2008) have only been able to attenuate, not completely abolish, CO sensitivity using point mutations. However, here we show that 1mM KCN effectively abolishes channel stimulation caused by the CO donor (also see Brazier et al., 2008 in this volume), but that it is completely unable to alter the inhibition evoked by the H$_2$S donor, (Fig. 2C). Indeed, both the current depression and the increase in Va$_{50}$ evoked by 10 mM NaSH were not statistically different in the absence or presence of 1 mM KCN. These data suggest non-competitive actions of these two gases at cyanide-sensitive (for CO) and cyanide-insensitive (for H$_2$S) centres in the BK$_{Ca}$ α-subunit.

Since regulation of BK$_{Ca}$ channel is crucial for arterial chemoreception (see (Kemp and Peers, 2007) for recent review) and H$_2$S is a BK$_{Ca}$ inhibitor, we used RT-PCR and immunohistochemistry to probe for the expression of the H$_2$S synthesising enzymes, CBS and CGL, in both the stable HEK 293 cell line and the carotid body. RT-PCR showed that both CBS and CGL transcripts were present in HEK 293 cells, with amplicons of the predicted sizes of 470 and 277 bp, respectively (Fig. 3A). CBS and CGL were both identified in HEK 293 cells using immunohistochemistry (Fig. 3B). Employing separate, rat-specific primers, amplicons of the predicted sizes for CBS and CGL (420 and 500 bp, respectively) were amplified from carotid body (Fig. 3C); sequencing demonstrated amplification of the correct products. Furthermore, CBS expression was also successfully detected by immunohistochemistry in carotid body (Fig. 3D). This study explored the novel proposal that H$_2$S may regulate the activity of a specific potassium channel. It has shown that NaSH inhibits the activity of BK$_{Ca}\alpha$ channels by reducing NPo and shifting the Va$_{50}$ value to positive voltages, without altering its conductance. Unlike the activation evoked by the CO donor, inhibition of BK$_{Ca}\alpha$ channels by NaSH is not sensitive to KCN, suggesting that the actions of H$_2$S and CO are non-competitive. Moreover, H$_2$S-producing enzymes are expressed in the rat carotid body, suggesting that they have the capacity to produce H$_2$S. Taken together, these data show that BK$_{Ca}\alpha$ is inhibited by H$_2$S and suggest that this endogenously produced gas may be involved in chemotransduction in carotid body.

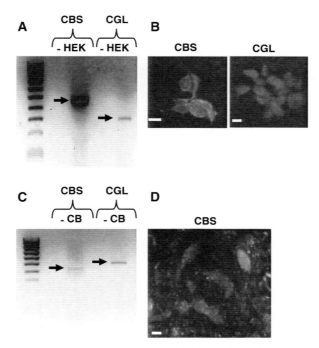

Fig. 3 Identification of H₂S producing enzymes in HEK 293 cells and rat carotid body.
(**A**) Reverse transcriptase-polymerase chain reaction (RT-PCR) amplification products run on a 2% agarose gel showing amplicons of the correct sizes predicted from the sequences of human cystathionine-beta-synthase (CBS – 470 bp) and cystathionine-gamma-lyase (CGL – 277 bp). On the left are shown size markers, in 100 bp increments. (**B**) Immunohistochemical localization of cystathionine-beta-synthase (CBS – *green*) and cystathionine-gamma-lyase (CGL – *red*) in HEK 293 cells. Nuclei are shown in blue. Scale bar = 20 μm. (**C**) RT-PCR products run on a 1% agarose gel showing amplicons of the correct sizes predicted from the sequences of rat CBS (420 bp) and CGL (500 bp). On the left are shown size markers, in 100 bp increments. (**D**) Immunohistochemical localization of cystathionine-beta-synthase (CBS – *green*) in 4 μm fixed sections of rat carotid body. Nuclei are shown in blue. Scale bar = 20 μm

Acknowledgments The authors would like to thank The British Heart Foundation.

References

Bolotina VM, Najibi S, Palacino JJ, Pagano PJ, & Cohen RA (1994). Nitric oxide directly activates calcium-dependent potassium channels in vascular smooth muscle. *Nature* **368**, 850–853.

Cheng Y, Ndisang JF, Tang G, Cao K, & Wang R (2004). Hydrogen sulfide-induced relaxation of resistance mesenteric artery beds of rats. *Am J Physiol Heart Circ Physiol* **287**, H2316–H2323.

Edwards G, Niederste-Hollenberg A, Schneider J, Noack T, & Weston AH (1994). Ion channel modulation by NS 1619, the putative BK$_{Ca}$ channel opener, in vascular smooth muscle. *Br J Pharmacol* **113**, 1538–1547.

Ghatta S, Nimmagadda D, Xu X, & O'Rourke ST (2006). Large-conductance, calcium-activated potassium channels: structural and functional implications. *Pharmacol Ther* **110**, 103–116.

Hou S, Xu R, Heinemann SH, & Hoshi T (2008). The RCK1 high-affinity Ca^{2+} sensor confers carbon monoxide sensitivity to Slo1 BK channels. *Proc Natl Acad Sci USA* **105**, 4039–4043.

Jaggar JH, Li A, Parfenova H, Liu J, Umstot ES, Dopico AM, & Leffler CW (2005). Heme is a carbon monoxide receptor for large-conductance Ca^{2+}-activated K^+ channels. *Circ Res* **97**, 805–812.

Kemp PJ & Peers C (2007). Oxygen sensing by ion channels. *Essays Biochem* **43**, 77–90.

Li L & Moore PK (2007). An overview of the biological significance of endogenous gases: new roles for old molecules. *Biochem Soc Trans* **35**, 1138–1141.

Peers C (1990). Hypoxic suppression of K^+ currents in type-I carotid-body cells – selective effect on the Ca^{2+}-activated K^+ current. *Neurosci Lett* **119**, 253–256.

Riesco-Fagundo AM, Perez-Garcia MT, Gonzalez C, & Lopez-Lopez JR (2001). O_2 modulates large-conductance Ca^{2+}-dependent K^+ channels of rat chemoreceptor cells by a membrane-restricted and CO-sensitive mechanism. *Circ Res* **89**, 430–436.

Wang R (2002). Two's company, three's a crowd: can H_2S be the third endogenous gaseous transmitter? *FASEB J* **16**, 1792–1798.

Wang R & Wu L (1997). The chemical modification of K_{Ca} channels by carbon monoxide in vascular smooth muscle cells. *J Biol Chem* **272**, 8222–8226.

Williams SE, Brazier SP, Baban N, Telezhkin V, Muller CT, Riccardi D, & Kemp PJ (2008). A structural motif in the C-terminal tail of slo1 confers carbon monoxide sensitivity to human BK_{Ca} channels. *Pflugers Arch* **456**, 561–572.

Williams SE, Wootton P, Mason HS, Bould J, Iles DE, Riccardi D, Peers C, & Kemp PJ (2004). Hemoxygenase-2 is an oxygen sensor for a calcium-sensitive potassium channel. *Science* **306**, 2093–2097.

DPPX Modifies TEA Sensitivity of the Kv4 Channels in Rabbit Carotid Body Chemoreceptor Cells

O. Colinas, F.D. Pérez-Carretero, E. Alonso, J.R. López-López and M.T. Pérez-García

Abstract Chemoreceptor cells from rabbit carotid body (CB) exhibit transient outward currents reversibly inhibited by low P_{O2}. Molecular and functional dissection of the components of these outward currents indicates that at least two different channels (Kv4.3 and Kv3.4) contribute to this current. Furthermore, several lines of evidence support the conclusion that Kv4 channel subfamily members (either Kv4.3 alone or Kv4.3/Kv4.1 heteromultimers) are the oxygen sensitive K channels (K_{O2}) in rabbit CB chemoreceptor cells. However, the pharmacological characterization of these currents shows that they are almost completely blocked by high external TEA concentrations, while Kv4 channels have been shown to be TEA-insensitive. We hypothesized that the expression of regulatory subunits in chemoreceptor cells could modify TEA sensitivity of Kv4 channels. Here, we explore the presence and functional contribution of DPPX to K_{O2} currents in rabbit CB chemoreceptor cells by using DPPX functional knockdown with siRNA. Our data suggest that DPPX proteins are integral components of K_{O2} currents, and that their association with Kv4 subunits modulate the pharmacological profile of the heteromultimers.

Keywords Carotid body · Chemoreceptor cells · Kv4 channels · DPPX · siRNA · Tetraethylammonium · Rabbit

1 Introduction

In rabbit carotid body chemoreceptor cells, genes of the Kv4 family (either Kv4.3 alone or Kv4.3/Kv4.1 heteromultimers) have been shown to represent the molecular correlate of the oxygen-sensitive voltage dependent K^+ current (K_{O2}) originally described in this preparation (Perez-Garcia et al., 2000; Sanchez et al., 2002; López-López et al., 2003).

O. Colinas (✉)
Departamento de Bioquímica y Biología Molecular y Fisiología e Instituto de Biología y Genética Molecular (IBGM), Universidad de Valladolid y Consejo Superior de Investigaciones Científicas (CSIC), c/Sanz y Forés s/n, 47003 Valladolid, Spain
e-mail: olaiacm@ibgm.uva.es

The variability observed in gating kinetics and conductance among the Kv4 mediated native current does not seem to affect the pharmacological profile of Kv4 currents, which are typically described as 4-AP sensitive and TEA-resistant. This holds true when characterizing Kv4 currents in heterologous expression systems (Pak et al., 1991; Jerng and Covarrubias, 1997) and also when studying native currents (Martina et al., 1998; Song et al., 1998), suggesting that the association of Kv4 pore-forming subunits with accessory subunits does not change the pharmacological properties of the heteromultimers. However, in rabbit CB chemoreceptor cells we have observed that transient outward K^+ currents are sensitive to 4-AP (López-López et al., 1993) and heteropodatoxin (HpTx-2), (Sanchez et al., 2002), but can also be blocked by external TEA application. While we have identified Kv3.4 as the high sensitive TEA component of the transient K^+ current (Kaab et al., 2005; Sanchez et al., 2002), we have no explanation for the blockade of the Kv4 component with low-millimolar TEA concentrations. In the search for an explanation of this perplexing observation, we have explored the possibility that some accessory subunit of Kv4 channels could determine the atypical pharmacological profile of the transient K^+ current in rabbit CB chemoreceptor cells. Among Kv4 auxiliary subunits, the structural properties of DPPX, with a large C-terminal extracellular domain (Wada et al., 1992), made conceivable the hypothesis that its association with Kv4 α subunits could modify the binding of TEA to the external side of the pore. We detected high expression levels of DPPX mRNA in rabbit CB chemoreceptor cells (when compared to cerebellar granule neurons), DPPX protein was found to co-express with Kv4.3 in this preparation, and its functional knockdown with siRNA decreased its sensitivity to extracellular TEA, demonstrating the physiological association in a native tissue.

2 Methods

2.1 Dissociation and Culture of Rabbit Carotid Body Cells

Adult New Zealand rabbits (1.5–2 kg) were anesthetized with intravenous application of sodium pentobarbital (40 mg/kg) through the lateral vein of the ear. After tracheostomy, carotid artery bifurcations were dissected out and animals were killed by intracardiac injection of sodium pentobarbital. The CBs were enzymatically dispersed as previously described (Pérez-García et al., 1992). Dispersed cells were plated onto poly-L-lysine-coated coverslips with 2 ml of growth medium, and maintained in culture at 37°C in a 5%CO_2 atmosphere up to 96 h.

2.2 Electrophysiological Methods

Ionic currents were recorded at room temperature (20–25°C) using the whole-cell configuration of the patch-clamp technique. Whole-cell current recordings and data acquisition from CB chemoreceptor cells were made as previously described (López-López et al., 1997; Sanchez et al., 2002).

2.3 siRNA Design and Construction

We selected a target sequence for designing siRNA against rabbit DPPX in a region with the highest identity with the human sequence. The siRNA sequence was 5'-ACACGAGGATGAAAGTGAA-3'. After using BLAST program to ensure specificity of the sequences across mammalian genomes, the target sequence was used to generate siRNA Expression Cassette (SEC) with the Silencer™ Express (siRNA Expression Cassette Kit, Ambion) according to the manufacturer's instructions. A negative control SEC with limited homology to the mammalian genomes was used as control.

2.4 RNA Isolation and RT-PCR Methods

The expression levels of DPPX mRNA in CB chemoreceptor cells were determined from total RNA extracted from pooled chemoreceptor cells kept in culture during 24–48 h. Control experiments were performed in cultured cerebellar granule cells obtained from rabbit cerebellum according to previously published protocols (Liu et al., 2007) and kept in culture for 5–7 days. Electrodes made from capillary glass baked overnight at 200°C were filled with 7 μl of RNAse-free water and the tip was broken to facilitate the aspiration of multiple cells. After collecting 25–50 cells (CB chemoreceptor cells or cerebellar granule cells), the contents of the pipette were ejected into a 0.2 ml eppendorf tube containing 1 μl of RNAsin (20 u/μl, Applied Biosystems) and kept at −80°C. RNA was reverse transcribed with MuLvRT (5000 u/ml) as previously described (Kaab et al., 2005).

Total mRNA from rabbit hippocampus or cerebellum using TRIzol Reagent was used as the calibrator.

The mRNA levels for DPPX were determined by quantitative real-time PCR (qPCR) on a Rotor-Gene 3000 instrument (Corbett Research) using ribosomal protein L18 (RPL18) expression levels as housekeeping gene. Amplification of cDNA was performed as previously described (Kaab et al., 2005). The PCR primers were designed using the Primer 3 website (http://frodo.wi.mit.edu/cgi-bin/primer3/primer3_www.cgi) and were selected to spam an intronic sequence and to recognize both human and rabbit DPPX. The primer sequences were 5' CACGAGGATGAAAGTGAACG 3' (forward) and 5' TGATGGACTGGATGTTGTCG 3' (reverse), and they amplify a 178 bp fragment.

The data were analyzed using the threshold cycle (Ct) relative quantification method, (Livak and Schmittgen, 2001) so that the fold change in expression ($2^{-\Delta\Delta Ct}$) was calculated from $\Delta\Delta Ct$ values obtained with the expression

$$\Delta\Delta Ct = (Ct_{DPPX} - Ct_{RPL18})_{CBcells} - (Ct_{DPPX} - Ct_{RPL18})_{Hippocampus}$$

Using the $2^{-\Delta\Delta Ct}$ method, the data in CB cells and G cells are presented as the fold change in gene expression normalized to RPL18 and relative to hippocampus.

2.5 Electroporation of Cultured Rabbit CB Chemoreceptor Cells

Single cell electroporation was performed using modified patch-clamp techniques, following previously published models (Rae and Levis, 2002). A home-made device consisting on an operational amplifier connected to a current to voltage converter allows to supply train pulses from −9 to +9 V through a patch pipette. The train pulses were produced using pClamp an Digidata 1200 as the pulse generator, which allows independent setting of the duration and frequency of each pulse and the total time of the train pulse delivery. The protocol used and the electroporation pipettes were made as previously described (Colinas et al., 2008). The pipette solution was the internal solution used for the recording experiments to which 33 ng/µl of a plasmid DNA encoding GFP alone or in combination with 7 ng/µl of DPPX or negative control SEC were added. GFP-transfected cells were recorded 24 to 48h after electroporation.

2.6 Immunocytochemical Methods

Immunocytochemistry on isolated CB chemoreceptor cells in culture was performed as described previously (Perez-Garcia et al., 2004).

3 Results

3.1 TEA Effect on Transient K^+ Currents From Rabbit CB Chemoreceptor Cells

We tested the effect of TEA on the native CB chemoreceptor cells by analyzing the reduction in the peak current amplitude in 200 ms depolarizing pulses to +40 mV in the presence of increasing TEA concentrations (from 0.1 µM to 100 mM) in the bath solution. The best fit to the data from 12 CB chemoreceptor cells was obtained with the three binding-site model with affinity constants obtained of 3.4 µM, 130 µM and 24 mM and relative amplitudes of 8.1, 62.64 and 12.26% respectively. In addition to this three components, there is a non-inactivating component that was unaffected by high concentrations of TEA representing a 17% of the total current. In all the cases, the best fit to the data from each individual cell was also obtained with the three binding-site model. The dose-response curve obtained in one cell is depicted in Fig. 1A. The relative amplitude of these three components in the example shown was 14%, 45% and 16% respectively. The TEA resistant fraction averaged 24%. The dissociation constants (Kd) are also indicated on the figure.

Transient outward currents in rabbit CB chemoreceptor cells have been shown to be comprised of Kv3.4 and Kv4.1/4.3 channels (Sanchez et al., 2002; Perez-Garcia

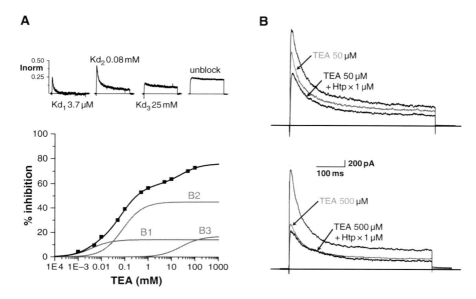

Fig. 1 TEA effect on transient K⁺ currents from rabbit chemoreceptor cells. (**A**) TEA dose–response curve obtained with data from one CB chemoreceptor cell. The continuous line through the data points shows the best fit obtained with a hyperbolic function with three binding sites. The functions representing the three components are also shown in the figure (B1, B2 and B3). The upper panels in the figure show normalized traces obtained by subtracting the traces in the presence of 5 μM, 100 μM and 25 mM TEA from the control traces, to get an estimation of the overall shape of the currents blocked by TEA concentrations closed to the obtained affinity constants (Kd indicated in each trace). The unblock trace is the current remaining in the presence of 100 mM TEA. (**B**) Traces show representative examples of depolarizing pulses to +40 mV in control or induced by the application of 50 μM TEA (*upper panel*) or 500 μM TEA (*bottom panel*) alone or in combination with 1 μM HpTx as indicated

et al., 2000), and both the pharmacological and kinetic profile are consistent identifying Kv3.4 as the high-affinity TEA component. However, the intermediate affinity (and to a lower extent the low-affinity) component is clearly a transient current representing the larger portion of the total outward current, that can be attributable to Kv4 currents, and that shows an atypical behavior with respect to TEA sensitivity. In order to explore more directly this extent, we investigated if selective blockers of Kv4 channels, such as heteropodatoxin-2 (Sanguinetti et al., 1997), inhibit the same channel population than submillimolar TEA concentrations. We have used 50 μM TEA to block completely the high TEA-sensitivity component, that we have previously described as the BDS-sensitive component (Kaab et al., 2005), and TEA 500 μM to block the intermediate TEA sensitivity component. In this set of experiments (see Fig. 1B), CB chemoreceptor cells were subjected to two consecutive applications of 1 μM HpTx, in the presence of 50 and 500 μM TEA respectively.

Perfusion of the cells with 50 μM TEA in the bath solution led to a 24.02±4.6% reduction of the peak current amplitude, and 1 μM HpTx increased inhibition up to a 46.2±5.4%. However, application of 500 μM TEA completely eliminated the HpTx-sensitive current, suggesting that both drugs are acting on the same channel population. These data prompted us to explore the possibility that some accessory subunit of Kv4 channels with an extracellular domain could modify TEA binding to Kv4 channels in rabbit chemoreceptor cells.

3.2 Presence of DPPX in Rabbit CB Chemoreceptor Cells

We have used immunocytochemistry in isolated chemoreceptor cells to determine the presence at the protein level of DPPX, a Kv4 accessory subunit with a large C-terminal extracellular domain (Wada et al., 1992). We found that all TH–positive cells were stained with anti-DPPX antibody and also observed a perfect correlation with double-immunocytochemistry with antiKv4.3 and anti-DPPX antibodies (Fig. 2A, B). To further confirm the presence of DPPX in rabbit CB chemoreceptor cells, qPCR experiments using SYBRgreen were performed in pooled chemoreceptor cells obtained from CB cultures, using RPL18 as housekeeping gene (Kaab et al., 2005). We have also measured DPPX mRNA levels in granule cells from cerebellum. We found high expression levels of DPPX mRNA in our preparation (rabbit CB chemoreceptor cells) where the quantification shows a 50-fold increase as compared to the control tissue (hippocampus), and almost ten times more than in granule cells (Fig. 2C). Altogether, these data suggested that DPPX could represent a good candidate to modulate Kv4 currents in CB chemoreceptor cells.

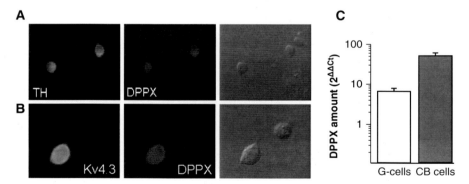

Fig. 2 DPPX expression in rabbit chemoreceptor cells. (**A**) Immunofluorescence labeling of DPPX shows the expression of DPPX in every TH-positive cell. (**B**) Double labeling of CB chemoreceptor cells with anti-Kv4.3 and anti-DPPX antibodies shows an almost perfect coexpression of these two proteins. (**C**) Real-time PCR showing the relative abundance of DPPX mRNA in CB chemoreceptor cells in primary culture. DPPX mRNA abundance in cerebellar granule cells (G cells) were determined for comparison. Each bar is the mean ± SEM of four to seven individual determinations

3.3 DPPX Contribution to Transient Outward K^+ Currents in Chemoreceptor Cells

In order to determine whether TEA sensitivity of transient outward currents in rabbit CB chemoreceptor cells was in fact due to the presence of Kv4/DPPX heteromultimers, we studied the effects of functional suppression of DPPX on the kinetics and pharmacology of the native currents by using siRNA against rabbit DPPX.

We introduced siRNA against rabbit DPPX in the chemoreceptor cells by using single-cell electroporation and analyzing the magnitude of the currents and their kinetics. We observed a slower time course of inactivation in the siRNA DPPX transfected cells, and the fit of the data to a biexponential decay function indicated that siRNA DPPX induced a decrease in the amplitude of the fast component of the current. Also, we found a significant reduction in the proportion of the transient component of the current, which is in agreement with the well-known chaperone function of DPP proteins on Kv4 currents (Jerng et al., 2005; Nadal et al., 2003).

Once we obtained functional evidence of DPPX down-regulation in chemoreceptor cells, we explored if there was some effect on the TEA sensitivity of the transient current. In each cell studied, we applied 4 different TEA concentrations (from 10 μM to 10 mM) and we normalized the reduction in the peak current amplitude observed in depolarizing pulses to +40 mV. The dose-response curve obtained in

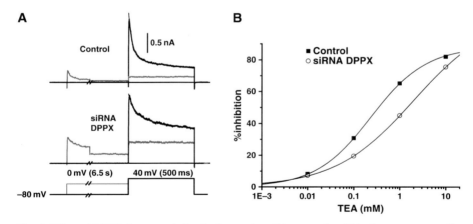

Fig. 3 Effect of DPPX down-regulation in the transient K^+ current in rabbit chemoreceptor cells. (**A**) Currents recorded from two individual CB chemoreceptor cells in control and after electroporation of siRNA DPPX, with a voltage protocol in which 500 ms depolarizing steps to +40 mV follow 6.5 s prepulses to two different potentials, −80 mV (*black trace*) and 0 mV (*grey trace*). The difference between the current amplitude at +40 mV in these two pulses is defined as the transient outward current. The *black thick line* shows the fit of the currents to a double-exponential function. (**B**) The effect of TEA in two individual CB chemoreceptor cells in control and after electroporation of siRNA DPPX is represented as percentage of inhibition, and was calculated from the reduction in the peak current amplitude in depolarizing pulses to +40 mV. The continuous lines show the fit of the data to a hyperbolic function with one binding site for TEA. The half inhibitory concentration in control cell is 240 μM and in siRNA DPPX transfected cell 2 mM

two individual CB chemoreceptor cells in control and after electroporation of siRNA DPPX is depicted in Fig. 3B. We found a significant reduction of TEA sensitivity in siRNA DPPX transfected cells in the intermediate range of TEA concentrations (0.1 and 1 mM) with no changes at lower and higher concentrations (Fig. 3B). The average fit of the data to a hyperbolic function shows that down-regulation of DPPX leads to a 3-fold decrease in the TEA affinity of the current (half inhibitory concentration of $189.76\pm13.44\,\mu M$ in control cells versus $540.96\pm40.84\,\mu M$ in siRNA DPPX transfected cells).

4 Discussion

The data presented in this work allow us to propose that the sensitivity to TEA of the Kv4 channels that largely contribute to the transient outward K^+ currents of rabbit CB chemoreceptor cells is related to the presence of DPPX accessory subunits in these cells. These findings provide an explanation to our previous observation regarding TEA sensitivity of rabbit CB K_{O2} and are also relevant to understand the contribution of DPPX to the functional properties of transient K^+ currents in this preparation.

The tissue-specific assembly of the Kv4 multiproteic complex could explain TEA-sensitivity of Kv4 channels in rabbit CB chemoreceptors. It has been proposed that both the relative abundance of DPPX as well as the cell-specific expression of the DPPX splice isoforms could confer Kv4 channels with differential modulation at various levels. In this regard, it is noteworthy that the expression levels of DPPX are ten times higher in CB chemoreceptor cells than in cerebellar granule cells, a preparation in which DPPX has been found to be prominently expressed (Nadal et al., 2003, 2006).

The data obtained allow us to conclude that many of the modulatory effects of DPPX on Kv4 currents described in heterologous expression system are also relevant for native currents. However, it must be taken into account that these effects are muffled because the transient K^+ current component does not reflect an homogeneous channel population (Sanchez et al., 2002; López-López et al., 2003; Kaab et al., 2005), and because siRNA against DPPX does not produce a complete knock out of DPPX subunit expression.

In spite of these limitations, we can extract relevant and unambiguous conclusions from the data obtained from siRNA DPPX transfected cells regarding the role of DPPX subunit in rabbit CB chemoreceptor cells: (1) DPPX has a chaperone role on Kv4 currents, as in the presence of siRNA DPPX there is a significant decrease in the current density of the transient component, (2) DPPX accelerates the time course of inactivation of the channels, as DPPX down-regulation decreases the proportion of the fast component of the current, and (3) DPPX contributes to TEA sensitivity of transient K^+ current, as siRNA DPPX transfected cells exhibited a significant decrease of their TEA sensitivity. With regards to this latter observation, it is interesting to point out that the change in TEA sensitivity was not observed along the whole range of TEA concentrations, but only for the high micromolar-low millimolar ones, which in all likelihood represent the Kv4 component of the outward K^+ current (component B2 in Fig. 1). We have previously described that the component blocked by lower TEA concentrations (1–50 μM) represents Kv3.4 channels (Sanchez et al., 2002; Kaab et al., 2005), and this component is not modified by siRNA DPPX, but is reduced by down-regulation of Kv3.4 channels (as under chronic hypoxia stimulation (Kaab et al., 2005), and selectively abolished with transfection of siRNA against rabbit Kv3.4 (unpublished observation). On the other hand, at high TEA concentrations the fraction of the current blocked is a non-inactivating one, that we have evidences indicating that is carried by members of the Kv1 family of channels (Perez-Garcia et al., 2000).

Altogether, this set of data indicates that the effect of siRNA DPPX is very selective on Kv4 currents and that DPPX association contributes to the atypical pharmacological profile of Kv4 currents in rabbit CB chemoreceptor cells.

Acknowledgments We thank Esperanza Alonso for excellent technical assistance. This work was supported by Ministerio de Sanidad y Consumo, Instituto de Salud Carlos III grants R006/009 (Red Heracles) and PI041044 (JRLL), Ministerio de Educación y Ciencia grant BFU2004-05551 (MTPG) and Junta de Castilla y León grant VA011C05.

References

Colinas, O., Perez-Carretero, F. D., Lopez-Lopez, J. R., & Perez-Garcia, M. T. (2008). A Role for DPPX Modulating External TEA Sensitivity of Kv4 Channels. *J General Physiol* 131, 455–471.
Jerng, H. H. & Covarrubias, M. (1997). K+ channel inactivation mediated by the concerted action of the cytoplasmic N- and C-terminal domains. *Biophysical J* 72, 163–174.
Jerng, H. H., Kunjilwar, K., & Pfaffinger, P. J. (2005). Multiprotein assembly of Kv4.2, KChIP3 and DPP10 produces ternary channel complexes with ISA-like properties. *J Physiol* 568, 767–788.
Kaab, S., Miguel-Velado, E., López-López, J. R., & Perez-Garcia, M. T. (2005). Down regulation of kv3.4 channels by chronic hypoxia increases acute oxygen sensitivity in rabbit carotid body. *J Physiol* 566, 395–408.
Liu, L. Y., Hoffman, G. E., Fei, X. W., Li, Z., Zhang, Z. H., & Mei, Y. A. (2007). Delayed rectifier outward K+ current mediates the migration of rat cerebellar granule cells stimulated by melatonin. *J Neurochem* 102, 333–344.
Livak, K. J. & Schmittgen, T. D. (2001). Analysis of relative gene expression data using real-time quantitative PCR and the 2(-Delta Delta C(T)) Method. *Methods* 25, 402–408.
López-López, J. R., De Luis, D. A., & Gonzalez, C. (1993). Properties of a transient K^+ current in chemoreceptor cells of rabbit carotid body. *J Physiol Lond* 460, 15–32.
López-López, J. R., Gonzalez, C., & Pérez-García, M. T. (1997). Properties of ionic currents from isolated adult rat carotid body chemoreceptor cells: effect of hypoxia. *J Physiol* 499, 429–441.
López-López, J. R., Perez-Garcia, M. T., Sanz-Alfayate, G., Obeso, A., & Gonzalez, C. (2003). Functional identification of Kvalpha subunits contributing to the O_2-sensitive K^+ current in rabbit carotid body chemoreceptor cells. *Adv Exp Med Biol* 536, 33–39.
Martina, M., Schultz, J. H., Ehmke, H., Monyer, H., & Jonas, P. (1998). Functional and molecular differences between voltage-gated K^+ channels of fast-spiking interneurons and pyramidal neurons of rat hippocampus. *J Neurosci* 18, 8111–8125.

Nadal, M. S., Amarillo, Y., de Miera, E. V.-S., & Rudy, B. (2006). Differential characterization of three alternative spliced isoforms of DPPX. *Brain Res* 1094, 1–12.

Nadal, M. S., Ozaita, A., Amarillo, Y., Vega-Saenz, d. M., Ma, Y., Mo, W., Goldberg, E. M., Misumi, Y., Ikehara, Y., Neubert, T. A., & Rudy, B. (2003). The CD26-related dipeptidyl aminopeptidase-like protein DPPX is a critical component of neuronal A-type K^+ channels. *Neuron* 37, 449–461.

Pak, M. D., Baker, K., Covarrubias, M., Butler, A., Ratcliffe, A., & Salkoff, L. (1991). mShal, a subfamily of A-type K^+ channel cloned from mammalian brain. *Proc Natl Acad Sci USA* 88, 4386–4390.

Perez-Garcia, M. T., Colinas, O., Miguel-Velado, E., Moreno-Dominguez, A., & López-López, J. R. (2004). Characterization of the Kv channels of mouse carotid body chemoreceptor cells and their role in oxygen sensing. *J Physiol-Lond* 557, 457–471.

Perez-Garcia, M. T., López-López, J. R., Riesco, A. M., Hoppe, U. C., Marban, E., Gonzalez, C., & Johns, D. C. (2000). Viral gene transfer of dominant-negative Kv4 construct suppresses an O_2-sensitive K^+ current in chemoreceptor cells. *J Neurosci* 20, 5689–5695.

Pérez-García, M. T., Obeso, A., López-López, J. R., Herreros, B., & Gonzalez, C. (1992). Characterization of cultured chemoreceptor cells dissociated from adult rabbit carotid body. *AJP - Lung Cell Mol Physiol* 263, C1152–C1159.

Rae, J. L. & Levis, R. A. (2002). Single-cell electroporation. *Pflugers Arch.* 443, 664–670.

Sanchez, D., López-López, J. R., Perez-Garcia, M. T., Sanz-Alfayate, G., Obeso, A., Ganfornina, M. D., & Gonzalez, C. (2002). Molecular identification of Kvalpha subunits that contribute to the oxygen-sensitive K(+) current of chemoreceptor cells of the rabbit carotid body. *J Physiol* 542, 369–382.

Sanguinetti, M. C., Johnson, J. H., Hammerland, L. G., Kelbaugh, P. R., Volkmann, R. A., Saccomano, N. A., & Mueller, A. L. (1997). Heteropodatoxins: peptides isolated from spider venom that block Kv4.2 potassium channels. *Mol Pharmacol* 51, 491–498.

Song, W. J., Tkatch, T., Baranauskas, G., Ichinohe, N., Kitai, S. T., & Surmeier, D. J. (1998). Somatodendritic depolarization-activated potassium currents in rat neostriatal cholinergic interneurons are predominantly of the A type and attributable to coexpression of Kv4.2 and Kv4.1 subunits. *J Neurosci* 18, 3124–3137.

Wada, K., Yokotani, N., Hunter, C., Doi, K., Wenthold, R. J., & Shimasaki, S. (1992). Differential expression of two distinct forms of mRNA encoding members of a dipeptidyl aminopeptidase family. *Proc Natl Acad Sci USA* 89, 197–201.

Sustained Hypoxia Enhances TASK-like Current Inhibition by Acute Hypoxia in Rat Carotid Body Type-I Cells

F. Ortiz, R. Iturriaga and R. Varas

Abstract Carotid body type-I cells respond to acute hypoxia with membrane depolarization and calcium-dependent neurotransmitter release. The inhibition of a TASK-like background potassium channels plays a key role in initiating this response. Chronic hypoxia enhances the carotid body chemosensory responses evoked by acute hypoxia, however the accurate mechanism by which chronic hypoxia increases carotid body reactivity is not clear. Therefore, we investigated the effects of chronic hypoxia upon TASK-like currents in isolated type-I cells. Carotid bodies were excised from anaesthetized newborn Sprague-Dawley rats and dissociated by collagenase-trypsin digestion. Isolated cells were maintained under 5% CO_2 in normoxic (21% O_2) or hypoxic (1–2% O_2) environment for 24 and 48 hours. Channel activity (NPo) was recorded using the cell-attached configuration of the patch-clamp technique. In normoxic and 24 hours hypoxic cultured cells, acute hypoxic stimuli decreases NPo approximately 70% with no effects on current amplitude. On the other hand, in cultured cells subjected to 48 hours of hypoxia, NPo decreases near to 90% in response to acute hypoxia. We concluded that continuous hypoxic exposure enhances the TASK-like channel activity inhibition in response to acute hypoxia. Our results provide a potential mechanism by which chronic hypoxia increases carotid body reactivity.

Keywords Chronic sustained hypoxia · Acute hypoxia · TASK-like channel

1 Introduction

The current model for acute hypoxic stimuli transduction states that hypoxia evokes a depolarization of carotid body (CB) type-I cells, leading to an intracellular Ca^{2+} increase and the subsequent release of excitatory neurotransmitter(s) (Gonzalez

R. Varas (✉)
Laboratory of Neurobiology, Facultad de Ciencias Biológicas, P. Universidad Católica de Chile, Santiago, Chile
e-mail: rvaras@bio.puc.cl

et al. 1994, Iturriaga et al. 2007). It is well known that hypoxia produces a fast and reversible inhibition of type-I cells K^+ currents, promoting the aforementioned depolarization (López-Barneo et al. 1988, López-Lopez and Gonzalez 1992, Wyatt and Peers 1995, Buckler 1997). In the neonatal rat CB model, it has been established that a key event in initiating the response to acute hypoxia is the inhibition of an oxygen-sensitive, TASK-like background K^+ current (Buckler 1997, Buckler et al. 2000).

The effects of chronic sustained hypoxia (CSH) on CB function, however, are complex and far from being elucidated (Bisgard 2000). It has been shown in several animal models and humans that subjects exposed to sustained hypoxia for hours to days (i.e. high altitude) developed a progressive increase in basal ventilation and showed augmented ventilatory responses to hypoxia (Vizek et al. 1987).

This phenomenon, called "ventilatory acclimatization", is characterized by enhanced CB chemosensory responses (Vizek et al. 1987). Currently, it is known that CSH produces structural and functional changes in the CB. Most of these reports have been focused on transmitters and modulators of CB transduction and only few of have tackled the study of type-I cell ionic currents in sustained hypoxia models (Bisgard 2000, Powell 2007). In this regard, some studies showed that sustained hypoxia increases type-I cell excitability and modifies the density of K^+ and Na^+ channels (Hempleman 1995, Carpenter et al. 1998, Kääb et al. 2005, Cáceres et al. 2007). However, the possible CSH effect on TASK-like K^+ current of type-I cells has not been studied. Since TASK-like K^+ current is thought to have a key role in sensing acute hypoxia and/or metabolic cell status (Duprat et al. 2007, Wyatt et al. 2007), we hypothesized that the inhibition of type-I TASK-like K^+ current evoked by acute hypoxia is promoted by sustained hypoxia, explained -at least in part- the CB hyperreactivity to acute hypoxia.

2 Methods

CBs were extracted from newborn male Sprague-Dawley rat pups (ten days old), previously anesthetized with ketamin/xylazine (75/7.5 mg/Kg, respectively). The CBs were placed on ice cold modified Hank's balanced salt solution (4°C, pH = 7.43). At the end of surgery the animals were killed by beheading whilst still anaesthetized. Then, the carotid bodies were enzymatic (trypsin/collagenase) and mechanical dissociated as previously described (Buckler 1997). Cellular suspension was plated out onto poly-D-lysine coated glass coverslip. The cells were cultured in normoxia (21% O_2; N24 and N48) and hypoxia (1–2% O_2; CH24 y CH48) conditions for 24 or 48 hours. Additionally, we studied acutely disassociated (AD, 2–4 hours after CB dissociation) cells.

Experiments were conducted on cells from each group using the cell-attached configuration of the patch clamp technique. Single and multiple channel recordings were performed using an Axopatch 200B amplifier. Membrane currents were filtered at 2 kHz and recorded at a sampling frequency of 20 kHz (using a Digidata 1200). Electrodes were fabricated from borosilicate glass capillaries and were fire-polished before use (electrode resistance was between 5 and 15 MΩ and seal resistances were $\geq 5\,G\Omega$). Cell-attached channel recordings were performed with no imposed pipette potential.

Cells were bathed with standard HCO_3^--buffered saline contained (in mM) 117 NaCl, 4.5 KCl, 23 NaHCO$_3$, 1.0 MgCl$_2$, 2.5 CaCl$_2$ and 11 glucose; bubbled with either 5% CO$_2$ and 95% air (normoxia, PO$_2$ ~140 mmHg) or 5% CO$_2$ and 95% N$_2$ (hypoxia, PO$_2$ ~5 mmHg). Pipette (extracellular) solutions contained (in mM) 140 KCl, 4 MgCl$_2$, 1 EGTA, 10 HEPES, 10 tetraethylammonium (TEA)-Cl and 5 4-aminopyridine (4-AP), both solutions were equilibrated at pH = 7.43. Experiments were conducted at 28–30°C.

Single channel recordings were analyzed with pClamp7.0 software (Molecular Devices). Channel activity is reported as the open probability times the number of channels in a given patch (NPo). Mean values are expressed as MEAN ± SEM. Non-parametric ANOVA (Kruskal-Wallis test), and Mann Whitney Test was used (statistical significant differences were set at $p < 0.05$).

3 Results

The main form of channel activity observed at resting conditions (see Methods) had current amplitude near 1 pA and rapid flickery openings, in concordance to what has been described as a TASK-like K$^+$ current, as is shown in a representative recording in Fig. 1. According to this, when control type-I cells were exposed to an acute hypoxic stimuli (PO$_2$ = 5 mmHg for 3 min) we observed a reduction in the channel activity (NPo, Figs. 2 and 3b), while single channel current amplitude was not affected (Figs. 2 and 3a).

Interestingly, the TASK-like K$^+$ current inhibition evoked by acute hypoxia in cells exposed to hypoxia for 48 hours is enhanced (when compared to control cells: AD, N24 and N48). Indeed, the percentage of NPo inhibition for control group was 67.8 ± 0.02 %, while this value rise to 88.9 ± 0.01% in the hypoxic group (Mann Whitney test, $p < 0.05$, Fig. 3b).

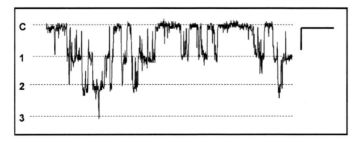

Fig. 1 Representative record of single channel currents in the cell-attached voltage-clamp configuration at 0 mV pipette potential (resting membrane potencial). Symmetric potassium, TEA (10 mM), and 4-AP (5 mM) conditions were maintained. Please note the high degree of flickering and the "burst" discharge. C: closed state; 1,2,3: open states 1,2 and 3. Scale bars: 1 pA, 20 ms

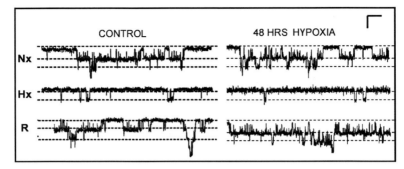

Fig. 2 Recording traces of the same membrane patch in the cell-attached configuration (pipette potential = 0 mV, symmetric K$^+$), in normoxia (Nx), acute hypoxia (Hx, PO$_2$ = 5 mmHg) and recovery (R, after 3 minutes of washing out hypoxic solution) conditions. *Left panel*, recording from a cell cultured in normoxia (control). *Right panel*, cell cultured in chronic hypoxia for 48 hours (see methods). Scale bars: 1 pA, 10 ms

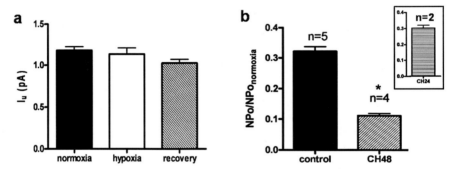

Fig. 3 (**a**) Average of single event current amplitude (I$_u$). Pipette potential = 30 mV, Kruskal-Wallis, p = 0.7986, n = 7. (**b**) Chronic hypoxia (48 hrs; CH48) enhances channel activity inhibition induced by acute hypoxia. Mann Whitney, $p < 0.05$. The CH24 is not statistically different to the control (inset)

4 Discussion

Several evidences strongly suggest that potentiation of CB chemosensory responses in chronically hypoxic animals is essential for ventilatory acclimatization (Vizek et al. 1987). Nevertheless, the mechanism by which this increased response to acute hypoxia is achieved is not clear.

Background K$^+$ current (mediated mainly by TASK-like K$^+$ channels) plays a primary role in carotid body type-I cells' chemoreception process since they control the cells' resting membrane potential and therefore their excitability (Buckler 1997). Particularly, in isolated rat type-I cells the membrane depolarization evoked by acute hypoxia appears to be largely mediated through the inhibition of this TASK-like

K^+ current (Buckler et al. 2000, Williams and Buckler 2004). In addition, TASK-like K^+ channels are also coupled to mitochondrial activity (Buckler and Vaughan-Jones 1998, Wyatt and Buckler 2004) and therefore they might also be involved in sensing type-I cells metabolic status (Williams and Buckler 2004, Varas et al. 2007).

Our data shows an enhanced inhibition of the TASK-like current in neonatal rat type-I cells cultured in hypoxia (1–2% O_2) for two days: while in control conditions acute hypoxia (PO_2 = 5 mmHg) evokes a ~70% inhibition of the TASK-like current, in type-I cells subjected to hypoxia for 48 hours, current inhibition reaches ~90% (Figs. 2 and 3). These results could be reflecting the well-known effect of CSH in type-I cells metabolism, which includes changes in AMP/ATP ratio, increased cAMP levels, among others (Delpiano and Acker 1991, Nurse et al. 1994, Kobayashi et al. 1998).

There are probably multiple mechanisms underlying the potentiation of CB responses during chronic hypoxia. In that regard, it is worth to note that the increased inhibition of TASK-like currents, in neonatal rat type-I cells subjected to chronic hypoxia reported here, provides a novel potential mechanism to explain the CB hyperreactivity.

5 Conclusion

Our data shows that the sustained hypoxia for 48 hours enhances the TASK-like current inhibition by acute hypoxia, suggesting a mechanism that could contribute to explain the carotid body increased chemosensory during chronic hypoxia.

Acknowledgments This research was supported by grant 1070854 from FONDECYT. F.O. is supported by CONICYT fellowship for PhD students.

References

Bisgard, G.E. 2000, Carotid body mechanisms in acclimatization to hypoxia, *Respir Physiol*, 121: 237–246.
Buckler, K. 1997, A novel oxygen-sensitive potassium current in rat carotid body type I cells, *J Physiol*, 498: 649–662.
Buckler, K., Vaughan-Jones, R. D. 1998, Effects of mitochondrial uncouplers on intracellular calcium, pH and membrane potential in rat carotid body type I cells, *J Physiol*, 513: 819–833.
Buckler, K., Williams, B., Honoré, E. 2000, An oxygen, acid and anaesthetic sensitive TASK-like background potassium channel in rat arterial chemoreceptor cells, *J Physiol*, 525: 135–142.
Cáceres, A.I., Obeso, A., Gonzalez, C., Rocher, A. 2007, Molecular identification and functional role of voltage-gated sodium channels in rat carotid body chemoreceptor cells. Regulation of expression by chronic hypoxia in vivo, *J Neurochem*, 102: 231–245.
Carpenter, E., Bee, D., Peers, C. 1998, Ionic currents in carotid body type I cells isolated from normoxic and chronically hypoxic adult rats, *Brain Res*, 811: 79–87.
Delpiano, M.A., Acker, H. 1991, Hypoxia increases the cyclic AMP content of the cat carotid body in vitro. *J Neurochem*, 57: 291–297.
Duprat, F., Lauritzen, I., Patel, A., Honoré, E. 2007, The TASK background K2P channels: chemo- and nutrient sensors, *Trends Neurosci*, 30: 573–580.

Gonzalez, C., Almaraz, L., Obeso, A., Rigual, R. 1994, Carotid body chemoreceptors: from natural stimuli to sensory discharges, *Physiol Rev*, 74: 829–898.

Hempleman, S.C. 1995, Sodium and potassium current in neonatal rat carotid body cells following chronic in vivo hypoxia, *Brain Res*, 699: 42–50.

Iturriaga, R., Varas, R., Alcayaga, J. 2007, Electrical and pharmacological properties of petrosal ganglion neurons that innervate the carotid body. *Resp Physiol & Neurobiol*, 157: 130–139.

Kääb, S., Miguel-Velado, E., López-Lopez, J.R., Pérez-García, M.T. 2005, Down regulation of Kv3.4 channels by chronic hypoxia increases acute oxygen sensitivity in rabbit carotid body, *J Physiol*, 566: 395–408.

Kobayashi, S., Beitner-Johnson, D., Conforti, L., Millhorn, D.E. 1998, Chronic hypoxia reduces adenosine A2A receptor-mediated inhibition of calcium current in rat PC12 cells via downregulation of protein kinase A. *J Physiol*, 512: 351–363.

López-Barneo, J., López-Lopez, J.R., Ureña, J., Gonzalez, C. 1988, Chemotransduction in the carotid body: Current modulated by PO_2 in type I chemoreceptor cells, *Science*, 241: 580–582.

López-Lopez, J.R., Gonzalez, C. 1992, Time course of K^+ current inhibition by low oxygen in chemoreceptor cells of adult rabbit carotid body. Effects of carbon monoxide, *FEBS Lett*, 299: 251–254.

Nurse, C. A., Jackson, A., Stea, A. 1994, Plasticity in cultured arterial chemoreceptors: effects of chronic hypoxia and cyclic AMP analogs. *Adv Exp Med Biol*, 360: 167–170.

Powell, F.L. 2007, The influence of chronic hypoxia upon chemoreception, *Resp Physiol Neuro*, 157: 154–161.

Varas, R., Wyatt, C.N., Buckler, K. 2007. Modulation of TASK-like background potassium channels in rat arterial chemoreceptor cells by intracellular ATP and other nucleotides. *J Physiol*, 583: 521–536.

Vizek, M., Pickett, C.K., Weil, J.V. 1987, Increased carotid body hypoxic sensitivity during acclimatization to hypobaric hypoxia, *J Appl Physiol*, 63: 2403–2410.

Williams, B.A., Buckler, K.J. 2004, Biophysical properties and metabolic regulation of a TASK-like potassium channel in rat carotid body type 1 cells. *Am J Physiol Lung Cell Mol Physiol*, 286: L221–L230.

Wyatt, C.N., Buckler, K. 2004, The effect of mitochondrial inhibitors on membrane currents in isolated neonatal rat carotid body type I cells, *J Physiol*, 556: 175–191.

Wyatt, C.N., Mustard, K.J., Pearson, S.A., Dallas, M.L., Atkinson, L., Kumar, P., Peers, C., Hardie, D.G., Evans, A.M. 2007, AMP-activated protein kinase mediates carotid body excitation by hypoxia, *J Biol Chem*, 282: 8092-8098

Wyatt, C.N., Peers, C. 1995, Ca^{2+}-activated K^+ channels in isolated type I cells of the neonatal rat carotid body, *J Physiol*, 483: 559–565.

Inhibition of L-Type Ca^{2+} Channels by Carbon Monoxide

M.L. Dallas, J.L. Scragg and C. Peers

Abstract Inhibition of K^+ channels in glomus cells underlies excitation of the carotid body by hypoxia. It has recently been proposed that hypoxic inhibition involves either activation of AMP activated protein kinase (AMPK) or inhibition of carbon monoxide (CO) production by heme oxygenase 2 (HO-2). In the vasculature, L-type Ca^{2+} channels are also O_2 sensitive. Here, we have investigated the possible involvement of either AMPK or CO in the hypoxic inhibition of L-type Ca^{2+} channels. Using whole-cell patch clamp recordings from HEK293 cells stably expressing the human cardiac α_{1C} Ca^{2+} channel subunit, we found that pre-treatment of cells with AICAR (to activate AMPK) was without effect on Ca^{2+} currents. CO, applied via the donor molecule CORM-2 caused reversible, voltage-independent Ca^{2+} channel inhibition of up to *ca.* 50%, whereas its inactive form (iCORM) was without significant effect. Effects of CO were prevented by the antioxidant MnTMPyP, but not by inhibition of NADPH oxidase (with either apocynin or diphenyleneiodonium), or xanthine oxidase (with allopurinol). Instead, inhibitors of complex III of the mitochondrial electron transport chain and a mitochondrial-targeted antioxidant (Mito Q), prevented the effects of CO. Our data suggest that hypoxic inhibition of L-type Ca^{2+} channels does not involve AMPK or CO. However, the known cardioprotective effects of HO-1 could arise from an inhibitory action of CO on L-type Ca^{2+} channels.

Keywords Ca^{2+} channel · L-type · Carbon monoxide · Reactive oxygen species · Mitochondria · Splice variant · Patch clamp · AMP kinase · Electron transport chain

1 Introduction

The membrane hypothesis for chemotransduction supports the concept that hypoxia (and other chemostimuli) excite the carotid body by inhibiting K^+ channels in glomus cells (Kumar, 2007). This leads to their depolarization, voltage-gated Ca^{2+}

C. Peers (✉)
Division of Cardiovascular and Neuronal Remodelling, Leeds Institute of Genetics, Health and Therapeutics University of Leeds, Leeds LS2 9JT, UK
e-mail: c.s.peers@leeds.ac.uk

entry and hence release of neurotransmitters onto afferent chemosensory fibres. The principal mechanisms underlying hypoxic inhibition of K^+ channels remains to be fully elucidated. However, two proposed mechanisms have received attention in recent years, both involving enzymes coupled to channel activity. Firstly, heme oxygenase-2 (HO-2) was found to co-localise with O_2 sensitive maxiK channels in a recombinant expression system. In the presence of O_2 and NADPH, HO-2 produces CO during the breakdown of heme. CO increases maxiK channel activity, and so during hypoxia, CO levels decline thereby reducing K^+ channel activity (Williams et al. 2004). This mechanism was confirmed in rat glomus cells, though may not be so prominent in other species (Ortega-Saenz et al. 2006). Secondly, the "metabolic fuel gauge" enzyme AMP activated protein kinase (AMPK (Hardie et al. 2006)) was proposed to be activated (possibly by a shift in the AMP:ATP ratio) in glomus cells during hypoxia (Evans et al. 2005). This enzyme was subsequently demonstrated to phosphorylate maxiK channels directly (Wyatt et al. 2007) and, in so doing, suppress their activity. Furthermore, AMPK activation caused glomus cell depolarization, raised $[Ca^{2+}]_i$ and excited afferent chemosensory fibres (Wyatt et al. 2007) and inhibition of AMPK activity suppressed hypoxic excitation of the carotid body.

Hypoxia also suppresses the activity of L-type Ca^{2+} channels in the vasculature (Franco-Obregon et al. 1995), an effect which may contribute to systemic hypoxic vasodilation. However, in contrast to the studies described above for hypoxic inhibition of maxiK channels, no mechanisms have yet been proposed to account for this action of hypoxia on Ca^{2+} channels. The present study was therefore designed to investigate whether hypoxic inhibition of recombinant L-type Ca^{2+} channels – which appears indistinguishable from effects seen on native channels (Fearon et al. 1997) – involves either activation of AMPK or altered CO production.

2 Methods

HEK293 cells stably transfected with the hHT splice variant of the human cardiac L-type Ca^{2+} channel α_{1C} subunit (Fearon et al. 2000) were cultured in growth medium comprising MEM with Earle's salts and L-glutamine, supplemented with 9% (v/v) fetal calf serum (Globepharm, Esher, Surrey, UK), 1% (v/v) non-essential amino acids, 50 μg/ml gentamicin, 100 units/ml penicillin G, 100 μg/ml streptomycin and 0.25 μg/ml amphotericin in a humidified atmosphere of air/CO_2 (19:1) at 37°C. All culture reagents were purchased from Gibco-BRL (Paisley, UK) unless otherwise stated.

HEK293 cells attached to coverslip fragments were placed in a perfused (2–4 ml/min) chamber and whole-cell patch-clamp recordings were made as previously described (Fearon et al. 2000). Perfusate contained (in mM): NaCl, 95; CsCl, 5; $MgCl_2$, 0.6; $BaCl_2$, 20; HEPES, 5; D-glucose, 10; TEA-Cl 20 (pH 7.4, 21–24°C). Patch electrodes (resistance 4–7 MΩ) contained (in mM): CsCl, 120; TEA-Cl, 20; $MgCl_2$, 2; EGTA, 10; HEPES, 10; ATP, 2 (pH adjusted to 7.2 with CsOH). Cells were clamped at –80 mV and whole cell capacitance determined from analogue compensation. Series resistance compensation of 70–90% was applied. Whole-cell currents were evoked by 100 ms step depolarizations to various test potentials (0.1 Hz) and leak-subtraction was applied as previously described (Fearon et al. 2000). Evoked currents were filtered at 1 kHz, digitized at 2 kHz and current amplitudes measured over the last 10–15 ms of each step depolarization since Ba^{2+} was used as the charge carrier, as they displayed little or no inactivation during step depolarizations. Currents showing notable run-down before CORM-2 application were discarded.

All voltage-clamp and analysis protocols were performed with the use of an Axopatch 200A amplifier/Digidata 1200 interface controlled by Clampex 9.0 software (Molecular Devices, Foster City, CA). Offline analysis was performed using the data analysis package Clampfit 9.0 (Molecular Devices, Foster City, CA). Results are presented as means ± S.E.M., and statistical analysis performed using unpaired Student's t-tests where $P < 0.05$ was considered statistically significant.

3 Results

3.1 AMPK Activation Does Not Alter Ca^{2+} Channels

To investigate the possible involvement of AMPK in the hypoxic modulation of Ca^{2+} channels, HEK293 cells were exposed to AICAR (1 mM) for at least 20 min at 37°C. AICAR is metabolised intracellularly to generate an AMP-like molecule, thereby activating AMPK (Evans et al. 2005). This approach has previously been employed to demonstrate AMPK regulation of maxiK channels (Wyatt et al. 2007). As shown in Fig. 1, this approach was completely without effect on Ca^{2+} channel currents throughout the range of activating test potentials studied. Thus, it appears unlikely that AMPK activation can account for hypoxic inhibition of L-type Ca^{2+} channels.

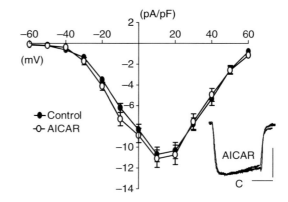

Fig. 1 Mean (with S.E.M. bars, n=8) current-density versus voltage relationships measured in HEK293 cells expressing α_{1C} subunits. Cells were either untreated, or exposed to AICAR (1 mM) for at least 20 min prior to recordings. Inset, example currents from the two cells groups. Currents evoked by step depolarizations from −80 mV to +10 mV. Scales bars; 100pA *vertical*, 50 ms *horizontal*

3.2 CO Inhibits Ca^{2+} Channels

Exposure of HEK293 cells to the established CO donor CORM-2 caused marked inhibition of currents, as exemplified in Fig. 2A. Inhibition was slowly reversible and not associated with changes in current kinetics (Fig. 2A). The effects of CORM-2 were concentration-dependent, with an IC_{50} of ca. 15 μM. Importantly, currents were largely unaffected (~10% inhibition) either by the vehicle, DMSO, or by the inactive breakdown product, iCORM-2 (Fig. 2B) indicating that the effects of CORM-2 were largely due to its release of CO, as has previously been established (Williams et al. 2004; Boczkowski et al. 2006; Chatterjee, 2007). In addition,

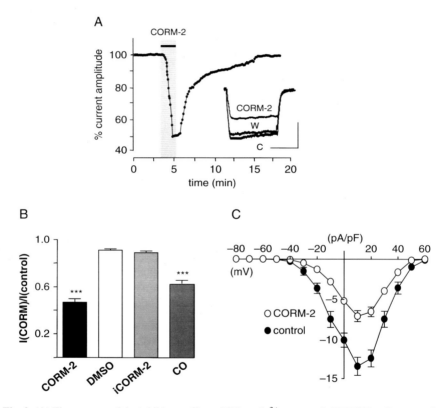

Fig. 2 (**A**) Time course of the inhibitory effect of CO on Ca^{2+} currents in HEK293 cells expressing $α_{1C}$ subunits. CORM-2 (30 μM) was applied for period indicated by bar. Each point is the current amplitude evoked by a depolarization from −80 to +10 mV (100 ms, 0.1 Hz). Inset, superimposed example currents (C; control, W; wash). Scale bars; 100pA vertical, 50 ms horizontal. (**B**) Bar graph showing mean (± s.e.m.) inhibitory effects of CORM-2, vehicle (DMSO), the inactive molecule (iCORM-2) and CO itself. ***$P < 0.001$ vs. the response to iCORM-2. (**C**) Mean (with s.e.m. bars, n=10) current-density versus voltage relationships measured in HEK293 cells expressing $α_{1C}$ subunits. Cells were either untreated (*closed circles*), or exposed to 30 μM CORM-2 (*open circles*)

we exposed cells to dissolved CO and this caused a similar degree of reversible inhibition (Fig. 2B). Full current-voltage relationships indicated that the effects of CO were apparent at all activating test potentials (Fig. 2C), indicating a lack of voltage-dependence.

3.3 CO Increases Mitochondrial ROS Production

CO can modulate a variety of signalling pathways to exert a diverse array of cellular effects (Wu and Wang 2005; Kim et al. 2006). We therefore took a pharmacological approach to investigate candidate mechanisms underlying the ability of CO to inhibit L-type Ca^{2+} channels. Results are summarized in Table 1.

In sum, we found that the inhibitory effects of CO were fully prevented by either the antioxidant MnTMPyP (100 μM) or by the reducing agent, dithiothreitol (DTT; 2 mM). Thus, CO modulation of Ca^{2+} currents appeared to involve reactive oxygen species (ROS). To identify the source of ROS, we inhibited xanthine oxidase with allopurinol (1 μM) and NADPH oxidase with apocynin (30 μM) and diphenyleneiodonium (3 μM). In the presence of these drugs, CO still inhibited Ca^{2+} currents significantly. Next, we examined mitochondria as the source of ROS and found that the mitochondria-targeted antioxidant, MitoQ (250 nM) prevented the actions of CO. Furthermore, both antimycin A (3 μM) and stigmatellin (1 μM), but not rotenone (2 μM), prevented the actions of CO, indicating that CO exerted its effects via ROS production specifically from complex III of the electron transport chain.

4 Discussion

The present work clearly indicates that hypoxia is unlikely to inhibit L-type Ca^{2+} channels either via AMPK activation or via inhibition of HO-2, as has been proposed for hypoxic modulation of maxiK channels (see Introduction). AMPK activation

Table 1 Pharmacological manipulation of the inhibition of L-type Ca^{2+} channels by CO (applied as the CO donor, CORM-2, at 30 μM). Inhibition is expressed as percentage ± S.E.M. taken from at least 6 recordings. Significance compared with CORM-2 vehicle alone

Compound	Inhibition of Ca^{2+} current (%)	Significance
None	53.2 ± 2.3	P < 0.001
MnTMPyP	14.5 ± 3.4	n.s.
DTT	10.7 ± 1.1	n.s.
Allopurinol	52.3 ± 0.2	P < 0.001
Diphenyleneiodonium	41.6 ± 4.2	P < 0.001
Apocynin	37.8 ± 2.1	P < 0.001
MitoQ	10.8 ± 3.6	n.s.
Rotenone	46.1 ± 3.0	P < 0.001
Antimycin A	13.3 ± 2.7	n.s.
Stigmatellin	15.3 ± 4.3	n.s.

was without effect on Ca^{2+} currents, and the HO-2 product, CO *inhibited* currents, whereas enhancement would be expected if it were to account for hypoxic inhibition. Thus, a novel, as yet undetermined mechanism must account for this action of hypoxia on Ca^{2+} channels. Nevertheless, our study has revealed a novel inhibitory effect of CO which occurs via an increase in ROS production specifically from mitochondria. Cellular ROS production can occur at numerous sites, including NADPH oxidases and xanthine oxidase, and CO can modulate NADPH oxidase (Taille et al. 2005). However, we show that mitochondria are the source of ROS increases evoked by CO. Thus, both a general antioxidant (MnTMPyP) and a mitochondrial targeted one (MitoQ), but not inhibition of NADPH or xanthine oxidases, prevented the actions of CO (Table 1). This is in accordance with previous studies (Taille et al. 2005; Zuckerbraun et al. 2007; D'Amico et al. 2006). CO binds to complex IV (cytochrome c oxidase) of the electron transport chain (see D'Amico et al. 2006), thereby presumably inhibiting its acceptance of electrons from complex III and so causing more to leak and form ROS. Indeed, the effect of CO was prevented by complex III (but not complex I) inhibition (Table 1).

Although these studies have not identified the mechanism of hypoxic inhibition of Ca^{2+} channels, the inhibition by CO is likely to be of physiological significance. Cardiac atrial and ventricular myocytes express HO-1 and HO-2 which both generate CO (together with biliverdin and Fe^{2+} by heme catabolism) and HO-1 levels can be increased by various stress factors (Ewing et al. 1994) including myocardial infarction (Lakkisto et al. 2002) and chronic exposure to hipoxia (Grilli et al., 2003). CO limits the cellular damage of ischemia/reperfusion injury in the heart (Clark et al. 2003). Indeed, greater cardiac damage is seen following ischemia/reperfusion injury in HO-1 knockout mice (Yet et al. 1999). Conversely, HO-1 over-expression in the heart reduces infarct size and other markers of damage following ischemia/reperfusion injury (Yet et al. 2001). CO also improves cardiac blood supply through coronary vessel dilation and reduces cardiac contractility (McGrath 1984). Our data provide a mechanism to account for these protective effects of CO, through inhibition of L-type Ca^{2+} channels via mitochondrial ROS production.

Acknowledgments This work was supported by the Wellcome Trust and the British Heart Foundation.

References

Boczkowski, J., Poderoso, J. J., & Motterlini, R. 2006, CO-metal interaction: Vital signaling from a lethal gas. *Trends Biochem Sci* 31: 614–621.
Chatterjee, P. K. 2007, Physiological activities of carbon monoxide-releasing molecules: Ca ira. *Br J Pharmacol* 150: 961–962.
Clark, J. E., Naughton, P., Shurey, S., Green, C. J., Johnson, T. R., Mann, B. E., Foresti, R., & Motterlini, R. 2003, Cardioprotective actions by a water-soluble carbon monoxide-releasing molecule. *Circ Res* 93: e2–e8.
D'Amico, G., Lam, F., Hagen, T., & Moncada, S. 2006, Inhibition of cellular respiration by endogenously produced carbon monoxide. *J Cell Sci* 119: 2291–2298.

Evans, A. M., Mustard, K. J., Wyatt, C. N., Peers, C., Dipp, M., Kumar, P., Kinnear, N. P., & Hardie, D. G. 2005, Does AMP-activated protein kinase couple inhibition of mitochondrial oxidative phosphorylation by hypoxia to calcium signaling in O_2-sensing cells? *J Biol Chem* 280: 41504–41511.

Ewing, J. F., Raju, V. S., & Maines, M. D. 1994, Induction of heart heme oxygenase-1 (HSP32) by hyperthermia: possible role in stress-mediated elevation of cyclic 3':5'-guanosine monophosphate. *J Pharmacol Exp Ther* 271: 408–414.

Fearon, I. M., Palmer, A. C. V., Balmforth, A. J., Ball, S. G., Mikala, G., Schwartz, A., & Peers, C. 1997, Hypoxia inhibits the recombinant α1C subunit of the human cardiac L-type Ca^{2+} channel. *J Physiol* 500: 551–556.

Fearon, I. M., Varadi, G., Koch, S., Isaacsohn, I., Ball, S. G., & Peers, C. 2000, Splice variants reveal the region involved in oxygen sensing by recombinant human L-type Ca^{2+} channels. *Circ Res* 87: 537–539.

Franco-Obregon, A., Urena, J., & Lopez-Barneo, J. 1995, Oxygen-sensitive calcium channels in vascular smooth muscle and their possible role in hypoxic arterial relaxation. *Proc Natl Acad Sci USA* 92: 4715–4719.

Grilli, A., De Lutiis, M. A., Patruno, A., Speranza, L., Gizzi, F., Taccardi, A. A., Di, N. P., De, C. R., Conti, P., & Felaco, M. 2003, Inducible nitric oxide synthase and heme oxygenase-1 in rat heart: direct effect of chronic exposure to hypoxia. *Ann Clin Lab Sci* 33: 208–215.

Hardie, D. G., Hawley, S. A., & Scott, J. W. 2006, AMP-activated protein kinase–development of the energy sensor concept. *J Physiol* 574: 7–15.

Kim, H. P., Ryter, S. W., & Choi, A. M. 2006, CO as a cellular signaling molecule. *Annu Rev Pharmacol Toxicol* 46: 411–449.

Kumar, P. 2007, Sensing hypoxia in the carotid body: from stimulus to response. *Essays Biochem* 43: 43–60.

Lakkisto, P., Palojoki, E., Backlund, T., Saraste, A., Tikkanen, I., Voipio-Pulkki, L. M., & Pulkki, K. 2002, Expression of heme oxygenase-1 in response to myocardial infarction in rats. *J Mol Cell Cardiol* 34: 1357–1365.

McGrath, J. J. 1984, The effects of carbon monoxide on the heart: an in vitro study. *Pharmacol Biochem Behav* 21 Suppl 1: 99–102.

Ortega-Saenz, P., Pascual, A., Gomez-Diaz, R., & Lopez-Barneo, J. 2006, Acute Oxygen Sensing in Heme Oxygenase-2 Null Mice. *J General Physiol* 128: 405–411.

Taille, C., El-Benna, J., Lanone, S., Boczkowski, J., & Motterlini, R. 2005, Mitochondrial respiratory chain and NAD(P)H oxidase are targets for the antiproliferative effect of carbon monoxide in human airway smooth muscle. *J Biol Chem* 280: 25350–25360.

Williams, S. E., Wootton, P., Mason, H. S., Bould, J., Iles, D. E., Riccardi, D., Peers, C., & Kemp, P. J. 2004, Hemoxygenase-2 is an Oxygen Sensor for a Calcium-Sensitive Potassium Channel. *Science* 306: 2093–2097.

Wu, L. & Wang, R. 2005, Carbon monoxide: endogenous production, physiological functions, and pharmacological applications. *Pharmacol Rev* 57: 585–630.

Wyatt, C. N., Mustard, K. J., Pearson, S. A., Dallas, M. L., Atkinson, L., Kumar, P., Peers, C., Hardie, D. G., & Evans, A. M. 2007, AMP-activated protein kinase mediates carotid body excitation by hypoxia. *J Biol Chem* 282: 8092–8098.

Yet, S. F., Perrella, M. A., Layne, M. D., Hsieh, C. M., Maemura, K., Kobzik, L., Wiesel, P., Christou, H., Kourembanas, S., & Lee, M. E. 1999, Hypoxia induces severe right ventricular dilatation and infarction in heme oxygenase-1 null mice. *J Clin Invest* 103: R23–R29.

Yet, S. F., Tian, R., Layne, M. D., Wang, Z. Y., Maemura, K., Solovyeva, M., Ith, B., Melo, L. G., Zhang, L., Ingwall, J. S., Dzau, V. J., Lee, M. E., & Perrella, M. A. 2001, Cardiac-specific expression of heme oxygenase-1 protects against ischemia and reperfusion injury in transgenic mice. *Circ Res* 89: 168–173.

Zuckerbraun, B. S., Chin, B. Y., Bilban, M., de Costa, d. J., Rao, J., Billiar, T. R., & Otterbein, L. E. 2007, Carbon monoxide signals via inhibition of cytochrome c oxidase and generation of mitochondrial reactive oxygen species. *FASEB J* 21: 1099–1106.

Effects of the Polyamine Spermine on Arterial Chemoreception

S. Cayzac, A. Rocher, A. Obeso, C. Gonzalez, P.J. Kemp and D. Riccardi

Abstract Polyamines modulate many biological functions. Here we report a novel inhibitory modulation by spermine of catecholamine release by the rat carotid body and have identified the molecular mechanism underpinning it. We used molecular (RT-PCR and confocal microscopy) and functional (i.e., neurotransmitter release, patch clamp recording and calcium imaging) approaches to test the involvement of: (i) voltage-dependent calcium channels, and; (ii) the extracellular calcium-sensing receptor, CaR, a G protein-coupled receptor which is also activated by polyamines. RT-PCR and immunohistochemistry of isolated carotid bodies revealed that only $Ca_v1.2$ and $Ca_v2.2$ were expressed in type 1 cells while $Ca_v1.3$, $Ca_v1.4$, $Ca_v2.1$, $Ca_v2.3$ and $Ca_v3.1$, $Ca_v3.2$ and $Ca_v3.3$, could not be detected. CaR expression was detected exclusively in the nerve endings. In isolated carotid bodies, the hypoxia-dependent (7% O_2 for 10 minutes) and depolarization-evoked catecholamine release were partially suppressed by pre- (and co)-incubation with 500 µM spermine. In dissociated type 1 glomus cells intracellular calcium concentration did not change following spermine treatment, but this polyamine did inhibit the depolarisation-evoked calcium influx. Whole-cell patch clamp recordings of HEK293 cells stably transfected with $Ca_v1.2$ demonstrated that spermine inhibits this calcium channel. Interestingly, this inhibition was not apparent if the extracellular solution contained a concentration of Ba^{2+} above 2 mM as the charge carrier. In conclusion, spermine attenuates catecholamine release by the carotid body principally via inhibition of $Ca_v1.2$. This mechanism may represent a negative feedback, which limits transmitter release during hypoxia.

Keywords Carotid body · Polyamine · Calcium sensing receptor · Spermine · Calcium channels · Catecholamine

P.J. Kemp (✉)
School of Biosciences, Cardiff University, Museum Avenue, Cardiff, CF10 3AX, U.K.
e-mail: kemp@cardiff.ac.uk

1 Introduction

Naturally occurring polyamines (putrescine, spermidine and spermine) are small organic molecules, which have been reported to modulate many cellular processes. Polyamines are necessary to maintain optimal cell growth (Tabor and Tabor, 1984) and are present in the plasma at micromolar concentrations (Chaisiri et al., 1979). However, tissue levels of polyamines are increased as a consequence of normal hormonal variations (Gilad et al., 2002), and also as during pathological states, such as malignant transformation (Casero and Marton, 2007), brain injury (Dogan et al., 1999) or neuronal ischemia (Li et al., 2007). Furthermore, synaptic vesicles contain spermine at concentrations as high as 2 mM (Masuko et al., 2003). Polyamines are charged at physiological pH (Heby, 1986) and, due to their polycationic nature, exhibit greater affinities than other mono- and di-valent cations for acidic components of proteins and have, therefore, been suggested to target ion channels and receptors involved in the trafficking of calcium ions (Li et al., 2007). Specifically, they can inhibit voltage-gated calcium channels (Gomez and Hellstrand, 1995; Lasater and Solessio, 2002) and activate the cell surface calcium-sensing receptor, CaR (Quinn et al., 1997).

The carotid body responds to a decrease in oxygen availability with a release of neurotransmitters by type 1 glomus cells. While a variety of molecular players have been postulated to form the basis of the oxygen sensing machinery (see, for instance, Kemp et al. (2006) for a recent review), all of these mechanisms converge onto an increase in intracellular calcium concentration, necessary for neurotransmitter release. This is achieved through activation of voltage-gated calcium channels (Conde et al., 2006a). Polyamines may be stored at high concentration in neurotransmitter vesicles, suggesting that they play some physiological role in the modulation of the carotid body response to hypoxia. The purpose of this study was to investigate the presence of known targets for spermine action, namely voltage-gated calcium channels (Gomez and Hellstrand, 1995; Lasater and Solessio, 2002) and CaR, in the carotid body, and to study the effects of spermine on catecholamine release by isolated carotid bodies. Polyamine effects on intracellular calcium homeostasis in dissociated type 1 cells and in recombinant systems overexpressing calcium channels were also investigated.

2 Methods

2.1 Surgery

Wistar rats (100–350 g) were anesthetized with sodium pentobarbital in accordance with U.K. Home Office regulations. The carotid artery bifurcations were excised, carotid bodies dissected and placed in ice-cold solution containing (in mM): 143 NaCl, 2 KCl, 2 CaCl$_2$, 1.1 MgCl$_2$, 5.5 glucose and 10 HEPES, pH 7.4, bubbled with 100% O$_2$.

2.2 RT-PCR and Immunohistochemistry

Total RNA was extracted from carotid bodies using RNeasy Micro kit (Qiagen, Crawley, U.K.) according to the manufacturer's instructions. As positive controls for calcium channel amplification, total RNA was extracted from eye and brain using Trizol (Invitrogen, Paisley, U.K.) according to the manufacturer's instructions. 1 μg of total RNA was reverse-transcribed at 50°C for 30 min using Superscript III (Invitrogen, Strathclyde, U.K.). 2 μl of cDNA were used for each PCR reaction using the Premix Ex-taq polymerase kit (Lonza, Basel, Switzerland) with published sequences of intron-spanning primers designed to amplify: $Ca_v1.2$-1.3-1.4 (L-type), $Ca_v2.1$ (P/Q-type), $Ca_v2.2$ (N-type), $Ca_v2.3$ (R-type), $Ca_v3.1$-3.2-3.3 (T-type), CaR, β-actin and tyrosine hydroxylase (TH). All PCR products amplified from carotid body were fully sequenced to demonstrate the presence of *bona fide* transcripts for the genes of interest.

For immunohistochemistry, rats were perfusion-fixed with 4% paraformaldehyde, the tissue dissected, embedded in OCT (Tissue Tek, Sakura Finetek, Torrance, USA) and 4 μm-thick cryosections were cut onto Superfrost slides. Primary antibodies were incubated over night at 4°C: $Ca_v1.2$ (1/200), $Ca_v1.3$ (1/200), $Ca_v2.1$ (1/200), $Ca_v2.2$ (1/100), $Ca_v2.3$ (1/200), TH (1/1000) and CaR (1/200). The Alexa 594- and FITC-conjugated secondary antibodies (1/1000) were applied for 1 h at room temperature.

2.3 Measurement of Neurotransmitter Release

This technique has been extensively described elsewhere (e.g. (Conde et al., 2006b)). Briefly, the catecholamine store of intact carotid bodies was radiolabelled by a 2 h incubation with 30 μM [^{3}H]tyrosine (48 Ci/mmol), 100 μM of 6-methyltetrahydropterine and 1 mM of ascorbic acid dissolved in a solution containing (in mM): 140 NaCl, 5 KCl, 2 $CaCl_2$, 1.1 $MgCl_2$, 5 glucose, 10 HEPES, pH 7.4. Unincorporated [^{3}H]tyrosine was removed by rinsing the carotid bodies in the same solution for 1 h. Carotid bodies were individually incubated at $37 \pm 1°C$ in bicarbonate-buffered solution containing (in mM): 116 NaCl, 24 $NaHCO_3$, 5 KCl, 2 $CaCl_2$, 1.1 $MgCl_2$, 5 glucose, 10 HEPES, pH 7.4, equilibrated with 20% O_2, 5% CO_2, 75% N_2. The solutions used were bubbled either with 20% O_2, 5% CO_2 and 75% N_2 (normoxia); or 7% O_2, 5% CO_2 and 88% N_2 (hypoxia). Where a "high potassium" solution was employed, NaCl and KCl concentrations were 86 and 35 mM, respectively. Spermine and R-568 (the latter prepared as a 10 mM stock solution in dimethyl sulfoxide), were diluted in the bicarbonate-buffered solution. Every 10 min, the bathing solution was collected and replaced by fresh solution. A solution containing ascorbic and acetic acid was added to each collected sample before the [^{3}H]-catecholamines were absorbed with alumina. The alumina was then washed extensively and eluted with 1N HCl before the [^{3}H]-catecholamine was quantified by liquid scintillation counting.

2.4 Ca^{2+} Imaging in Type 1 Cells

Carotid bodies were dissociated with trypsin (1 mg/ml) and collagenase (2.5 mg/ml) and cultured for 24 h. carotid body cells were loaded with 4 μM fura-2 AM for 40 min in a HEPES-buffered solution containing (in mM): 125 NaCl, 4 KCl, 1 CaCl$_2$, 1 MgSO$_4$, 1 NaH$_2$PO$_4$, 20 HEPES, 6 glucose (pH 7.4) supplemented with 0.1% (w/v) bovine serum albumin and then washed for 15 min in experimental HEPES-buffered physiological solution to be used for the experiments, which contained (in mM): 125 NaCl, 5 KCl, 0.5 CaCl$_2$, 0.5 MgCl$_2$, 5 HEPES and 10 glucose, pH 7.4. For the "high potassium solution", NaCl was lowered to 115 mM and KCl increased to 15 mM. All the recordings were performed at 26 ± 1°C. Individual cell types were identified by *post-hoc* immunostaining using a tyrosine hydroxylase antibody.

2.5 Whole-Cell Recordings of Ca$_v$1.2-HEK293 Cells

HEK293 stably expressing the human Ca$_v$1.2 (Ca$_v$1.2-HEK293) were a gift from Prof. C Peers (University of Leeds, U.K). Bath solution contained (in mM): 135 NaCl, 5 KCl, 1.2 CaCl$_2$, 1.2 MgCl$_2$, 5 HEPES and 10 glucose, pH 7.4. Pipette solution contained (mM): 120 CsCl$_2$, 20 TEA-Cl, 2 MgCl$_2$, 10 EGTA, 10 HEPES, 2 Na-ATP, pH 7.2 with CsOH. After the whole-cell configuration had been achieved, the bath solution was changed to a solution containing (in mM): 113 NaCl, 5 CsCl, 0.6 MgCl$_2$, 2 BaCl$_2$, 5 HEPES, 10 glucose and 20 TEA-Cl, pH 7.4. Spermine, R-568 and nifedipine were diluted in the Ba^{2+} solution. Cells were voltage-clamped at −70 mV and then ramped at 0.1 Hz from −100 mV to +100 mV during 200 ms before being stepped from −80 mM to + 15 mV during 50 ms. Experiments were conducted at 21°C and all currents were leak-subtracted off-line.

3 Results and Discussion

RT-PCR showed that only Ca$_v$1.2 and Ca$_v$2.2 could be amplified from carotid body RNA (not shown). Consistent with these results, immunohistochemistry shows that both Ca$_v$1.2 (green – Fig. 1A) and Ca$_v$2.2 (green – Fig. 1D) were detected in type 1, tyrosine hydoxylase-positive cells (red - Fig. 1B and E). In addition, Ca$_v$1.2 was also expressed in the nerve fibres (arrows – Fig. 1C). These immunohistochemistry data were fully in agreement with the results obtained from RT-PCR amplifications of specific voltage-gated calcium channel mRNAs (not shown). In contrast, CaR could not be amplified by RT-PCR in the carotid body (data not shown) while immunostaining revealed the receptor to be present exclusively in the nerve terminals (green – Fig. 1G and I). This pattern of expression of CaR has also been observed in the adult (Ruat et al., 1995) and developing (Vizard et al., 2008) brain. The lack of CaR PCR products from carotid body RNA suggests that receptor

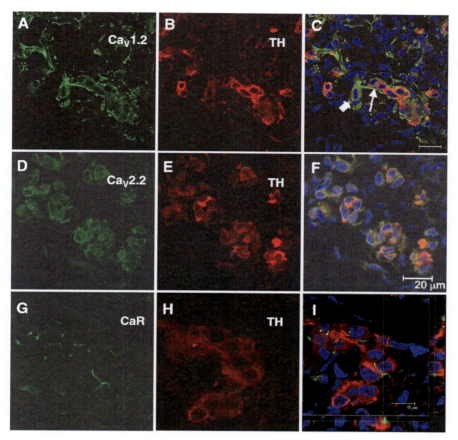

Fig. 1 Expression of voltage-gated calcium channels and CaR in the carotid body. Confocal images of Ca$_v$1.2 (**A**), Ca$_v$2.2 (**D**) and CaR (**G**) immunoreactivities in cryosections from rat carotid body. Type 1 cells are identified by tyrosine hydroxylase (TH, *red*) staining (**B**, **E** and **H**) and the overlay of the two images with the nuclear staining (DAPI, *blue*) are shown in panels C, F and I. Ca$_v$1.2-positive staining is found at the cell surface of type 1 cells (*block arrow*) and in the nerve terminals (*arrow*). Ca$_v$2.2 is found in type 1 cells while CaR immunoreactivity is selectively present in nerve terminals. Note that in panel I, the right and lower margins are z-stacks through the section along the white dotted lines, as indicated in the main panel. Scale bar = 20 µm (**C** and **F**) or 10 µm (**I**)

transcripts are predominantly expressed in the soma of neurons, presumably of the petrosal ganglion.

Measurement of neurotransmitter release from isolated carotid bodies in vitro showed that application of 500 µM spermine did not affect baseline catecholamine secretion but significantly inhibited the hypoxia-evoked release by 30% ($p < 0.05$, n = 6 – Fig. 2A) and the depolarization-evoked release by 60% ($p < 0.01$, n = 6 – Fig. 2B). That spermine inhibited the secretion evoked by depolarisation indicated that the O_2 sensing mechanism was not affected per se. As catecholamine secretion

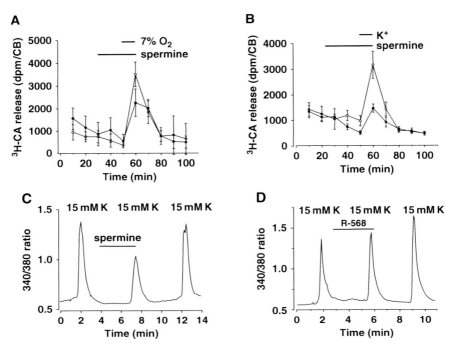

Fig. 2 Spermine inhibition of catecholamine secretion from isolated carotid bodies and [Ca^{2+}]$_i$ rises in type 1 cells. (**A**) Mean ($\pm$S.E.M.) tritiated catecholamine (^{3}H-CA) secretion induced by bubbling the tubes with 7% oxygen (hypoxia) in untreated (○) and spermine-treated (500 μM - ●); n = 6 carotid bodies. (**B**) Mean ($\pm$S.E.M.) tritiated catecholamine (^{3}H-CA) secretion induced by raising the extracellular potassium chloride concentration isoosmotically to 35 mM K$^+$ in untreated (○) and spermine-treated (500 μM - ●); n = 6 carotid bodies. (**C**) Exemplar recording of fura 2 fluorescence ratio from a cluster of carotid body type 1 cells. Voltage-dependent Ca^{2+} influx was stimulated by isoosmotic addition of 15 mM potassium chloride in the presence or absence of 200 μM spermine, as indicated (representative of 22 cell clusters from 5 isolations). (**D**) Typical recording of fura 2 fluorescence ratio from a cluster of carotid body type 1 cells. Voltage-dependent Ca^{2+} influx was stimulated by isoosmotic addition of 15 mM potassium chloride in the presence or absence of 100 nM of the CaR positive allosteric modulator, R-586, as indicated (representative of 19 cell clusters from 4 isolations)

is mediated by, and strictly depend upon, a rise in Ca^{2+}$_i$ (Gonzalez et al., 1994), the effect of spermine on [Ca^{2+}]$_i$ homeostasis was investigated in type 1 cells. 200 μM spermine inhibited the Ca^{2+} influx induced by high K$^+$ by 34% (p < 0.01, N = 5 – Fig. 2C). The CaR positive allosteric modulator, R-568, did not significantly affect Ca^{2+} influx induced by high K$^+$ (p > 0.5, n = 4 – Fig. 2D). Consistent with the molecular biology observations, these experiments strongly suggest that the target of the inhibitory action of spermine is the voltage-dependent Ca^{2+} channels.

Since 60 to 80% of the hypoxia-evoked Ca^{2+} influx is mediated by L-type Ca^{2+} channels (e Silva and Lewis, 1995), we investigated the effect of spermine on the only L-type channel which we could detect in type 1 cells, namely Ca$_v$1.2. 200 μM

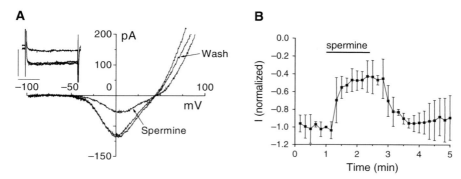

Fig. 3 Inhibitory effect of spermine on whole cell $Ca_v1.2$ currents. (**A**) Typical example of the effect of 200 µM spermine on $Ca_v1.2$ currents. The effect of spermine was reversible. The insert shows traces evoked by a + 15 mV step. Scale bars are 100 pA and 40 ms. (**B**) Mean (± S.E.M.) time-course of spermine inhibition of $Ca_v1.2$ currents evoked by repetitive steps to + 15 mV, n = 3

spermine inhibited whole cell currents from HEK293 cells stably expressing $Ca_v1.2$, by 53 ± 5% (n = 3; Fig. 3).

In conclusion, our data show that spermine exerts an inhibitory influence on carotid body chemosensitivity. This effect is due, at least in part, to an inhibition of the $Ca_v1.2$. The effect of spermine on $Ca_v2.2$, which is also expressed in carotid body glomus cells, is yet to be determined. Spermine is stored at millimolar concentrations in neuronal synaptic vesicles (Masuko et al., 2003) and is co-secreted with neurotransmitters (Fage et al., 1992). Although no data are currently available in the carotid body, it seems reasonable to assume that type 1 cells will also co-secrete spermine. If this is the case, spermine co-secretion during hypoxia may act to limit the effect of hypoxic depolarization by inhibiting L-type calcium channels. Such an inhibitory mechanism may work in parallel to that proposed for the effect of GABA where, through $GABA_B$ autoreceptors, background potassium currents are re-activated (Fearon et al., 2003). Such auto-feedback mechanisms may act to limit neurotransmitter release by type 1 cells during hypoxia.

Acknowledgments We thank the British Heart Foundation (to PJK), Amgen, Inc. (to DR),. BFU2007-61848 (DGICYT), CIBER CB06/06/0050 (FISS-ICiii) and JCyL-GR242 for funding this work.

References

Casero, R. A., Jr. & Marton, L. J. (2007) Targeting polyamine metabolism and function in cancer and other hyperproliferative diseases. *Nat Rev Drug Discov*, 6, 373–90.
Chaisiri, P., Harper, M. E. & Griffiths, K. (1979) Plasma spermine concentrations of patients with benign and malignant tumours of the breast or prostate. *Clin Chim Acta*, 92, 273–82.

Conde, S. V., Caceres, A. I., Vicario, I., Rocher, A., Obeso, A. & Gonzalez, C. (2006a) An overview on the homeostasis of Ca^{2+} in chemoreceptor cells of the rabbit and rat carotid bodies. *Adv Exp Med Biol*, 580, 215–22; discussion 351–9.

Conde, S. V., Obeso, A., Vicario, I., Rigual, R., Rocher, A. & Gonzalez, C. (2006b) Caffeine inhibition of rat carotid body chemoreceptors is mediated by A2A and A2B adenosine receptors. *J Neurochem*, 98, 616–28.

Dogan, A., Rao, A. M., Hatcher, J., Rao, V. L., Baskaya, M. K. & Dempsey, R. J. (1999) Effects of MDL 72527, a specific inhibitor of polyamine oxidase, on brain edema, ischemic injury volume, and tissue polyamine levels in rats after temporary middle cerebral artery occlusion. *J Neurochem*, 72, 765–70.

E Silva, M. J. & Lewis, D. L. (1995) L- and N-type Ca^{2+} channels in adult rat carotid body chemoreceptor type I cells. *J Physiol*, 489(Pt 3), 689–99.

Fage, D., Voltz, C., Scatton, B. & Carter, C. (1992) Selective release of spermine and spermidine from the rat striatum by N-methyl-D-aspartate receptor activation in vivo. *J Neurochem*, 58, 2170–5.

Fearon, I. M., Zhang, M., Vollmer, C. & Nurse, C. A. (2003) GABA mediates autoreceptor feedback inhibition in the rat carotid body via presynaptic GABAB receptors and TASK-1. *J Physiol*, 553, 83–94.

Gilad, V. H., Halperin, R., Chen-Levy, Z. & Gilad, G. M. (2002) Cyclic changes of plasma spermine concentrations in women. *Life Sci*, 72, 135–41.

Gomez, M. & Hellstrand, P. (1995) Effects of polyamines on voltage-activated calcium channels in guinea-pig intestinal smooth muscle. *Pflugers Arch*, 430, 501–7.

Gonzalez, C., Almaraz, L., Obeso, A. & Rigual, R. (1994) Carotid body chemoreceptors: from natural stimuli to sensory discharges. *Physiol Rev*, 74, 829–98.

Heby, O. (1986) Putrescine, spermidine and spermine. *NIPS*, 1, 3.

Kemp, P. J., Peers, C., Riccardi, L., Iles, D. E., Mason, H. S., Wootton, P. & Williams, S. E. (2006) In search of the acute oxygen sensor: functional proteomics and acute regulation of large-conductance, calcium-activated potassium channels by hemeoxygenase-2. *Adv Exp Med Biol*, 580, 137–46; discussion 351–9.

Lasater, E. M. & Solessio, E. (2002) Regulation of voltage-sensitive Ca^{2+} channels in bipolar cells by divalent cations and polyamines. *Adv Exp Med Biol*, 514, 275–89.

Li, J., Doyle, K. M. & Tatlisumak, T. (2007) Polyamines in the brain: distribution, biological interactions, and their potential therapeutic role in brain ischaemia. *Curr Med Chem*, 14, 1807–13.

Masuko, T., Kusama-Eguchi, K., Sakata, K., Kusama, T., Chaki, S., Okuyama, S., Williams, K., Kashiwagi, K. & Igarashi, K. (2003) Polyamine transport, accumulation, and release in brain. *J Neurochem*, 84, 610–7.

Quinn, S. J., Ye, C. P., Diaz, R., Kifor, O., Bai, M., Vassilev, P. & Brown, E. (1997) The Ca2+-sensing receptor: a target for polyamines. *Am J Physiol*, 273, C1315–23.

Ruat, M., Molliver, M. E., Snowman, A. M. & Snyder, S. H. (1995) Calcium sensing receptor: molecular cloning in rat and localization to nerve terminals. *Proc Natl Acad Sci USA*, 92, 3161–5.

Tabor, C. W. & Tabor, H. (1984) Polyamines. *Annu Rev Biochem*, 53, 749–90.

Vizard, T. N., O'keeffe, G. W., Gutierrez, H., Kos, C. H., Riccardi, D. & Davies, A. M. (2008) Regulation of axonal and dendritic growth by the extracellular calcium-sensing receptor. *Nat Neurosci*, 11, 285–91.

RT-PCR and Pharmacological Analysis of L-and T-Type Calcium Channels in Rat Carotid Body

A.I. Cáceres, E. Gonzalez-Obeso, C. Gonzalez and A. Rocher

Abstract Mechanisms involved in carotid body (CB) chemoreceptor cells O_2-sensing and responses are not fully understood. So far, it is known that hypoxia depolarizes chemoreceptor cells via O_2-sensitive K^+-channel inhibition; calcium influx via voltage-gated channels and neurotransmitter secretion follow. Presence of high voltage activated (HVA) calcium channels in rat CB chemoreceptor cells is well documented, but the presence of low voltage activated (LVH) or T-type calcium channels has not been reported to date. The fact that O_2-sensitive PC12 cells express T-type channels and that they are inducible by chronic hypoxia (CH) lead us to hypothesize they could be present and play a role in the genesis of the hypoxic response in rat CB chemoreceptor cells. We have analyzed the expression of the three isoforms of T-type calcium channels (α1G, α1H and α1I) and the isoforms α1C and α1D of L-type calcium channels in rat CB by RT-PCR. We found that rat CB expresses α1G and α1C subunits. After chronic hypoxic treatment of adult rats (10% O_2, 8 days), expression of α1G seems to be down-regulated whereas α1C expression is up-regulated. Functionally, it was found that the release of catecholamine induced by hypoxia and high external K^+ from CB chemoreceptor cells was fully sensitive to L-type channel inhibition (nisoldipine, 2 μM), while specific inhibition of T-channels (mibefradil, 2 μM) inhibited exclusively hypoxia-induced release (50%). As a whole, present findings demonstrate the presence of T-type as well as L-type calcium channels in rat CB and suggest a selective participation of the T-type channels in the hypoxic activation of chemoreceptor cells.

Keywords T-type Ca channel · High-threshold Ca channel · Mibefradil · Chemoreceptor cells · Hypoxia

A. Rocher (✉)
Departamento de Bioquímica y Biología Molecular y Fisiología/Instituto de Biología y Genética Molecular.CIBER de Enfermedades Respiratorias (ISCiii) Facultad de Medicina. Universidad de Valladolid - CSIC. 47005 Valladolid, Spain
e-mail: rocher@ibgm.uva.es

1 Introduction

Voltage-dependent Ca^{2+} channels (VDCC) have been subdivided on the basis of their electrophysiological and pharmacological properties into low voltage-activated (LVA) or T-type channels and high voltage-activated (HVA) channels that include L-, N-, P/Q-, and R-types (Hofmann et al., 1999; Catterall et al., 2005). The HVA channels possess similar biophysical properties and mainly have been characterized by their different sensitivities to pharmacological agents and inhibitory toxins. HVA channels are complexes composed of a pore-forming α_1-subunit together with modulatory β-, α_2/δ- and y-subunits (Tsien et al., 1988; Catterall et al., 2005). The $\alpha 1$ subunit confers fundamental properties to a specific Ca^{2+} channel and is considered to be the molecular signature defining the different subtypes: $\alpha 1C$ and $\alpha 1D$ forms the L-type channel (also named Cav1.2 and Cav1.3) and $\alpha 1G$, $\alpha 1H$ and $\alpha 1I$ compose the T-type channels: Cav3.1, Cav3.2 and Cav3.3 (Perez-Reyes, 1999).

In CB chemoreceptor cells, VDCC are clearly involved in the release of neurotransmitters from type I cells to afferent fibbers (Gonzalez et al., 1994). Although a number of studies describe several HVA Ca^{2+} channel types present in rabbit and rat CB chemoreceptor cells in an electrophysiological perspective (see Lopez-Lopez & Peers, 1997), in rabbit CB only the L- and P/Q- subtypes participate in exocytotic responses (Rocher et al., 2005). In the rat, it is known that L-type supports 75% of the increase in $[Ca^{2+}]i$ (Buckler and Vaughan-Jones, 1994) and of the neurotransmitter secretion elicited by hypoxia (Conde et al., 2006), but to date neither the molecular identity of the HVA α_1-subunits nor the presence of LVA Ca^{2+} channels have been studied in detail. The fact that O_2-sensitive PC12 cells express T-type channels and that they are inducible by chronic hypoxia (CH) lead us to hypothesize they could be present and play a role in the genesis of the hypoxic response in rat CB chemoreceptor cells (Del Toro et al., 2003).

In the present work we have analyzed: first, the expression of the three isoforms of T-type calcium channels ($\alpha 1G$, $\alpha 1H$ and $\alpha 1I$) and the isoforms $\alpha 1C$ and $\alpha 1D$ of L-type calcium channels in rat CB by reverse transcription (RT) followed by cDNA PCR amplification. Second, by studying the effects of pharmacological inhibition, we have compared the participation of T-type and L-type calcium channels in the secretory response of rat CB to hypoxic (7%O_2) and depolarizing stimulus (35 mM K^+).

2 Methods

2.1 Animals and Surgery

Experiments were performed in intact CB of adult Wistar rats anaesthetized with sodium pentobarbital, 60 mg/Kg (i.p.). After tracheostomy, a block of tissue containing the carotid artery bifurcation was removed and placed in a lucite chamber filled with ice-cold 100% O_2 Hepes-buffered Tyrode and the CBs were cleaned of surrounding tissues under a dissecting microscope.

2.2 RNA Isolation and cDNA Amplification

Total RNA was prepared from fresh rat CB using the Trizol method (Gibco, BRL) and transcribed into cDNA as described in detail previously (Cáceres et al., 2007). Thereafter cDNA was stored at -80^aC until it was used for PCR amplification. The RT reaction was performed in the absence of reverse transcriptase or RNA as control to exclude genomic DNA contamination of the RNA preparation and possible amplification of genomic DNA. PCR amplification was performed according to manufacture's instruction, using specific primers (see Table 1) that amplify fragments of each isoform of the VDCC. Aliquots of first-strand cDNA were amplified by PCR in a 25 μl reaction mix containing buffer, dNTP, $MgCl_2$, Taq polymerase and primers at appropriated concentrations, as described previously (Cáceres et al., 2007). Before PCR, samples were heated to 94°C for 3 min. Each PCR cycle consisted of denaturation at 94°C for 30 sec, annealing at 55°C for 30 sec and elongation at 72°C for 1 min. After PCR, samples were heated to 72°C for 10 min. PCR products were collected at 30 cycles (before saturation) and analyzed on 1.5% agarose gels containing ethidium bromide. The sizes of the amplified fragments were as calculated from the published sequences.

Table 1 mRNA primer sequences

Gene Bank	Target mRNA	Primers sequences 5'-3'	Tm (°C)	(%) CG	Product (bp)
NM_030875	$Ca_v3.1$ (α1G)	Fw: ACCTGCCTGACACTCTGCAG	61	60	857 bp
		Rv: GCTGGCCTCAGCGCAGTCGG	67	75	
NM_012647	$Ca_v3.2$ (α1H)	Fw: GGCGTGGTGGTGGAGAACTT	61	60	469 bp
		Rv: GATGATGGTGGGATTGAT	51	44	
NM_013119	$Ca_v3.3$ (α1I)	Fw: GAGTTAGACAAGCTCCCAGA	57	50	902 bp
		Rv: CAGTTGAGGAAGATAAAGGC	55	47	
NM_012517	Cav 1.2 (α1C)	Fw: GTGAGGCTGAGCGAAGAAGT	60	55	233 bp
		Rv: TGAGAGATGTCTCCCCCTTG	60	55	
NM_017298	Cav 1.3 (α1D)	Fw: TTGACTCCCTCATCGTAA	52	44	837 bp
		Rv: GGTTGCCTTCAGTCTTG	52	53	
NM_031144	β-actin	Fw: AAGATCCTGACCGAGCGTGG	62	54	327 bp
		Rv: AGCACTGTGTTGGCATAGAGG	61	60	

2.3 Measurement of the Release of CA

To label the CA stores, 12 CBs /experiment were incubated during 2 h in small vials containing 0.5 ml of 100% O_2 equilibrated Hepes-buffered Tyrode (in mM: NaCl, 140; KCl, 5; $CaCl_2$, 2; $MgCl_2$, 1.1; glucose, 5.5; HEPES, 10) and placed in a metabolic shaker at 37°C. The incubating solution contained ^{3}H-tyrosine (30 μM) with high specific activity (50 Ci/mmol), 100 μM 6-methyl-tetrahydropterine and 1 mM ascorbic acid as cofactors for tyrosine hydroxylase and dopamine-β-hydroxylase, respectively. At the end of the labeling period individual CB were transferred to a glass vial containing 4 ml of precursor-free Tyrode bicarbonate solution (24 mM NaCl was substituted by 24 mM NaHCO$_3$), and kept at 37°C for the rest of the experiment. Solutions were continuously bubbled with a gas mixture saturated with water vapor of composition 20% O$_2$/5% CO$_2$/75% N$_2$, except when hypoxia was applied (7%O$_2$). After an hour washing period, incubating solutions were collected every 10 min and saved for analyses in ^{3}H-CA content. The analysis of ^{3}H-catechols present in the collected solutions included: adsorption to alumina at alkaline pH, extensive washing of the alumina with distilled water, bulk elution of all ^{3}H-catechols with 1 ml of 1N HCl and liquid scintillation counting of the eluates. At the end of the experiments, the CBs were homogenized at 0–4°C and the homogenates analyzed for their ^{3}H-CA content.

3 Results

3.1 PCR Analysis of Calcium Channel α1-Subunits

Selected primers amplify DNA fragments that correspond in length to those predicted from the cDNAs of the corresponding VDCC α1-subunits (see Table 1). In Figure 1A we analyze T-type Ca^{2+} channel expression and we observed that the predominant subunit in rat CB is Cav3.1 (α1G) that seems to be down-regulated during chronic hypoxic treatment, while in adrenal medulla Cav3.2 (α1H) is predominant. As it was expected Cav3.3 (α1I) only expresses in brain, which is positive for all the subunits (Perez-Reyes, 1999). Figure 1B shows that normoxic and hypoxic CB expresses Cav1C isoform of L-type calcium channels and lack of isoform Cav1D. Furthermore there is an appreciable up-regulation of α1C isoform in chronic hypoxically rat CB when compared with β-actin expression.

3.2 CA Release Experiments

This group of experiments was directed to explore the identity of VDCC supporting the exocytotic release of CA in rat CB on response to hypoxia and high K^+_e. After labeling endogenous deposits of CA by incubating the organs in a medium containing 30 μM 3,5-[^{3}H]tyrosine, CBs were superfused with control solutions or with stimuli solutions (7%O$_2$ or 35 mM K^+_e). These CB superfusion solutions were renewed every 10 min and analyzed for their content in [^{3}H]-CA as described previously (Rocher et al., 2005).

Figure 2A and B represent prototype experiments showing the mean secretory response to two cycles of 7%O$_2$ stimulation, quantified as [^{3}H]CA release to the medium in successive 10 min periods. In experimental CBs, the Ca^{2+} channels inhibitors nisoldipine (a blocker of L-type Ca^{2+} channels; part A) or mibefradil (a blocker of T Ca^{2+} channels; part B) were present 10 min prior and during the application of the second stimulus. Similar experiments were performed stimulus being an unspecific depolarizing stimulus (35 mM K$^+$). Figure 2C summarizes mibefradil and nisoldipine effect on [^{3}H]CA release induced by 7%O$_2$ and

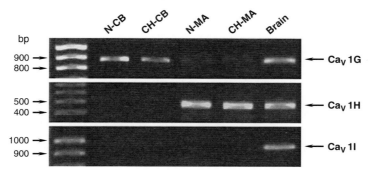
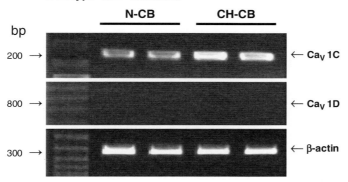

Fig. 1 RT-PCR analysis of Ca^{2+} channel α1-subunits in RNA from whole rat CB. **A** shows gene expression analysis of LVA channel α1G (Cav3.1), α1H (Cav3.2) and α1I (Cav3.3) subunits in normoxic (N) and chronically hypoxic rat CBs (CH) compared with adrenal medulla (MA), and brain as positive control. **B** shows gene expression analysis of the HVA channel α1C (Cav2.1) and α1D (Cav2.2), and β-actin as housekeeping gene

35 mM K^+, expressed as % total CB content. Data are means ± SEM for 8–12 CBs (**$p < 0.01$; ***$p < 0.01$). Note that nisoldipine inhibited the release by about 95% both, hypoxia and high K^+-induced, whereas mibefradil specifically inhibits 50% hypoxia induced release without effect on high external K^+ evoked release.

4 Discussion

The main findings of this study on Ca^{2+} channels in rat CB are as follows:

1. We report, for the first time, the presence of T-type calcium channels in rat CB, specifically the α1G (Cav3.1) subunit, and the α1C subunit (Cav1.2) of L-type Ca^{2+} channel.

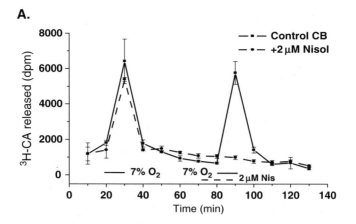
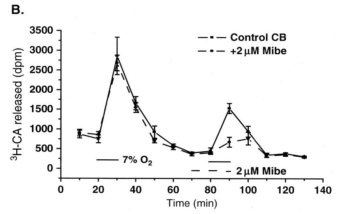
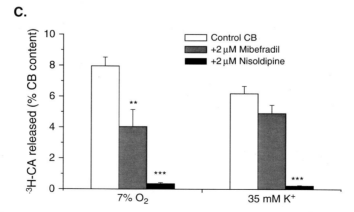

Fig. 2 Effects of HVA calcium channels (**Part A**) and LVA calcium channel blockers (**Part B**) on the release of ^{3}H-CA induced by hypoxia (7%O_2). **Part C** summarizes participation of L-type and T-type calcium channels in the secretory response to hypoxia and high external K^+. Data are means ± SE of 6-8 CB (**p < 0.01; ***p < 0.001)

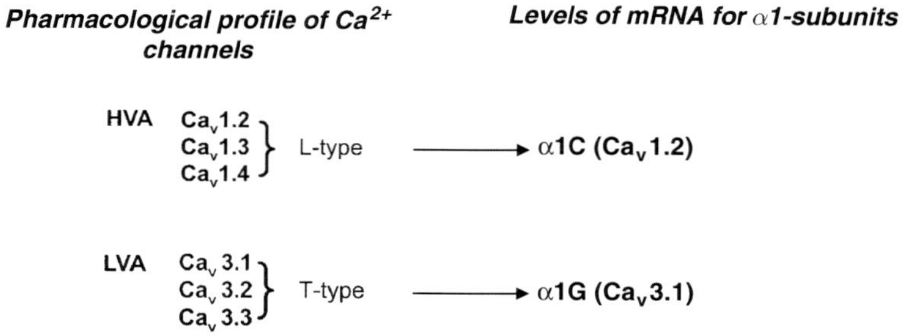

Pharmacological profile of Ca²⁺ channels

Levels of mRNA for α1-subunits

HVA Ca$_v$1.2
     Ca$_v$1.3 } L-type ⎯⎯⎯→ α1C (Ca$_v$1.2)
     Ca$_v$1.4

LVA Ca$_v$3.1
     Ca$_v$3.2 } T-type ⎯⎯⎯→ α1G (Ca$_v$3.1)
     Ca$_v$3.3

Fig. 3 Scheme summarizing, in rat CB chemoreceptor cell, the proposed distribution of HVA and LVA Ca²⁺ channel types from functional studies and the a₁-subunit mRNA expression pattern from RT-PCR

2. After in vivo chronic hypoxia exposure, α1G subunit seems to be down-regulated whereas α1C expression is up-regulated; the mechanism and functional significance of these modifications of expression is unknown.
3. Secretory response induced by hypoxia and high external K⁺ from chemoreceptor cells is fully dependent (more than 95%) of Ca²⁺ entry through L-type channels.
4. Finally, we demonstrate specific participation of T-type Ca²⁺ channels in rat CB hypoxic activation because their pharmacological inhibition with mibefradil blocks 50% the hypoxia-induced secretory response without affecting high K⁺ induced secretion.

Figure 3 summarizes our present findings. In rabbit and rat CB chemoreceptor cells it is known that the release of ³H-CA by chemoreceptor cells in response to hypoxic and high external K⁺ stimulus is reduced by around 95–98% in nominally Ca²⁺ solutions (Obeso et al., 1992; Vicario et al., 2000). In both species, Ca²⁺ enters during hypoxic and high K⁺ stimulation via VDCC, because Cd²⁺, a blocker of VDCC abolishes totally CA secretion (Rocher et al., 2005; Conde et al., 2006). However, whereas in rabbit CB other types of VDCC more than L-type channels also participate in the release of NT (Rocher et al., 2005), here we demonstrate that in rat CB only the L-subtype provides Ca²⁺ for exocytotic responses because they are totally sensitive to L-type channel blockers (dihidropyridines).

Our data using the release of NT as an index of Ca²⁺ dynamics in rat CB are compatible with the findings of Buckler and Vaughan-Jones (1994) showing that about 70% of the increase in [Ca²⁺]i was suppressed by blocker of L-type Ca²⁺ channels. Our results suggest that Ca²⁺ influx through L-type calcium channels is the only effective for NT secretion. The fact that mibefradil inhibits by 50% the hypoxic response but does not affect those to high external K⁺ demonstrate, first, that the effect is not unspecific on L-type channels (it would produce a similar effect on both stimuli) and second, because calcium entry through T-type channels does not contribute to secretory responses, and because they activate to very low voltage, they could represent an amplification step to the depolarization triggering sensory transduction in the rat CB. Because physiological properties of the calcium channels are determined by the α1 subunit, changes of this subunit, as suggested by PCR analysis, may give rise or attenuation to the current density in chronic hypoxia (Hofmann et al., 1999). Preliminary results show that inhibitory effect of mibefradil disappears in chronically hypoxic CBs whereas nisoldipine effect remain unaltered; we interpret this finding to suggest that the possible change in T-type calcium channel expression (down-regulation) induced by chronic hypoxia is responsible of the lack of inhibitory effect. Future studies will be necessary to clarify significance of these results.

Acknowledgments We are grateful to Mª Llanos Bravo for technical assistance. This work was supported by grants BFU2007-61848 (MEC, Spain), CIBER CB06/06/0050 (FISS-ICiii) and by JCyL grants: GR242, VA104A08 and SAN673/VA12/08.

References

Buckler KJ, Vaughan-Jones RD. 1994. Effects of hypoxia on membrane potential and intracellular calcium in rat neonatal carotid body type I cells. *J Physiol* 476: 423–428.

Cáceres AI, Obeso A, González C, Rocher A. 2007. Molecular identification and functional role of voltage-gated Na channels in rat carotid body chemoreceptor cells. Regulation of expression by chronic hypoxia in vivo. *J Neurochem* 102: 231–245.

Catterall WA Perez-Reyes E, Snutch TP, Striessnig J. 2005. Nomenclature and Structure-function Relationships of voltage-gated Ca^{2+}-channels. *Pharmacol* 57: 411–425.

Conde SV, Cáceres AI, Vicario I, Rocher A, Obeso A, Gonzalez C. 2006. An overview on the homeostasis of Ca^{2+} in chemoreceptor cells of the rabbit and rat carotid bodies. *Adv Exp Med Biol* 580: 215–222.

Del Toro R, Levitsky KL, Lopez-Barneo J, Chiara MD. 2003. Induction of T-type calcium channel gene expression by chronic hypoxia.*J Biol Chem* 278: 22316–11324.

Gonzalez C, Almaraz L, Obeso A, Rigual R. 1994. Carotid body chemoreceptors: From natural stimuli to sensory discharges. *Physiol Rev* 74: 829–898.

Hofmann F, Lacinova L, Klugbauer N. 1999. Voltage-dependent calcium channels: from structure to function. *Rev. Physiol Biochem Pharmacol* 139: 33–87.

Lopez-Lopez J.R, Peers, C. 1997. Electrical properties of chemoreceptor cells. In: *The Carotid Body Chemoreceptors* (Ed. C. González) Springer, NY. pp. 65–77.

Obeso A, Rocher A, Fidone S, Gonzalez C. 1992. The role of dihydropyridine-sensitive Ca^{2+} channels in stimulus-evoked catecholamine release from chemoreceptor cells of the carotid body. *Neuroscience* 47: 463–472.

Perez-Reyes E. 1999. Three for T: molecular analysis of the low voltage-activated calcium channel family. *Cell Mol Life Sci* 56: 660–669.

Rocher A, Geijo-Barrientos E, Cáceres AI, Rigual R, Gonzalez, C, Almaraz L. 2005. Role of voltage-dependent calcium channels in stimulus-secretion coupling in rabbit carotid body chemoreceptor cells. *J Physiol* 562:407–420.

Tsien RW, Lipscombe D, Madison DV, Bley KR, Fox AP. 1988. Multiple types of neuronal calcium channels and their selective modulation. *Trends Neurosci* 11: 431–438.

Vicario I, Obeso A, Rocher A, Lopez-Lopez JR, Gonzalez C. 2000. Intracellular Ca^{2+} stores in chemoreceptor cells of the rabbit carotid body: significance for chemoreception. *Am J Physiol Cell Physiol* 279: C51–C61.

Functional Characterization of Phosphodiesterases 4 in the Rat Carotid Body: Effect of Oxygen Concentrations

A.R. Nunes, J.R. Batuca and E.C. Monteiro

Abstract The non-specific cAMP phosphodiesterase (PDE) inhibitor isobutylmethylxanthine (IBMX) has been used to manipulate cAMP levels in carotid body (CB) preparations but the characterization of different PDE isoforms in CB has never been performed. PDE4 is one of the PDE families that uses cAMP as a specific substrate and changes its activity and affinity for drug inhibitors according to the degree of its phosphorylation. We investigated the effects of hypoxia on cAMP accumulation induced by different PDE4 inhibitors in the CB based on the hypothesis that acute changes in O_2 could interfere with their affinity.

Concentration-response curves for the effects of the PDE4 selective inhibitors, rolipram and Ro 20-1724 and IBMX on cAMP were obtained in CBs, removed from rats and incubated in normoxia (20%O_2) or hypoxia (5%O_2).

No differences were found between cAMP concentrations in normoxic and hypoxic conditions in the absence of PDE inhibitors. In both conditions, the E_{max} calculated for IBMX was similar to that of the specific PDE4 inhibitors. Hypoxia shifted the concentration response curves to the left with the following rank order of potency IBMX> RO 20-1724=rolipram and increased E_{max} by about 25%.

This pharmacological approach supports the hypothesis that there is PDE4 activity in CBs that is enhanced by acute hypoxia although the low potency of the PDE4 inhibitors to increase cAMP do not support an important role for PDE4 activation in the O_2-sensing machinery at the CB.

Keywords Carotid body · Hypoxia · phosphodiesterases · cAMP · IBMX · Rolipram · Ro 20-1724

1 Introduction

cAMP is an ubiquitous intracellular second messenger common to metabotropic receptors coupled to transmembrane adenylate cyclase and hydrolysed by cyclic

E.C. Monteiro (✉)
Department of Pharmacology, Faculty of Medical Sciences, New University of Lisbon,
Campo Mártires da Pátria, 130, Lisbon 1169-056, Portugal
e-mail: emilia.monteiro@fcm.unl.pt

nucleotide phosphodiesterases (PDE). cAMP concentrations in the CB are regulated by the stimulatory effects of adenosine mediated by A_{2A} and A_{2B} receptors (Chen et al. 1997; Conde et al. 2006; Monteiro et al. 1996), by beta-adrenoceptor activity (Mir et al. 1984) and by dopamine D_2 and D_1 receptors (Almaraz et al. 1991; Batuca et al. 2003; Mir et al. 1984). Some evidence also suggests a modulatory role for cAMP in the O_2-sensing machinery in the CB (Gonzalez et al. 1994) but it has also been shown that cyclic nucleotides alone do not modify K^+ current inhibition by hypoxia in isolated type I carotid body cells (Hatton and Peers 1996).

The non-specific cAMP phosphodiesterase (PDE) inhibitor isobutylmethylxanthine (IBMX) has been used to manipulate cAMP levels in CB preparations but the characterization of different PDE isoforms in CB, has never been performed. Eleven kinetically distinct PDE isoform families that differ in their affinity for subtracts (cAMP and/or cGMP), pharmacological inhibitors and their response to modulators have so far been identified (Bender and Beavo 2006).

PDE4 is one of the PDE families that uses cAMP as a specific substrate, its activity is not regulated by Ca^{2+} but changes its activity and affinity for drug inhibitors according to the degree of its phosphorylation (Bender and Beavo 2006). Chronic hypoxia (several days) regulates PDE4 activity in pulmonary arteries (Millen et al. 2006) as well as the activity of cAMP-PDE in rat blood (Spoto et al. 1998) but little is known about the effect of acute hypoxia. Increases in the activity of a Ca^{2+}-dependent PDE in CBs of rats submitted in vivo to 40 min of hypoxia have been reported (Hanbauer and Lovenberg 1977).

We investigated the effects of hypoxia on cAMP accumulation induced by different PDE4 inhibitors at the CB based on the hypothesis that acute changes on O_2 could interfere with their affinity.

2 Methods

2.1 Surgical Procedures

The experiments were performed on CBs, isolated from adult (3 months old) male and female Wistar rats. Rats were obtained from the Faculty of Medical Sciences animal house, kept at a constant temperature (21°C) with a regular light (08–20 h) and dark (20–08 h) cycle, with food and water *ad libitum*. Experiments were carried out in accordance with the Portuguese regulations for the protection of the animals. Surgical procedures for isolation of the tissues have been described previously (Batuca et al. 2003; Conde and Monteiro 2004). In brief, rats were anaesthetised with sodium pentobarbital (60 mg.Kg^{-1} i.p.), then tracheostomised to allow spontaneous breathing during the surgical procedure. After the CBs had been removed the animals were killed by an intracardiac injection of a lethal dose of pentobarbital, in accordance with the EU directives for the use of experimental animals (Portuguese laws no. 1005/92 and 1131/97). The tissues removed were further prepared as follows.

The CBs removed were pre-incubated to allow recovery from the surgical manipulation in a 37°C shaker bath for 15 min in a medium containing: NaCl 116 mM, NaHCO$_3$ 24 mM, KCl 5 mM, CaCl$_2$ 2 mM, MgCl$_2$ 1.1 mM, Hepes 10 mM, Glucose 5.5 mM and adjusted to pH 7.40 (Perez-Garcia et al. 1990) and equilibrated with 20%O$_2$ (PO$_2 \approx$ 142 mmHg). After the pre-incubation period, all the CBs were incubated in a fresh incubation medium, equilibrated in normoxia or hypoxia (5%O$_2$, PO$_2$ = 33 mmHg) for 30 min in the absence or the presence of PDE inhibitors. Only one PDE inhibitor at one concentration was tested in each experiment.

2.2 Cyclic Nucleotide Extraction and Quantification

After incubation, CBs were immersed in cold 6% (w/V) trichloroacetic acid (TCA, 600 μL) for 10 min, weighed on an electrobalance (mc 215 Sartorius), homogenized using a Potter glass homogeniser at 2–8°C and centrifuged at 12000 g for 10 min at 4°C. The supernatants were washed four times in 3 ml of water saturated with diethyl ether solution (50:50), collected for lyophilisation (Christ Alpha 1–2 B. Brawn, Biotech International) and stored at −20°C until cAMP quantification by enzyme immunoassay, using an EIA commercial kit (RPN 2255 GE Healthcare Bio-Sciences AB).

2.3 Data Analysis and Statistical Procedures

cAMP content was expressed in picomoles per milligram tissue (pmol/mg tissue). Data were evaluated using Graph Pad Prism (GraphPad Software Inc., version 4, San Diego, CA, USA) and expressed as means ± S.E.M. Statistical differences between normoxic and hypoxic basal cAMP levels, were assessed by the Mann-Whitney nonparametric test. Comparison between the effects of hypoxia on different PDE inhibitors in CBs was performed by two-way analysis of variance with Bonferroni's post test. Concentration-response curves for the effects of inhibitors were analysed using sigmoidal curve-fitting analysis, extrapolating the E$_{max}$ (mean maximal effect (%) observed in cAMP levels) and the EC$_{50}$ value (concentration of inhibitor required to elicit 50% of maximum effect observed) for each inhibitor.

3 Results

In the absence of PDE inhibitors, the basal cAMP content observed in CBs incubated with 20%O$_2$ was 0.75 ± 0.08 pmol/mg (n=15) and hypoxia did not significantly modify cAMP concentrations (0.63 ± 0.05 pmol/mg, n=15).

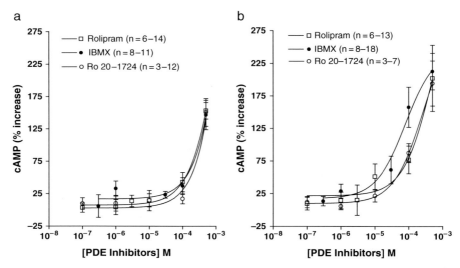

Fig. 1 Effects of rolipram, Ro 20-1724 and IBMX on cAMP levels in rat carotid bodies in (**a**) normoxia (20%O_2) (0% effect- 0.75 ± 0.08 pmol cAMP/mg) and (**b**) hypoxia (5%O_2) (0% effect- 0.63 ± 0.05 pmol cAMP/mg). Data values represent means ± s.e.m

Figure 1 shows the comparison between the effects of PDE inhibitors in CBs incubated in normoxic (Fig. 1a) and hypoxic (Fig. 1b) conditions. In normoxia only high concentrations of inhibitors (500 μM) caused clear effects on cAMP levels (Fig. 1a) but hypoxia increased the effects of the PDE4 inhibitors (rolipram and Ro 20-1724) as well as the non-specific inhibitor IBMX (Fig. 1b). The EC_{50} and E_{max} values obtained from these concentration-response curves are represented in Table 1. The E_{max} values found in hypoxia were similar for the three inhibitors and higher than those obtained in normoxia (Table 1 and Fig. 2). Similar potencies, quantified as EC_{50}, for IBMX, Ro 20-1724 and rolipram were found in normoxia (Table 1). Hypoxia increased the potency of these drugs with a more pronounced effect on IBMX (Table 1). Although higher levels of cAMP were found in the CBs incubated in hypoxia with IBMX, Ro 20-1724 and rolipram than those obtained in normoxia, only the differences in experiments with 100 μM of IBMX and Ro 20-1724 achieved statistical significance (Fig. 2).

4 Discussion

This pharmacological approach provided evidence that PDE4 is active at the CB and that acute hypoxia modifies the effects of PDE4 inhibitor.

An initial work showing an increase in the activity of a Ca^{2+}-dependent PDE in CBs of rats submitted in vivo to 40 min of hypoxia appeared in 1977 (Hanbauer and Lovenberg 1977) but no further studies are known and the characterisation of PDE isoforms has never been performed at the CB. From the eleven PDE isoform

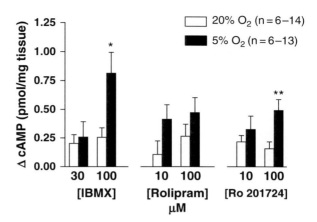

Fig. 2 Effects of hypoxia in the presence of selective (rolipram and Ro 20-1724) and non-selective (IBMX) PDE inhibitors on cAMP levels in rat carotid bodies. *P < 0.05 and **P < 0.01 two way ANOVA with Bonferroni's post test. Values represent means ± s.e.m

families (PDE1-PDE11) identified and cloned within the superfamily of PDE only three types are known to hydrolyse cAMP selectively (PDE4, PDE7 and PDE8) and five hydrolyse both cAMP and cGMP (PDE1, PDE2, PDE3, PDE10 and PDE11). This work was focused on PDE that use cAMP as a specific substract and that are not modulated by cGMP.

PDE4 isoezymes are characterised by high affinity and specificity for cAMP, low K_M (1.2–10 μM), insensitivity to Ca^{2+} and cGMP, selective inhibition by the archetypal rolipram and Ro 20-1724 (IC_{50} of 1 μM and 2 μM, respectively) (Bender and Beavo 2006). No specific inhibitors are available for PDE8 but it is known that both PDE7 and 8 are insensitive to rolipram (Lugnier 2006). IBMX is a commonly used tool for causing non-specific inhibition of PDE1 to PDE5 but is devoid of effects on PDE7, PDE8 and PDE9 (Lugnier 2006).

In this study we observed concentration-dependent effects of Ro 20-1724, rolipram and IBMX in CBs, and concluded that PDE4 is present in the CB. The efficacy of the non-selective inhibitor, IBMX, to increase cAMP was similar to that of the PDE4 selective inhibitors Ro 20-1724 and rolipram in CBs suggesting that probably no other PDE is relevant in terms of hydrolysing specifically cAMP in this tissue.

Table 1 Comparison between the efficacies and potencies of PDE inhibitors in normoxia and hypoxia

IBMX (0.3-500 μM)				Rolipram (0.1-500 μM)				Ro 20-1724 (0.1-500 μM)			
20%O_2		5%O_2		20%O_2		5%O_2		20%O_2		5%O_2	
EC50 (μM)	Emax (%)	EC50 (μM)	Emax (%)	EC50 (μM)	Emax (%)	EC50 (μM)	Emax (%)	EC50 (μM)	Emax (%)	EC50 (μM)	Emax (%)
211.8	146.5± 22.8	50.4	212.4± 27.9	203.7	152.9± 12.8	145.2	202.0± 50.8	260.6	147.0± 41.6	117.8	194.1± 34.5

Emax= Mean maximal effect (%) observed in cAMP levels. EC50= Drug concentration that produces 50% of the maximal effect.

cAMP concentrations were used to investigate the effects of PDE inhibitors and to characterise PDE in different tissues. Since cAMP accumulation depends basically on the equilibrium between adenylate cyclase activation and degradation by PDE, this is not a direct, specific method to characterise the enzyme isoforms in comparison, for instance, with enzymatic activity and/or protein expression methodologies. However, from the point of view of drug effects, quantifying cAMP provided functional and integrated results useful for choosing the most appropriate tool and its concentrations to manipulate cAMP.

Since whole CBs were used in this study the effects on cAMP accumulation can be attributed to PDE4 inhibition in chemosensitive elements (type I and II cells and CSN afferent endings) as well as vessels and autonomic nerve endings.

In normoxic conditions the EC_{50} calculated for all the PDE4 inhibitors studied were in general higher than the IC_{50} values described in the literature (rolipram = 1 μM; Ro 20-1724= 2 μM and IBMX= 1-50 μM (Bender and Beavo 2006)) but were similar to the concentrations of IBMX and Ro 20-1724 ($>3 \times 10^{-5}$M) needed to increase cAMP in isolated cardiomyocytes (Katano and Endoh 1990). Since both IBMX and the specific inhibitors, rolipram and Ro 20-1724, showed similar EC_{50} values (respectively 208, 203.7 and 262.4 μM) the low potencies should be attributed to low PDE4 activity in the CBs instead of any particular drug effect.

The effect of chronic hypoxia (several days) on PDE activity has been addressed in other preparations (Millen et al. 2006; Spoto et al. 1998) but this work showed for the first time that acute hypoxia increases the effect of PDE inhibitors on cAMP accumulation. With the methodological approach used in the present work we cannot exclude that the effects of hypoxia include changes in adenylate cyclase activity. In the absence of PDE inhibitors hypoxia did not increase basal cAMP values. In the assumption that hypoxia activates transmembrane adenylate cyclase, this means that hypoxia should also increase cAMP degradation. The hypothesis that hypoxia preferentially regulates degradation than cAMP production is in keeping with the general pharmacological principle that regulation of any ligand or second messenger degradation can often make a more rapid and larger percentage change in concentration than comparable regulation of the rates of synthesis. This is particularly true in the case of cAMP/cGMP where PDE activity is at least one order of magnitude higher than adenylate and guanylate cyclases in all tissues (see e.g. (Bender and Beavo 2006)).

Acute hypoxia increased the efficacy as well as the potency of the inhibitors suggesting that probably an increase in both PDE activity and drug affinities are involved. Complex cAMP driven mechanisms through protein kinase A and extracellular signal regulated kinase mediated phosphorylation can modify PDE4 specific isoforms activity and subsequently alter sensitivity to selective inhibitors (Bender and Beavo 2006). The hypothesis that acute hypoxia may cause PDE4 activation by PKA- or ERK mediated phosphorylation could be advanced.

In conclusion, this pharmacological approach supports the hypothesis that there is PDE4 activity in CBs that is enhanced by acute hypoxia although the low potency of the PDE4 inhibitors to increase cAMP do not support an important role for PDE4 activation in the O_2-sensing machinery at the CB.

Acknowledgments This work was financially supported by CEPR/FCT. The authors are grateful to the Department of Microbiology, Faculty of Medical Sciences for the valuable help in sample lyophilisation.

References

Almaraz, L., Perez-Garcia, M. T., & Gonzalez, C. 1991, Presence of D1 receptors in the rabbit carotid body, *Neurosci. Lett.*, 132: 259–262.

Batuca, J. R., Monteiro, T. C., & Monteiro, E. C. 2003, Contribution of dopamine D2 receptors for the cAMP levels at the carotid body, *Adv. Exp. Med. Biol.*, 536: 367–373.

Bender, A. T. & Beavo, J. A. 2006, Cyclic nucleotide phosphodiesterases: molecular regulation to clinical use, *Pharmacol. Rev.*, 58: 488–520.

Chen, J., Dinger, B., & Fidone, S. J. 1997, cAMP production in rabbit carotid body: role of adenosine, *J. Appl. Physiol.*, 82: 1771–1775.

Conde, S. V. & Monteiro, E. C. 2004, Hypoxia induces adenosine release from the rat carotid body, *J. Neurochem.*, 89: 1148–1156.

Conde, S. V., Obeso, A., Vicario, I., Rigual, R., Rocher, A., & Gonzalez, C. 2006, Caffeine inhibition of rat carotid body chemoreceptors is mediated by A_2A and A_2B adenosine receptors, *J. Neurochem.*, 98: 616–628.

Gonzalez, C., Almaraz, L., Obeso, A., & Rigual, R. 1994, Carotid body chemoreceptors: from natural stimuli to sensory discharges, *Physiol. Rev.*, 74: 829–898.

Hanbauer, I. & Lovenberg, W. 1977, Presence of a calcium2+-dependent activator of cyclic-nucleotide phosphodiesterase in rat carotid body: effects of hypoxia, *Neuroscience*, 2: 603–607.

Hatton, C. J. & Peers, C. 1996, Hypoxic inhibition of K^+ currents in isolated rat type I carotid body cells: evidence against the involvement of cyclic nucleotides, *Pflugers Arch.*, 433: 129–135.

Katano, Y. & Endoh, M. 1990, Differential effects of Ro 20-1724 and isobutylmethylxanthine on the basal force of contraction and beta-adrenoceptor-mediated response in the rat ventricular myocardium, *Biochem. Biophys. Res. Commun.*, 167: 123–129.

Lugnier, C. 2006, Cyclic nucleotide phosphodiesterase (PDE) superfamily: a new target for the development of specific therapeutic agents, *Pharmacol. Ther.*, 109: 366–398.

Millen, J., MacLean, M. R., & Houslay, M. D. 2006, Hypoxia-induced remodelling of PDE4 isoform expression and cAMP handling in human pulmonary artery smooth muscle cells, *Eur. J. Cell Biol.*, 85: 679–691.

Mir, A. K., McQueen, D. S., Pallot, D. J., & Nahorski, S. R. 1984, Direct biochemical and neuropharmacological identification of dopamine D2-receptors in the rabbit carotid body, *Brain Res.*, 291: 273–283.

Monteiro, E. C., Vera-Cruz, P., Monteiro, T. C., & Silva e Sousa, M. A. 1996, Adenosine increases the cAMP content of the rat carotid body in vitro, *Adv. Exp. Med. Biol.*, 410: 299–303.

Perez-Garcia, M. T., Almaraz, L., & Gonzalez, C. 1990, Effects of different types of stimulation on cyclic AMP content in the rabbit carotid body: functional significance, *J. Neurochem.*, 55: 1287–1293.

Spoto, G., Di Giulio, C., Contento, A., & Di Stilio, M. 1998, Hypoxic and hyperoxic effect on blood phosphodiesterase activity in young and old rats, *Life Sci.*, 63: L349–L353.

Calcium Sensitivity for Hypoxia in PGNs with PC-12 Cells in Co-Culture

G.P. Patel, S.M. Baby, A. Roy and S. Lahiri

Abstract Calcium sensitivity of petrosal ganglion neurons (PGNs) to chemical stimuli with and without PC-12 cells in co-culture instead of glomus is not known – the idea being that two types of unusual cells could form synapse and provide a model for studies of chemotransduction. Calcium levels in the PGNs were measured in the presence of different chemical stimuli in the bath medium. Remarkably, the PGNs alone were not sensitive to hypoxia (10 torr), PCO (∼300 torr in normoxa) nor to ATP (100 μM) but they developed the sensitivity to these stimuli in synaptic contact with PC-12 cells. The sharp rise in calcium level was suppressed (2/3) by suramin (100 μM), a purinergic blocker, and the remaining 1/3 was blocked by hexomethonium, a cholinergic blocker. Taken together, these observations suggest that PGNs developed neurotransmission when in contact with PC-12 cells, as if the latter substituting for glomus cells, thus providing a model for chemotransduction studies. The reason for the insensitivity of PGNs alone to the chemical stimuli is unknown at this time.

Keywords Co-culture of PGNs and PC-12 cells · Hypoxia · High PCO · PGNs · PC-12 cells · Calcium sensitivity

1 Introduction

Alcayaga et al. (2003) showed that natural stimuli of the carotid body (CB) cell have no effect on PG neurons (PGNs), but the effects on PGNs can be recorded only in co-culture with carotid body cells. The synaptic contacts between CB cells and PGNs appear to be necessary for the generation of chemosensory activity. However,

S. Lahiri (✉)
Department of Physiology, University of Pennsylvania Medical Center, Philadelphia, PA, 19104, USA
e-mail: lahiri@mail.med.upenn.edu

no absolute correlation between the sensory activity and the electrical properties of the neurons has been found (Zhong et al., 1997; Zhang et al., 2000). Thus, there remains an uncertainty as to whether the PGNs respond with or without glomus cells connection.

We tested the possibility using another preparation, i.e. co-culturing PGNs with PC-12 cells, because PC-12 cells have been shown to have properties similar to glomus cells (Seta and Millhorn, 2004). We found that PGNs developed sensitivity to hypoxia only in presence of PC-12 cells, arguing that synaptic contact between PC-12 cells and PGNs was an absolute necessity.

In the co-culture of rats PGNs and CB cells, basal discharge and hypoxia-induced action potential are partially blocked by suramin (purinargic blocker) or hexomethonium (cholinergic blocker) but completely blocked by simultaneous application of both blockers (Zhang et al., 2000). Thus functional synapse and PGNs and glomus cells is a real possibility. Like glomus cell, PC-12 cells are also oxygen sensitive and are excitable. Both originate from neural crest and stimulated by hypoxia which is translated into an increase in intercellular calcium and neurotransmitter release.

Several molecules including acetylcholine, ATP, dopamine etc. have been proposed to participate in the synaptic transmission (Gonzalez et al., 1994; Iturriaga and Alcayaga, 2004; Lahiri et al., 2006; Nurse, 2005). To investigate the synaptic mechanism we have developed and tested a co-culture model of PC-12 cells and PGNs for chemical synapse. Results of this investigation have been presented in this paper.

2 Methods

PGNs were co-cultured with or without PC-12 cells. Intercellular calcium of PGNs was measured with Fura-2 (Roy et al., 2004). The idea being that calcium was increased in the PGNs when in contact with PC-12 cells only.

3 Results and Discussion

The summarized results are presented in the following table (Table 1). The scenario that the PGNs, which were in contact with the PC-12 cells, responded to hypoxia, high PCO, and the application of ATP (Figs. 1 and 2); whereas the PGNs which were cultured without PC-12 cells remained quiescent to the same stimuli (Fig. 2). This results point to the importance of PC-12 cells in the co-culture and are consistence with those of Alcayaga et al. (2003).

A co-culture between PGNs and glomus cells has been used before with hypoxia with similar results (Nurse, 2005).

The reason for the lack of response of the PGNs is not obvious. It is hypothesized that the receptors for the response were absent in PGNs.

This effect of CO resembles those of hypoxia.

Table 1 Summary responses in isolated PGN and co-cultures of PGN +PC12 cells

Stimuli	Effects	Effects
	[Ca2+]i	[Ca2+]i
Hypoxia	↑	No response
ATP	↑	No response
CO	↑	No response
CO + Suramin	↓	No response
CO + Suramin + Hexamethonium	↓	No response

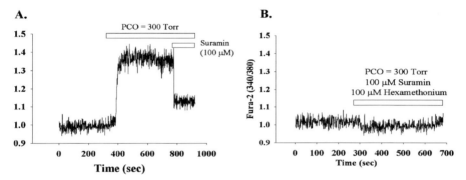

Fig. 1 Example of the effect of CO (Pco = 300 Torr) on intracellular [Ca^{2+}] response of PGN co-cultured with PC-12 cells, **A**. Increase in intracellular Ca^{2+} of PGN co-cultured with PC-12 cells during CO is partially inhibited (75%) by suramin (100 μM). **B**. Combined application of suramin and hexamethonium completely inhibited the intracellular [Ca^{2+}] response of PGN

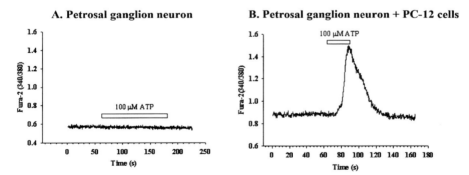

Fig. 2 Effect of ATP on intracellular [Ca^{2+}] of PGNs cultured without (A) and with (B) PC-12 cells. Increase in intracellular Ca^{2+} of PGN co-cultured with PC-12 cells with ATP suggests transduction of the stimulus through functional synapse (**B**). However, PGNs without PC-12 connection remain quiescent to ATP stimulus (**A**). This suggests that ATP, acting via postsynaptic purinoceptors, excites the PGN

Acknowledgments Supported by the grant: NIH 43413-15.

References

Alcayaga, J., Varas, R., Iturriaga, R., 2003. Petrosal ganglion responses in vitro. In: Lahiri, S., Semenza, G.L., Prabhakar, N.R. (Eds.), Oxygen Sensing. Marcel Dekker, Inc., New York, pp. 671–683.

Gonzalez, C., Almaraz, L., Obeso, A., Rigual, R., 1994. Carotid body chemoreceptors: from natural stimuli to sensory discharges. Physiol. Rev. 74, 829–898.

Iturriaga, R., Alcayaga, J., 2004. Neurotransmission in the carotid body: transmitters and modulators between glomus cells and petrosal ganglion nerve terminals. Brain Res. Brain Res. Rev. 47, 46–53.

Lahiri, S., Roy, A., Baby, S.M., Semenza, G.L., Prabhakar, N.R., 2006. Oxygen sensing in the body. Prog. Biophys. Mol. Biol. 91, 249–286.

Nurse, C.A., 2005. Neurotransmission and neuromodulation in the chemosensory carotid body. Autonom. Neurosci. Basic Clin. 120, 1–9.

Roy, A., Li, J., Baby, S.M., Mokashi, A., Buerk, D.G., Lahiri, S., 2004. Effects of iron-chelators on ion-channels and HIF-1α in the carotid body. Respir. Physiol. Neurobiol. 141, 115–123.

Seta, K.A., Millhorn, D.E., 2004. Functional genomics approach to hypoxia signaling. J. Appl. Physiol. 96, 765–773.

Zhang, M., Zhong, H., Vollmer, C., Nurse, C.A., 2000. Co-release of ATP and Ach mediates hypoxic signaling at rat carotid body chemoreceptors. J. Physiol. 525, 143–158.

Zhong, H., Zhang, M., Nurse, C.A., 1997. Synapse formation and hypoxic signaling in co-cultures of rat petrosal neurons and carotid body type 1 cells. J. Physiol. 503, 599–612.

Modification of Relative Gene Expression Ratio Obtained from Real Time qPCR with Whole Carotid Body by Using Mathematical Equations

J.H. Kim, I. Kim and J.L. Carroll

Abstract Quantitative real time PCR (qPCR) is a common tool used to compare the relative gene expression between treated/untreated cells, different types of tissues, or immature/mature organs. When homogeneous cells are used for qPCR, the Ct number of a tested gene solely represents the quantity of gene expression in cells. However, when a heterogeneous tissue is used for qPCR, the Ct number of a tested gene should be modified depending on several factors: the percentage of each cell type in the sample tissue, the cell type where the target gene is expressed, and the cell type in which the target gene is regulated. The carotid body (CB) is mainly composed of three types of cells: type I (chemoreceptor) cells, type II cells, and other types of cells. Therefore, the relative gene expression ratio obtained from qPCR data using whole CB could be modified by applying one of the following 19 different cases: (1) the target gene is expressed in only one type of cell (3 cases), (2) the gene is expressed in two types of cells and increased in only one or both cell types (9 cases), and (3) the gene is expressed in all three types of cells and increased in only one, two, or all three cell types (7 cases). For example, in the case that the target gene is expressed in all three types of cells and the gene is increased in only a cell comprising 10% of whole CB, the gene expression ratio in that cell will be 9 times as that derived from whole CB. Thus, once the percentage of each cell type in whole CB is observed, the cell type of interest gene (E-gene) expression is identified, and the cell type that regulates E-gene expression by treatment is identified. Thus, the corresponding mathematical equation out of 19 cases could be applied to modify the gene expression ratios measured by qPCR.

Keywords Mathematical equation · qRT-PCR · Gene expression · Ratio · CB · Type I cell · Type II cell · Cell type · Percentage · Modification

J.H. Kim (✉)
Dept of Systems Engineering, University of Arkansas at Little Rock, Little Rock, AR, USA
e-mail: jhkim@ualr.edu

1 Introduction

Quantitative real time reverse transcription PCR (qRT-PCR) is the most suitable and sensitive method for the detection and quantification of gene expression levels, in particular for low abundance mRNA, from limited tissue samples, and to elucidate small changes in mRNA expression levels (Bustin 2000, Pfaffl and Hageleit 2001). The relative expression is based on the expression ratio of the target gene versus a reference gene and is adequate for most purposes to investigate physiological changes in gene expression levels (Pfaffl et al. 2002). However, when the heterogeneous tissues are used for qRT-PCR, the relative gene expression ratio could be considered to re-evaluate depending on several factors: the percentage of each cell type which sample tissue consists of, the cell type expressing target genes, and the cell type in which the target gene is regulated.

For example, when whole rat carotid body (CB) is used for the gene expression studies, the relative gene expression ratio of qRT-PCR could be re-evaluated. Carotid body is a small organ sensing blood O_2, CO_2, and pH level. CB consists of four principle components: cell clusters, blood vessels, connective tissue, and nerve fibers (Izal-Azcarate et al. 2008) and is mainly composed with chemoreceptor glomus cells (type I cells) and sustentacular cells (type II cells) at ~60% of total CB volume (Gonzalez et al. 1995, Lopez-Barneo et al. 2001). When we assume that CB consists of three types of cells: type I cells, type II cells, and other types of cell, we find 19 different cases, as shown in methods. We try to generate mathematical equations to modify the relative gene expression ratio of qRT-PCR for all possible cases. In cases which the target gene is only expressed in one type of cell, the ratio won't be changed, as shown in case 1–3. However, for other cases the relative gene expression ratios for qRT-PCR could be modified with the proper mathematical equations generated for 16 cases (case 4–19). Therefore, if we know what percentage each cell type is occupied, which cell types express target genes, and which cell types in which the target gene is regulated, the relative gene expression ratio of qRT-PCR can be easily re-calculated by applying the corresponding mathematical equations.

2 Methods

In order to study the developmental gene expression ratio of postnatal 1 day and 14days old CB having three different cell types, the following notations will be used;

Nc_1, d_1 or Nc_1, d_{14} : number of E-gene in cell 1 (type I cell) at day 1 or 14
Nc_2, d_1 or Nc_2, d_{14}: number of E-gene in cell 2 (type II cell) at day 1 or 14
Nc_3, d_1 or Nc_3, d_{14} : number of E-gene in cell 3 (other type) at day 1 or 14
Nd_1 or Nd_{14} : total number of E-gene in whole CB cells at day 1 or 14

Pc_1, d_1 or Pc_1, d_{14} : the percentage of cell 1 in whole CB cells at day 1 or 14
Pc_2, d_1 or Pc_2, d_{14} : the percentage of cell 2 in whole CB cells at day 1 or 14

Pc$_3$, d$_1$ or Pc$_3$, d$_{14}$: the percentage of cell 3 in whole CB cells at day 1 or 14
Thus, Pc$_1$, d$_1$ + Pc$_2$, d$_1$ + Pc$_3$, d$_1$ = Pc$_1$, d$_{14}$ + Pc$_2$, d$_{14}$ + Pc$_3$, d$_{14}$ = 1

The number of cell 1 at day 1
= Pc$_1$, d$_1$ * total number of cell in whole CB at day 1
The number of cell 1 at day 14
= Pc$_1$, d$_{14}$ * total number of cell in whole CB at day 14
The number of cell 2 at day 1
= Pc$_2$, d$_1$ * total number of cell in whole CB at day 1
The number of cell 2 at day 14
= Pc$_2$, d$_{14}$ * total number of cell in whole CB at day 14
The number of cell 3 at day 1
= Pc$_3$, d$_1$ * total number of cell in whole CB at day 1
The number of cell 3 at day 14
= Pc$_3$, d$_{14}$ * total number of cell in whole CB at day 14

It is assumed that each cell has the same number of E-gene.
The total number of E-gene at day 1 = Nd$_1$
The total number of E-gene at day 14 = Nd$_{14}$
Then, at day 1

Nc$_1$, d$_1$ = Pc$_1$, d$_1$ * Nd$_1$ →→→ Nd$_1$ = Nc$_1$, d$_1$/Pc$_1$, d$_1$ (1)
Nc$_2$, d$_1$ = Pc$_2$, d$_1$ * Nd$_1$ →→→ Nd$_1$ = Nc$_2$, d$_1$/Pc$_2$, d$_1$ (2)
Nc$_3$, d$_1$ = Pc$_3$, d$_1$ * Nd$_1$ →→→ Nd$_1$ = Nc$_3$, d$_1$/Pc$_3$, d$_1$ (3)
Hence, from (1) and (2), Nc$_1$, d$_1$/Pc$_1$, d$_1$ = Nc$_2$, d$_1$/Pc$_2$, d$_1$
→→→ Nc$_1$, d$_1$ = Nc$_2$, d$_1$/Pc$_2$, d1 * Pc$_1$, d$_1$ = Pc$_1$, d$_1$/Pc$_2$, d$_1$ * Nc$_2$, d$_1$ (4)
→→→ Nc$_2$, d$_1$ = Nc$_1$, d$_1$/Pc$_1$, d1 * Pc$_2$, d$_1$ = Pc$_2$, d$_1$/Pc$_1$, d$_1$ *Nc$_1$, d$_1$ (5)
From (2) and (3), Nc$_3$, d$_1$/Pc$_3$, d$_1$ = Nc$_2$, d$_1$/Pc$_2$, d
→→→ Nc$_2$, d$_1$ = Nc$_3$, d$_1$/Pc$_3$, d1 * Pc$_2$, d$_1$ = Pc$_2$, d$_1$/Pc$_3$, d$_1$ * Nc$_3$, d$_1$ (6)
→→→ Nc$_3$, d$_1$ = Nc$_2$, d$_1$/Pc$_2$, d$_1$ * Pc$_3$, d = Pc$_3$, d$_1$/Pc$_2$, d$_1$ * Nc$_2$, d$_1$ (7)
From (1) and (3), Nc$_3$, d$_1$/Pc$_3$, d$_1$ = Nc$_1$, d$_1$/Pc$_1$, d$_1$
→→→ Nc$_1$, d$_1$ = Nc$_3$, d$_1$/Pc$_3$, d$_1$ * Pc$_1$, d$_1$ = Pc$_1$, d$_1$/Pc$_3$, d$_1$ * Nc$_3$, d$_1$ (8)
→→→ Nc$_3$, d$_1$ = Nc$_1$, d$_1$/Pc$_1$, d$_1$ * Pc$_3$, d$_1$ = Pc$_3$, d$_1$/Pc$_1$, d$_1$ * Nc$_1$, d$_1$ (9)

The total number of E-gene at day 1 = Nd$_1$ = Nc$_1$, d$_1$ + Nc$_2$, d$_1$ + Nc$_3$, d$_1$
from (5) and (9)=Nc$_1$, d$_1$+Pc$_2$, d$_1$/Pc$_1$, d$_1^*$Nc$_1$, d$_1$ + Pc$_3$, d$_1$/Pc$_1$, d$_1$ * Nc$_1$, d$_1$
= (1 + Pc$_2$, d$_1$/Pc$_1$, d$_1$ + Pc$_3$, d$_1$/Pc$_1$, d$_1$) * Nc$_1$, d$_1$
Nd$_1$ = Nc$_1$, d$_1$ + Nc$_2$, d$_1$ + Nc$_3$, d$_1$
from (4) and (7)
= Pc$_1$, d$_1$/Pc$_2$, d$_1$ * Nc$_2$, d$_1$ + Nc$_2$, d$_1$ + Pc$_3$, d$_1$/Pc$_2$, d$_1$ * Nc$_2$, d$_1$
= (Pc$_1$, d$_1$/Pc$_2$, d$_1$ + 1 + Pc$_3$, d$_1$/Pc$_2$, d$_1$) * Nc$_2$, d$_1$
Nd$_1$ = Nc$_1$, d$_1$ + Nc$_2$, d$_1$ + Nc$_3$, d$_1$

from (8) and (9)
$= Pc_2, d_1/Pc_3, d_1 * Nc_3, d_1 + Pc_1, d_1/Pc_3, d_1 * Nc_3, d_1 + Nc_3, d_1$
$= (Pc_1, d_1/Pc_3, d_1 + Pc_3, d_1/Pc_3, d_1 + 1) * Nc_3, d_1$

The E-gene expression at day 14 is α times (calculated from qRT-PCR using whole CB) as that at day 1,
then, $Nd_{14} = \alpha * Nd_1$
that is, $Nc_1, d_{14} + Nc_2, d_{14} + Nc_3, d_{14} = \alpha * (Nc_1, d_1 + Nc_2, d_1 + Nc_3, d_1)$.

3 Results

We generate the mathematical equations using the percentage of each cell type to modify the gene expression ratio of qRT-PCR in the following 19 different cases.

3.1 Interest Gene is Expressed in Only One Type of Cells

Case 1: If gene is expressed only at cell 1:

then, $Nc_2, d_{14} = Nc_3, d_{14} = Nc_2, d_1 = Nc_3, d_1 = 0$
hence, $Nc_1, d_{14} + Nc_2, d_{14} + Nc_3, d_{14} = \alpha * (Nc_1, d_1 + Nc_2, d_1 + Nc_3, d_1)$
$Nc_1, d_{14} = \alpha * Nc_1, d_1$ (10)

Therefore, the gene expression ratio gotten from whole CB is the same as the ratio gotten from only cell 1.

Case 2: If gene is expressed only at cell 2 :

Similarly to Case 1, $\quad Nc_2, d_{14} = \alpha * Nc_2, d_1 \quad$ (11)

Case 3: If gene is expressed only at cell 3 :

Similarly to Case 1, $\quad Nc_3, d_{14} = \alpha * Nc_3, d_1 \quad$ (12)

3.2 Interest Gene is Expressed in Cell 1 and Cell 2

Case 4: The number of gene at only cell 1 is increased,

then $Nc_2, d_{14} = Nc_2, d_1$ and $Nc_3, d_{14} = Nc_3, d_1 = 0$
hence, $Nc_1, d_{14} + Nc_2, d_{14} + Nc_3, d_{14} = \alpha * (Nc_1, d_1 + Nc_2, d_1 + Nc_3, d_1)$
$Nc_1, d_{14} + Nc_2, d_1 = \alpha * (Nc_1, d_1 + Nc_2, d_1)$
since $Nc_2, d_1 = Pc_2, d_1/Pc_1, d_1 * Nc_1, d_1$ (from Eq. (4))
$Nc_1, d_{14} + Pc_2, d_1/Pc_1, d_1 * Nc_1, d_1 = \alpha * (Nc_1, d_1 + Pc_2, d_1/Pc_1, d_1 * Nc_1, d_1)$
$Nc_1, d_{14} = \alpha * (Nc_1, d_1 + Pc_2, d_1/Pc_1, d_1 * Nc_1, d_1) - Pc_2, d_1/Pc_1, d_1 * Nc_1, d_1$
$= \{\alpha * (1 + Pc_2, d_1/Pc_1, d_1) - Pc_2, d_1/Pc_1, d_1\} * Nc_1, d_1$
$$Nc_1, d_{14}/Nc_1, d_1 = \alpha * (1 + Pc_2, d_1/Pc_1, d_1) - Pc_2, d_1/Pc_1, d_1 \quad (13)$$

Therefore, the gene expression ratio at only cell 1 will be re-calculated based on the factor given at Eq. (13) with the gene expression ratio (α) at whole CB.

Case 5: The number of gene at only cell 2 is increased,
Similar to Case 4,

$$Nc_2, d_{14}/Nc_2, d_1 = \alpha*(1 + Pc_1, d_1/Pc_2, d_1) - Pc_1, d_1/Pc_2, d_1 \quad (14)$$

Case 6: The number of gene at cell 1 and cell 2 is increased,

then $Nc_3, d_{14} = Nc_3, d_1 = 0$
$Nc_1, d_{14} + Nc_2, d_{14} + Nc_3, d_{14} = \alpha * (Nc_1, d_1 + Nc_2, d_1 + Nc_3, d_1)$
hence, $Nc_1, d_{14} + Nc_2, d_{14} = \alpha * (Nc_1, d_1 + Nc_2, d_1)$
since $Nc_2, d_1 = Pc_2, d_1/Pc_1, d_1 * Nc_1, d_1 (from Eq. (4))$ and
$Nc_2, d_{14} = Pc_2, d_{14}/Pc_1, d_{14} * Nc_1, d_{14}$
$(1 + Pc_2, d_{14}/Pc_1, d_{14})*Nc_1, d_{14} = \alpha * (1 + Pc_2, d_1/Pc_1, d_1) * Nc_1, d_1$
$Nc_1, d_{14}/Nc_1, d_1$
$$= \alpha*(1 + Pc_2, d_1/Pc_1, d_1)/(1 + Pc_2, d_{14}/Pc_1, d_{14}) \quad (15)$$
Since $Nc_1, d_1 = Pc_1, d_1/Pc_2, d_1 * Nc_2, d_1$, and
$Nc_1, d_{14} = Pc_1, d_{14}/Pc_2, d_{14} * Nc_2, d_{14}$
$Nc_1, d_{14}/Nc_1, d_1 = (Pc_1, d_{14}/Pc_2, d_{14} * Nc_2, d_{14})/(Pc_1, d_1/Pc_2, d_1 * Nc_2, d_1)$
thus, $Nc_2, d_{14}/Nc_2, d_1$
$= \{(Pc_2, d_1 * Pc_2, d_{14})/(Pc_1, d_1^*Pc_1, d_{14})\}*(Nc_1, d_{14}/Nc_1, d_1)$
$$Nc_2, d_{14}/Nc_2, d_1 = \{(Pc_1, d_1^*Pc_2, d_{14})/(Pc_2, d_1^*Pc_1, d_{14})\}*\{\alpha*(1 + Pc_2, d_1/Pc_1, d_1)/ (1 + Pc_2, d_{14}/Pc_1, d_{14})\} \quad (16)$$

3.3 Interest Gene is Expressed in Cell 1 and Cell 3

Case 7: The number of gene at only cell 1 is increased,

Similar to case 4,

$$Nc_1, d_{14}/Nc_1, d_1 = \alpha^* (1 + Pc_3, d_1/Pc_1, d_1) - Pc_3, d_1/Pc_1, d_1 \quad (17)$$

Case 8: The number of gene at only cell 3 is increased,
Similar to case 4,

$$Nc_3, d_{14}/Nc_3, d_1 = \alpha^*(1 + Pc_1, d_1/Pc_3, d_1) - Pc_1, d_1/Pc_3, d_1 \quad (18)$$

Case 9: The number of gene at cell 1 and cell 3 is increased,
Similar to Case 6,

$$Nc_1, d_{14}/Nc_1, d_1$$
$$= \alpha^*(1 + Pc_3, d_1/Pc_1, d_1)/(1 + Pc_3, d_{14}/Pc_1, d_{14}) \quad (19)$$
$$Nc_3, d_{14}/Nc_3, d_1 = \{(Pc_1, d_1^*Pc_3, d_{14})/(Pc_3, d_1^*Pc_1, d_{14})\}^*\alpha^*(1+Pc_3, d_1/Pc_1, d_1)$$
$$/(1 + Pc_3, d_{14}/Pc_1, d_{14}) \quad (20)$$

3.4 Interest Gene is Expressed in Cell 2 and Cell 3

Case 10: The number of gene at only cell 2 is increased,
Similar to case 4,

$$Nc_2, d_{14}/Nc_2, d_1 = \alpha^*(1 + Pc_3, d_1/Pc_2, d_1) - Pc_3, d_1/Pc_2, d_1 \quad (21)$$

Case 11: The number of gene at only cell 3 is increased,
Similar to case 4,

$$Nc_3, d_{14}/Nc_3, d_1 = \alpha^*(1 + Pc_2, d_1/Pc_3, d_1) - Pc_2, d_1/Pc_3, d_1 \quad (22)$$

Case 12: The number of gene at cell 2 and cell 3 is increased,
Similar to Case 6,

$$Nc_2, d_{14}/Nc_2, d_1 = \alpha^*(1 + Pc_3, d_1/Pc_2, d_1)/(1 + Pc_3, d_{14}/Pc_2, d_{14}) \quad (23)$$
$$Nc_3, d_{14}/Nc_3, d_1 = \{(Pc_2, d_1^*Pc_3, d_{14})/(Pc_3, d_1^*Pc_2, d_{14})\}^*\{\alpha^*(1 + Pc_3, d_1$$
$$/Pc_2, d_1)/(1 + Pc_3, d_{14}/Pc_2, d_{14})\} \quad (24)$$

3.5 Interest Gene is Expressed in All 3 Cells

Case 13: The number of gene at only cell 1 is increased,

then $Nc_2, d_{14} = Nc_2, d_1$ and $Nc_3, d_{14} = Nc_3, d_1$

hence, $Nc_1, d_{14} + Nc_2, d_{14} + Nc_3, d_{14} = \alpha^*(Nc_1, d_1 + Nc_2, d_1 + Nc_3, d_1)$

$Nc_1, d_{14} + Nc_2, d_1 + Nc_3, d_1 = \alpha^*(Nc_1, d_1 + Nc_2, d_1 + Nc_3, d_1)$

since, $Nc_2, d_1 = Pc_2, d_1/Pc_1, d_1^*Nc_1, d_1$ and $Nc_3, d_1 = Pc_3, d_1/Pc_1, d_1^*Nc_1, d_1$

$Nc_1, d_{14} + Pc_2, d_1/Pc_1, d_1^*Nc_1, d_1 + Pc_3, d_1/Pc_1, d_1^*Nc_1, d_1$
$= \alpha^*(Nc_1, d_1 + Pc_2, d_1/Pc_1, d_1^*Nc_1, d_1 + Pc_3, d_1/Pc_1, d_1^*Nc_1, d_1)$

then, $Nc_1, d_{14} = \{\alpha^*(1 + Pc_2, d_1/Pc_1, d_1 + Pc_3, d_1/Pc_1, d_1) - (Pc_2, d_1/Pc_1, d_1 + Pc_3, d_1/Pc_1, d_1)\}^*Nc_1, d_1$

$$Nc_1, d_{14}/Nc_1, d_1 = \alpha^*(1 + Pc_2, d_1/Pc_1, d_1 + Pc_3, d_1/Pc_1, d_1) - (Pc_2, d_1/Pc_1, d_1 + Pc_3, d_1/Pc_1, d_1) \qquad (25)$$

Case 14: The number of gene at only cell 2 is increased,
Similarly to Case 13,

$$Nc_2, d_{14}/Nc_2, d_1 = \alpha^*(1 + Pc_1, d_1/Pc_2, d_1 + Pc_3, d_1/Pc_2, d_1) - (Pc_1, d_1/Pc_2, d_1 + Pc_3, d_1/Pc_2, d_1) \qquad (26)$$

Case 15: The number of gene at only cell 3 is increased,
Similarly to Case 13,

$$Nc_3, d_{14}/Nc_3, d_1 = \alpha^*(1 + Pc_1, d_1/Pc_3, d_1 + Pc_2, d_1/Pc_3, d_1) - (Pc_1, d_1/Pc_3, d_1 + Pc_2, d_1/Pc_3, d_1) \qquad (27)$$

Case 16: The number of gene at cell 1 and cell 2 (not at cell 3) is increased,

then $Nc_3, d_{14} = Nc_3, d_1$

hence, $Nc_1, d_{14} + Nc_2, d_{14} + Nc_3, d_{14} = \alpha^*(Nc_1, d_1 + Nc_2, d_1 + Nc_3, d_1)$

$Nc_1, d_{14} + Nc_2, d_{14} + Nc_3, d_1 = \alpha^*(Nc_1, d_1 + Nc_2, d_1 + Nc_3, d_1)$

since $Nc_2, d_1 = Pc_2, d_1/Pc_1, d_1^*Nc_1, d_1$, $Nc_3, d_1 = Pc_3, d_1/Pc_1, d_1^*Nc_1, d_1$,

and $Nc_2, d_{14} = Pc_2, d_{14}/Pc_1, d_{14}^*Nc_1, d_{14}$

$Nc_1, d_{14} + Pc_2, d_{14}/Pc_1, d_{14}^*Nc_1, d_{14} + Pc_3, d_1/Pc_1, d_1^*Nc_1, d_1$
$= \alpha^*(Nc_1, d_1 + Pc_2, d_1/Pc_1, d_1^*Nc_1, d_1 + Pc_3, d_1/Pc_1, d_1^*Nc_1, d_1)^*(1 + Pc_2, d_{14}/Pc_1, d_{14})^*Nc_1, d_{14}$

$= \{\alpha^*(1 + Pc_2, d_1/Pc_1, d_1 + Pc_3, d_1/Pc_1, d_1) - Pc_3, d_1/Pc_1, d_1\}^*Nc_1, d_1$

$$Nc_1, d_{14}/Nc_1, d_1 = \{\alpha^*(1 + Pc_2, d_1/Pc_1, d_1 + Pc_3, d_1/Pc_1, d_1) - Pc_3, d_1/Pc_1, d_1\}/(1 + Pc_2, d_{14}/Pc_1, d_{14}) \qquad (28)$$

Since $Nc_1, d_1 = Pc_1, d_1/Pc_2, d_1^*Nc_2, d_1$, and $Nc_1, d_{14} = Pc_1, d_{14}/Pc_2, d_{14}^*Nc_2, d_{14}$

$Nc_1, d_{14}/Nc_1, d_1 = (Pc_1, d_{14}/Pc_2, d_{14}^*Nc_2, d_{14})/(Pc_1, d_1/Pc_2, d_1^*Nc_2, d_1)$

thus, $Nc_2, d_{14}/Nc_2, d_1 = \{(Pc_1, d_1^*Pc_2, d_{14})/(Pc_2, d_1^*Pc_1, d_{14})\}^*(Nc_1, d_{14}/Nc_1, d_1)$

$Nc_2, d_{14}/Nc_2, d_1 = [\{(Pc_1, d_1^*Pc_2, d_{14})/(Pc_2, d_1^*Pc_1, d_{14})\}^*$

$\{\alpha^*(1 + Pc_2, d_1/Pc_1, d_1 + Pc_3, d_1/Pc_1, d_1) - Pc_3, d_1/Pc_1, d_1\}]$

$/(1 + Pc_2, d_{14}/Pc_1, d_{14})$ \hfill (29)

Case 17: The number of gene at cell 1 and cell 3 is increased,
Similar to Case 16,

$Nc_1, d_{14}/Nc_1, d_1 = \{\alpha^*(1 + Pc_2, d_1/Pc_1, d_1 + Pc_3, d_1/Pc_1, d_1) - Pc_2, d_1/Pc_1,$

$d_1\}/(1 + Pc_3, d_{14}/Pc_1, d_{14})$ \hfill (30)

$Nc_3, d_{14}/Nc_3, d_1 = [\{(Pc_1, d_1^*Pc_3, d_{14})/(Pc_3, d_1^*Pc_1, d_{14})\}^*\{\alpha^*(1$

$+ Pc_2, d_1/Pc_1, d_1 + Pc_3, d_1/Pc_1, d_1) - Pc_2, d_1/Pc_1, d_1\}]$

$/(1 + Pc_3, d_{14}/Pc_1, d_{14})$ \hfill (31)

Case 18: The number of gene at cell 2 and cell 3 is increased,
Similar to Case 16,

$Nc_2, d_{14}/Nc_2, d_1 = \{\alpha^*(1 + Pc_1, d_1/Pc_2, d_1 + Pc_3, d_1/Pc_2, d_1) - Pc_1, d_1/Pc_2,$

$d_1\}/(1 + Pc_3, d_{14}/Pc_2, d_{14})$ \hfill (32)

$Nc_3, d_{14}/Nc_3, d_1 = [\{(Pc_2, d_1^*Pc_3, d_{14})/(Pc_3, d_1^*Pc_2, d_{14})\}^*\{\alpha^*$

$(1 + Pc_1, d_1/Pc_2, d_1 + Pc_3, d_1/Pc_2, d_1) - Pc_1, d_1/Pc_2, d_1\}]/$

$(1 + Pc_3, d_{14}/Pc_2, d_{14})$ \hfill (33)

Case 19: The number of gene at all three cells is increased,

$Nc_1, d_{14} + Nc_2, d_{14} + Nc_3, d_{14} = \alpha^*(Nc_1, d_1 + Nc_2, d_1 + Nc_3, d_1)$

$Nc_1, d_{14} + Pc_2, d_{14}/Pc_1, d_{14}^*Nc_1, d_{14} + Pc_3, d_{14}/Pc_1, d_{14}^*Nc_1, d_{14}$

$= \alpha^*(Nc_1, d_1 + Pc_2, d_1/Pc_1, d_1 * Nc_1, d_1 + Pc_3, d_1/Pc_1, d_1^*Nc_1, d_1)$

$(1 + Pc_2, d_{14}/Pc_1, d_{14} + Pc_3, d_{14}/Pc_1, d_{14})^*Nc_1, d_{14}$

$= \alpha^*(1 + Pc_2, d_1/Pc_1, d_1 + Pc_3, d_1/Pc_1, d_1)^*Nc_1, d_1$

$Nc_1, d_{14}/Nc_1, d_1 = \alpha * (1 + Pc_2, d_1/Pc_1, d_1 + Pc_3, d_1/Pc_1, d_1)/(1 + Pc_2, d_{14}$

$/Pc_1, d_{14} + Pc_3, d_{14}/Pc_1, d_{14})$ \hfill (34)

Since $Nc_1, d_1 = Pc_1, d_1/Pc_2, d_1^*Nc_2, d_1$, and

$Nc_1, d_{14} = Pc_1, d_{14}/Pc_2, d_{14}^*Nc_2, d_{14}$

$Nc_1, d_{14}/Nc_1, d_1 = (Pc_1, d_{14}/Pc_2, d_{14}^*Nc_2, d_{14})/(Pc_1, d_1/Pc_2, d_1^*Nc_2, d_1)$

thus, $Nc_2, d_{14}/Nc_2, d_1 = (Pc_1, d_1^*Pc_2, d_{14})/(Pc_2, d_1^*Pc_1, d_{14})\}^*$
$(Nc_1, d_{14}/Nc_1, d_1)$
$Nc_2, d_{14}/Nc_2, d_1$
$= [\{(Pc_1, d_1^*Pc_2, d_{14})/(Pc_2, d_1^*Pc_1, d_{14})\}^*\alpha^*(1 + Pc_2, d_1/Pc_1, d_1 +$
$Pc_3, d_1/Pc_1, d_1)/(1 + Pc_2, d_{14}/Pc_1, d_{14} + Pc_3, d_{14}/Pc_1, d_{14})$ (35)

Since $Nc_1, d_1 = Pc_1, d_1/Pc_3, d_1 * Nc_3, d_1$, and
$Nc_1, d_{14} = Pc_1, d_{14}/Pc_3, d_{14} * Nc_3, d_{14}$
$Nc_1, d_{14}/Nc_1, d_1 = (Pc_1, d_{14}/Pc_3, d_{14}^*Nc_3, d_{14})/(Pc_1, d_1/Pc_3, d_1^*Nc_3, d_1)$
thus, $Nc_3, d_{14}/Nc_3, d_1 = \{(Pc_1, d_1^*Pc_3, d_{14})/(Pc_3, d_1^*Pc_1, d_{14})\}^*(Nc_1, d_{14}/Nc_1, d_1)$
$Nc_3, d_{14}/Nc_3, d_1 =$
$[\{(Pc_1, d_1^*Pc_3, d_{14})/(Pc_3, d_1^*Pc_1, d_{14})\}^*\alpha^*(1 + Pc_2, d_1/Pc_1, d_1 +$
$Pc_3, d_1/Pc_1, d_1)/(1 + Pc_2, d_{14}/Pc_1, d_{14} + Pc_3, d_{14}/Pc_1, d_{14})$ (36)

Here are two examples of how the E-gene expression ratio of qRT-PCT could be changed by applying above mathematical equations in each case under two different conditions, day 1 and day 14 CB have the same or different percentage of types of cell (Table 1). As shown in Table 1, when the E-gene expression at day 14 is 10 times as that at day 1 ($\alpha = 10$), the ratio remained the same value in some cases (case 1–3), but the ratio 10 was changed into the ratio 91 in case 14.

4 Discussion

The ideal qRT-PCR for CB sample having heterogeneous cell types will be using one type of single or few cells. But in present time, especially detecting low abundant genes in single or few cells with qRT-PCR still has technical difficulties. Therefore, we try to find the way to modify the value of gene expression ratio of qRT-PCR by utilizing a mathematical tool. As shown in Table 1, we can see how much the ratio can be modified depending on the cases. Although the ratio of qRT-PCR won't be changed in the case which gene is expressed in only one type of cells (case 1–3), 10 times of expression ratio can be changed as much as 91 times of ratio depending on the case (case 14). Nevertheless our proposed mathematical equations are derived under the assumption that the tissue consists of three types of cells, other mathematical equations for tissues composed of various number of cell type can be derived in the same way as above mathematical equations.

Choosing one of the proposed equations can be determined, if we find out following three factors by biological experiments: (1) the cell types expressing target genes, (2) the cell type the target genes are regulated in, and (3) the percentage of each cell type. It might be plausible to identify and estimate the cell types expressing the target gene and the percentage of each cell type by using immunostaining method, if the antibodies of target genes and cell type specific markers are available. Also, it might be possible to identify the cell type in which the target genes

Table 1 Two examples, day 1 and day 14 have the same percentage of cells or different percentage of cells, were examined how the expression ratio can be changed by above mathematical equations depending on the cases under assumption E-gene expression at day 14 is 10 times ($\alpha = 10$) higher than that at day 1

Different cases	$Pc_1, d_1 = 50\%$, $Pc_2, d_1 = 10\%$, $Pc_3, d_1 = 40\%$, $Pc_1, d_{14} = 50\%$, $Pc_2, d_{14} = 10\%$, $Pc_3, d_{14} = 40\%$	$Pc_1, d_1 = 50\%$, $Pc_2, d_1 = 10\%$, $Pc_3, d_1 = 40\%$, $Pc_1, d_{14} = 30\%$, $Pc_2, d_{14} = 30\%$, $Pc_3, d_{14} = 40\%$

I. Interest gene is expressed only one type of cell

Case 1	From Eq. (10), $Nc_1, d_{14}/Nc_1, d_1 = 10$	From Eq. (10), $Nc_1, d_{14}/Nc_1, d_1 = 10$
Case 2	From Eq. (11), $Nc_2, d_{14}/Nc_2, d_1 = 10$	From Eq. (11), $Nc_2, d_{14}/Nc_2, d_1 = 10$
Case 3	From Eq. (12), $Nc_3, d_{14}/Nc_3, d_1 = 10$	From Eq. (12), $Nc_3, d_{14}/Nc_3, d_1 = 10$

II. Interest gene is expressed in two types of cells and increased in only one or both types of cell

Case 4	From Eq. (13), $Nc_1, d_{14}/Nc_1, d_1 = 11.8$	From Eq. (13), $Nc_1, d_{14}/Nc_1, d_1 = 11.8$
Case 5	From Eq. (14), $Nc_2, d_{14}/Nc_2, d_1 = 55$	From Eq. (14), $Nc_2, d_{14}/Nc_2, d_1 = 55$
Case 6	From Eq. (15), $Nc_1, d_{14}/Nc_1, d_1 = 10$	From Eq. (15), $Nc_1, d_{14}/Nc_1, d_1 = 6$
	From Eq. (16), $Nc_2, d_{14}/Nc_2, d_1 = 10$	From Eq. (16), $Nc_2, d_{14}/Nc_2, d_1 = 30$
Case 7	From Eq. (17), $Nc_1, d_{14}/Nc_1, d_1 = 17.2$	From Eq. (17), $Nc_1, d_{14}/Nc_1, d_1 = 17.2$
Case 8	From Eq. (18), $Nc_3, d_{14}/Nc_3, d_1 = 21.25$	From Eq. (18), $Nc_3, d_{14}/Nc_3, d_1 = 21.25$
Case 9	From Eq. (19), $Nc_1, d_{14}/Nc_1, d_1 = 10$	From Eq. (19), $Nc_1, d_{14}/Nc_1, d_1 = 7.7$
	From Eq. (20), $Nc_3, d_{14}/Nc_3, d_1 = 10$	From Eq. (20), $Nc_3, d_{14}/Nc_3, d_1 = 12.9$
Case 10	From Eq. (21), $Nc_2, d_{14}/Nc_2, d_1 = 36$	From Eq. (21), $Nc_2, d_{14}/Nc_2, d_1 = 36$
Case 11	From Eq. (22), $Nc_3, d_{14}/Nc_3, d_1 = 12.25$	From Eq. (22), $Nc_3, d_{14}/Nc_3, d_1 = 12.25$
Case 12	From Eq. (23), $Nc_2, d_{14}/Nc_2, d_1 = 10$	From Eq. (23), $Nc_2, d_{14}/Nc_2, d_1 = 21.4$
	From Eq. (24), $Nc_3, d_{14}/Nc_3, d_1 = 10$	From Eq. (24), $Nc_3, d_{14}/Nc_3, d_1 = 7.1$

III. Interest gene is expressed in three types of cells and increased in only one, two, or three types of cell

Case 13	From Eq. (25), $Nc_1, d_{14}/Nc_1, d_1 = 19$	From Eq. (25), $Nc_1, d_{14}/Nc_1, d_1 = 19$
Case 14	From Eq. (26), $Nc_2, d_{14}/Nc_2, d_1 = 91$	From Eq. (26), $Nc_2, d_{14}/Nc_2, d_1 = 91$
Case 15	From Eq. (27), $Nc_3, d_{14}/Nc_3, d_1 = 23.5$	From Eq. (27), $Nc_3, d_{14}/Nc_3, d_1 = 23.5$
Case 16	From Eq. (28), $Nc_1, d_{14}/Nc_1, d_1 = 16$	From Eq. (28), $Nc_1, d_{14}/Nc_1, d_1 = 9.6$
	From Eq. (29), $Nc_2, d_{14}/Nc_2, d_1 = 16$	From Eq. (29), $Nc_2, d_{14}/Nc_2, d_1 = 48$
Case 17	From Eq. (30), $Nc_1, d_{14}/Nc_1, d_1 = 11$	From Eq. (30), $Nc_1, d_{14}/Nc_1, d_1 = 8.5$
	From Eq. (31), $Nc_3, d_{14}/Nc_3, d_1 = 11$	From Eq. (31), $Nc_3, d_{14}/Nc_3, d_1 = 14.1$
Case 18	From Eq. (32), $Nc_2, d_{14}/Nc_2, d_1 = 19$	From Eq. (32), $Nc_2, d_{14}/Nc_2, d_1 = 40.7$
	From Eq. (33), $Nc_3, d_{14}/Nc_3, d_1 = 19$	From Eq. (33), $Nc_3, d_{14}/Nc_3, d_1 = 13.6$
Case 19	From Eq. (34), $Nc_1, d_{14}/Nc_1, d_1 = 10$	From Eq. (34), $Nc_1, d_{14}/Nc_1, d_1 = 6$
	From Eq. (35), $Nc_2, d_{14}/Nc_2, d_1 = 10$	From Eq. (35), $Nc_2, d_{14}/Nc_2, d_1 = 30$
	From Eq. (33), $Nc_3, d_{14}/Nc_3, d_1 = 10$	From Eq. (33), $Nc_3, d_{14}/Nc_3, d_1 = 10$

are regulated by running conventional PCR with single or few homogeneous cells. Then, one of the proposed mathematical equations can be applied to re-calculate the relative gene expression ratio derived from qRT-PCR.

In summary, the gene expression ratio of qRT-PCR from heterogeneous tissue which composed of three types of cells, can be modified by one of the proper mathematical equations depending on cell types expressing target genes, which cell type the target genes are regulated in, and the percentage of each cell type.

Acknowledgments This study was supported by ABI grant from Arkansas Children's Hospital and CUMG grant from University of Arkansas for Medical Sciences.

References

Bustin, S.A. 2000, Absolute quantification of mRNA using real-time reverse transcription polymerase chain reaction assays, *J. Mol. Endocrinol.* 25:169–193.
Gonzalez, C., Vicario, I., Almaraz, L., & Rigual, R. 1995, Oxygen sensing in the carotid body. *Biol. Signals* 4:245–256.
Izal-Azcarate, A., Belzunegui, S., Sebastian, W.S., Garrido-Gil, P., Vazquez-Claverie, M., Lopez, B., Marcilla, I., & Luquin, M.A. 2008, Immunohistochemical characterization of the rat carotid body, *Respir. Physiol. Neurobiol.* 161:95–99.
Lopez-Barneo, J., Pardal, R., & Ortega-Saenz, P. 2001, Cellular mechanism of oxygen sensing, *Annu. Rev. Physiol.* 63:259–287.
Pfaffl, M.W. and Hageleit, M. 2001, Validities of mRNA quantification using recombinanat RNA and recom binant DNA external calibration curves in real-time RT-OCR, *Biotechnol. Lett.* 23:275–282.
Pfaffl, M.W., Horgan, G.W., & Dempfle, L. 2002, Relative expression software tool (REST) for group-wise comparison and statistical analysis of relative expression results in real-time PCR, *Nucleic Acids Res.* 30:e36.

Neurotransmitters in Carotid Body Function: The Case of Dopamine – *Invited Article*

R. Iturriaga, J. Alcayaga and C. Gonzalez

Abstract The carotid body (CB) is the main peripheral chemoreceptor. The present model of CB chemoreception states that glomus (type I) cells are the primary receptors, which are synaptically connected to the nerve terminals of the petrosal ganglion neurons. In response to hypoxia, hypercapnia and acidosis, glomus cells release one (or more) transmitter(s) which, acting on the nerve terminals of chemosensory neurons, increases the afferent discharge. Among several molecules present in glomus cells, dopamine, acetylcholine and 5′-adenosine-triphosphate have been proposed to be the excitatory transmitters in the CB. Beside these putative excitatory transmitters, other molecules modulate the chemosensory process through direct actions on glomus cells and/or by producing tonic effects on CB blood vessels. In this review, we focus on the role played by dopamine in the CB chemoreception, with emphasis on the open question if the reported differences on its actions on the generation of afferent chemosensory activity reflect true species differences. The available data suggest that dopamine may play a modulatory role within the cat CB, while in the rabbit CB, dopamine is an excitatory transmitter. Therefore, the reported differences on the actions of exogenously applied dopamine and its participation on the generation of afferent chemosensory activity appear to reflect true species differences.

Keywords Carotid body · Petrosal ganglion neurons · Afferent activity · Neurotransmitters · Dopamine · ATP · Rat rabbit · Hypoxia · Hypercapnia · Acidosis

1 Introduction

The carotid body (CB) is the main chemoreceptor organ that senses the levels of arterial PO_2, PCO_2 and pH, playing a crucial role in cardioventilatory homeostasis. The basic morphological unit of the CB is the glomoid, constituted by clusters

---

R. Iturriaga (✉)
Lab. Neurobiología, Facultad de Ciencias Biológicas, P. Universidad Católica de Chile
e-mail: riturriaga@bio.puc.cl

of glomus (type I) cells synaptically connected between them and with the nerve terminals of the petrosal ganglion (PG) neurons, and surrounded by sustentacular (type II) cells. Hypoxia, hypercapnia and acidosis increase the rate of discharge in the afferent axons of the chemosensory PG neurons (Gonzalez et al., 1994). The most accepted model for CB chemoreception proposes that glomus cells are the primary loci for the transduction of chemical stimuli. As a result of the transduction of natural stimuli the glomus cell depolarizes through a fast and reversible inhibition of one or more K^+ currents (López–Barneo et al., 1988; Buckler, 1997), which in turn raises the intracellular $[Ca^{2+}]$ (Buckler and Vaughan-Jones, 1994), eliciting the release of one -or more- transmitters or modulators (Gonzalez et al., 1994; Iturriaga and Alcayaga, 2004; Nurse, 2005). Among several molecules present in glomus cells, dopamine (DA), acetylcholine (ACh) and 5′-adenosine-triphosphate (ATP) have been proposed to be the excitatory transmitters in the CB (Gonzalez et al., 2003; Iturriaga and Alcayaga, 2004; Nurse, 2005). Beside these putative excitatory transmitters, other molecules such as nitric oxide and endothelin-1 modulate the chemosensory process through direct actions on glomus cells and/or by producing tonic effects on CB blood vessels (Iturriaga and Alcayaga, 2004). In this short review, we will focus on the role played by DA in CB chemoreception, with emphasis on the open question if the reported differences on its actions on the generation of afferent chemosensory activity reflect true species differences.

2 Dopamine Actions in the Carotid Body Afferent Activity

Dopamine meets most of the criteria to be considered an excitatory transmitter in the CB (Gonzalez et al., 1981, 1994, 2003). Dopamine is the predominant catecholamine synthesized, taken-up and stored in dense-cored vesicles of glomus cells, and released by the CB natural stimuli. The suggestion that DA acts as an excitatory transmitter in the CB was strongly supported by the pioneer studies of Gonzalez, Fidone and colleagues. Using radiometric methods they found that hypoxia increased, in a Ca^{2+} dependent manner, the release of DA from the rabbit CB and isolated glomus cells (Fidone and Gonzalez, 1982; Fidone et al., 1982a,b; Gonzalez and Fidone, 1977; Pérez García et al., 1992). These observations were later confirmed in CBs from cats and rats (Rigual et al., 1986; Vicario et al., 2000). In addition, Fidone et al. (1982b) found that after the incubation with [^{3}H]-tyrosine for 2–3 hours, the amount of radiolabeled catecholamines (mainly DA) released from the rabbit CB superfused in vitro was proportional to the intensity of the hypoxic challenge. Figure 1 taken from Fidone et al. (1982b) shows that chemosensory activity recorded from the rabbit carotid sinus nerve in response to hypoxia is highly correlated with the increases in [^{3}H]DA + [^{3}H]DOPAC release. More recently, using voltammetric methods, Donnelly (1996) and Iturriaga et al. (1996) confirmed that hypoxia releases DA from the rat and cat CB. However, Donnelly (1996) and Iturriaga et al. (1996) found that repeated hypoxic stimulation of the cat and rat CB produces highly conserved increases in chemosensory discharge, while the amplitude of the DA efflux was progressively reduced. These observations

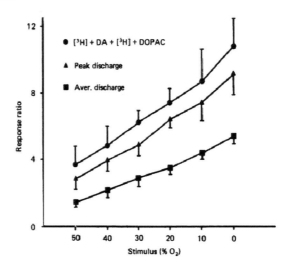

Fig. 1 Relationship between total [³H]DA release ([³H]DA + [³H]DOPAC), peak and average chemoreceptor discharge during hypoxic stimulation in a rabbit CB. [³H]DA release is expressed as a percent of the release during the control period prior to the stimulus (i.e. [stimulated – control]/control). (Taken from Fidone et al. 1982b with permission of Journal of Physiology, Blackwell Publishing, UK)

show a dissociation between chemosensory excitation and DA efflux. However, an attenuation of catecholamine release in response to repetitive stimulation has also been observed in other catecholaminergic systems (Artalejo et al., 1985; García et al., 1976).

To consider DA as an excitatory transmitter in the junction between glomus cells and PG neurons, DA needs to mimic the effect of the natural stimuli on carotid chemosensory activity. In the rabbit, the application of DA to the CB produces chemosensory excitation, but in most other species DA produces chemosensory inhibition. In the cat, intracarotid and intravenous injections of DA evoke transient inhibition of chemosensory discharges, while continuous intravenous infusion of DA in situ decrease the CB responsiveness to hypoxia and hypercapnia (Llados and Zapata, 1978; Iturriaga et al., 1994). The inhibitory effect produced by DA on chemosensory activity is blocked or reversed into excitation after blockade of DA type 2 (D_2) dopamine receptors. In addition, the D_2 receptor antagonist domperidone produces a sustained increase in basal chemosensory discharge, and enhances the responses to hypoxia (Iturriaga et al., 1994). In the superfused cat CB in vitro, DA produces a transient inhibition of chemosensory activity, but repeated injections of DA result in desensitization of the inhibitory actions, and even late excitatory effects in response to large doses (Zapata, 1975).

3 Dopamine Actions in the Petrosal Ganglion

The inhibitory effects of DA on CB chemosensory activity have been attributed to postsynaptic effects mediated by D_2 receptors located in the nerve terminals of the PG neurons that innervate the CB (Bairam et al., 1996; Czyzk-Krezska, 1992). However, the chronic sensory denervation of the CB reduces the density of D_2 receptors in the CB by 50%, indicating that half of these D_2 receptors are located in the CB parenchyma or stroma (Dinger et al., 1981). Moreover, other studies have found evidence for the presence of D_1 and D_2 receptors in the CB (Bairam et al., 1998, Verna et al., 1985). Thus, it is possible that the exogenous application of DA to the CB may produce chemosensory inhibition by acting on dopaminergic receptors located in the CB blood vessels or on presynaptic receptors located in glomus or sustentacular cells. Since the presence of dopaminergic neurons (Katz and Black, 1986) as well as mRNA for D_2 receptors (Bairam et al., 1996; Czyzyk-Krzeska et al., 1992) have been shown in a population of rat and rabbit PG neurons, Alcayaga et al. (1999, 2003) studied the responses elicited by the application of DA at the perikarya of cat PG neurons in an isolated preparation. This preparation of the PG allows the recording of antidromic discharges from the carotid sinus nerve evoked by the application of putative transmitters and drugs to the PG surface (Alcayaga et al., 1998). Application of DA to the isolated cat PG in vitro does not produce any noticeable effect on the basal neural activity recorded from the carotid sinus nerve (Alcayaga et al., 1999, 2003). However, as is shown in Fig. 2, DA

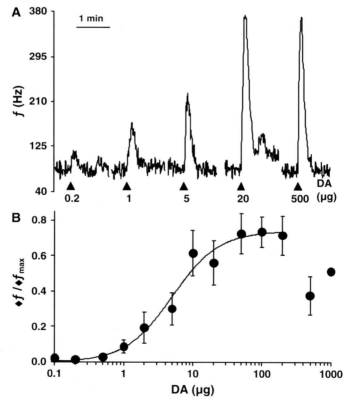

Fig. 2 (**A**) Effects of DA on rabbit carotid sinus nerve frequency of discharge in a representative recording from a single PG preparation. (**B**) Dose-response relationship in 9 isolated PG preparations

applied to the isolated rabbit PG evokes a dose-dependent increase in the carotid sinus nerve frequency of discharge. Moreover, this excitatory effect is blocked by the D_2 receptor antagonist spiperone, indicating that DA excites rabbit PG neurons projecting through the carotid sinus nerve (Alcayaga et al., 2006).

As was mentioned, applications of DA to the isolated cat PG did not modify the frequency of discharge from the carotid sinus nerve (Alcayaga et al., 1999). However, when DA is applied before ACh, produces a dose-dependent modification of the neural responses evoked by ACh and ATP. For a given dose of ACh, low doses of DA potentiate the responses induced by ACh, while large doses inhibit the responses elicited by ACh (Alcayaga et al., 1999) and ATP (Alcayaga et al., 2003). The inhibitory effect of DA on ACh-induced responses was partly reversed by the D_2 receptor blocker spiperone (Alcayaga et al., 1999).

The reported differences in the effects of exogenously applied DA and its possible participation in the generation of afferent chemosensory activity appear to reflect true species differences. In the rat and cat CB, the excitatory response seems to rely on several neurotransmitters, with DA acting as a key modulator of the chemosensory activity. On the other hand, chemosensory activity in the rabbit appears to be generated with an important participation of dopaminergic synapses (Iturriaga et al., 2007).

4 Conclusion

In summary, the effects of DA and D_2 receptor antagonist on cat CB chemoreception and PG neurons suggest a modulatory role for DA within the cat CB. However, in the rabbit PG in vitro, DA applied to the ganglion induces a dose-dependent increase in the carotid nerve discharge frequency, suggesting an excitatory action of DA on rabbit PG neurons projecting through the carotid nerve. Thus, the reported differences on the actions of exogenously applied DA and its participation on the generation of afferent chemosensory activity appear to reflect true species differences.

For several decades, it was proposed that a unique molecule was responsible for the excitatory transmission between glomus cells and the terminals of PG neurons. The precise nature of this unique transmitter was matter of active debate. The high degree of co-localization for amine-synthesizing enzymes (tyrosine hydroxylase, dopamine-β-hydroxylase and choline acetyltransferase) and peptides found in the glomus cells of several species suggest that several molecules may participate in the chemosensory process in the CB. However, DA plays a fundamental role in this process.

Acknowledgments This work was supported in part by grants 1070854 from the National Fund for Scientific and Technological Development of Chile (FONDECYT) and BFU2007-61848 (DGICYT), CIBER CB06/06/0050 (FISS-ICiii) and JCyL-GR242 form Spanish Institutions.

References

Alcayaga, J., Iturriaga, R., Varas, R., Arroyo, J., & Zapata, P. 1998. Selective activation of carotid nerve fibers by acetylcholine applied to the cat petrosal ganglion in vitro. *Brain Res* 786: 47–54.

Alcayaga, J., Retamal, M., Cerpa, V., Arroyo, J., & Zapata, P, 2003. Dopamine inhibits ATP-induced responses in the cat petrosal ganglion in vitro. *Brain Res* 966: 283–287.

Alcayaga, J., Soto, R., Vargas, R., Oritz, F., Arroyo, J., & Iturriaga, R. 2006. Carotid body transmitters actions on rabbit petrosal ganglion in vitro. *Adv Exp Med Biol* 580: 331–337.

Alcayaga, J., Varas, R., Arroyo, J., Iturriaga, R., & Zapata, P. 1999. Dopamine modulates carotid nerve responses induced by acetylcholine on cat petrosal ganglion in vitro. *Brain Res* 831: 97–103.

Artalejo, A.R., García, A.G., Montiel, C., & Sánchez García, P.A. 1985. Dopaminergic receptor modulates catecholamine release from the cat adrenal gland. *J Physiol* 362: 359–368.

Bairam, A., Dauphin, C., Rousseau, F., & Khandjian, E.W. 1996. Dopamine D_2 receptor mRNA isoforms expression in the carotid body and petrosal ganglion of developing rabbits. *Adv Exp Med Biol* 410: 285–289.

Bairam, A., Frenette, J., Dauphin, C., Carroll, J.L., & Khandjian, E.W. 1998. Expression of dopamine D1-receptor mRNA in the carotid body of adult rabbits, cats and rats. *Neurosci Res* 31: 147–154.

Buckler, K. 1997. A novel oxygen-sensitive potassium current in rat carotid body type I cells. *J Physiol* 498: 649–662.

Buckler, K.J., & Vaughan-Jones, R.D. 1994. Effects of hypoxia on membrane potential and intracellular calcium in rat neonatal carotid body type I cells. *J Physiol* 476: 423–428.

Czyzyk-Krzeska, M.F., Lawson, E.E., & Millhorn, D.F. 1992. Expression of D_2-dopamine receptor messenger RNA in the arterial chemoreceptor afferent pathway. *J Autonom Nervous Syst* 41: 31–39.

Donnelly, D.F. 1996. Chemoreceptornerve excitation may be not proportional to catecholamine secretion. *J Appl Physiol* 81: 2330–2337.

Dinger, B., Gonzalez, C., Yoshizaki, K., & Fidone, S. 1981. [^{3}H]Spiroperidol binding in normal and denervated carotid bodies. *Neurosci Let,* 21: 51–55.

Fidone, S., & Gonzalez, S. 1982. Catecholamine synthesis in rabbit carotid body in vitro. *J Physiol* 333: 69–79.

Fidone, S., Gonzalez, C., & Yoshizaki, K. 1982a. Effects of hypoxia on catecholamine synthesis in rabbit carotid body in vitro. *J Physiol* 333: 81–91.

Fidone, S.J., Gonzalez, C., & Yoshizaki, K. 1982b. Effects of low oxygen on the release of dopamine from the rabbit carotid body in vitro. *J Physiol* 333: 93–110.

García, A.G., Kirpekar, S.M., & Sánchez García, P. 1976. Release of noradrenaline from the cat spleen by nerve stimulation and potassium. *J Physiol* 261: 301–317.

Gonzalez, C., & Fidone, S. 1977. Increased release of ^{3}H dopamine during low O_2 stimulation of rabbit carotid body in vitro. *Neurosci Lett* 6: 95–99.

Gonzalez, C., Almaraz, L., Obeso, A., & Rigual, R. 1994. Carotid body chemoreceptors: from natural stimuli to sensory discharges. *Physiol Rev* 74: 829–898.

Gonzalez, C., Kwok, Y., Gibb, J., & Fidone, S. 1981. Physiological and pharmacologic effects on TH activity in rabbit and cat carotid body. *Am J Physiol* 240: R38–R43.

Gonzalez, C., Rocher, A., & Zapata, P. 2003. Quimiorreceptores arteriales: mecanismos celulares y moleculares de las funciones adaptativa y homeostática del cuerpo carotídeo. *Rev Neurol* 36: 239–254.

Iturriaga, R., & Alcayaga, J. 2004. Neurotransmission in the carotid body: Transmitters and modulators between glomus cells and petrosal ganglion nerve terminals. *Brain Res Rev* 47: 46–53.

Iturriaga, R., Alcayaga, J., & Zapata, P. 1996. Dissociation of hypoxia-induced chemosensory responses and catecholamine efflux in the cat carotid body superfused in vitro. *J Physiol* 497: 551–564.

Iturriaga, R., Larraín, C., & Zapata, P. 1994. Effects of dopamine blockade upon carotid chemosensory activity and its hypoxia-induced excitation. *Brain Res* 663: 145–164.

Iturriaga, R., Varas, R., & Alcayaga, J. 2007. Electrical and pharmacological propierties of petrosal ganglion neurons that innervate the carotid body. *Resp Physiol & Neurobiol* 157: 130–139.

Katz, D.M., & Black, I.B. 1986. Expression and regulation of catecholaminergic traits in primary sensory neurons: relationship to target innervation in vivo. *J Neurosci* 6: 983–989.

Llados, F., & Zapata, P. 1978. Effects of dopamine analogues and antagonists on carotid body chemosensory in situ. *J Physiol* 274: 487-499.

López–Barneo, J., Lopez-Lopez, J.R., Ureña, J., & Gonzalez, C. 1988. Chemotransduction in the carotid body: Current modulated by PO_2 in type I chemoreceptor cells. *Science* 241: 580–582.

Nurse, C.A. 2005. Neurotransmission and neuromodulation in the chemosensory carotid body. *Auton Neurosci* 120: 1–9.

Pérez García, M.T., Obeso, A., López-López, J.R., Herreros, B., & Gonzalez, C. 1992. Characterization of chemoreceptor cells in primary culture isolated from adult rabbit carotid body. *Am J Physiol* 263: C1152–C1159.

Rigual, R., Gonzalez, E., Gonzalez, C., & Fidone, S. 1986. Synthesis and release of catecholamines by the cat carotid body in vitro: effects of hypoxic stimulation. *Brain Res* 374: 101–109.

Verna, A., Schamel, A., Le Moine, C., & Bloch, B. 1985. Localization of dopamine D_2 receptor mRNA in glomus cells of the rabbit carotid body by in situ hybridization. *J Neurocytol* 24: 265–270.

Vicario, I., Rigual, R., Obeso, A., & Gonzalez, C. 2000. Characterization of the synthesis and release of catecholamine in the rat carotid body in vitro. *Am J Physiol* 278: C490–C409.

Zapata, P. 1975. Effects of dopamine on carotid chemo and baroreceptors in vitro. *J Physiol* 244: 235–251.

Adenosine in Peripheral Chemoreception: New Insights into a Historically Overlooked Molecule – *Invited Article*

S.V. Conde, E.C. Monteiro, A. Obeso and C. Gonzalez

Abstract In the present article we review in a concise manner the literature on the general biology of adenosine signalling. In the first section we describe briefly the historical aspects of adenosine research. In the second section is presented the biochemical characteristics of this nucleoside, namely its metabolism and regulation, and its physiological actions. In the third section we have succinctly described the role of adenosine and its metabolism in hypoxia. The final section is devoted to the role of adenosine in chemoreception in the carotid body, providing a review of the literature on the presence of adenosine receptors in the carotid body; on the effects of adenosine at presynaptic level in carotid body chemoreceptor cells, as well as, its metabolism and regulation; and at postsynaptic level in carotid sinus nerve activity. Additionally, a review on the effects of adenosine in ventilation was done. This review discusses evidence for a key role of adenosine in the hypoxic response of carotid body and emphasizes new research likely to be important in the future.

Keywords Adenosine · ATP · cAMP · Carotid body · Adenosine transport · Adenosine receptors · Caffeine · Afferent discharge · Ventilation · Nucleotidase · Presynaptic · Postsynaptic.

1 Historical Aspects

Adenosine is a product of ATP metabolism, which can be recycled to resynthesise ATP itself. Adenosine is not stored or released as a classical neurotransmitter and is an ubiquitous substance, being released by almost all cells in response to hypoxia. In addition, adenosine is involved in key pathways such as puringeric nucleic acid base synthesis, amino acid metabolism and modulation of cellular metabolic status and in fact adenosine analogs are commercially available as anti-viral and anti-neoplasic drugs. The concept of adenosine as an extracellular mediator was first suggested

S.V. Conde (✉)
Department of Pharmacology, Faculty of Medical Sciences, New University of Lisbon,
Campo Mártires da Pátria, 130, 1169-056 Lisbon, Portugal
e-mail: silvia.conde@fcm.unl.pt

in 1929 by Drury and Szent-Györgyi (1929) where they observed that adenosine could affect neuronal function. These effects of adenosine have become more relevant after the work of Sattin and Rall (1970) on the effects of adenosine on cAMP production in cortical slices conjunctly with the findings that the stimulation of neuronal preparations induces the release of adenosine (Pull and McIlwain 1972) and that exogenously application of adenosine modulates neuromuscular transmission (Ginsborg and Hirst 1972). The first indication that ATP and adenosine exert their actions in different receptors come from the findings that ATP was more potent than AMP and adenosine in causing contraction in guinea-pig ileum and uterus (Gillespie 1934). Many decades later, in 1978, separate membrane receptors for adenosine and ATP were identified (Burnstock 1978) and at about the same time, two subtypes of adenosine receptors, A_1 and A_2, were recognized (Van Calker et al. 1979). New adenosine receptors subtypes were further identified and four subtypes, A_1, A_{2A}, A_{2B} and A_3 are known on the basis of their molecular structures showing different tissue localization and pharmacological profile (Fredholm et al. 1994).

2 Biochemical Characteristics of Adenosine and Adenosine Functions

2.1 Endogenous Ligand: Level, Metabolism and Regulation

Adenosine modulates the activity of several systems at presynaptic level (inhibiting or facilitating neurotransmitter release), at post-synaptic or at non-synaptic level (e.g. modulating blood flow or regulating of T cells function).

Extracellular adenosine comes from the extracellular production through ATP catabolism via 5′-ectonucleotidases, as well as by its intracellular production and release by nucleoside transport systems. Intracellular adenosine production from AMP is mediated by an intracellular 5′-nucleotidase or by the hydrolysis of S-adenosylhomocysteine (SAH) (Fig. 1) (Fredholm et al. 2001). Beyond ATP catabolism, another source of extracellular adenosine is cAMP that can be released into the extracellular media by a probenecid-sensitive transporter (Rosenberg et al. 1994) and converted by extracellular phosphodiesterases (PDE) in AMP and then by a 5′-ectonucleotidase into adenosine (see Fredholm et al. 2001).

Intra- and extra-cellular concentrations of adenosine are kept in equilibrium by means of nucleoside transporters. Two families constitute the nucleoside transport system: a Na^+ dependent nucleoside transport system which carriers nucleosides against a concentration gradient and the Na^+ independent family of nucleoside transporters, which is bi-directional, (equilibrative nucleoside transport system, ENT) formed of two different groups, differentiated by their sensitivity to nitrobenzylthioinosine (Griffith and Jarvis 1996; Podgorska et al. 2005). Two enzymes constitute the most important pathways of adenosine removal: adenosine kinase (AK) and adenosine deaminase (ADA). ADA is present mostly intracellularly, however it is also found in some extracellular compartments, being this enzyme

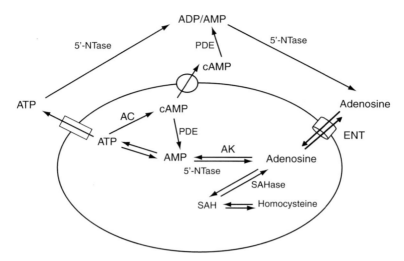

Fig. 1 Extra- and intracellular adenosine metabolism and the transporters that contribute to its release, uptake and production. 5′NTase, 5′-nucleotidase; AC, adenylyl cyclase; AK, adenosine kinase; ENT, equilibrative nucleoside transporter; PDE, phosphodiesterase; SAHase, S-adenosyl homocystein hydrolase

of particularly importance when the adenosine concentrations are high (Arch and Newsholme 1978).

As a whole, these enzymatic transport pathways allow the conversion of minor changes in intracellular ATP into disproportionably large changes in extracellular adenosine concentrations.

2.2 Physiological Actions of Adenosine

Adenosine receptors are G protein coupled receptors. A_1 receptors are coupled preferentially to $G_{i1/2/3}$, nevertheless they can also be coupled to G_o. A_{2A} and A_{2B} receptors preferentially activate G protein from the G_s family, but also can activate G_{olf} and G_q, respectively. A_3 receptors activate mainly $G_{i/o}$ proteins but also can activate G_q (Fredholm et al. 2001). After activation of G proteins, enzymes and ion channels are affected. A_1 receptors mediate the inhibition of adenylyl cyclase, the activation of several types of K^+ channels (Trussell and Jackson 1985), the inactivation of N, P and Q-type Ca^{2+} channels (Wu and Saggau 1994), the activation of phospholipase C, etc. (reviewed in Fredholm et al. 2001). A_3 receptors seem to mediate the same effectors than A_1 receptors. The majority of effects of A_{2A} and A_{2B} are due to the activation of adenylyl cyclase and the generation of cyclic AMP, but other independent cAMP actions, including mobilization of intracellular calcium, have also been described (Fredholm et al. 2001). The distinct adenosine receptors, A_1, A_{2A}, A_{2B} and A_3 are activated by different endogenous adenosine concentrations (affinity for adenosine: $A_1 > A_{2A} > A_{2B} > A_3$) (Fredholm et al. 1994). The disposable endogenous adenosine to activate these receptors equilibrates with the density of adenosine receptors at the site of action help to control the distinct physiological responses to this nucleotide.

Adenosine regulates many physiological processes, particularly in excitable tissues such as heart and brain. Many of the actions of adenosine either reduce the activity of excitable tissues (e.g. by slowing the heart rate) or increase the delivery of metabolic substrates (e.g. by inducing vasodilatation) and, thus, help to couple the rate of energy expenditure to the energy supply (see Dunwiddie 1999). Most often, this adenosine-mediated inhibition of cell metabolism is mimicked by A_1 receptor agonists, which leads to the idea that adenosine-induced inhibition of cell metabolism is mediated by A_1 adenosine receptors. The adenosine-mediated excitation is thought to be due to an action on A_2 adenosine receptors (reviewed in Cunha 2001). The A_3 adenosine receptor is the most enigmatic having 2 personalities often come into direct conflict, for example A_3 receptors can be neuroprotective but also can contribute to neurodegeneration (Gessi et al. 2008).

The "retaliatory" action of adenosine is particularly evident in pathological conditions associated to hypoxia/ischemia, being this nucleoside neuroprotective agent against hypoxic or ischemic events (see Latini and Pedata 2001).

3 Adenosine and Hypoxia

Hypoxia changes cellular metabolism and causes accumulation of extracellular adenosine (see Latini and Pedata 2001). The accumulation of adenosine does not require time and energy-consuming protein synthesis and therefore ensures a rapid response to hypoxia. This accumulation is in part justified by hypoxic-mediated regulation of enzymes that are involved in adenosine metabolism – adenosine kinase (Decking et al. 1997) and 5′-nucleotidase (Kobayashi et al. 2000a; Li et al. 2006). In fact, the release of adenosine evoked by hypoxia in CNS is completely derived from extracellular nucleotide degradation (Koos et al. 1997) and increased by EHNA, an inhibitor of adenosine deaminase (Zetterstrom et al. 1982). The majority of the protective effects of adenosine against hypoxia, like in ischemia-reperfusion injury in cardiac tissue and in neuronal tissue, are mediated by A_1 receptors. A_1 receptors acts in order to reduce neurotoxicity limiting Ca^{2+} entry and to reduce metabolic demand which would preserve ATP stores (Matherne et al. 1997). Alternatively, some of the protective effects could be mediated by other adenosine receptors. For example, through A_2 adenosine receptors, adenosine regulates several pathological processes related with systemic consequences of chronic hypoxia: vasodilation (Hinschen et al. 2003), inhibition of cardiac fibroblasts (Dubey et al. 2001) and vascular smooth muscle (Dubey et al. 2002) growth, stimulation of endothelial cells growth (Grant et al. 1999), angiogenesis (Feoktistov et al. 2002) and mast cell activation (Feoktistov & Biaggioni 1998). Additionally, it is known that HIF-1α contributes to the increase of adenosine extracellular concentrations during hypoxia by up-regulation of 5′-ectonucleotidase (Synnestvedt et al. 2002) and repression of equilibrative nucleoside transport (ENT) (Eltzschig et al. 2005) and that A_{2B} adenosine receptor promoters have a functional binding site for HIF (Kong et al. 2006), demonstrating that adenosine signalling is amplified by HIF-1α during hypoxia. Some evidence also support the inverse interaction: VEGF expression in macrophages is induced by adenosine A_{2A} receptor and involves the transcriptional regulation of the VEGF promoter by HIF-1α through the hypoxia response element; NECA strongly increase HIF-1α mRNA levels through an A_{2A} dependent manner (Ramanathan et al. 2007); and caffeine, an antagonist of adenosine receptors, inhibits adenosine induced HIF-1α accumulation in cancer cells, being the observed increase in HIF-1α and VEGF due to an activation of A_3 receptors (Merigui et al. 2007). These last data on interactions between HIF-1α and adenosine in macrophages, endothelial and cancer cells demonstrate that adenosine is involved in HIF-1α mediating hypoxic signalling.

4 Adenosine and Carotid Body

In 1981, McQueen and Ribeiro presented for the first time, in the British Pharmacological Society, that adenosine was capable of stimulating the carotid sinus nerve (CSN) chemosensory activity (McQueen and Ribeiro 1981). After that pioneer study, much experimental evidence has been obtained supporting the idea of an important role for adenosine in arterial chemoreception (see Gonzalez et al. 1994).

4.1 Adenosine in the Carotid Body: Metabolism and Regulation

Adenosine is released in normoxic conditions by the whole CB (Conde and Monteiro 2004), and its release increases in response to moderate hypoxia. The origin of this extracellular accumulation of adenosine caused by hypoxia was attributed to the CB chemoreceptor cells because moderate hypoxia is not a strong enough stimulus to induce the release of adenosine in CB related non-chemosensitive tissues (Conde and Monteiro 2004).

It is known that ATP is present in the CB in the rat (Buttigieg and Nurse 2004; Conde and Monteiro 2006a), in the cat (Acker and Starlinger 1984; Obeso et al. 1986) and in the rabbit (Verna et al. 1990) being released in response to acute hypoxia (Buttigieg and Nurse 2004; Conde and Monteiro 2006a). Approximately 40% of extracellular adenosine came from the extracellular catabolism of ATP, both under normoxic and moderate hypoxic conditions (Conde and Monteiro 2004). Nevertheless, moderate hypoxia also triggers adenosine efflux through NBTI-sensitive equilibrative nucleoside transport system (Conde and Monteiro 2004). Other source of extracellular adenosine could be cAMP. Although several studies have been produced on the role of cAMP in chemotransduction, the contribution of this cyclic nucleotide to adenosine production has never been investigated in the CB.

Indirect evidences of adenosine degradation by adenosine deaminase at the carotid body were produced: concentrations of adenosine released from CB are higher in the presence of the adenosine deaminase inhibitor, EHNA (Conde and

Monteiro 2002) and EHNA by itself mimics the excitatory effect of adenosine on ventilation mediated by CB activation (Monteiro and Ribeiro 1989).

4.2 Adenosine Receptors in the Carotid Body

The A_{2A} and A_{2B} receptors are the predominant receptor subtypes that have been localized in the rat carotid body. A_{2A} mRNA expression is developmentally regulated in the carotid body (Gauda et al. 2000) as it happens with tyrosine hydroxylase (TH) and dopamine D_2-receptor mRNAs (Gauda et al. 1996). Immunocytochemical studies have demonstrated the presence of A_{2A} receptors in all cells that express TH in the CB (Kobayashi et al. 2000b; Gauda et al. 2000). In the same manner rat chemoreceptor cells express A_{2B} receptors (Conde et al. 2006). It should be noted, however, that for every cell identified as a chemoreceptor by their positive staining to tyrosine hydroxylase, there are many other cells in the primary cultures of the CB that also express A_{2B} receptors (Conde et al. 2006). This evidence is in agreement with previous PCR studies of Kobayashi et al. (2000b) that show a high level of A_{2B} receptor expression in whole CB homogenates. The pharmacological decomposition of the effects of caffeine on the CSN chemosensory activity allowed to distinguish the A_{2A} and A_{2B} receptor contribution to this effect and demonstrated that apart from their localization on chemoreceptor cells A_{2A} receptors are also present post-synaptically on CSN (Conde et al. 2006).

The presence and function of A_1 receptors at the CB is more controversial and apparently species dependent. Rocher and co-workers demonstrated the presence of A_1 receptors in the rabbit CB chemoreceptor cells by showing that A_1 agonists and antagonists are capable of modulate Ca^{2+} currents in these cells (Rocher et al. 1999). However, A_1 receptors appear to be absent in rat CB (Kobayashi et al. 2000b; Gauda et al. 2000) being localized in petrosal ganglion neurons that also express TH mRNA (Gauda 2002).

Regarding A_3 adenosine receptors, Kobayashi et al. (2000b) have reported the absence of mRNA for this adenosine receptor in the rat CB.

4.3 Effects of Adenosine on Ventilation: The Role of the CB

Adenosine has an excitatory role on ventilation. In the rat, intracarotid injections of adenosine and its analogues increase in a dose-dependent manner ventilation, an effect abolished by CSN section and mediated by A_2 receptors (Monteiro and Ribeiro 1987). A_{2A} receptors seems to be responsible, at least in part, by the excitatory effect of adenosine in ventilation ($\approx 55\%$), since CGS21680, which is a selective A_{2A} agonist at low doses, increase respiration in rats by 31% (Fig. 2). This excitatory effect of adenosine on ventilation was also supported by the finding that in unanesthetized monkeys, caffeine, an unspecific adenosine receptor antagonist, significantly attenuated the CB-mediated hyperventilation that occurred while the animals are breathing 10% O_2 (Howell and Landrum 1995). In contrast, caffeine

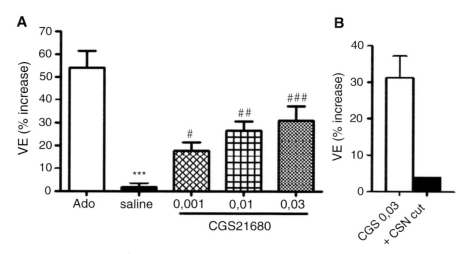

Fig. 2 Effects of adenosine (Ado, 100 nmol) and the A_{2A} agonist, CGS21680 (0.001; 0.01 and 0.03 nmol), on minute ventilation (VE) in rats before (A) and after (B) carotid sinus nerve (CSN) section. *Vertical bars* represent means ± SEM. ∗∗∗p< 0,001 adenosine (n=4) vs. saline (n=8); # p < 0,05; ## p<0,01; ### p<0,001 saline vs. ≠ CGS21680 concentrations (n=7); One-Way Anova

stimulates ventilation and decrease apneic episodes in premature infants (Aranda and Turmen 1979; Bairam et al. 1987) these results being explained by a dominance of central chemoreceptors in infants due to the immatureness of the CB in the early stage of life (see Gauda et al. 2004). The excitatory effect of exogenous adenosine observed in the rat in vivo was mimicked by drugs that increase levels of endogenous adenosine, EHNA and dipyridamole, inhibitors of adenosine deamination and uptake, respectively (Monteiro and Ribeiro 1989). The importance of these effects of adenosine on ventilation described in rats was strongly reinforced by the results obtained in humans. The adverse effects of intravenous administration of adenosine commercially available as anti arrhytmic are hyperventilation, dyspnoea and chest discomfort (Watt and Routledge 1985; Reid et al. 1987; Uematsu et al. 2000).

This effect of adenosine in man was attributed to the activation of CB chemoreceptors because: (1) adenosine, when applied intravascularly does not cross the blood brain barrier (Berne et al. 1974); (2) its effect on ventilation is proportionally higher the closer to the CB the bolus of adenosine is administered (Watt et al. 1987), and; (3) adenosine and its antagonists modify the hyperventilatory responses to hypoxia that is completely mediated by the CB (Maxwell et al. 1986, 1987).

4.4 Effects of Adenosine on Carotid Sinus Nerve Activity

Adenosine applied exogenously stimulates the CSN chemosensory activity in the cat both in vivo (McQueen and Ribeiro 1983) and in vitro (Runold et al. 1990), as well as in the rat (Vandier et al. 1999). The effect of adenosine on CSN chemosensory activity was mimicked by its analogues: NECA, L-PIA, D-PIA and CADO and inhibited by theophylline and 8-phenyltheophylline (8-PT), this being compatible with the involvement of an adenosine receptor of the A_2 subtype (McQueen and Ribeiro 1983, 1986). Additionally McQueen and Ribeiro (1986) also demonstrated that chemoexcitation produced by hypoxia (10% O_2, 4 min) was substantially reduced in the presence of 8-PT suggesting the involvement of adenosine in chemosensory excitation acting directly on nerve endings or as a modulator. Another finding that corroborates this suggestion is the fact that dypiridamole increases CSN discharge in normoxic conditions which suggests that increases in endogenous adenosine cause chemoexcitation (McQueen and Ribeiro 1983). Recently, Conde et al. (2006), demonstrates that caffeine, a non-selective adenosine receptor antagonist, inhibits in 55% the CSN chemosensory activity elicited by hypoxia this effect being due to a mixed action on A_{2B} presynaptic receptors (25%) and A_{2A} postsynaptic receptors (30%) (Conde et al. 2006). These last results concerning the effect of adenosine on A_{2A} postsynaptic receptors are in agreement with the excitatory effect of CGS21680 on ventilation in rats (Fig. 2). All these findings demonstrate that chemosensory activity in CSN elicited by hypoxia is partially controlled by adenosine but contrast with the results obtained by Bairam et al. (1997) in which they observe that caffeine did not modify CSN chemosensory activity in adult cats breathing 8% O_2.

4.5 Cellular Actions of Adenosine in the Carotid Body

A_{2A} and A_{2B} adenosine receptors are positively coupled to adenylyl cyclase and thereby their activation should modify cAMP levels, intracellular Ca^{2+} levels, depolarize cells, etc. The chemoexcitatory effect of adenosine in the CB appears to be mediated, at least in part, via modifications in cAMP levels (Fig. 3). Adenosine and its analogues increase cAMP content in the rat CB (Monteiro et al. 1996; Conde et al. 2008), and the increase in cAMP levels (e.g. by the administration of cAMP analogues resistant to enzymatic degradation) potentiates the response to hypoxia (Perez-Garcia et al. 1991). It is also known that dipyridamole, potentiates the augmentation of cAMP produced by hypoxia; this hypoxic response is blocked nearly completely by A_2 antagonists (Chen et al. 1997).In fact, A_{2B} receptor activation, are the responsible for the majority of cAMP increase in the CB (Conde et al. 2008). This increase in cAMP produced by adenosine could modulate K^+ channels, as dibutyryl cAMP, a cAMP analogue, applied exogenously to isolated rabbit chemoreceptor cells decreases the amplitude of 4-aminopyridine-sensitive K^+ current (Lopez-Lopez et al. 1993). However, the effect of cAMP on K^+ currents is not clarified since Hatton and Peers (1996) demonstrated that 8-bromo-cAMP and dibutyryl cAMP did not modify K^+ currents in isolated rat chemoreceptor cells. Still, it is known that adenosine, itself, decreases the amplitude of these K^+ currents in rat isolated chemoreceptor cells being this effect voltage independent and largely Ca^{2+} independent, although a small but significant component of the current was Ca^{2+} dependent (Vandier et al. 1999) (Fig. 3). Regarding the effects of adenosine

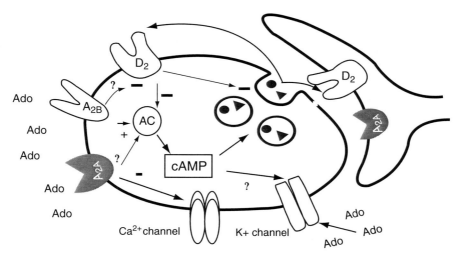

Fig. 3 Representation of some of adenosine cellular actions in rat carotid body. AC-adenylyl cyclase; Ado-adenosine. Adapted from Conde et al. 2008

in Ca^{2+} currents, it is known that adenosine inhibits L-type Ca^{2+} current, as well as, catecholamines release in rabbit carotid body chemoreceptor cells, an effect that has been attributed to A_1 receptors since agonists and antagonists of this receptor are capable of modulate Ca^{2+} currents (Rocher et al. 1999). Nevertheless, in the rat chemoreceptor cells adenosine stimulates catecholamines release through an interaction between A_{2B} and D_2 receptors (Conde et al. 2008) (Fig. 3). These results are in agreement with those obtained in the anaesthetized rat where adenosine injected into the carotid artery potentiates the inhibitory action of dopamine on ventilation (Monteiro and Ribeiro 2000). Additionally, it was observed that adenosine inhibits the voltage-dependent Ca^{2+} currents in rat chemoreceptor cells being this inhibition abolished by ZM241385 in a concentration that is specific for A_{2A} receptors and that this nucleoside did not modify $[Ca^{2+}]_i$ in cells exposed to normoxia but attenuating the increase in $[Ca^{2+}]_i$ produced by hypoxia (Kobayashi et al. 2000b). In contrast, Xu et al. (2006) have described that adenosine via A_{2A} receptors triggers a small increase in intracellular Ca^{2+} levels in rat chemoreceptor cells, but the increase in the intracellular Ca^{2+} observed by these authors seems to be insufficient to reach the threshold to evoke a release of neurotransmitters (Vicario et al. 2000).

It was suggested that the effects of adenosine on ventilation could be exclusively attributed to its action as a modulator of the release of the classical neurotransmitters, enhancing the release of what is accepted as a stimulatory neurotransmitter in the cat carotid body, ACh, and depressing the hypoxia-induced release of DA a transmitter which is accepted to be inhibitory (Fitzgerald et al. 2004). This is one possible interpretation particularly based on the work of Fitzgerald (Fitzgerald et al. 2004), however the findings in here described, as the presence of adenosine receptors post-synpatically and the fact that ACh via nicotinic receptors with α4 subunits modulates the release of adenosine and ATP from rat chemoreceptor cells (Conde and Monteiro 2006b) meaning that adenosine is also involved in the chemoexcitation produced by ACh in the carotid body lead us in the opposite direction. Any way, the excitatory role of adenosine in CB chemosensory activity is consensual and can be due to its direct action at postsynaptic sensory nerve endings as well as at presynaptic site in chemoreceptor cells.

4.6 Role of Adenosine in Chronic Hypoxia

Chronic hypoxia is extremely important to health having significant clinical implications. Chronic hypoxia has several consequences at cellular level (tumour cell growth, angiogenesis, remodelling) and at systemic level (hypertension), however most studies have been focus on their consequences at the cellular level. Pathologically, several cardiorespiratory diseases result in chronic sustained or intermittent hypoxia. These include e.g. sleep apnea, congestive heart failure, emphysema, chronic obstructive pulmonary disease and sudden infant death syndrome. It is known that in both chronic intermittent and sustained hypoxia CB sensitivity is altered (Peng and Prabhakar 2004; Powell 2007). A lot of research has been performed in order to investigate the neurotransmitter(s) involved in this altered CB sensitivity nevertheless, none of the neurotransmitters extensively studied seem to be the responsible for this altered sensitivity (Powell 2007). Adenosine A_{2A} receptors are expressed in rat CB chemoreceptor cells (Gauda et al. 2000) and are up-regulated by chronic sustained hypoxia in rat phaeochromocytoma (PC12) cells, an O_2-sensitive cell line (Kobayashi et al. 2000a). It has also been shown that chronic sustained hypoxia enhances adenosine release in PC12 cells by altering adenosine metabolism and membrane transport (Kobayashi et al. 2000a) suggesting that adenosine has a role in chronic sustained hypoxia adaptive responses. Nevertheless, information about the functional significance of adenosine in chronic hypoxia in the CB is lacking. After chronic exposure to 10% O_2 a moderate down-regulation of adenosine A_{2B} receptor was observed in mice CB (Ganfornina et al. 2005). Recently, it has been shown that 8-SPT, an antagonist of adenosine receptors, attenuates the increase in respiratory frequency elicited by acute hypoxic tests in rats exposed to an atmosphere of 12% O_2 for 7 days (Walsh and Marshall 2006). These authors suggested that the increased respiration observed during acute hypoxic tests in chronically hypoxic rats is due to the activation of A_2 receptors by stimulus-induced released adenosine. Therefore, it could be postulated that adenosine have a role in the increased sensitivity of CB during chronic hypoxia as well as in acclimatization however, more experiments related with this theme are necessary.

5 Summary

Despite the fact that the cellular mechanisms of adenosine action are not totally clarified, the adenosine net effect in the CB is excitatory and is clearly related with the response to acute hypoxia and probably with the increased CB sensitivity observed during chronic hypoxia. This effect of adenosine as an excitatory mediator in the CB could be due to a direct action on sensory nerve endings as well as to its actions as modulator on the release of other neurotransmitters.

Acknowledgments This work was financially supported by CEDOC/FCT of Portugal and by grants BFU2007-61848, CIBER CB06/06/0050 (FISS-ICiii) and by JCYL GR 242 of Spain.

References

Acker H. & Starlinger H. 1984, Adenosine triphosphate content in the cat carotid body under different arterial O_2 and CO_2 conditions. *Neurosci Lett* 50: 175–179.
Aranda J.V. & Turmen T. 1979, Metylxanthines in apnea of prematurity. *Clin Prerinatol* 6: 87–108.
Arch J.R. & Newsholme E.A. 1978, Activities and some properties of 5′-nucleotidase, adenosine kinase and adenosine deaminase in tissues from vertebrates and invertebrates in relation to the control of the concentration and the physiological role of adenosine. *Biochem J* 174: 965–977.
Bairam A., Boutroy M.J., Badonnel Y. & Vert P. 1987, Theophylline versus caffeine: comparative effects in treatment of idiophatic apnea in the preterm infant. *J Pediatr* 110: 636–639.
Bairam A., De Grandpré P., Dauphin C. & Marchal F. 1997, Effects of caffeine on carotid sinus nerve chemosensory discharge in kittens and cats. *J Appl Physiol* 82: 413–418.
Berne R.M., Rubio R. & Curnish R.R. 1974, Release of adenosine from ischaemic brain. Effect of cerebral vascular resistance and incorporation into cerebral adenine nucleotides. *Circ Res* 35: 262–271.
Burnstock G. 1978, A basis for distinguishing two types of purinergic receptor. In: *Cell Membrane Receptors for Drugs and Hormones: A Multidisciplinary Approach*. Eds. Straub R.W. & Bolis L., pp. 107–118. Raven Press, New York.
Buttigieg J. & Nurse C.A. 2004, Detection of hypoxia-evoked ATP release from chemoreceptor cells of the rat carotid body. *Biochem Biophy Res Com* 322: 82–87.
Chen J., Dinger B. & Fidone S.J. 1997, cAMP production in rabbit carotid body: role of adenosine. *J Appl Physiol* 82: 1771–1775.
Conde S.V., Gonzalez C., Batuca J.R., Monteiro E.C. & Obeso A. 2008, An antagonistic interaction between A2B adenosine and D2 dopamine receptors modulates the function of rat carotid body chemoreceptor cells. *J Neurochem* 107(5): 1369–1381.
Conde S.V. & Monteiro E.C. 2002, Importance of adenosine to the specificity of chemoreceptor cells response to hypoxia. *Eur J Neurosci* http://fens2002.bordeaux.inserm.fr/pages/posters/LetC.html
Conde S.V. & Monteiro E.C. 2004, Hypoxia induces the release of adenosine from the rat carotid body. *J Neurochem* 89: 1148–1156.
Conde S.V. & Monteiro E.C. 2006a, Profiles for ATP and adenosine release at the rat carotid body in response to O_2 concentrations. *Adv Exp Med Biol* 580: 179–184.
Conde S.V. & Monteiro E.C. 2006b, Activation of nicotinic ACh receptors with $\alpha 4$ subunits induces adenosine release at the rat carotid body. *Br J Pharmacol* 147: 783–789.
Conde S.V., Obeso A., Vicario I., Rigual R., Rocher A. & Gonzalez C. 2006, Caffeine inhibition of rat carotid body chemoreceptors is mediated by A2A and A2B adenosine receptors. *J Neurochem* 98: 616–628.
Cunha RA 2001, Adenosine as a neuromodulator and as a homeostatic regulator in the nervous system: different roles, different sources and different receptors. *Neurochem Int* 38: 107–125.
Decking U.K., Schlieper G., Kroll K. & Schrader J. 1997, Hypoxia-induced inhibition of adenosine kinase potentiates cardiac adenosine release. *Circ Res* 81: 154–164.

Drury A.N. & Szent-Györgyi A. 1929, The physiological activity of adenine compounds with especial reference to their action upon the mammalian heart. *J Physiol* 68: 213–237.
Dubey R.K., Gillespie D.G. & Jackson E.K. 2002, A2B adenosine receptors stimulate growth of porcine and rat arterial endothelial cells. *Hypertension* 39: 530–535.
Dubey R.K., Gillespie D.G., Zacharia L.C., Mi Z. & Jackson E.K. 2001, A(2b) receptors mediate the antimitogenic effects of adenosine in cardiac fibroblasts. *Hypertension* 37: 716–721.
Dunwiddie T.V. 1999, Adenosine and suppression of seizures. In: *Jasper's Basic Mechanisms of Epilepsies*. Eds. Delgado-Escuetta A., Wilson W.A., Olsen R.W., & Porter R.J., pp. 1001–1010. Lippicott Williams & Wilkins, Philadephia.
Eltzschig H.K., Abdulla P., Hoffman E., Hamilton K.E., Daniels D., Schonfeld C., Loffler M., Reyes G., Duszenko M., Karhausen J., Robinson A., Westerman K.A., Coe I.R. & Colgan S.P. (2005) HIF-1-dependent repression of equilibrative nucleoside transporter (ENT) in hypoxia. *J Exp Med* 202: 1493–1505.
Feoktistov I. & Biaggioni I. 1998, Pharmacological characterization of adenosine A2B receptors: studies in human mast cells co-expressing A2A and A2B adenosine receptor subtypes. *Biochem Pharmacol* 55: 627–633.
Feoktistov I., Goldstein A.E., Ryzhov S., Zeng D., Belardinelli L., Voyno-Yasenetskaya T. & Biaggioni I. 2002, Differential expression of adenosine receptors in human endothelial cells: role of A2B receptors in angiogenic factor regulation. *Circ Res* 90: 531–538.
Fitzgerald R.S., Shirahata M., Wang H.Y., Balbir A. & Chang I. 2004, The impact of adenosine on the release of acetylcholine, dopamine, and norepinephrine from the cat carotid body. *Neurosci Lett* 367: 304–308.
Fredholm B.B., Abbracchio M.P., Burnstock G., Daly J.W., Harden T.K., Jacobson K.A., Leff P. & Williams M. 1994, Nomenclature and classification of purinoreceptors. *Pharmacol Rev* 46: 143–156
Fredholm B.B., Ijzerman A.P., Jacobson K.A., Klotz K. & Linden J. 2001, Nomenclature and classification of adenosine receptors. *Pharmacol Rev* 53: 527–552.
Ganfornina M.D., Perez-Garcia M.T., Gutierrez G., Miguel-Velado E., Lopez-Lopez J.R., Marin A., Sanchez D. & Gonzalez C. 2005, Comparative gene expression profile of mouse carotid body and adrenal medulla under physiological hypoxia. *J Physiol* 566: 491–503.
Gauda E.B. 2002, Gene expression in peripheral arterial chemoreceptors. *Micro Res Tech* 59: 153–167.
Gauda E.B., Bamford O. & Gerfen C.R. 1996, Developmental expression of tyrosine hydroxylase, D2-dopamine receptor and substance P genes in the carotid body of the rat. *Neuroscience* 75: 969–977.
Gauda E.B., McLemore G.L., Tolosa J, Marston-Nelson J. & Kwak D. 2004, Maturation of pheripheral arterial chemoreceptors in relation to neonatal apnoea. *Semin Neonatol* 9: 181–194.
Gauda E.B., Northington F.J., Linden J. & Rosin D.L. 2000, Differential expression of A(2a), A(1)-adenosine and D(2)-dopamine receptor genes in rat peripheral arterial chemoreceptors during postnatal development. *Brain Res* 28: 1–10.
Gessi S., Merighi S., VaraniK., Leung E., Lennan S.M. & Borea P.A. 2008, The A3 adenosine receptor: An enigmatic player in cell biology. *Pharmacol & Therap* 117: 123–140.
Gillespie J.H. 1934, The biological significance of the linkages in adenosine triphosphoric acid. *J Physiol* 80: 345–359.
Ginsborg B.L. & Hirst G.D.S. 1972, The effect of adenosine on the release of the transmitter from the phrenic nerve of the rat. *J Physiol* 221:629–645.
Gonzalez, C., Almaraz, L., Obeso, A., & Rigual, R. 1994. Carotid body chemoreceptors: from natural stimuli to sensory discharges. Physiol Rev 74: 829–898
Grant M.B., Tarnuzzer R.W., Caballero S., Ozeck M.J., Davis M.I., Spoerri P.E., Feoktistov I., Biaggioni I., Shryock J.C. & Belardinelli L. (1999) Adenosine receptor activation induces vascular endothelial growth factor in human retinal endothelial cells. *Circ Res* 85: 699–706.

Griffith D.A. & Jarvis S.M. 1996, Nucleoside and nucleobase transport systems of mammalian cells. *Biochim Biophys Acta* 1286: 153–181.

Hatton C.J. & Peers C. 1996, Hypoxic inhibition of K$^+$ currents in isolated rat type I carotid body cells: evidence against the involvement of cyclic nucleotides. *Pflugers Arch* 433: 129–135.

Hinschen A.K., Rose'Meyer R.B. & Headrick J.P. 2003, Adenosine receptor subtypes mediating coronary vasodilation in rat hearts. *J Cardiovasc Pharmacol* 41: 73–80.

Howell L.L. & Landrum A.M. 1995, Attenuation of hypoxia-induced increases in ventilation by adenosine antagonists in rhesus monkeys. *Life Sci* 57: 773–783.

Kobayashi S., Conforti L. & Millhorn D.E. 2000b, Gene expression and function of A$_{2A}$ receptor in the rat carotid body. *Am J Physiol Lung Cell Mol Physiol* 279: L273–L282.

Kobayashi S., Zimmermann H. & Millhorn D. 2000a, Chronic hypoxia enhances adenosine release in rat PC12 cells by altering adenosine metabolism and membrane transport. *J Neurochem* 74: 621–632.

Kong T., Westerman K.A., Faigle M., Eltzschig H.K. & Colgan S.P. 2006, HIF-dependent induction of adenosine A2B receptor in hypoxia. *FASEB J* 20: 2242–2250.

Koos B.J., Kruger L. & Murray T.F. 1997, Source of extracellular brain adenosine during hypoxia in fetal sheep. *Brain Res* 778: 439–442.

Latini S. & Pedata F. 2001, Adenosine in the central nervous system: release mechanisms and extracellular concentrations. *J Neurochem* 79:463–484.

Li X, Zhou T, Zhi X, Zhao F, Yin L. & Zhou P. 2006, Effect of hypoxia/reoxygenation on CD73 (ecto-5'-nucleotidase) in mouse microvessel endothelial cell lines. *Microvasc Res.* 72: 48–53.

Lopez-Lopez J.R., De Luis D.A. & Gonzalez C. 1993, Properties of a transient K+ current in chemoreceptor cells of rabbit carotid body. *J Physiol* 460: 15–32.

Matherne G.P., Linden J, Byford A.M., Gauthier N.S. & Headrick J.P. 1997, Transgenic A1 adenosine receptor overexpression increases myocardiac resistence to ischemia. *Proc Natl Acad Sci USA* 94: 6541–6546.

Maxwell DL, Fuller RW, Nolop KB, Dixon CMS & Hughes MB 1986, Effects of adenosine on ventilatory responses to hypoxia and hypercapnia in humans, *J Appl Physiol* 61: 1762–1766.

Maxwell D.L., Fuller R.W., Conradson T-B., Dixon C.M.S., Aber V., Hughes M.B. & Barnes P.J. 1987, Contrasting effects of two xanthines, theophylline and enprofylline, on the cardiorespiratory stimulation of infused adenosine in man, *Acta Physiol Scan* 131: 459–465.

McQueen D.S. & Ribeiro J.A. 1981, Effect of adenosine on carotid chemoreceptor activity in the cat. *Br J Pharmacol* 74: 129–136.

McQueen D.S. & Ribeiro J.A. 1983, On the specificity and type of receptor involved in carotid body chemoreceptor activation by adenosine in the cat. *Br J Pharmacol* 80: 347–354.

McQueen D.S. & Ribeiro J.A. 1986, Pharmacological characterization of the receptor involved in chemoexcitation induced by adenosine. *Br J Pharmacol* 88: 615–620.

Merigui S., Benini A., Mirandola P., Gessi S., Varani K., Simioni C., Leung E., Maclennan S., Baraldi P.G. & Borea P.A. 2007, Caffeine inhibits adenosine-induced accumulation of hypoxia-inducible factor-1aplha, vascular endothelial growth factor, and interleukin-8 expression in hypoxic human colon cancer cells. *Mol Pharmacol* 72: 395–406.

Monteiro E.C. & Ribeiro J.A. 1987, Ventilatory effects of adenosine mediated by carotid body chemoreceptors in the rat. *Naunyn-Schmiedeberg's Arch Pharmacol* 335: 143–148.

Monteiro E.C. & Ribeiro J.A. 1989, Adenosine deaminase and adenosine uptake inhibitors facilitate ventilation in rats. *Naunyn-Schmiedeberg's Arch Pharmacol* 340: 230–238.

Monteiro E.C. & Ribeiro J.A. 2000, Adenosine-dopamine interactions and ventilation mediated through carotid body chemoreceptors. *Adv Exp Med Biol* 475: 671–684.

Monteiro E. C., Vera-Cruz P., Monteiro T. E. & Silva e Sousa M. A. 1996, Adenosine increases the cAMP content of the rat carotid body in vitro. *Adv Exp Med Biol* 410: 299–303.

Obeso A., Almaraz L. & Gonzalez C. 1986, Effects of 2-deoxy-D-glucose on in vitro cat carotid body. *Brain Res* 371: 25–36.

Peng Y.J. & Prabhakar N.R. 2004, Effect of two paradigms of chronic intermittent hypoxia on carotid body sensory activity. *J Appl Physiol* 96: 1236–1242.

Perez-Garcia M.T., Almaraz L. & Gonzalez C. 1991, Cyclic AMP modulates differentially the release of dopamine induced by hypoxia and other stimuli and increases dopamine synthesis in the rabbit carotid body. *J Neurochem* 57: 1992–2000.

Podgorska M., Kocbuch K. & Pawelczyk T. 2005, Recent advances in studies on biochemical and structural properties of equilibrative and concentrative nucleoside transporters. *Act Biochem Pol* 52: 749–758.

Powell F.L. 2007, The influence of chronic hypoxia upon chemoreception. *Respir Physiol Neurobiol* 157: 154–161.

Pull I. & McIlwain H. 1972, Adenine derivatives as neurohumoral agents in the brain. The quantities liberated on excitation of superfused cerebral tissues. *Biochem J.* 130: 975–981

Ramanathan M., Pinhal-Enfield G., Hao I. & Leibovich S.J. 2007, Synergistic Up-regulation of vascular endothelial growth factor (VEGF) expression in macrophages by adenosine A2A receptor agonists and endotoxin involves transcriptional regulation via the hypoxia response element in the VEGF promoter. *Mol Biol Cell* 18: 14–23.

Reid P.G., Watt A.H., Routledge P.A. & Peter Smith A. 1987, Intravenous infusion of adenosine but not inosine stimulates ventilation in man. *Br J Clin Pharmacol* 23: 331–338.

Rocher A., Gonzalez C. & Almaraz L. 1999, Adenosine inhibits L-type Ca^{2+} currents and catecholamine release in the rabbit carotid body chemoreceptor cells. *Eur J Neurosci* 11: 673–681.

Rosenberg P.A., Knowles R., Knowles K.P. & Li Y. 1994, Beta-adrenergic receptor-mediated regulation of extracellular adenosine in cerebral cortex in culture. *J Neurosci* 14: 2953–2965.

Runold M., Cherniak N.S. & Prabhakar N.R. 1990, Effect of adenosine on isolated and superfused cat carotid body activity. *Neurosci Lett* 113: 111–114.

Sattin A. & Rall T.W. 1970, The effect of adenosine and adenine nucleotides on the cyclic adenosine 3', 5'-phosphate content of guinea pig cerebral cortex slices. *Mol Pharmacol* 6: 13–23.

Synnestvedt K., Furuta G.T., Comerford K.M., Louis N., Karhausen J., Eltzschig H.K., Hansen K.R., Thompson L.F. & Colgan S.P. 2002, Ecto-5'-nucleotidase (CD73) regulation by hypoxia-inducible factor-1 mediates permeability changes in intestinal epithelia. *J Clin Invest* 110: 993–1002.

Trussell L.O. & Jackson M.B. 1985, Adenosine-activated potassium conductance in cultured striatal neurons. *Proc Natl Acad Sci USA*. 82: 4857–4861.

Uematsu T., Kozawa O., Matsuno H., Niwa M., Yoshikoshi H., Oh-uchi M., Cono K., Nagashima S. & Kanamaru M. 2000, Pharmacokinetics and tolerability of intravenous infusion of adenosine (SUNY4001) in healthy volunteers. *Br J Clin Pharmacol* 50: 177–181.

Vandier C., Conway A.F., Landauer R.C. & Kumar P. 1999, Presynaptic action of adenosine on a 4-aminopyridine-sensitive current in the rat carotid body. *J Physiol* 515: 419–429.

Van Calker D., Muller M. & Hamprecht B. 1979, Adenosine regulates the accumulation of cyclic AMP in cultured brain cells. *J Neurochem* 33: 999–1005.

Verna A., Talib N., Roumy M. & Pradet A. 1990, Effects of metabolic inhibitors and hypoxia on the ATP, ADP and AMP content of the rabbit carotid body in vitro: the metabolic hypothesis in question. *Neurosci Lett* 116: 156–161.

Vicario I., Obeso A., Rocher A., Lopez-Lopez J.R. & Gonzalez C. 2000, Intracellular Ca^{2+} stores in chemoreceptor cells of the rabbit carotid body: significance for chemoreception. *Am J Physiol Cell Physiol* 279: C51–C61.

Walsh M.P. & Marshall J.M. 2006, The role of adenosine in the early respiratory and cardiovascular changes evoked by chronic hypoxia in the rat. *J Physiol* 575: 277–289.

Watt A.H. & Routledge P.A. 1985, Adenosine stimulates respiration in man, *Br J Pharmacol* 20: 503–506.

Watt A.H., Reid P.G., Stephens M.R. & Routledge P.A. 1987, Adenosine-induced respiratory stimulation in man depends on site of infusion. Evidence for an action on the carotid body? *Br J Clin Pharmacol* 23: 486–490.

Wu L.G. & Saggau P. 1994, Adenosine inhibits evoked synaptic transmission primarily by reducing presynaptic calcium influx in area CA1 of hippocampus. *Neuron* 12: 1139–1148.

Xu F., Xu J., Tse F.W. & Tse A. 2006, Adenosine stimulates depolarization and rise in cytoplasmatic Ca^{2+} concentration in type I cells of rat carotid bodies. *Am J Physiol Cell Physiol* 290: C1592–C1598.

Zetterstrom T., Vernet L., Ungerstedt U., Tossman U., Jonzon B. & Fredholm B.B. 1982, Purine levels in the intact brain, studies with an implated perfused hollow fibre. *Neurosci Lett* 29: 111–115.

The A_{2B}-D_2 Receptor Interaction that Controls Carotid Body Catecholamines Release Locates Between the Last Two Steps of Hypoxic Transduction Cascade

S.V. Conde, A. Obeso, E.C. Monteiro and C. Gonzalez

Abstract We have recently demonstrated that adenosine controls the release of catecholamines (CA) from carotid body (CB) acting on A_{2B} receptors. Here, we have investigated the hypothesis that this control is exerted via an interaction between adenosine A_{2B} and dopamine D_2 receptors present in chemoreceptor cells and if it is, the location of this interaction on the CB hypoxic transduction cascade. Experiments were performed *in vitro* in CB from 3 months rats. The effect of adenosine A_{2B} and dopamine D_2 receptor agonists applied alone or conjunctly, was studied on the basal and evoked release (10% O_2 and ionomycin) of CA from CB. We have observed that the inhibitory action of propylnorapomorphine, a D_2 selective agonist, on the normoxic and 10%O_2-evoked release of CA was abolished by NECA, an A_2 agonist, meaning that an interaction between the D_2 and A_{2B} receptors controls the release of CA from CB. Further, propylnorapomorphine inhibits the release of CA evoked by ionomycin, being this effect totally reversed by NECA. The present results provide direct pharmacological evidence that A_{2B} and D_2 receptors interact to modulate the release of CA from rat CB between the steps of Ca^{2+} entry and increase in intracellular free Ca^{2+}, and the activation of exocytosis and neurotransmitter release, of the stimulus-secretion coupling process.

Keywords Adenosine · Dopamine · A2b adenosine receptors · D2 dopamine receptors · Dopamine agonists · Adenosine agonists · Dopamine antagonists · Adenosine antagonists

---

S.V. Conde (✉)
Departamento de Bioquímica y Biología Molecular y Fisiología, Universidad de Valladolid, Facultad de Medicina. Instituto de Biología y Genética Molecular, CSIC. Ciber de Enfermedades Respiratorias, CIBERES, Instituto de Salud Carlos III. 47005 Valladolid, Spain; Department of Pharmacology, Faculty of Medical Sciences, New University of Lisbon, Campo Mártires da Pátria, 130, 1169-056 Lisbon, Portugal
e-mail: silvia.conde@fcm.unl.pt

1 Introduction

Carotid bodies are major peripheral chemoreceptor organs sensing changes in blood O_2 responding by generating action potentials at the carotid sinus nerve (CSN), which are integrated in the brainstem to induce a hyperventilatory compensatory response. Hypoxia, the physiological CB stimulus increases the release of dopamine (DA) (Vicario et al. 2000) and adenosine (Conde and Monteiro 2004) from rat CB. Adenosine is an excitatory neurotransmitter at the CB, increasing CSN electrical activity and ventilation in several species (McQueen and Ribeiro 1983; Monteiro and Ribeiro 1987; Runold et al. 1990) including humans (Watt and Routledge 1985; Uematsu et al. 2000). Recently we have demonstrated that adenosine mediates up to 60% of the low PO_2-induced CSN activity (Conde et al. 2006). About half of this effect was postsynaptic and mediated by A_{2A} receptors and the rest was presynaptic mediated by A_{2B} receptors and associated to an augmentation of the release CA from chemoreceptor cells (Conde et al. 2006).

Ribeiro and McQueen (1983) in their pioneer study for the effects of adenosine in chemoreception found that adenosine potentiated the inhibitory effect of DA on cat CSN activity. Similarly, Monteiro and Ribeiro (2000) in in vivo experiments in the rat observed that the inhibitory effect of D_2 dopamine agonists on ventilation was attenuated or even reversed by the blockage of A_2 receptors. We postulated that this modulatory effect could occur due to an interaction between adenosine A_{2B} and dopamine D_2 receptors present in glomus cells. Interactions between A_{2A}-D_2 receptors are well described in the central nervous system (Ferré et al. 1997; Fuxe et al. 2007) and these findings prompted us to investigate if interactions between A_{2B} and D_2 receptors could be involved in the modulation by adenosine of CA release in the CB.

We have previously suggested that A_{2B} receptors operate at the CB in the last steps of the stimulus-secretion coupling process, since caffeine inhibits identically the release of CA from CB in response to high K^+ (a non-specific depolarising stimulus) as in response to hypoxia (Conde et al. 2006). These results lead us to investigate with more detail the location of the interaction between D_2 and A_{2B} receptors in the CB hypoxic transduction cascade.

2 Methods

2.1 Animals and Surgical Procedures

Experiments were performed in Wistar adult rats of both sexes (250–350 g) obtained from *vivarium* of the Faculty of Medicine of the University of Valladolid. The Institutional Committee of the University of Valladolid for Animal Care and Use approved the protocols. Rats were anaesthetized with sodium pentobarbital (60 mg/kg i.p.), tracheostomized and the carotid arteries were dissected past the carotid bifurcation. The carotid bifurcation was placed in a Lucite chamber in ice-cold 95% O_2-equilibrated Tyrode (in mM: NaCl 140; KCl 5; $CaCl_2$ 2; $MgCl_2$ 1.1;

HEPES 10; glucose 5.5, pH 7.40 and the CB was cleaned free of CSN and nearby connective tissue (Vicario et al. 2000). In all instances animals were killed by an intracardiac overdose of sodium pentobarbital.

2.2 Labelling of Catecholamines Stores: Release of ^{3}H-CA

In brief, to label CA stores of the chemoreceptor cells the CBs (8-12/experiment) were incubated during 2 hours in Tyrode solution containing the natural precursor ^{3}H-tyrosine (30 μM) (specific activity of 45 Ci/mmol) and 6-methyltetrahydropterine (100 μM) and ascorbic acid (1 mM), cofactors of tyrosine hydroxylase and dopamine-β-hydroxylase, respectively. After the labelling period, individual CBs were transferred to vials containing 4 ml of precursor-free Tyrode-bicarbonate solution (composition as above except for the substitution of 24 mM of NaCl by 24 mM of NaHCO$_3$). Solutions were continuously bubbled with 20%O$_2$/5%CO$_2$/75%N$_2$ saturated with water vapour, except when applied hypoxic stimuli. The solutions of the initial incubation periods (3×20 min) were discarded to washout the precursor and the readily releasable pool of labelled CA and thereafter renewed at fixed times (every 10 min) and collected for subsequent analysis of their ^{3}H-CA content. Specific protocols for stimulus and drug applications are provided in the Results. Stimulus included hypoxia 10% O$_2$-equilibrated solutions (PO$_2$ ≈ 66 mmHg) and ionomycin. The drugs used were: 5′-(N-ethylcarboxamido)adenosine (NECA, an adenosine A$_2$ agonist; 10 and 100 μM), propylnorapomorphine (a D$_2$ dopamine selective agonist, 200 nM) and ionomycin (an ionophore, 5 μM). Analysis of ^{3}H-CA CB release and content was performed as previously described by Vicario et al. (2000). Release data are presented as percentage of ^{3}H-CA of the tissue content; this presentation corrects for the potential variations in CB size and/or rate of ^{3}H-CA synthesis.

Data were evaluated using a Graph Pad Prism Software, version 4 and were presented as mean ± SEM. The significance of the differences between the means was calculated by One-Way ANOVA with Dunnett's multiple comparison test. P values of 0.05 or less were considered to represent significant differences.

3 Results

3.1 Propylnorapomorphine, a D$_2$ Dopamine Agonist, Inhibits the Basal and Evoked Release of ^{3}H-CA from Carotid Body and its Reversal by NECA

After the initial washout period (see Methods) both basal-normoxic (Fig. 1A) and 10% O$_2$ (Fig. 1B) release of ^{3}H-CA follows a monotonic decay. Figure 1C and D show that addition of propylnorapomorphine (200 nM), a D$_2$ selective agonist, to the incubating solutions produced a marked inhibition of the release of ^{3}H-CA both

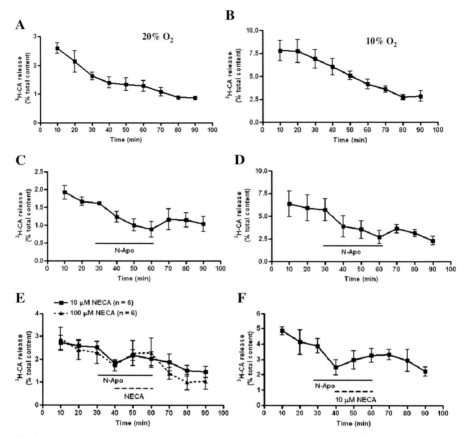

Fig. 1 Effect of propylnorapomorphine (N-Apo, 200 nM), a D_2 dopamine agonist, on the release of ^{3}H-CA from the carotid body in normoxic conditions (20% O_2) and in response to moderate hypoxia (10% O_2) in the absence (**C** and **D**) and in the presence of NECA (**E** and **F**). (**A** and **B**) Time courses for the release of ^{3}H-CA from CB in normoxia and moderate hypoxia, respectively. (**C** and **D**) Effect of 200 nM of N-Apo on the release of ^{3}H-CA from CB in normoxia and moderate hypoxia, respectively. (**E** and **F**) shows the reversal of the inhibitory effect of N-Apo on the release of ^{3}H-CA from CB in normoxia and moderate hypoxia, respectively by NECA. Note that the partial reversion of inhibitory effect of N-Apo on ^{3}H-CA release obtained with 10 μM of NECA is totally abolished with 100 μM of NECA Propylnorapomorphine was applied for 30 min between 30 and 60 min and NECA for 20 min between 40 and 60 min of the experimental protocol. Values represent means ± S.E.M

in normoxia and moderate hypoxia conditions, respectively, that recovered after the elimination of the drug.

In order to investigate if an interaction exists between A_{2B} and D_2 receptors that modulate the release of ^{3}H-CA from CB we tested NECA on the effect of propylnorapomorphine, on the release of this neurotransmitter from rat CB in normoxia and moderate hypoxia. We have observed that NECA, an A_2 agonist, when applied in a 10 μM concentration attenuated the effect of the D_2 agonist on the release of ^{3}H-CA from CB in normoxia (Fig. 1E) and in moderate hypoxia (Fig. 1F).

We have observed that NECA when applied in a 10 μM concentration attenuated the effect of the D_2 agonist on the release of ^{3}H-CA from CB in normoxia (Fig. 1E) and in moderate hypoxia (Fig. 1F). It can also be observed that the inhibitory effect of the D_2 agonist on the basal release of ^{3}H-CA is totally abolished if the concentration of NECA is increased by 10x (100 μM). These results suggest that an interaction between A_{2B} and D_2 receptors modulate the release of ^{3}H-CA from CB chemoreceptor cells both in normoxic and moderate hypoxic conditions.

3.2 Effect of Ionomycin on the Release of ^{3}H-CA by Carotid Body Modified by Propylnorapomorphine and NECA

We have previously suggested that A_{2B} receptors operate at the CB in the last steps of the stimulus-secretion coupling process, since caffeine produces identical effects in the response elicited by high K^+ and hypoxia (Conde et al. 2006). In order to investigate at which step of the stimulus-secretion process the interaction between A_{2B} and D_2 are involved, we studied this interaction on the release of ^{3}H-CA evoked by ionomycin. Ionomycin alters cell permeability producing an increase in intracellular Ca^{2+} and, as a consequence, induce a Ca^{2+} dependent vesicular release of CA (Dedkova et al. 2000). This effect can be observed both in Fig. 2A on the time course

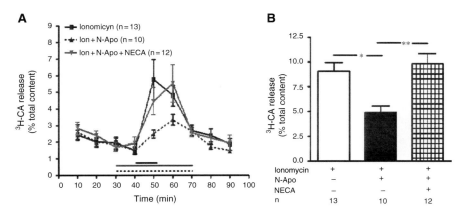

Fig. 2 Inhibitory effect of propylnorapomorphine, a D_2 agonist, on the ionomycin evoked release of ^{3}H-CA from CB and reversion of the inhibitory effect by NECA. Panel **A** shows the time course for the effects of 200 nM propylnorapomorphine (N-Apo) and the reversion of this effect by 10 μM of NECA on the release of ^{3}H-CA evoked by 5 μM ionomycin. Ionomycin was applied for 10 min between 40 and 50 min of the experiment. N-Apo and NECA were applied 10 min prior and 20 min after stimulation with ionomycin. 0% corresponds to 1.946 ± 0.36, 1.49 ± 0.235 and 1.914 ± 0.36 % of ^{3}H-CA total content respectively for ionomycin in the absence and in the presence of N-APO and N-APO plus NECA. Panel **B** shows means ± S.E.M. of the effect of N-Apo in the absence and in the presence of NECA on ionomycin evoked release of ^{3}H-CA from CB. *$P<0.05$, **$P<0.01$; One-Way ANOVA with Dunnett's multi-comparison test

of the ^{3}H-CA release as in the magnitude of the increase, in Fig. 2B, where the application of 5 µM of ionomycin increased the release of ^{3}H-CA from rat CB by 8.78 ± 1.10%. The application of 200 nM of propylnorapomorphine decreased the ^{3}H-CA release from CB evoked by ionomycin by 44% (Fig. 2B). When we tested the effect of NECA on the inhibitory effect of propylnorapomorphine on ^{3}H-CA release evoked by ionomycin we observed that the application of 10 µM of NECA completely reversed the inhibition (Fig. 2A and B). These results suggest that A_{2B} and D_2 receptors interact to modulate the release of ^{3}H-CA from rat CB between the steps of Ca^{2+} entry and increase in intracellular free Ca^{2+}, and the activation of exocytosis and neurotransmitter release, of the stimulus-secretion coupling process.

4 Discussion

Using a pharmacological approach we have demonstrated the existence of an interaction between A_{2B} and D_2 receptors that modulate the release of CA from rat CB, being located between the last two steps of hypoxic transduction cascade, that is, between the steps of Ca^{2+} entry and increase in intracellular free Ca^{2+}, and the activation of exocytosis and neurotransmitter release.

At the onset of the discussion we want to state that we have attributed all the effects of NECA to an effect on A_{2B} receptors, because we have previously observed that NECA increased the release of CA both in normoxia and moderate hypoxia, an effect that is not mimicked by A_1 or A_{2A} agonists (Conde et al. 2006).

The nature of D_2-A_{2B} receptor effects on chemoreceptor cells should be defined as an antagonistic, because the effects of activation of D_2 receptors with propylnorapomorphine diminished with increasing concentrations of NECA, which activates A_{2B} receptors. With the results herein described and the protocols used in this work we cannot establish if the interaction is a functional coupling between A_{2B} and D_2 receptors, or at adenylyl cyclase level, or at both sites. Nevertheless, we can suggest that a functional coupling between A_{2B} and D_2 receptors in chemoreceptor cells could exist, as described in the central nervous system (mostly in the striatum) between D_2 and A_{2A} receptors (Ferré et al. 1997; Fuxe et al. 2007).

The interactions previously described between dopamine and adenosine in the CB observed when monitoring CSN activity in the cat (Ribeiro and McQueen 1983) and ventilation in the rat (Monteiro and Ribeiro 2000) have defined D_2 as the dopamine receptor involved, but could not identify the subtype of A_2 adenosine receptor participating in the interaction. Our data indicate that the subtype of adenosine receptor is the A_{2B} adenosine receptor. We cannot exclude the possibility that interactions between D_2 and A_{2A} receptors also exist in the CB. Nevertheless, the interaction herein observed is in agreement with the potentiation of the inhibitory effect of dopamine by adenosine observed in ventilation in the rat (Monteiro and Ribeiro 2000).

A final consideration relates to the fact that D_2 interactions, particularly in the central nervous system, are most generally described with A_{2A} receptors. Nevertheless, the interaction between catecholamine autoreceptors and A_{2B} receptors is not unique to chemoreceptor cells. For example, Talaia et al. (2005) found that adenosine acting via A_{2B} receptors heightens the release of noradrenalin from the sympathetic endings innervating the vas deferens, this effect being annuled by yohimbine, a blocker of presynaptic α2-adrenoreceptors.

Ionomycin is a calcium ionophore, and it is assumed that these substances directly facilitate the transport of Ca^{2+} across the plasma membrane (Himmel et al. 1990; Dedkova et al. 2000). It is generally accepted that in chemoreceptor cells the stimulus-secretion coupling in response to hypoxic stimulus occurs according to following steps (Gonzalez et al. 1992, 1994): (1) O_2-sensing at an O_2-sensor → (2) activation of coupling mechanisms with K^+ channels → (3) change in kinetics of these K^+ channels resulting in a decrease in their opening probability → (4) cell depolarization → (5) activation voltage operated channels → (6) Ca^{2+} entry and increase in intracellular free Ca^{2+} → (7) activation of exocytosis and neurotransmitter release. The Ca^{2+} signal induced by ionomycin that produces an increase in Ca^{2+} intracellular concentrations, originating the release of neurotransmitters eliminates an effect of the interaction between D_2 and A_{2B} receptors on the entire hypoxic transduction cascade until step 6 in our study. The fact that NECA abolishes the inhibitory effect of propylnorapomorphine on the ionomycin evoked release of ^{3}H-CA supports the above suggestion and indicates that the interaction between the A_{2B} and D_2 receptors present in CB chemoreceptor cells modulating CA release is between the step 6 and step 7 of the stimulus-secretion coupling process. Nevertheless, clarification of the intimate mechanisms of interaction and action on the exocytotic process would require additional experiments.

To summarise, we can assume that an antagonistic interaction between A_{2B} and D_2 receptors exists in the CB that modulates the release of CA from chemoreceptor cells being located between the two last steps of hypoxic stimulus-secretion coupling.

Acknowledgments We want to thank Mª Llanos Bravo for technical assistance. The work was supported by CEDOC/FCT (Portugal) and by grants BFU2007-61848, CIBER CB06/06/0050 (FISS-ICiii) and by JCYL GR 242 of Spain.

References

Conde S.V. & Monteiro E.C. 2004, Hypoxia induces the release of adenosine from the rat carotid body. *J Neurochem*, 89: 1148–1156.
Conde S.V., Obeso A., Vicario I., Rigual R., Rocher A., & Gonzalez C. 2006, Caffeine inhibition of rat carotid body chemoreceptors is mediated by A2A and A2B adenosine receptors. *J Neurochem*, 98: 616–628.
Dedkova E.N., Sigova A.A., & Zinchenko V.P. 2000, Mechanism of action of calcium ionophores on intact cells: ionophore resistant cells. *Membr Cell Biol*, 13: 357–368.
Ferré S., Fredholm B.B., Morelli M., Popoli P., & Fuxe K. 1997, Adenosine-dopamine receptor-receptor interactions as an integrative mechanism in the basal ganglia. *Trends Neurosci*, 20: 482–487.
Fuxe K., Canals M., Torvinen M., Marcellino D., Terasmaa A., Genedani S., Leo G., Guidolin D., Diaz-Cabiale Z., Rivera A., Lundstrom L., Langel U., Narvaez J., Tanganelli S., Lluis C.,

Ferré S., Woods A., Franco R., & Agnati L.F. 2007, Intramembrane receptor–receptor interactions: a novel principle in molecular medicine. *J Neural Transm*, 114: 49–75.

Gonzalez C., Almaraz L., Obeso A., & Rigual R. 1992, Oxygen and acid chemoreception in the carotid body chemoreceptors. *Trends Neurosci*, 15: 146–153.

Gonzalez, C., Almaraz, L., Obeso, A., & Rigual, R. 1994, Carotid body chemoreceptors: from natural stimuli to sensory discharges. *Physiol Rev*, 74: 829–898.

Himmel H.M., Riehle R., Sticker K. & Siess M. 1990, Effects of the divalent cation ionophore ionomycin on the performance of isolated guinea-pig atria. *Basic Res Cardiol*, 85: 247–256.

McQueen D.S. 1983, Pharmacological aspects of putative transmitters in the carotid body. In *Physiology of the Peripheral Arterial Chemoreceptors*, eds. Acker, H. & O'Regan, R.G., Amsterdam, Elsevier Science, pp. 149–155.

Monteiro E.C. & Ribeiro J.A. 1987, Ventilatory effects of adenosine mediated by carotid body chemoreceptors in the rat. *Naunyn-Schmiedeberg's Arch Pharmacol*, 335: 143–148.

Monteiro E.C. & Ribeiro J.A. 2000, Adenosine-dopamine interactions and ventilation mediated through carotid body chemoreceptors. *Adv Exp Med Biol*, 475: 671–684.

Runold M., Cherniak N.S. & Prabhakar N.R. 1990, Effect of adenosine on isolated and superfused cat carotid body activity. *Neurosci Lett*, 113: 111–114.

Talaia C., Queiroz G., Quintas C. & Gonçalves J. 2005, Interaction between A2B-receptors and alpha2-adrenoreceptors on the modulation of noradrenaline release in the rat vas deferens: possible involvement of a group 2 adenylyl cyclase isoform. *Neurochem Int*, 47: 418–429.

Uematsu T., Kozawa O., Matsuno H., Niwa M., Yoshikoshi H., Oh-uchi M., Cono K., Nagashima S., & Kanamaru M. 2000, Pharmacokinetics and tolerability of intravenous infusion of adenosine (SUNY4001) in healthy volunteers. *Br J Clin Pharmacol*, 50: 177–181.

Vicario I., Rigual R., Obeso A., & Gonzalez C. 2000, Characterization of the synthesis and release of catecholamine in the rat carotid body in vitro. *Am. J.Physiol. Cell Physiol.*, 278: C490–C499.

Watt A.H. & Routledge P.A. 1985, Adenosine stimulates respiration in man, *Br J Pharmacol*, 20: 503–506.

Benzodiazepines and GABA-GABA$_A$ Receptor System in the Cat Carotid Body

A. Igarashi, N. Zadzilka and M. Shirahata

Abstract Benzodiazepines (BZs) suppress ventilation possibly by augmenting the GABA$_A$ receptor activity in the respiratory control system, but precise sites of action are not well understood. The goals of this study were: (1) to identify GABA$_A$ receptor subunits in the carotid body (CB) and petrosal ganglion (PG); (2) to test if BZs exert their effects through the GABA$_A$ receptor in the CB chemosensory unit. Tissues were taken from euthanized adult cats. RNA was extracted from the brain, and cDNA sequences of several GABA$_A$ receptor subunits were determined. Subsequent RT-PCR analysis demonstrated the gene expression of α2, α3, β3, and γ2 subunits in the CB and the PG. Immunoreactivity for GABA and for GABA$_A$ receptor β3 and γ2 subunits was detected in chemosensory glomus cells (GCs) in the CB and neurons in the PG. The functional aspects of the GABA-GABA$_A$ receptor system in the CB was studied by measuring CB neural output using in vitro perfusion setup. Two BZs, midazolam and diazepam, decreased the CB neural response to hypoxia. With continuous application of bicuculline, a GABA$_A$ receptor antagonist, the effects of BZs were abolished. In conclusion, the GABA-GABA$_A$ receptor system is functioning in the CB chemosensory system. BZs inhibit CB neural response to hypoxia by enhancing GABA$_A$ receptor activity.

Keywords Hypoxia · Carotid body · GABA · GABA$_A$ receptor · Benzodiazepine · Bicuculline · Immunohistochemistry · RT-PCR · Diazepam

1 Introduction

BZs are widely prescribed as sleep medicines and as sedatives. It has been known that sedative doses of BZs impair the ventilatory response to hypoxia significantly (Alexander and Gross 1988). The decreased respiratory response to hypoxia had

A. Igarashi (✉)
Division of Physiology, Department of Environmental Health Sciences, The Johns Hopkins University, Yamagata Prefectural Shinjo Hospital, Department of Anesthesiology
e-mail: igarashi-ayu@umin.ac.jp

been long considered due to the suppression of the respiratory center in the central nervous system. However, our preliminary study (Igarashi and Shirahata 1997) and a recent study (Kim et al. 2006) suggest that BZs also reduced the CB neural response to hypoxia.

BZs bind to the BZs recognition site on $GABA_A$ receptors, and potentiate the action of GABA, the natural agonist of $GABA_A$ receptors. Without GABA, BZs have no effect on the $GABA_A$ receptor. Thus we hypothesized that GABA and $GABA_A$ receptors exist in the carotid chemoreceptor system and mediate the effect of BZs. GABA and $GABA_B$ receptors were previously identified in mouse and rat GCs (Oomori et al. 1994, Fearon et al. 2003). A possible presence of $GABA_A$ receptors in the cat CB was suggested, but its functional role remained unclear (Pokorski et al. 1994). The aims of our study were to identify the $GABA_A$ receptors in the CB chemosensory unit, and to investigate modulation of the GABA-$GABA_A$ receptor system by BZs.

2 Methods

Adult cats were euthanized with overdose of ketamine and pentobarbital, decapitated, and CBs, PGs and cerebral cortex were excised. The tissues were immediately immersed in the TRIzol (Invitrogen) for RT-PCR studies; zinc fixative (Beckstead 1994) for immunohistochemistry. For nerve recording experiments the carotid bifurcations were immersed in ice-cold Leibovitz's medium (L15, Sigma Chemical Co.) containing bicarbonate (20 mM; equilibrated with 5% CO_2/air). All animal experiments were carried out in accordance of the National Institute of Health Guide for the care and Use of Laboratory Animals. The protocols were approved by the Animal Care and Use Committee of the Johns Hopkins University.

2.1 RT-PCR Analysis

We determined the cat DNA sequences for $GABA_A$ receptors. Based on our data in the mouse (Fujii et al. 2008), our targets were $\alpha 2$, $\alpha 3$, $\beta 3$ and $\gamma 2$ subunits. The tissue preparation was described before (Shirahata et al. 2004). Primers for PCR reactions included coding regions of these subunits in the human. The PCR products were separated by gel electrophoresis and the appropriate bands were excised. DNA was purified, and sequenced at the DNA Analyzing Facility at Johns Hopkins University.

To determine the expression of these genes in the cat CBs and PGs, cDNAs were prepared as described before (Shirahata et al. 2004). PCR was performed with the primers designed based on the cat DNA sequences for $GABA_A$ receptor subunits, and gel electrophoresis was performed.

2.2 Immunohistochemistry

Immunohistochemistry in the paraffin embedded tissue was performed as previously described (Yamaguchi et al. 2003) with a slight modification. The primary antibodies were: anti-GABA, 1/300; anti-GABA$_A$ receptor β3 subunit, 1/50, anti- GABA$_A$ receptor γ2 subunit, 1/100 (Abcam). Each antibody was visualized by fluorescein-conjugated anti-rabbit IgG (1/100, Vector). As negative control for the primary antibodies, rabbit primary antibody isotype control (Zymed) was used at appropriate concentrations.

2.3 Carotid Sinus Nerve (CSN) Recording

Preparation of the nerve recording was similar to our previous study (Igarashi et al. 2002). In short, branches of the common carotid artery were ligated except the feeding artery of the CB. The CB was perfused via the common carotid artery with Krebs solution (Krebs; 35 to 36°C) using a peristaltic pump (3–4 ml/min; mean perfusion pressure≅100 mmHg). Krebs was equilibrated in reservoirs with either 5%CO_2/air (normoxia) or 5%CO_2/95%N_2 (hypoxia). PO_2, PCO_2, and the perfusion pressure of the perfusate were continuously monitored. Desheathed CSN was placed on bipolar silver-silver chloride electrodes and lift into mineral oil which covered the surface of the Krebs. The signals were preamplified (Grass P15D), monitored with an oscilloscope (Tektronix 5113), led to a 60 Hz notch filter, amplified, and fed to an integrator (Grass 7P3). All these parameters were recorded continuously with a pen-recorder (Grass 7D).

Midazolam and diazepam (Sigma Chemical Co.) were initially dissolved with dimethylsulfoxide and diluted with Krebs (10, 20 μM for diazepam; 3.3 μM for midazolam). Bicuculline methiodide (Research Biochemical International) was prepared with ethanol, diluted with distilled water (1 M), and further diluted in Krebs at 10 μM.

A control CB neural response to hypoxia was obtained by switching the Krebs from normoxia to hypoxia for 4 minutes. Subsequently, normoxic Krebs containing midazolam (3.3 μM) or diazepam (10 or 20 μM) was administered to the CB for 15 minutes, then the perfusate was switched to hypoxic Krebs containing a BZ for 4 minutes. To examine the effect of bicuculline, after obtaining control neural response to hypoxia and 10 minutes normoxic perfusion with bicuculline 10 μM, hypoxic response of CSN with bicuculline was obtained. Subsequently, the CB was perfused with Krebs containing midazolam 3.3 μM or diazepam 20 μM in the presence of bicuculline for 10 minutes, and the hypoxic neural response was obtained in a similar manner as described above. Between each hypoxic challenge, normoxic perfusion was resumed for 10 minutes.

The difference between the CSN activity during the normoxia and the peak CSN activity during hypoxia was defined as the CB neural response to hypoxia (ΔCSNA).

All values were expressed as means ± SEM, and analyzed with Wilcoxon's signed rank test or one way ANOVA. $p < 0.05$ was considered statistically significant.

3 Result

3.1 RT-PCR Analysis

The nucleotide sequences of the cat $GABA_A$ receptor α2, α3, β3, and γ2 subunits were more than 93% similar to those of human (gene bank accession number: EU543280, EU580545, EU553272, EU545645, respectively). The CB expressed $GABA_A$ receptor α2, α3 and β3 subunits genes. Gamma2 subunit gene was only weakly detected. PG expressed all genes tested, but α3 and β3 subunits appeared to be dominant (Fig. 1A).

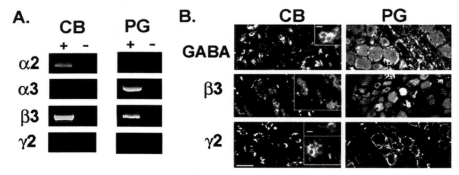

Fig. 1 Gene (A) and protein (B) expression of $GABA_A$ receptor subunits in the CB and the PG. (A) Gel electrophoresis showing bands of PCR products in which cDNAs were used as templates (+). In negative control (−), RNA samples were used as templates. (B) Immunoreactivity for GABA, $GABA_A$ receptor β3 subunits, $GABA_A$ receptor γ2 subunits. Photomicrographs of higher magnifications are presented in insets showing immunostained glomus cells (bar in the inset = 10 μm). Negative control did not show any staining (data not shown). Bar = 50 μm

3.2 Immunohistochemistry

The immunoreactivity for GABA, $GABA_A$ receptor β3 and γ2 subunits was mostly distributed in the cytoplasm of GCs. The immunoreactivity for GABA was observed both in the cell bodies and the nerve fibers in the PG. Both $GABA_A$ receptor β3 and γ2 immunoreactivity was found in the cell bodies (Fig. 1B).

3.3 CSN Recording

Midazolam (3.3 μM) or diazepam (10–20 μM) did not change the baseline CSN activity. Midazolam (3.3 μM) and diazepam (10 μM) siginificantly decreased

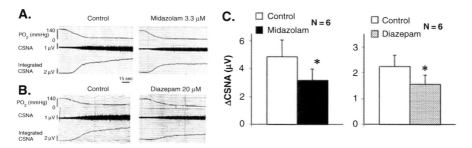

Fig. 2 The CSN responses to hypoxic Krebs perfusion before and with BZs. CSNA = carotid sinus nerve activity. (**A**) The effect of midazolam 3.3 μM. (**B**) Hypoxic responses of a carotid body before and with diazepam 20 μM. (**C**) Data summarized the effects of midazolam (3.3 μM) and diazepam (10 μM) on the hypoxic response of a carotid body. ΔCSNA is the difference between the values of CSNA during normoxia and those during hypoxia. * − significantly different from own control. The composition of Krebs was (in mM): NaCl 118, KCl 4.50, $MgSO_4 \cdot 7H_2O$ 1.20, Na_2PO_4 1.20, EDTA 0.0016, $CaCl_2$ 2.40, $NaHCO_3$ 22, glucose 10.0 and choline chloride 30 μM equilibrated in reservoirs with either 5%CO_2/air (normoxic Krebs) or 5%CO_2/95%N_2 (hypoxic Krebs)

CSN activity to hypoxia (Fig. 2; 34% and 29% decrease, respectively). Perfusion of the CB with bicuculline 10 μM did not alter the baseline CSN activity or the response to hypoxia. However, after the perfusion with bicuculline, the effect of midazolam (3.3 μM) to the CB response to hypoxia was inhibited (Fig. 3). A similar phenomenon was observed using diazepam 20 μM (data not presented).

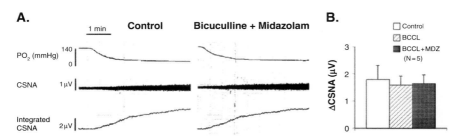

Fig. 3 The effect of bicuculline on the hypoxic response of the carotid body with or without midazolum (3.3 μM). Bicuculline did not significantly change hypoxia-induced increase in CSNA. Bucuculline blocked the effect of midazolam, showing no significant change in the hypoxic response of the carotid body with combined administration of bicuculline and midazolam

4 Discussion

In the present study, we demonstrated that: (1) gene expressions of $GABA_A$ receptor in the cat CB (α2, α3, β3, and γ2 subunits) and the PG (α2, α3, β3, and γ2 subunits); (2) immunoreactivities for GABA and $GABA_A$ receptor β3 and γ2 subunits in GCs, in the cell bodies and neural fibers in the PG; (3) the inhibitory effect of midazolam and diazepam on the chemoreceptor neural response to hypoxia, which was blocked by $GABA_A$ receptor antagonist bicuculline. These data confirmed our hypothesis that the $GABA-GABA_A$ receptor system is present in the CB; and that the activation of $GABA_A$ receptors in chemoreceptor afferent nerve endings, at least in part, accounts for the inhibition of the hypoxic ventilatory response after BZ administration.

BZs work through the BZ binding site in $GABA_A$ receptors which are ligand-gated chloride channels (Atack 2005). Potentiation of the GABAergic inhibitory system in the central nervous system by BZs induces anticonvulsant, anesthetic, anxiolytic, and sedative-hypnotic effects, which make BZs useful therapeutics for many medical conditions. However, activation of the GABAergic system in the respiratory neurons in the nucleus tractus solitalius leads to depression of ventilation (Hedner et al. 1984), and decreased respiratory response to hypoxia and hypercapnia (Alexander and Gross 1988, Forster et al. 1980). The current study shows that midazolam and diazepam also inhibits the neural response of the CB to hypoxia and that the inhibition was antagonized by $GABA_A$ receptor antagonist bicuculline. These data suggest that the depression of the hypoxic response by midazolam and diazepam is partly mediated via $GABA_A$ receptors in the CB chemosensory unit.

GABA is a major inhibitory neurotransmitter in the brain. Two major types of receptors, $GABA_A$ and $GABA_B$ are identified. The receptors are hetero-pentamers composed of two α, two β, and one γ. Up to date, eighteen members of $GABA_A$ receptor subunits are identified. A major type of subunit composition is $\alpha1\beta2\gamma2$ in the brain (Möhler et al. 2002, Atack 2005). Our RT-PCR analysis and immunohistochemical studies demonstrated that GABA and $GABA_A$ receptors are present in GCs and in chemoreceptor afferent neurons. Our study further showed that the subunit composition in the CB chemosensory unit is distinct. Alpha2, $\beta3$ and $\gamma2$ subunits are predominant in the CB, and $\alpha3$, $\beta3$, and $\gamma2$ are dominant in the PG. We can not deny the possibility that other subunits may also exist in the cat CB chemosensory unit. However, major subunits in the mouse CB are $\alpha3$, $\beta3$, and $\gamma2$ (Fujii et al. 2008). Interestingly, some general anesthetics, such as etomidate and propofol precipitate their respiratory depressant action through $\beta3$ containing $GABA_A$ receptors (Zeller et al. 2005). Since the subunit composition of $GABA_A$ receptors determines the affinity of agonists and antagonists (Möhler et al. 2002, Atack 2005), investigating the molecular target of BZs in the chemoreceptor would be beneficial to develop new BZs without undesired effect such as respiratory depression.

In our study, during normoxia, midazolam and diazepam did not change the basal chemoreceptor nerve activity. In addition, bicuculline did not influence the basal activity. It is possible that the release of GABA is not significant and GABA is not actively influencing $GABA_A$ receptor activity during normoxia. During hypoxia we observed BZs' inhibited neural activity, suggesting an increased release of GABA and $GABA_A$ receptor activation. However, bicuculline did not increase CSN activity during hypoxia. It is possible that the modification of the CB neural activity by

GABA-GABA$_A$ interaction is present but relatively small, and without a further inhibitory action caused by BZs, the excitatory neurotransmitters dominate the CSN response. Further studies are needed to know how much extent and under what circumstances the GABA-GABA$_A$ system control the CB neural activity.

Acknowledgments This study was supported by HL72293.

References

Alexander C.M., Gross J.B. 1988, Sedative doses of midazolam depress hypoxic ventilatory responses in humans. *Anesth Analg*, 67: 377–382.

Atack J.R. 2005, The benzodiazepine binding site of GABA(A) receptors as a target for the development of novel anxiolytics. *Expert Opin Investig Drugs*, 14: 601–618.

Beckstead J.H. 1994, A simple technique for preservation of fixation-sensitive antigen in paraffin-embedded tissues. *J Histochem Cytochem*, 42: 1127–1134.

Fearon I.M., Zhang M., Vollmer C., Nurse C.A. 2003, GABA mediates autoreceptor feedback inhibition in the rat carotid body via presynaptic GABAB receptors and TASK-1. *J Physiol*, 553: 83–94.

Forster A., Gardaz J.P., Suter P.M., Gemperle M. 1980, Respiratory depression by midazolam and diazepam. *Anesthesiology*, 53: 494–497.

Fujii K., Lieu T.M., Wei E., Shirahata M. 2008, Gene expression profile of neurotransmitters, receptors and ion channels in the mouse carotid body. *XVIIth International Society for Arterial Chemoreception Meeting Abstract*, p. 54.

Hedner J., Hedner T., Wessberg P., Jonason J. 1984, An analysisi of the mechanism by which γ-aminobutyric acid depresses ventilation in the rat. *J Appl Physiol Respirat Environ Exercise Physiol*, 56: 849–856.

Igarashi A., Shirahata M. 1997, Benzodiazepines depress the carotid body chemoreceptor response to hypoxia in the cat. *Society for Neuroscience Abstract*, 23: 432.

Igarashi A., Amagasa S., Horikawa H., Shirahata M. 2002, Vecuronium directly inhibits hypoxic neurotransmission of the rat carotid body. *Anesth Analg*, 94: 117–122.

Kim C., Shvarev Y., Takeda S., Sakamoto A., Lindahl S.G.E., Eriksson L.I. 2006, Midazolam depresses carotid body chemoreceptor activity. *Acta Anaesthesiol Scand*, 50: 144–149.

Möhler H., Fritschy J.M., Rudolph U. 2002, A new benzodiazepine pharmacology. *J Pharmacol Exp Ther*, 300: 2–8.

Oomori Y., Nakaya K., Tanaka H., Iuchi H., Ishikawa K., SatohY., Ono K. 1994, Immunohistochemical and histochemoical evidence for the presence of noradrenaline, serotonine and gamma-aminobutyric acid in chief cells of the mouse carotid body. *Cell Tissue Res*, 278: 249–254.

Pokorski M., Paulev P.E., Szereda-Przestaszewska M. 1994, Endogenous benzodiazepine system and regulation of respiration in the cat. *Respir Physiol*, 97: 33–45.

Shirahata M., Hirasawa S., Okumura M., Mendoza J.A., Okumura A., Balbir A., Fitzgerald R.S., 2004, Identification of M1 and M2 muscarinic acetylcholine receptors in the cat carotid body chemosensory system. *Neuroscience*, 128: 635–644.

Yamaguchi S., Balbir A., Schofield B., Coram J., Tankersley C.G., Fitzgerald R.S., O'Donnell C.P., Shirahata M. 2003, Structural and functional differences of the carotid body between DBA/2J and A/J strains of mice. *J Appl Physiol*, 94: 1536–1542.

Zeller A., Arras M., Lazaris A., Jurd R., Rudolph U. 2005, Distinct molecular targets for the central respiratory and cardiac actions of the general anesthetics etomidate and propofol. *FASEB J*, 19: 1677–1679.

Evidence for Histamine as a New Modulator of Carotid Body Chemoreception

R. Del Rio, E.A. Moya, J. Alcayaga and R. Iturriaga

Abstract It has been proposed that histamine is an excitatory transmitter between the glomus cells of the carotid body (CB) and the nerve endings of the petrosal ganglion (PG) neurons. The histamine biosynthetic pathway and the presence of histamine H1, H2 and H3 receptors have been reported in the CB. Thus, histamine meets some of the criteria to be regarded as a transmitter. However, there is no evidence that glomus cells contain histamine, or whether its application produces chemosensory excitation. Therefore, we studied its immunocytochemical localization on cat CB and its effects on chemosensory activity. Using perfused and superfused in vitro CB and PG preparations, we assessed the effects of histamine hydrochloride on chemosensory discharges and of histamine H1, H2 and H3 receptor blockers. We found the presence of histamine immunoreactivity in dense-core vesicles in glomus cells. In an in vitro CB preparation we performed pharmacological experiments to characterize histamine effects. The application of histamine hydrochloride (0.5–1,000 µg) to the CB produces a dose-dependent increase in the carotid sinus nerve activity. The H1 receptor blockade with pyrilamine 500 nM produces partial decrease of the histamine-induced response, whereas the H2 receptor blockade (ranitidine 100 µM) fail to abolish the histamine excitatory effects. Antagonism of the H3 receptor results in an increase in carotid body chemosensory activity. On the other hand, application of histamine to the isolated PG had no effect on the carotid nerve discharge. Our results suggest that histamine is a modulator of the carotid body chemoreception through H1 and H3 receptor activation.

Keywords Carotid body · Histamine · Chemoreception · Histamine receptor · Pyrilamine

R. Iturriaga (✉)
Laboratorio de Neurobiología, Facultad de Ciencias Biológicas, P. Universidad Católica de Chile
e-mail: riturriaga@bio.puc.cl

1 Introduction

The carotid body (CB) is the main arterial chemoreceptor, which senses the arterial blood levels of PO_2, PCO_2 and pH, contributing to the ventilatory and cardiovascular homeostasis (Gonzalez et al. 1994). The current model of CB chemoreception states that hypoxia depolarizes chemoreceptor (glomus cells) cells, increasing intracellular $[Ca^{2+}]$ and releasing molecules such as acetylcholine (ACh), adenosine nucleotides, dopamine and some peptides, which may act as excitatory transmitters or modulators between glomus cells and the petrosal ganglion (PG) nerve terminals, which in turn increase the chemosensory frequency of discharges in the carotid sinus nerve (Iturriaga and Alcayaga 2004; Nurse 2005).

Koerner et al. (2004) found that hypoxia increases the release of histamine from the isolated rat CB, while the histamine biosynthesis enzyme histidine decarboxylase and the vesicular monoamine transporters VMAT1 and VMAT2 are present in glomus cells. More recently, Lazarov et al. (2006) found the expression of H1 and H3 receptors in the rat glomus cells and PG neurons. Thus, histamine meets some of the criteria to be considered a transmitter in the CB (Eyzaguirre and Zapata 1984). However, it is not known if histamine is present in glomus cells or is restricted to mast cells within the CB. On the other hand, Landgren et al. (1954) did not find any changes of cat CB chemosensory activity following intracarotid injections of histamine. Since histamine produces marked vascular effects in cats (Champion and Kadowitz 1997), which may affect the CB chemosensory process is seems to be necessary to eliminate the systemic effects induced by histamine using in vitro CB preparations. Lazarov et al. (2006) found that the application of H1 and H3 specific agonists to the rat CB caused a mild increased of phrenic nerve activity, suggesting that histamine may excite the CB. Accordingly, to test if histamine may act as an excitatory transmitter between glomus cells and PG neurons, we studied the immunolocalization of histamine in the cat CB, and the effects of exogenous histamine application on CB chemosensory discharges using in vitro perfused or superfused CB preparations (Alcayaga et al. 1988; Iturriaga et al. 1991), and in an isolated preparation of the PG (Alcayaga et al. 1998).

2 Methods

2.1 Animals

Experiments were performed on 10 male adult cats (2.0–4.7 kg) anesthetized with sodium pentobarbitone (40 mg kg^{-1} ip), followed by additional doses (12 mg iv) to maintain a level of surgical anesthesia. The experimental protocol was approved by the Bioethical Committee of the Facultad de Ciencias Biológicas of the P. Universidad Católica de Chile.

2.2 Immunohistochemical Studies

The presence of positive immunoreactivity for histamine (HA-ir) was studied in 4 CBs removed from 2 anesthetized cats. The CBs were fixed by immersion in 2.5% glutaraldehyde for 12 hours at 4°C, dehydrated in ethanol and included in paraffin. The CBs were cut into 5 μm-thick sections and mounted on silanized slides. Deparaffinized slices were immunostained with specific antiserum for histamine (1:100, Mab AHA-1, Fujiwara et al. 1997) with ABC Vectastain kit (Vector Lab., USA). Finally the sections were washed and peroxidase was visualized using diaminobenzidine. The slides were counterstained with hematoxylin and permanently mounted. The omission of the primary antibody was used as negative control.

2.3 In Vitro Recording of Chemosensory Discharges

The effects of 0.5–1.000 μg histamine hydrochloride on CB chemosensory discharges were assessed using in vitro perfused or superfused CB preparations, which allow the separation of vascular effects. Briefly, the CB with the carotid sinus nerve (CSN) was excised from cats and superfused in vitro with modified Tyrode solution, pH 7.40 at $38.5 \pm 0.5°C$. To study the participation of the blood vessels in the CB chemosensory process, we used the arterially perfused CB preparation, which conserves its functional vascularization (Belmonte and Eyzaguirre 1974; Iturriaga et al. 1991). To assess the effect of histamine on PG neurons, the PG with the CSN was removed from cats and placed in a chamber as previously described (Alcayaga et al. 1998). The isolated PG was superfused with Hank's balanced solution at $38.0 \pm 0.5°C$, pH 7.42. Chemosensory discharges in CB and PG were recorded from the CSN placed on pair platinum electrodes and lifted into mineral oil. The neural signals were preamplified and amplified, filtered (10 Hz-1 kHz; notch filter, 50 Hz), and fed to an electronic amplitude discriminator to measure the frequency of CSN discharges (f_x) expressed in Hz. The f_x signal was digitized with an analog-digital board DIGITADA 1200 (Axon Instruments, USA).

Histamine hydrochloride was applied to the superfused CB and to the isolated PG in 20 μl boluses with a microdispenser, whose tip was placed about 1 mm from the surface of the organs. In the perfused CB, histamine was applied in 200 μl boluses into the arterial line. The maximal responsiveness of the CB and PG preparations were tested using acetylcholine hydrochloride (ACh). Histamine receptor antagonists were applied at constant concentrations through the perfusion or superfusion medium to the CB. We used the H1 blocker pyrilamine (500 nM), the H2 blocker ranitidine (100 μM), and the H3 blocker thioperamide (30 μM). All drugs were obtained from Sigma (USA).

Data was expressed as mean ± S.E.M. The change in frequency discharge (Δf) was calculated as the difference between the maximal frequency achieved during a single response and the mean basal activity, computed in a 30-s interval prior to an evoked response. Data was standardized to the maximal response attained in (max Δf) and fitted to the following logistic expression: $\Delta f / \max \Delta f = 1/(1 + \{ED_{50}/D\}^S)$, where D = dose, ED_{50} = mean effective dose, S = Hill slope factor determining the steepness of each curve. The correlation coefficients for all adjusted curves were >0.90 ($P < 0.01$) for all conditions studied.

3 Results

3.1 Immunolocalization of Histamine in the Cat Carotid Body

We found a strong expression of histamine in the cat CB (Fig. 1). Histamine positive immunoreactivity (HA-ir) staining was present in clusters of cells within the CB, which has been described as glomus cells in light microscope studies. In fact, the positive HA-ir staining was found in clusters of round or ovoid cells, with a diameter of about 10 μm and a prominent nucleus (Fig. 1). Note that not all of the glomus cell clusters expressed HA-ir. The HA-ir staining was also detected in mast cells, mainly located in the interlobular connective tissues and around blood vessels (not shown). Negative controls, omitting the primary antibody were consistently devoid of immunoreactive staining (see *inset* in Fig. 1).

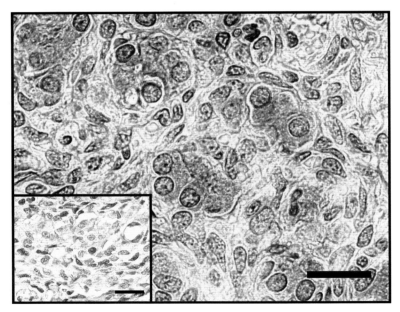

Fig. 1 Immunolocalization of histamine in the cat CB. High magnification photomicrograph of the CB showing HA –ir in glomus cells clusters, bar, 20 μm. *Inset*, negative control performed by omission of the primary antibody show no positive immunostaining

3.2 Effects of Exogenous Histamine on Chemosensory Discharges

In the perfused cat CB preparation, the application of histamine hydrochloride (0.5–1000 μg) produced a dose-dependent increase of f_x (Fig. 2). Similarly, in the superfused cat CB preparation, the application of histamine increases f_x in a dose-dependent manner (Fig. 2). However, the sensitivity and reactivity of the dose-responses curves were different in the perfused and superfused CB preparations. Transient desensitization was observed when repeated applications of histamine were applied to the superfused CB (not shown). Thus, 10–20 min intervals between applications of increasing doses of histamine were used to obtaining the dose-response curves. Figure 2 summarizes the effects of exogenous application of histamine hydrochloride (0.5–1000 μg) on f_x in perfused and superfused cat CB preparations, and to the isolated PG. In the perfused CB preparation, histamine at doses of 500 μg elicited maximal chemosensory responses, similar to those attained with nicotine. The ED_{50} for the histamine-induced chemosensory response on the perfused CB was 143.6 ± 0.7 μg ($n = 4$ CBs), while in the superfused CB preparation, the ED_{50} for the excitatory effect was 695.2 ± 0.8 μg ($n = 3$ CBs). Note that large doses of histamine (1000 μg) were ineffective to produce any neural response when were applied to the isolated PG ($n = 2$).

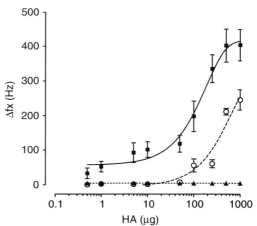

Fig. 2 Dose-response curves for the effects of histamine (HA) on chemosensory discharges in 4 perfused CB (■), 3 superfused CB (○) and 2 isolated PG (▲). Dose response curve are statistically different ($P < 0.001$, Two-way (ANOVA)

3.3 Pharmacological Effects of Histamine Antagonists on the Chemosensory Excitation Induced by Histamine

The application of the antagonist pyrilamine (500 nM) to the superfused CB reduced the excitatory response induced by 250 μg histamine ($P < 0.05$, $n = 4$), but not the chemosensory response induced by ACh (Fig. 3A). The H2 receptors blocker ranitidine (100 μM) did not modify the response elicited by histamine (Fig. 3B).

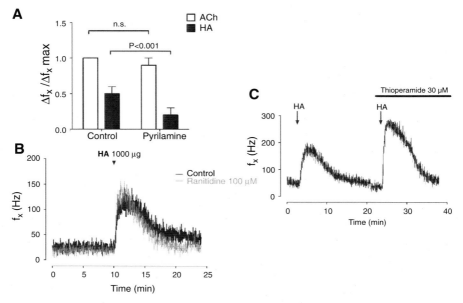

Fig. 3 Effects of specific histamine receptor antagonist on the chemosensory responses evoked by histamine. (**A**) Effect of 500 nM pyrilamine, H1 antagonist ($P<0.001$, Bonferroni after 2-way ANOVA). (**B**) Effect of 100 μM ranitidine, H2 antagonist on the response induced by histamine (HA). (**C**) Effect of 30 μM thioperamide, H3 antagonist, on the increase of chemosensory discharges elicits by HA

On the contrary, superfusion with Tyrode containing 30 μM thioperamide, a H3 antagonist, enhanced the chemosensory responses elicited by 250 μg histamine (Fig. 3C).

4 Discussion

Present results provide new information regarding the localization of histamine in glomus cells of the cat CB. In addition, we found that exogenous application of histamine to the CB produced a dose-response chemosensory excitation. In the superfused CB preparation, 500 nM pyrilamine, reduced the chemosensory excitation induced by histamine. On the contrary, the specific H2 receptor antagonist ranitidine (100 μM) was ineffective to modify the histamine-induced chemosensory response. Thus, present results indicate that glomus cells present positive immunoreactivity for histamine, and its application produces CB chemosensory excitation.

Since histamine has profound vascular effects in the cat (Champion and Kadowitz 1997), which may modify the chemosensory response, we study the effects of histamine on CB chemosensory discharges using perfused and superfused preparations in vitro of the cat CB. In both preparations of the CB, histamine produced a potent dose-dependent excitatory effect on chemosensory discharges. The excitatory response induced by histamine in the perfused cat CB is not due to reduction of the PO_2 in the CB parenchyma, and a consequently increase of chemosensory discharges. Indeed, using a superfused CB preparation, lacking the direct vascular effects, we found that histamine increased the chemosensory discharges, indicating that histamine excite CB chemoreceptors through non-vascular actions.

Koerner et al. (2004) proposed that histamine work as an excitatory transmitter between glomus cells of the CB and nerve terminals of the PG neurons. In addition to the findings of Koerner et al. (2004), we found that histamine immunoreactivity is present in the cytoplasm of glomus cells, and that exogenous application of histamine mimics the effects induced by natural stimuli. However, since histamine did not elicit any excitatory effect when was applied on PG chemosensory neurons, as other putative transmitters do (Iturriaga et al. 2007), our results support the idea that histamine acts at the presynaptic level in the CB (i.e. glomus cells), but not at the level of the somata of PG neurons.

The superfusion with Tyrode plus the H1 receptor antagonist pyrilamine reduced the chemosensory response induced by histamine while the H2 receptor antagonist ranitidine did not modified the excitatory chemosensory response elicited by histamine. On the contrary, the H3 antagonist thioperamide enhanced the chemosensory responses elicited by histamine. Thus, our data suggest that histamine may work as an excitatory modulator on the chemosensory process at the level of the glomus cells acting on H1 and H3 receptors.

Acknowledgments This work was supported by grant 1070854 from the National Fund for Scientific and Technological Development of Chile (FONDECYT).

References

Alcayaga, J., Iturriaga, R. & Zapata, P. 1988, Flow-dependent chemosensory activity in the carotid body super fused in vitro. *Brain Res*, 455: 31–37.

Alcayaga, J., Iturriaga, R., Varas, R., Arroyo, J. & Zapata, P. 1998, Selective activation of carotid nerve fibers by acetylcholine applied to the cat petrosal ganglion in vitro. *Brain Res*, 786: 47–54.

Belmonte, C. & Eyzaguirre, C. 1974, Efferent influences on carotid body chemoreceptors. *J Neurophysiol*, 37: 1131–1143.

Champion, H.C. & Kadowitz, P.J. 1997, NO release and the opening of K+ATP channels mediate vasodilator responses to histamine in the cat. *Am J Physiol*, 273: H928–H937.

Eyzaguirre, C. & Zapata, P. 1984, Perspectives in carotid body research. *J Appl Physiol*, 57: 931–957.

Fujiwara, K., Kitagawa, T., Inoue, Y. & Alonso, G. 1997, Monoclonal antibodies against glutaraldehyde-conjugated histamine: application to immunocytochemistry. *Histochem Cell Biol*, 107: 39–45.

Gonzalez, C., Almaraz, L., Obeso, A. & Rigual, R. 1994, Carotid body chemoreceptors: from natural stimuli to sensory discharges, *Physiol Rev*, 74: 829–898.

Iturriaga, R., Rumsey, W.L., Mokashi, A., Spergel, D., Wilson, D.F. & Lahiri, S. 1991, In vitro perfused-superfused cat carotid body for physiological and pharmacological studies. *J Appl Physiol*, 70: 1393–1400.

Iturriaga, R. & Alcayaga, J. 2004, Neurotransmission in the carotid body: Transmitters and modulators between glomus cells and petrosal ganglion nerve terminals. *Brain Res Rev*, 47: 46–53.

Iturriaga, R., Varas, R. & Alcayaga, J. 2007, Electrical and pharmacological propierties of petrosal ganglion neurons that innervate the carotid body. *Resp Physiol & Neurobiol*, 157: 130–139.

Koerner, P., Hesslinger, C., Schaefermeyer, A., Prinz, C. & Gratzl, M. 2004, Evidence for histamine as a transmitter in rat carotid body sensor cells. *J Neurochem*, 91: 493–500.

Landgren, S., Liljestrand, G. & Zotterman, Y. 1954, Impulse activity in the carotid sinus nerve following intracarotid injections of sodium-iodo-acetate, histamine hydrochloride, lergitin, and some purine and barbituric acid derivatives. *Acta Physiol Scand*, 30: 149–160.

Lazarov, N., Rozloznik, M., Reindl, S., Rey-Ares, V., Dutschmann, M. & Gratzl, M. 2006, Expression of histamine receptors and effect of histamine in the rat carotid body chemoafferent pathway. *Eur J Neurosci*, 24: 3431–3444.

Nurse, C.A. 2005, Neurotransmission and neuromodulation in the chemosensory carotid body. *Auton Neurosci*, 120: 1–9.

Fluoresceinated Peanut Agglutinin (PNA) is a Marker for Live O₂ Sensing Glomus Cells in Rat Carotid Body

I. Kim, D.J. Yang, D.F. Donnelly and J.L. Carroll

Abstract Experiments using live dissociated carotid body (CB) cells for patch clamping, $[Ca^{++}]_i$ or other measurements require positive identification of the cell being recorded. At present, cell morphology is usually employed, but several cell types within the carotid body evidence similar morphologic characteristics. Therefore, we sought to develop a method utilizing a vital dye to identify glomus cells before and during experiments that require live cells, such as patch clamp studies. It was previously reported that the binding sites for peanut agglutinin (PNA) were highly expressed by all neuroendocrine-derivatives of the sympathoadrenal neural crest, including glomus cells, small, intensely fluorescent cells, PC-12 cells, and adrenal chromaffin cells in situ (Katz et al. 1995). By utilizing the binding characteristics of galactose-specific lectin peanut agglutinin (PNA) on the outer cell membrane, we tested the possibility that the fluoresceinated PNA may preferentially bind to CB glomus cells. The results to date show: (1) Rhodamine tagged PNA (Rhod-PNA) binds to the live dissociated glomus cells in less than one hour incubation and can be visualized in superfused cells; (2) Rhod-PNA labeled cells are perfectly matched with tyrosine hydroxylase (TH) positive glomus cells; (3) Rhod-PNA did not interfere with Fura-2 for Ca^{++} imaging; (4) Rhod-PNA bound to glomus cells in $[Ca^{++}]_i$ studies does not affect O₂ response of glomus cells. Thus fluoresceinated PNA may be a useful marker for live CB glomus studies, without adversely affecting their physiologic response.

Keywords PNA · Peanut agglutinin · Lectin · CB · Glomus cells · Marker · O₂ sensing · TH · Ca^{++} imaging · Double immunostaining

I. Kim (✉)
Department of Pediatrics, University of Arkansas for Medical Sciences, Little Rock, USA
e-mail: kiminsook@uams.edu

1 Introduction

The carotid body (CB) is a small neural crest derived neuroendocrine organ sensing blood O_2, CO_2, and pH level. CB chemoreceptor glomus cells increase $[Ca^{++}]_i$ in response to low oxygen levels and their O_2 responses are fully developed during first two weeks of postnatal life (Carroll and Kim 2005) in rats.

The CB consists of four principle components: cell clusters, blood vessels, connective tissue, and nerve fibers (Izal-Azcarate et al. 2008). Chemoreceptor glomus cells and sustentacular cells comprise ~60% of total CB volume (Gonzalez et al. 1994, 1995, Lopez-Barneo et al. 2001). After enzymatic dissociation, CB cells consist of several types: glomus cells, sustentacular cells, fibroblasts etc. Usually, glomus cells are identified based on cell size, shape, and occurrence in clusters. However, clear identification is uncertain since different cells may have similar morphologic characteristics. It is known that tyrosine hydroxylase (TH) and synaptophysine are markers for glomus cells and glial fibrillary acid protein (GFAP), S-100 protein and vimentin (Kameda 1996, Kameda 2005, Yamamoto et al. 2006) are markers for sustentacular cells. However, these markers are for intracellular proteins and are only useful in permeabilized cells. Experiments using live dissociated carotid body (CB) cells for patch clamping, $[Ca^{++}]_i$ or other measurements require methods for live CB cell identification. In addition, live cell tagging must not interfere with the normal physiological responses of the cells. Therefore, we sought to develop a method to identify live dissociated glomus cells before and during experiments.

It was reported that the binding sites for peanut agglutinin (PNA) were highly expressed by all neuroendocrine-derivatives of the sympathyo-adrenal neural crest, including glomus cells, small, intensely fluorescent cells, PC-12 cells, and adrenal chromaffin cells in situ (Katz et al. 1995). Since galactose-specific lectin peanut agglutinin (PNA) (Lotan et al. 1977) is bound on the outer cell membrane, we investigated the possibility that the fluoresceinated PNA may bind to the membrane surface of CB glomus cells Although fluoresceinated PNA may bind to CB cells, several questions must be addressed: (1) whether the fluoresceinated-PNA bound CB cells are exclusively TH positive cells, (2) whether the fluoresceinated-PNA bound CB cells are viable after PNA treatment, and (3) whether PNA binding causes any adverse physiological effects on CB O_2 sensing.

In order to address the first concern, cultured dissociated CB cells were treated with rhodamine labeled PNA (Rhod-PNA) or biotin labeled PNA (Biotin-PNA) for one hour, and double-stained with TH antibodies. Secondly, the calcium responsiveness of Fura-2 labeled, Rhod-PNA treated cells was tested using high $[K^+]_o$ extracellular solution (20 mM K^+). Thirdly, alterations in physiologic responsiveness was tested in Fura-2, Rhod-PNA treated CB cells by stimulation with hypoxia or anoxia.

The results suggest that the fluoresceinated PNA is an efficient cell surface marker for live CB glomus cells with no adverse physiological effects.

2 Methods

2.1 Preparation of Dissociated CB Cells and Immunocytochemistry

Carotid bodies (CB) isolated from newborn Sparague-Dawley rats were dissociated with enzyme mixture consisting trypsin and collagenase (~0.5 mg/ml), and plated on poly-D-lysine coated glass coverslips. After culturing for 3–5 hours, the cells were treated with biotinylated-PNA or Rhod-PNA in complete CB medium (25 μg/ml: Vector) for 1 hour in cell culture incubator at 37°C. Then, the CB cells were fixed with 4% paraformaldehyde and treated with 0.4% triton X-100 in PBS. After treating with blocking solution (5% non-fat dry milk), cells were labeled with Alexa 546 tagged streptoavidin (1:200) for 2 hours at room temperature (RT). For double labeling, anti-TH (1:1000) was treated for 5–6 hours and labeled with Alexa flour 488 tagged secondary antibodies (1:500) for 2 hours at RT. After a final wash, the coverslips were mounted on a glass slide with Prolong Gold antifade reagent with DAPI (Invitrogen). Cells are observed under Nikon TE-2000 at 400X and the images were taken and merged with NIS Element software.

2.2 Labeling Live CB Cells with Rhodamine-PNA (Rhod-PNA) and Intracellular Ca^{++} ($[Ca^{++}]_i$) Measurement

For labeling live CB glomus cells with Rhod-PNA, CB cells were treated with Rhod-PNA in complete CB medium (25 μg/ml:Vector) for 1 hour, 37°C. Rhod-PNA treated CB cells were washed several times with warm PBS to remove unbound Rhod-PNA. $[Ca^{++}]_i$ was measured by quantitative fluorescence imaging using fura-2. Rhod-PNA treated cells were loaded with fura-2, by incubation for 30 minutes at 37°C (Molecular Probes). Cells were exposed to 20mM K^+ extracellular solution, hypoxia (0% O_2), or anoxia (0% O_2 with 0.5 mM dithionite). Fura-2 fluorescent emission was measured at 510 nm in response to alternating excitation at 340 and 380 nm. Images were acquired and stored using a Nikon TE2000 microscope and CCD (CoolSNAP HQ2) camera under Metafluor software (Molecular Devices). For each coverslip, the background light levels were determined and subtracted, from each image before measurement of the fluorescence intensity ratio at 340 nm/380 nm.

3 Results

3.1 Identification of PNA Bound Cells as Glomus Cells

To assess initially the binding ability of PNA to dissociated CB cells, Biotin-PNA or Rhod-PNA was added into cultured CB cells at 37°C. Within one hour, staining was positive for biotin- PNA and Rhod-PNA. After 3 hours incubation, cells were fixed

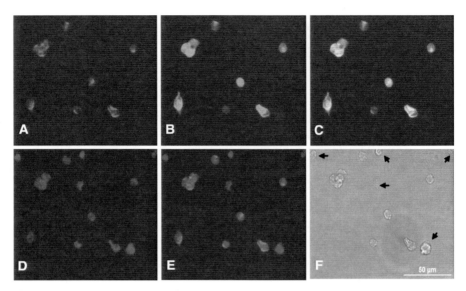

Fig. 1 Dissociated CB cells were incubated with biotinylated PNA and labeled with Alexa 546 tagged streptoavidin (**A**) and double labeled by TH with Alexa 488 secondary antibody (**B**). PNA and TH are overlaid (**C**). The nucleus are stained with DAPI contained mounting medium (**D**) and PNA stained image is overlaid over DAPI (**E**). The CB cells labeled with PNA are exactly matched with cell labeled with TH. Thus PNA only labels type I cells and the cells marked with arrows on DIC (**F**) are not type I cells. A scale bar (50 μm) in DIC (**F**)

in 4% paraformaldehyde and visualized with Alexa 546 streptoavidin. As shown in Fig. 1A CB cells from a P16 rat demonstrated positive binding for biotin-PNA. In order to identify the cell type that was labeled, cells were double labeled with glomus cell marker, TH. As shown in Fig. 1A–C, the cells labeled with biotin-PNA or TH were identical. This suggests that PNA specifically labels glomus cells. Biotin-PNA labeling on glomus cells is cell specific, because PNA did not labeled every CB cell as shown in Fig. 1E. The cells labeled only with DAPI in Fig. 1E were stained with neither biotin-PNA nor TH antibody and marked with arrows in Fig. 1F.

3.2 Identification of PNA Labeled Cells as O_2 Sensing Glomus Cells

Biotin-PNA and Rhod-PNA did not appear to affect viability or cell responsiveness. The CB cells treated with Rhod – PNA still responded to. hypoxia in Ca^{++} imaging studies. In order to identify the PNA positive cells as functioning glomus cells, PNA treated CB cells were tested $[Ca^{++}]_i$ response to acute hypoxia (0% O_2) or anoxia (0% O_2 + 0.5 mM dithionite). Both clusters (Fig. 2D) and single cells (Fig. 2E) on which Rhod-PNA was bound increased $[Ca^{++}]_i$ in response to 20 mM K^+, 0% O_2 hypoxia, and anoxia as shown in Fig. 2. Non-PNA labeled cells did not show any

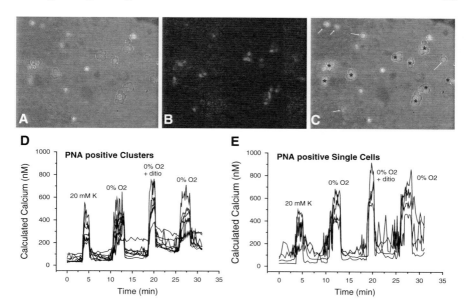

Fig. 2 Dissociated 16 days old CB cells were treated with Rho-PNA for 1 hr. (**A**) DIC of dissociated CB cells, (**B**) Rhod-PNA labeled CB cells, (**C**) Rhod-PNA labeled cells were overlaid on DIC image. Intracellular Ca_2^+ increases in response to 20 mM K, 0% O_2, and anoxia (with 0.5 mM dithionite [dithio]) in Rhod-PNA positive clusters marked asterisk (∗) on **C** (**D**) and in Rhod PNA positive single cells marked yellow arrows in **C** (**E**)

$[Ca^{++}]_i$ changes in response to hypoxia or anoxia. Thus, PNA labeled cells are O_2 sensing CB glomus cells.

3.3 No Adverse Physiological Effect of PNA Binding on O_2 Sensing

To find out whether PNA binding adversely affects O_2 sensing ability, the response to 20 mM K^+, hypoxia, and anoxia in PNA bound CB cells were compared to Non-PNA treated CB glomus cells. The averaged $[Ca^{++}]_i$ increase in response to 20 mM K^+ and 0% hypoxia showed that there are no significant differences between PNA treated and not treated CB cells. Thus, adding fluoresceinated PNA to identify glomus cells before/during physiological experiments requiring real time and live CB cells does not have any adverse physiological effects.

4 Discussion

Identification of cell types, and, in particular, glomus cells is essential for studying O_2 sensing mechanisms. The present results demonstrate that fluoresceinated PNA, Biotin-PNA and Rhod-PNA, bind to CB glomus cells within one hour of exposure.

By co-immunostaining for TH, fluoresceinated PNA-bound cells were identified as exclusively glomus chemoreceptor cells. In addition, PNA binding produced no detectable alteration in the calcium response to 20 mM K^+, 0% O_2 hypoxia, and anoxia.

Therefore these preliminary results suggest that the fluoresceinated PNA is an efficient cell surface marker for live CB glomus cells to identify the live dissociated CB glomus chemoreceptor cells before and during experiments.

Acknowledgments This study was supported by grants from the National Institutes of Health (RO1 HL54621) and CUMG grant from University of Arkansas for Medical Sciences.

References

Carroll, J.L., & Kim, I. 2005, Postnatal development of carotid body glomus cell O_2 sensitivity, *Respir. Physiol. Neurobiol.* 149:201–215.
Gonzalez, C., Almaraz, L., Obeso, A., Rigual, R. 1994, Carotid body chemoreceptors: from natural stimuli to sensory discharges. *Physiol Rev.* 74: 829–98.
Gonzalez, C., Lopez-Lopez, J.R., Obeso, A., Perez-Garcia, M.T., & Rocher, A. 1995, Cellular mechanisms of oxygen chemoreception in the carotid body, *Respir. Physiol.* 102:137–147.
Izal-Azcarate, A., Belzunegui, S., Sebastian, W.S., Garrido-Gil, P., Vazquez-Claverie, M., Lopez, B., Marcilla, I., & Luquin, M.A. 2008, Immunohistochemical characterization of the rat carotid body, *Respir. Physiol. Neurobiol.* 161:95–99.
Kameda, Y. 1996, Immunoelectron microscopic localization of vimentin in sustentacular cells of the carotid body and the adrenal medulla of guinea pigs, *J. Histochem. Cytochem.* 44: 1439–1449.
Kameda, Y. 2005, Mash1 is required for glomus cell formation in the mouse carotid body, *Dev. Biol.* 283:128–139.
Katz D.M., White M.E., & Hall A.K. 1995, Lectin binding distinguishes between neuroendocrine and neuronal derivatives of the sympathoadrenal neural crest. *J. Neurobiol.* 26:241–252.
Lopez-Barneo, J., Pardal, R., & Ortega-Saenz, P. 2001, Cellular mechanism of oxygen sensing, *Annu. Rev. Physiol.* 63:259–287.
Lotan, R., Beattie, G., Hubbell, W., & Nicolson, G.L. 1977, Activities of lectins and their immobilized derivatives in detergent solutions. Implications on the use of lectin affinity chromatography for the purification of membrane glycoproteins, *Biochemistry* 16:1787–1794.
Yamamoto, Y., & Taniguchi, K. 2006, Expression of tandem P domain K+ channel, TREK-1, in the rat carotid body, *J. Histochem. Cytochem.* 54:467–472.

Neuroglobin in Aging Carotid Bodies

V. Verratti, C. Di Giulio, G. Bianchi, M. Cacchio, G. Petruccelli, C. Di Giulio, L. Artese, S. Lahiri and R. Iturriaga

Abstract Neuroglobin (Ngb) is a member of the vertebrate globin family expressed particularly in the brain and in the retina. Ngb is concentrated in the mitochondria-containing areas of neurons, and its distribution is correlated with oxygen consumption rates. Previously we have shown that Ngb is expressed in carotid body (CB) tissues. Considering that hypoxia and aging may be linked through a series of adaptive and protective mechanisms (e.g. reduction in mitochondrial numbers), we investigate the role of Ngb during aging and hypoxia. Two groups of six rats (age-matched 3 and 24 months old) were kept in room air as a control groups, the others two groups were kept in a Plexiglas chamber for 12 days in chronic hypoxia (10–12% inspired oxygen). The presence of Ngb in the CB tissue was detected by immunohistochemistry using a polyclonal antibody. Ngb immunoreactivity was significantly higher in CB tissues from young rats exposed to chronic intermittent hypoxia, whereas CB tissues from old rats did not show any significant increase in Ngb levels after hypoxia. Similar to hemoglobin, Ngb may act as a respiratory protein by reversibly binding gaseous ligands NO and O_2 and could act as a NO scavenger and participate in detoxification of Reactive Oxygen Species (ROS) generated under hypoxic conditions.

Keywords Carotid body · Neuroglobin · Aging · Oxygen-sensing

1 Introduction

Neuroglobin (Ngb) is a 151-amino acid protein identified as a member of the vertebrate globin family (Burmester et al. 2000, Mammen et al. 2002). Ngb is predominantly expressed in nerve cells, particularly in the brain and in the retina, but it is also expressed in other tissues (Zhu et al. 2002, Burmester et al. 2004). The Ngb protein has three-on-three α-helical globin folds and a hexa-coordinated heme-Fe

V. Verratti (✉)
Department of Basic and Applied Medical Sciences, "G. d'Annunzio" University, Chieti, Italy
e-mail: vittorelibero@hotmail.it

complex with affinity for oxygen and CO (Vallone et al. 2004). The physiological role of Ngb is not well understood, but the protein is thought to participate in processes such as oxygen transport, oxygen storage, and nitric oxide detoxification. Similar to hemoglobin, Ngb may act as a respiratory protein by reversibly binding gaseous ligands (nitric oxide and oxygen) via the Fe-containing porphyrin ring. Consistent with this notion, Ngb is concentrated in the mitochondria-containing areas of neurons, and its distribution is correlated with oxygen consumption rates (Pesce et al. 2003, Moens and Dewilde 2000, Sun et al. 2003). Ngb levels are augmented during ischemia and hypoxia (Sun et al. 2001), further suggesting that Ngb enhances oxygen supply to neural components and may contribute to neuronal survival. The main oxygen chemoreceptor in the arterial blood system is the CB, which is characterized by high blood flow and oxygen consumption (Gonzalez et al. 1994, Prabhakar et al. 2001, Lahiri et al. 2002). We previously showed that Ngb is highly expressed in CB tissues under hypoxic compared to normoxic conditions (Di Giulio et al. 2006). This observation that in the CB, Ngb may work as a "neuronal globin protein" by providing oxygen to the respiratory chain of CB cells, and/or functioning as a sensor for the detection of cellular oxygen levels. Since aging is correlated with both decreased oxygen supply to cells and decreased oxygen demand by tissues (Di Giulio et al. 2003), it is likely that hypoxia and aging may be linked through a series of adaptive and protective mechanisms e.g. reduction in mitochondrial numbers. Considering that arteriosclerosis induces a state of hypoxia by stimulating extracellular matrix deposition and thus increasing the distance between cells and blood vessels and decreasing the oxygen diffusion gradient, we herein evaluated whether Ngb could play a role in the aging CB, and whether Ngb expression responds to hypoxia in aging CB tissues.

2 Methods

Four groups of male Wistar rats (250 g) were used. Two groups of six rats (age-matched 3 and 24 months old) were kept in room air (21% O_2) as a control groups, the other two groups were kept in a Plexiglas chamber for 12 days in chronic hypoxia (10–12% O_2). The chamber temperature and CO_2 were kept in physiological ranges. The rats were anaesthetized with Nembutal (40 mg/kg ip) and the CBs were dissected. The carotid body tissues were immersed overnight in ice cold 4% paraformaldehyde in 0.1 M phosphate buffered saline (PBS). Tissues were then rinsed in 15% sucrose PBS (1 h) and stored at 4°C in 30% sucrose PBS (2 h).Ten micrometer thick sections ($n = 6$ for each sample) were serially cut using a cryomicrotome (Reichert-Jung Frigocut 2800), thaw-mounted onto microscope slides, fixed by immersion in acetone at 4°C for 5 min, and air-dried. Slides were stored at 4°C until use.

The presence of Ngb in the CB tissue was detected by immunohistochemistry using a polyclonal antibody (E-16, Santa Cruz Biotech Antibodies). Slides were preincubated in PBS for 5 min and then with Ngb of goat origin which was diluted

1:100 in PBS and applied for 30 min. at 37°C. Slices were then washed twice in PBS for 5 min and in tris-HCl buffer, pH 7.6 for 10 min. A second antibody, goat antirabbit IgG, was added for 10 min. and slides were again washed in PBS. The immunohistochemical staining was analyzed using an histochemistry software package (Image-proplus 4.5) for the densitometric analysis. Five random fields were chosen for each CB and in each field an evaluation of the area was performed. The Ngb immunoreactivity was measured in optic integrated units (I.O.I.). Statistical comparison between young hypoxic group and the corresponding normoxic age-matched group, was performed using t-test (for unpaired data) with values of $p < 1$ considered significant. The same analysis was performed for the old group.

3 Results

The intensity of the Ngb immunoreactivity was significantly higher ($p < 0.01$) in CB tissues from young rats exposed to chronic intermittent hypoxia (Fig. 1). On the contrary, we did not find any significant increase in the Ngb level after hypoxic exposure in the CBs from old rats (Fig. 2).

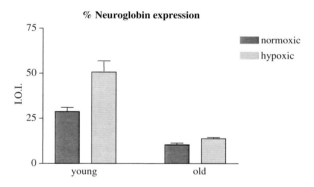

Fig. 1 Immunohistochemical analysis of Ngb in the young and old rat CB. Quantitative analyses are expressed as changes in integrated optical intensity (I.O.I.). Results are means ± S.D., $n = 5$ for all groups. Normoxic young vs. hypoxic young *$p < 0.01$

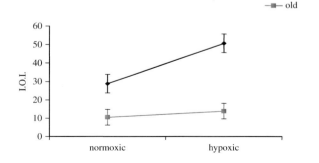

Fig. 2 Integrated Optical Intensity of Ngb in young and old Carotid Body submitted to hypoxia. Normoxic young vs. hypoxic young *$p < 0.01$

4 Discussion

Aging is characterized by a reduction in the general homeostatic adaptation to metabolic requirements (Di Giulio et al. 2003). The aged CB shows an increase in extracellular matrix deposition, a reduction in the number and volume of glomus cells and a reduction in the volume and number of mitochondria. We found that Ngb is present in CB tissues from both young and old rats under normoxic conditions. In addition, chronic hypoxia exposure for 12 days increased the Ngb levels in the CB tissues from young rats, but not old rats.

The presence of Ngb in the CB may play important physiological roles. As a respiratory protein, Ngb may act as a nitric oxide scavenger and/or participate in detoxification of reactive oxygen species generated under hypoxic conditions. It is also possible that Ngb in the CB of young rats may act as an oxygen sensor, detecting cellular oxygen concentrations. The increase of Ngb levels in the CB induced by chronic hypoxia is consistent with our previously studies (Di Giulio et al. 2006). This observation suggests that a number of CB factors involved with growth and neurotransmission are released in the CB. In terms of a possible link between Ngb and Hypoxic Inducible Factor (HIF), Ngb appears to play a protective role for glomus cell plasticity during aging, considering that aging an hypoxia are linked in some way at the cellular level.

Present results showing that Ngb levels in the CB from young increased following chronic hypoxia suggest that Ngb may enhance oxygen supply to the CB and may contribute to cell survival, as it was found in neuronal tissue (Sun et al. 2001). Thus, it is plausible that the observed age-related decrease in the response of Ngb to hypoxia could explain the reduction of the homeostatic responses to stress with age. However, further studies are needed to establish the role played by Ngb in the CB during aging, and to determine whether the age-related decrease in the hypoxic responsiveness noted herein is related to age and to the oxygen-sensitive Ngb-based mechanisms in the CB.

References

Burmester, T., Haberkamp, M., Mitz, S., Roesner, A., Schmidt, M., Ebner, B., Gerlach, F., Fuchs, C., Hankeln, T. 2004, Neuroglobin and cytoglobin: genes, proteins and evolution, *IUBMB Life*, 56: 703–707.

Burmester, T., Weich, B., Reinhardt, S., Hankeln T. 2000, A vertebrate globin expressed in the brain, *Nature*, 28: 407520–407523.

Di Giulio, C., Bianchi, G., Cacchio, M., Artese, L., Piccirilli, M., Verratti, V., Iturriaga, R. 2006, Neuroglobin, a new oxygen binding protein is present in the carotid body and increases after chronic intermittent hypoxia. In *The Arterial Chemoreceptors. Advances in Experimental Medicine and Biology*, eds. Y. Hayashida, C. Gonzalez and H. Kondo , Springer, New York, pp. 15–20.

Di Giulio, C., Cacchio, M., Bianchi, G., Rapino, C. and Di Ilio, C. 2003, Carotid body as a model for aging studies: is there a link between oxygen and aging? *J Appl Physiol*, 95: 1755–1758.

Gonzalez, C., Almaraz, L., Obeso, A., Rigual, R. 1994, Carotid body chemoreceptors: from natural stimuli to sensory discharges, *Physiol Rev*, 74(4): 829–898. Review.

Lahiri, S., Di Giulio, C. and Roy, A. 2002, Lesson from chronic intermittent and sustained hypoxia, *Respir Physiol Neurobiol*, 130: 223–233. Review.

Mammen, P.P.A., Shelton, J.M., Goetsch, S.C., Williams, S.C., Richardson, J.A., Garry, M.G., Garry, D.J. 2002, Neuroglobin, a novel member of the globin family, is expressed in focal regions of brain, *J Histochem Cytochem*, 50(12): 1591–1598.

Moens, L., Dewilde, S. 2000, Globins in the brain, *Nature*, 28;407(6803): 461–62.

Pesce, A., Dewilde, S., Nardini, M., Moens, L., Ascenzi, P., Hankeln, T., Burmester, T., Bolognesi, M. 2003, Human brain neuroglobin structure reveals a distinct mode of controlling oxygen affinity, *Structure*, 11(9): 1087–1095.

Prabhakar, N.R., Fields, R.D., Baker, T., Fletcher, E.C. 2001, Intermittent hypoxia: cell to system, *Am Physiol Lung Cell Mol Physiol*, 281: 524–528.

Sun, Y., Jin, K., Mao, X.O., Zhu, Y., Greenberg, D.A. 2001, Neuroglobin is up-regulated by a protects neurons from hypoxic-ischemic injury, *Proc Natl Acad Sci USA*, 98: 15306–15311.

Sun, Y., Jin, K., Peel, A., Mao, X.O., Xie, L., Greenberg, D.A. 2003, Neuroglobin protects the brain from experimental stroke in vivo, *Proc Natl Acad Sci USA*, 100(6): 3497–3500.

Vallone, B., Nienhaus, K., Matthes, A., Brunori, M., Nienhaus, G.U. 2004, The structure of carbonmonoxy neuroglobin reveals a heme-sliding mechanism for control of ligand affinity, *Proc Natl Acad Sci USA*, 101(50): 17351–17356.

Zhu, Y., Sun, Y., Jin, K., Greenberg DA. 2002, Hemin induces neuroglobin expression in neural cells, *Blood*, 100: 2494–2498.

Oxygen Sensing and the Activation of the Hypoxia Inducible Factor 1 (HIF-1) – *Invited Article*

Joachim Fandrey and Max Gassmann

Abstract For mammals, oxygen sensing is fundamental to survive. An adequate response to reduced oxygen tension, herein termed hypoxia, requires an instantaneous adaptation of the respiratory and the circulatory systems. While the glomus caroticum as well as the pulmonary and systemic vasculature and potentially also the airway chemoreceptors enable a corresponding response within seconds, changes in gene expression require minutes to hours. Hypoxia-induced gene expression depends on the activation of several transcription factors. Hypoxia-inducible factor-1 (HIF-1) has been identified as the key transcription factor complex that coordinates gene expression during hypoxia. To understand how abundance and activation of HIF-1 is regulated is of fundamental importance as it may open new therapeutic avenues to treat ischemic diseases or cancer where HIF-1 appears to be a key component of the pathophysiology.

Keywords O_2 sensing · Prolyl hydroxylases · PHD · FIH-1 · Erythropoietin

1 Introduction

The key regulatory hormone of red blood cell production, erythropoietin (EPO), served as a paradigm of O_2-regulated gene expression (Fandrey 2004). EPO plasma levels rise in response to hypobaric hypoxia, e.g. high altitude exposure, hypoxic hypoxia (e. g. at lowered inspiratory O_2 concentrations) or anaemic hypoxia. The common denominator of all these stimulators is tissue hypoxia in the interstitial cells of the kidney where most of the EPO is expressed in the adult. Deciphering the mechanisms of hypoxic induction of EPO expression has led to the identification of the key transcription factor complex of hypoxia-induced gene expression that accordingly was termed hypoxia-inducible factor-1 (HIF-1; Semenza and Wang 1992). HIF-1 is now recognized to control more than 100 genes directly

---

J. Fandrey (✉)
Institut fuer Physiologie, Universitaet Duisburg-Essen, Essen, Germany
e-mail: joachim.fandrey@uni-due.de

causing different degrees of increased expression depending on the severity of hypoxia and the cellular context. Constitutive activation of HIF-1 in endothelial cells and subsequent gene array analysis even suggested that more than 2% of all human genes may be directly or indirectly regulated by HIF-1 in this cell type (Manalo et al. 2005). HIF-1 target genes typically function to restore O_2 and energy homeostasis. This includes control of respiration and anaerobic glycolysis, vascular tone, angiogenesis and vascular remodelling, erythropoiesis and iron metabolism, cell growth and death, but also cellular O_2 sensing itself (Höpfl et al. 2004).

2 Hypoxia-Inducible Factor-1 (HIF-1)

HIF-1 is a heterodimer composed of a roughly 120 kDa α-subunit and a 91–94 kDa β-subunit. Both HIF-1 subunits are members of the basic helix-loop-helix (bHLH) – containing PAS domain family of transcription factors. PAS is an acronym of the three founding members of this transcription factor family Period (PER), Aryl hydrocarbon receptor nuclear translocator (ARNT) and Single-minded (SIM) (Höpfl et al. 2004). Both the HIF-α- and β-subunit contain 2 PAS domains, designated PAS-A and PAS-B. The bHLH and PAS domains mediate heterodimer formation between the α- and β-subunit which is necessary for DNA binding by the bHLH domain (Jiang et al. 1996). When the HIF-1 complex was analysed and the two heterodimerizing partners HIFα/β were cloned it turned out that HIF-1β was already known as ARNT, the dimerizing partner for the aryl hydrocarbon (e.g. dioxin) receptor. In contrast, HIF-1α isolated and cloned from binding to the EPO enhancer under hypoxic condition was previously unknown and turned out to be the oxygen sensitive subunit of the HIF complex (Wang et al. 1995).

Three HIF-α subunits have been reported in human cells with five splice variants identified so far (Fandrey et al. 2006). While both HIF-1α and -2α dimerize with HIF-1β to drive hypoxia-induced HIF-1 target gene expression the function of HIF-3α is less obvious. Some studies indicate that at least HIF-3α splice forms may act as antagonists of HIF-1α/-2α by scavenging the β-subunit and binding to DNA without having a transactivation domain (TAD) at its C-terminus to bind additional co-activators (for references Fandrey et al. 2006 and see below). We will herein focus on HIF-1α regulation as the paradigm of O_2-dependent activation of HIF-α subunits. In contrast to hypoxic regulation of HIF-αs, the β-subunit is constitutively expressed and found in the nucleus. Its abundance and activity is not affected by the O_2-tension.

3 Hypoxia-Induced Nuclear Accumulation of HIF-1

When exposed to hypoxia, HIF-1α accumulates in the cell and enters the nucleus via importins (Depping et al. 2008). Subsequently, HIF-1α dimerizes with abundantly nuclear HIF-1β to bind HIF DNA at HIF binding sites (HBS) within regulatory

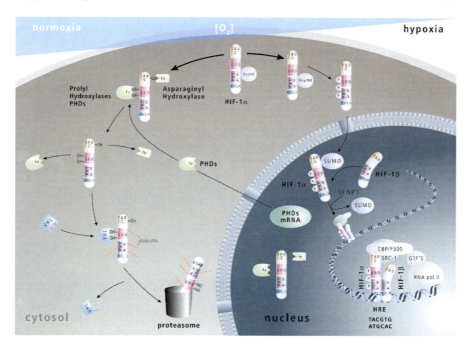

Fig. 1 Regulation of Hypoxia Inducible Factor 1 (HIF-1) depends on the O_2 lability of its HIF-1α subunit. Under normoxia (*left half* of the figure) prolyl hydroxylases (PHDs) are active to hydroxylate HIF-1α for subsequent recognition by the von Hippel-Lindau protein (VHL) which is part of an E3-ubiquitin ligase. Ubiquitinated HIF-1α is degraded in the proteasomes. Under hypoxia (*right half* of the figure) HIF-1α is stable and translocates into the nucleus where it might be further modified, e. g. by sumoylation. After dimerization with HIF-1β the HIF-1 complex binds to the hypoxia response element (HRE) and recruits further coactivators via CBP/P300. Among the HIF-1 target genes PHD2 and PHD3 which become increasingly expressed after HIF-1 activation. Active PHDs form a negative feedback loop on the cytoplasm but also the nucleus

DNA elements termed hypoxia regulatory elements (HRE) of hypoxia-inducible genes (Fig. 1).

For a better understanding of the mechanisms how HIF-1 activates its target genes, it is important to define when HIF-1α and HIF-1β dimerize. Two alternatives are possible: HIF-1α could immediately dimerize to its β-subunit upon its release from the nuclear import machinery. Alternatively, dimerization could depend on DNA binding of each partner via their bHLH-N-terminal end and, subsequently, be followed by dimerization of DNA-bound partners. To address this question, HIF-1α and −β were fluophore-labelled by construction of fusion proteins of HIF-1α and cyano-fluorescent protein (CFP) and HIF-1β and yellow-fluorescent protein (YFP), respectively. Cells were transiently transfected and nuclear mobility of HIF-1α/β fusion proteins was determined by fluorescence recovery after photo bleaching (FRAP) (Wotzlaw et al. 2007). For FRAP, designated areas within the nucleus (where the two subunits localize) were bleached to destroy the fluophore part by intensive laser light. Subsequently, fluorescence recovered over time by diffusion of non-bleached, intact HIF-1α/β fusion proteins into the bleached area. By scrutinizing the nuclei we noticed a heterogeneous distribution of fluophore-labelled HIF-α/β fusion proteins resembling the distribution of endogenous HIF-1α and HIF-1β. We had described this distribution before when we visualized the three-dimensional distribution of both HIF-1 subunits and binding partners by 2-photon-laser microscopy of immunostained cells (Berchner-Pfannschmidt et al. 2004). Interestingly, the heterogeneous speckle-like distribution of HIF-1α/β was not observed when a single fusion protein partner of the HIF complex was transfected. This suggests that a certain ratio of HIF-1α/β partners is required to allow the typical heterogeneous distribution and that the speckle-like structures resemble HIF-1 dimers. During the FRAP experiments both, HIF-1α and HIF-1β fluorescence recovered exactly at the same spots within the nucleus, indicated that the distribution within the nucleus is not at random but that both subunits are directed exactly to the same spots where they were localized before bleaching.

FRAP also allows the semi-quantitative evaluation of the speed by which proteins move. When we compared the time for HIF-1α and –β to recover fluorescence we noticed a significantly delayed recovery for HIF-α that indicates that HIF-1β moves faster within the nucleus than HIF-1α (Wotzlaw et al. 2007). As mentioned out above, FRAP only allows a semi-quantitative evaluation of protein movements. This depends on multiple factors, such as local protein concentration (as a driving force for diffusion), but also protein size and mass and binding to other cellular structures or proteins. In summary, the important finding of these experiments was the significant difference in mobility between HIF-1α and HIF-1β that clearly indicates both dimerizing partners do not form a complex in the nucleus instantaneously, since they would then have been forced to move at the same speed. Our finding is in good agreement with the studies that have reported further modification of HIF-1α within the nucleus, e. g. by sumoylation (Cheng et al. 2007), a process that is difficult to envision for a HIF-1α/β complex and most probably occurs with the undimerized α-subunit.

4 O$_2$ Sensors that Control HIF-1 Abundance and Activity

Upon hypoxia, HIF-1α accumulates instantaneously in the nucleus (Jewell et al. 2001). Interestingly, this is not due to de novo protein synthesis. Instead, oxygen-dependent degradation is halted under hypoxia. The seminal finding that HIF-1α abundance was independent of hypoxia in the absence of the von Hippel-Lindau protein indicated normoxic instability instead of de novo synthesis of HIF-1α upon hypoxia (Maxwell et al. 1999). Additional reports indicated the continuous expression of the *hif-1α* gene with no increase in HIF-1α mRNA and no enhancement of translation upon hypoxia. Thus, regulation was left to the posttranslational level and modifications that could confer VHL-dependent degradation at high pO$_2$ (Fandrey et al. 2006).

Several groups independently of each other identified hydroxylation of proline residues 402 and 564 in human HIF-1α as the key oxygen dependent posttranslational modifications which tags HIF-1α for recognition by VHL. VHL is part of an E3-ubiquitin ligase that poly-ubiquitinates HIF-1α and sends it to proteasomal degradation (Fig. 1). Oxygen lability of HIF-1α protein allows the instantaneous response to hypoxia at no extra energy expenses (Höpfl et al. 2004). While the continuous synthesis at high oxygen levels only provides HIF-1α protein for immediate subsequent prolyl hydroxylation and degradation, stabilization under hypoxia – when ATP might become scarce – does not need any additional energy. Thus, HIF-1α accumulates because it is stabilized and no longer is degraded under hypoxia.

Prolyl hydroxylation of HIF-1α is oxygen-dependent because the enzymes containing a prolyl hydroxylase domain (PHD) need oxygen as a co-substrate for enzymatic activity. So far, four isoforms of PHDs have been identified: PHD1, PHD2, PHD3 and PH4, the latter of which the role with respect to O_2 sensing is less clear (Fandrey et al. 2006). For PHD1 to 3, the hypoxic decrease in activity causes increased accumulation of HIF-1α. Among the PHDs, PHD2 is the most active and probably also most important O_2 sensor, at least for acute hypoxic responses (Berra et al. 2003). Upon hypoxia, enzymatic activity of PHDs ceases and HIF-1α levels rise instantaneously. Upon the return of oxygen, termed reoxygenation, PHDs very rapidly regain activity to hydroxylate HIF-1α causing a half-life of the protein of only 2–5 min in the presence of oxygen (Jewell et al. 2001). The group of PHD O_2 sensors is completed by an asparagyl hydroxylase which O_2-dependently hydroxylates the asparagine 803 of human HIF-1α (Lando et al. 2002). The asparagyl hydroxylase had been identified before and named factor inhibiting HIF-1 (FIH-1) due to its function to inhibit HIF-1 activity (for references Fandrey et al. 2006). Its asparagyl hydroxylase activity under normoxia prevents the recruitment of transcriptional co-activators, such as p300/CBP to the C-terminal transactivation domain (C-TAD) of HIF-1α. Thus, in addition to abundance HIF-1 activity is also repressed under normoxia. With hypoxia, enzyme activity decreases and allows the binding of co-activator proteins to HIF-1α to assemble the active HIF-1 complex.

5 Localization and Abundance of Prolyl Hydroxylases (PHDS)

Studies to localize the cellular O_2 sensors, both the PHDs and FIH-1, revealed a heterogeneous distribution within the cell (Metzen et al. 2003). While FIH-1 was predominantly cytoplasmic, PHDs showed a specific localization for each isoenzyme. PHD1 was only found in the nucleus, PHD2 predominantly in the cytoplasm and PHD3 in both the nucleus and the cytoplasm. It is not yet clear whether the different localization has functional implications, but all PHDs most likely recruit different proteins for full activity. In addition, interacting partners may influence the abundance of the PHDs differently (Fandrey et al. 2006).

PHDs are non-equilibrium enzymes and therefore the amount of enzyme determines its activity. Thus, oxygen sensing of cells can be modulated by induction of PHD proteins (Berchner-Pfannschmidt et al., 2007; Stiehl et al. 2006). It was soon recognized that PHD2 and PHD3 were themselves HIF-1 target genes and were induced by hypoxia (Fandrey et al. 2006; Metzen et al. 2005; Pescador et al. 2005). PHD1 and FIH-1, in contrast, are not affected by low oxygen tension. The HIF-1 dependent induction of cellular O_2 sensors forms a negative feedback loop which may limit HIF-1α accumulation under prolonged hypoxia. In many studies a transient increase of HIF-1α protein was observed with peak amounts around 4 to 8 hours depending on the cell type and severity of hypoxia. Subsequently, HIF-1α levels fell again, coinciding with increased PHD2 and PHD3 expression (Berchner-Pfannschmidt et al. 2007; Berra et al. 2003).

Other stimuli of HIF-1 affect this negative feedback loop and thus O_2 sensing such as the transient release of nitric oxide (NO) from cells or NO-releasing chemical compounds (Berchner-Pfannschmidt et al. 2007). NO acts like hypoxia and inhibits PHD activity by a yet unresolved mechanism (Berchner-Pfannschmidt et al. 2007). As long as NO is present, HIF-1α accumulates and causes HIF-1-dependent gene expression. Among these HIF-1 target genes is PHD2 which accumulates in NO-treated cells under normoxia and hypoxia. When the NO release ceases, newly formed PHD2 becomes active and impinges on O_2 sensitivity of the cells. As a consequence hypoxic HIF-1α accumulation is reduced due to the higher PHD2 activity in NO-treated cells, HIF-1α half-life is shortened and, finally, NO-treated cells become less sensitive to a certain degree of hypoxia (Berchner-Pfannschmidt et al. 2007). Collectively, the inflammatory mediator NO, modulates the hypoxic response. This may have wide implications in an inflammatory setting where macrophages and granulocytes need to function properly in a heavily hypoxic environment but also stromal cells have to cope with reduced oxygenation.

6 Assembly of the HIF Complex

To fully understand O_2 sensing with respect to hypoxia-induced gene expression, assembly of the HIF-1 complex needs to be elucidated. Genes, such as EPO, the paradigm of the HIF-regulated gene, require increased amounts of the HIF-1 complex bound to HREs. In addition, co-activators bridging between HIF-1 and other DNA-bound transcription factors are needed to ensure tissue-specific expression, i. e. under physiologic conditions in the kidneys and the liver (Fandrey 2004).

Full length HIF-1α and HIF-1β subunits hardly go into solution, a fact that has so far prevented successful crystallization of both proteins for X-ray structure analysis (Card et al. 2005). A way to analyse the assembly of the HIF-1 complex, however, is to determine protein-protein interaction between the different partners of the complex in living cells. To this end we have constructed fusion proteins of HIF-1α and CFP in HIF-1β and YFP respectively. Fluophore proteins were orientated both to a C- or N-terminal end of the respective HIF-1 subunit partner (Wotzlaw et al. 2007). Both fluophores, CFP and YFP, make a perfect couple for fluorescence resonance energy transfer (FRET) measurements. This method depends on the radiation-less transfer of energy from one fluophore to another. If the fluophores are sufficiently close (< 10 nm) and the emission spectrum of the donor (here CFP) overlaps with the excitation spectrum of the acceptor (here YFP) the efficiency of the energy transfer from the donor to the acceptor depends on the distance of both molecules. Thus, by determining the FRET efficiency one can calculate the distance between the donor and acceptor. We applied FRET for CFP-HIF-1α and YFP-HIF-1β, both labelled at the N-terminus of the HIF-1 subunits (Wotzlaw et al. 2007). Both constructs very efficiently activated a HIF-1 reporter gene, which indicates that the fluophores did not obstruct the binding of the HIF-1 dimer to DNA. FRET measurements were then performed in living cells in a specially designed chamber on the microscope table which allows to maintain stable CO_2 and O_2 levels over several hours and to induce well defined levels of hypoxia. Calculations based on the different FRET efficiencies revealed a mean distance of 6.6 nm for the N-termini of the HIF-1 partners. When HIF-1α and HIF-1β were labelled at their C-terminal ends, their distance was 7.2 nm (Fig. 2). This is well conceivable with the fact, that

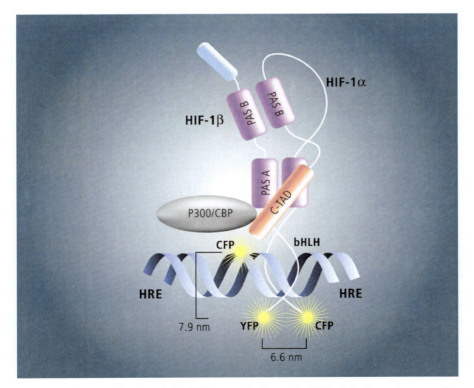

Fig. 2 Putative organization of the HIF-1 complex bound to DNA at the hypoxia responsive element (HRE). The antiparallel orientation of the PAS B segments of HIF-1α and HIF-1β allows the C-terminal end of HIF-1α with the coactivator P300/CBP to come close to the N-terminal end of the HIF-1 complex and the DNA. This proximity may allow interaction with other DNA bound proteins

co-activators like p300/CBP bind to the C-terminus of HIF-1α and may cause a slightly greater distance than at the N-terminus (Höpfl et al. 2004). HIF-1α and HIF-1β labelled on opposite ends were initially thought to serve as a negative control for FRET. Since the models of the HIF-1 complex so far suggested that the N- and the C-terminus would be further apart than 10 nm and therefore should not be close enough for FRET. Unexpectedly, however, a positive FRET response was recorded and a distance of 7.9 nm between the C-terminal CFP of HIF-1α and the N-terminal YFP of HIF-1β was calculated (Wotzlaw et al. 2007).

While these data were obtained, Card et al. successfully crystallized parts of the PAS-B domains of HIF-2α and HIF-1β (Card et al. 2005). Although no information on the organization of the whole HIF-1 complex was obtained in this study, X-ray structure analysis of the PAS-B domains alone indicated an antiparallel orientation of these two parts of HIF-1 subunits (Card et al. 2005). This finding helps to understand our FRET data because it indicates that the antiparallel orientation of the PAS domains causes an organization of the HIF-1 complex that brings the C-terminus close to the DNA where the N-terminal bHLH domains of HIF-1 bind. The N-terminus binds adapter proteins like p300/CBP that form a scaffold to recruit other transcription factors (Fig. 2). Thus, the data from the crystallization and the FRET studies provide first insights into the organization of the HIF-1 complex in living cells and may provide a tool of utmost relevance to better understand HIF-1 activation and function.

7 Outlook

The interest to understand hypoxia induced gene expression arose from increased red blood cell production after high altitude exposure almost 150 years ago. Following back the signalling cascade from the EPO gene to HIF-1 led to the discovery of cellular O_2 sensors, the PHDs and FIH-1 that are in control of O_2 dependent gene expression. We now understand that HIF-1 is a key regulator for O_2 dependent gene expression in physiology, pathophysiology of ischemic disease and tumour biology. This broad-ranged importance has immediately stimulated groups around the world to try to manipulate the HIF-1 system with the intention to treat human diseases. While it may be appropriate to increase HIF-1 in ischemia it may be beneficial to reduce HIF-1 in tumours. But even this needs to be properly studied and is not yet decided. Understanding of the underlying process of oxygen sensing, however, will undoubtedly help to provide the necessary knowledge for a targeted therapeutic approach.

Acknowledgments The authors thank Ms. Marianne Mathys (Audiovisual Services, University of Zurich) for the artwork. This project received financial support from by the European Commission under the 6th Framework Programme (Contract No: LSHM-CT-2005-018725, PULMOTENSION). This publication reflects only the authors' views and the European Community is in no way liable for any use that may be made of the information contained therein.

References

Berchner-Pfannschmidt, U., Wotzlaw, C., Merten, E., Acker, H., & Fandrey, J. 2004, "Visualization of the three-dimensional organization of hypoxia-inducible factor-1alpha and interacting cofactors in subnuclear structures", *Biol. Chem.*, vol. 385, no. 3–4, pp. 231–237.

Berchner-Pfannschmidt, U., Yamac, H., Trinidad, B., & Fandrey, J. 2007, "Nitric oxide modulates oxygen sensing by hypoxia-inducible factor 1-dependent induction of prolyl hydroxylase 2", *J. Biol. Chem.*, vol. 282, no. 3, pp. 1788–1796.

Berra, E., Benizri, E., Ginouves, A., Volmat, V., Roux, D., & Pouyssegur, J. 2003, "HIF prolyl-hydroxylase 2 is the key oxygen sensor setting low steady-state levels of HIF-1alpha in normoxia", *EMBO J.*, vol. 22, no. 16, pp. 4082–4090.

Card, P. B., Erbel, P. J. A., & Gardner, K. H. 2005, "Structural basis of ARNT PAS-B Dimerization: Use of a common beta-sheet interface for hetero- and homodimerization", *J. Mol. Biol.*, vol. 353, no. 3, pp. 664–677.

Cheng, J., Kang, X., Zhang, S., & Yeh, E. T. H. 2007, "SUMO-specific protease 1 is essential for stabilization of HIF1α during Hypoxia", *Cell*, vol. 131, no. 3, pp. 584–595.

Depping, R., Steinhoff, A., Schindler, S. G., Friedrich, B., Fagerlund, R., Metzen, E., Hartmann, E., & Kohler, M. 2008, "Nuclear translocation of hypoxia-inducible factors (HIFs): Involvement of the classical importin α/β pathway", *Biochimica et Biophysica Acta (BBA) – Mol. Cell Res.*, vol. 1783, no. 3, pp. 394–404.

Fandrey, J. 2004, "Oxygen-dependent and tissue-specific regulation of erythropoietin gene expression", *AJP – Regulatory, Integrative and Comparative Physiology*, vol. 286, no. 6, p. R977–R988.

Fandrey, J., Gorr, T. A., & Gassmann, M. 2006, "Regulating cellular oxygen sensing by hydroxylation", *Cardiovasc. Res.*, vol. 71, no. 4, pp. 642–651.

Höpfl, G., Ogunshola, O., & Gassmann, M. 2004, "HIFs and tumors – causes and consequences", *AJP – Regul. Integr. Comp. Physiol.*, vol. 286, no. 4, p. R608–R623.

Jewell, U. R., Kvietikova, I., Scheid, A., Bauer, C., Wenger, R. H., & Gassmann, M. 2001, "Induction of HIF-1alpha in response to hypoxia is instantaneous", *FASEB J.*, vol. 15, no. 7, pp. 1312–1314.

Jiang, B. H., Rue, E., Wang, G. L., Roe, R., & Semenza, G. L. 1996, "Dimerization, DNA binding, and transactivation properties of hypoxia-inducible factor 1", *J. Biol. Chem.*, vol. 271, no. 30, pp. 17771–17778.

Lando, D., Peet, D. J., Gorman, J. J., Whelan, D. A., Whitelaw, M. L., & Bruick, R. K. 2002, "FIH-1 is an asparaginyl hydroxylase enzyme that regulates the transcriptional activity of hypoxia-inducible factor", *Genes Develop.*, vol. 16, no. 12, pp. 1466–1471.

Manalo, D. J., Rowan, A., Lavoie, T., Natarajan, L., Kelly, B. D., Ye, S. Q., Garcia, J. G. N., & Semenza, G. L. 2005, "Transcriptional regulation of vascular endothelial cell responses to hypoxia by HIF-1", *Blood*, vol. 105, no. 2, pp. 659–669.

Maxwell, P. H., Wiesener, M. S., Chang, G. W., Clifford, S. C., Vaux, E. C., Cockman, M. E., Wykoff, C. C., Pugh, C. W., Maher, E. R., & Ratcliffe, P. J. 1999, "The tumour suppressor protein VHL targets hypoxia-inducible factors for oxygen-dependent proteolysis", *Nature*, vol. 399, no. 6733, pp. 271–275.

Metzen, E., Berchner-Pfannschmidt, U., Stengel, P., Marxsen, J. H., Stolze, I., Klinger, M., Huang, W. Q., Wotzlaw, C., Hellwig-Burgel, T., Jelkmann, W., Acker, H., & Fandrey, J. 2003, "Intracellular localisation of human HIF-1 alpha hydroxylases: implications for oxygen sensing", *J. Cell. Sci.*, vol. 116, no. Pt 7, pp. 1319–1326.

Metzen, E., Stiehl, D. P., Doege, K., Marxsen, J. H., Hellwig-Burgel, T., & Jelkmann, W. 2005, "Regulation of the prolyl hydroxylase domain protein 2 (phd2/egln-1) gene: identification of a functional hypoxia-responsive element", *Biochem. J.*, vol. 387, no. Pt 3, pp. 711–717.

Pescador, N., Cuevas, Y., Naranjo, S., Alcaide, M., Villar, D., Landazuri, M. O., & Del Peso, L. 2005, "Identification of a functional hypoxia-responsive element that regulates the expression of the egl nine homologue 3 (egln3/phd3) gene", *Biochem. J.*, vol. 390, no. Pt 1, pp. 189–197.

Semenza, G. L. & Wang, G. L. 1992, "A nuclear factor induced by hypoxia via de novo protein synthesis binds to the human erythropoietin gene enhancer at a site required for transcriptional activation", *Mol. Cell. Biol.*, vol. 12, no. 12, pp. 5447–5454.

Stiehl, D. P., Wirthner, R., Koditz, J., Spielmann, P., Camenisch, G., & Wenger, R. H. 2006, "Increased prolyl-4-hydroxylase domain (PHD) proteins compensate for decreased oxygen levels: evidence for an autoregulatory oxygen-sensing system", *J. Biol. Chem.*, p. M601719200.

Wang, G. L., Jiang, B. H., Rue, E. A., & Semenza, G. L. 1995, "Hypoxia-inducible factor 1 is a basic-helix-loop-helix-PAS heterodimer regulated by cellular O_2 tension", *Proc. Natl. Acad. Sci. USA*, vol. 92, no. 12, pp. 5510–5514.

Wotzlaw, C., Otto, T., Berchner-Pfannschmidt, U., Metzen, E., Acker, H., & Fandrey, J. 2007, "Optical analysis of the HIF-1 complex in living cells by FRET and FRAP", *FASEB J.*, vol. 21, no. 3, pp. 700–707.

Upregulation of Erythropoietin and its Receptor Expression in the Rat Carotid Body During Chronic and Intermittent Hypoxia

S.Y. Lam, G.L. Tipoe and M.L. Fung

Abstract The carotid body (CB) plays important roles in cardiorespiratory changes in intermittent hypoxia (IH). Erythropoietin (EPO), a hypoxia-inducible factor (HIF)-1 target gene, is present in the chemoreceptive type-I cells in the CB but its expression and role in IH resembling sleep apnoeic conditions are not known. We hypothesized that IH upregulates the expression of EPO and its receptor (EPOr) in the rat CB. The CB expressions of EPO and EPOr were examined in rats breathing 10% O_2 (in isobaric chamber for CH, 24 hour/day) or in IH (cyclic between air and 5% O_2 per minute, 8 hour/day) for 3–28 days. Immunohistochemical studies revealed that the EPO and EPOr proteins were localized in CB glomic clusters. The proportional amount of cells with positive staining of EPO immunoreactivities was significantly increased in both IH and CH groups when compared with the normoxic control. The EPO expression was more markedly increased in the CH than that of the IH groups throughout the time course, reaching a peak level at day 14. The positive EPOr immunostaining was increased significantly in the 3-day CH group. By day 14, the EPOr expression elevated considerably at peak levels in both IH and CH rats, whereas the elevation was greater in the CH rats. These results suggest an upregulation of EPO and its receptor expression in the rat CB under IH and CH conditions, presumably mediated by the activation of HIF-1 pathway. The increased EPO binding to its receptor might play a role in the enhancement of CB excitability during the early pathogenesis in patients with sleep-disordered breathing.

Keywords Carotid body · Erythropoietin · Chronic hypoxia · Intermittent hypoxia

M.L. Fung (✉)
Departments of Physiology, The University of Hong Kong, Faculty of Medicine Building, Pokfulam, Hong Kong SAR, China
e-mail: fungml@hkucc.hku.hk

1 Introduction

Carotid body (CB) plays important roles in cardiorespiratory changes in both chronic hypoxia (CH) and intermittent hypoxia (IH) (Gonzalez et al. 1994; Prabhakar 2001). During CH, the hypertrophy and hyperplasia of CB glomus cells contribute to the CB enlargement as observed in high altitude dwellings and patients with chronic obstructive pulmonary disease (Bee and Howard 1993; Heath et al. 1970). Under IH conditions, hypoxic chemosensitivity and the ventilatory chemoreflex were significantly augmented in the rat CB (Peng et al. 2004). Also, IH enhances the hypoxia-evoked neurotransmitter release from the CB as well as PC12 cells (Kim et al. 2004).

Hypoxia-inducible factor (HIF)-1α plays a physiological role in CH and IH. The expression of HIF-1α and the transcriptional regulation of an array of HIF-regulated genes in the rat CB is an essential molecular mechanism for the structural remodelling and functional changes of the organ under hypoxic conditions (Lam et al. 2008; Prabhakar 2001). Upregulation of the HIF-regulated genes such as vascular endothelial growth factor (VEGF), VEGF receptors, endothelin-1 and tyrosine hydroxylase are crucial for the CB response to hypoxic conditions (Chen et al. 2002; Fung 2003; Lam et al. 2008; Tipoe and Fung 2003).

Erythropoietin (EPO), a HIF-regulated gene, despite its novel erythropoietic function in enhancing red blood cells production, has been shown to modulate the ventilatory response to reduced oxygen supply mediated by activation of CB chemoreceptors. In fact, the CB was proposed to be an endocrine gland participating in the control of erythropoiesis (Tramezzani et al. 1971). Recent study has demonstrated that exogenous EPO could activate the CB chemoreceptor during CH. In addition, EPO receptor (EPOr) was found to be present in the chemoreceptive type-I cells in the CB (Soliz et al. 2005). Furthermore, high EPO plasma levels could modulate the CB chemotransduction in CH (Soliz et al. 2007). However, the expression and role of EPO under IH resembling sleep apnoeic conditions are unclear. It is hypothesized that the expression of EPO and its receptor (EPOr) in the rat CB could be regulated by CH or IH. The aim of the present study was to examine the CB expressions of EPO and EPOr in rats breathing 10% O_2 for CH or in cyclic O_2 levels between 5–21% for IH for 3–28 days. Our results suggest an upregulation of EPO and its receptor expression in the rat CB under IH and CH conditions, presumably mediated by the activation of HIF-1 pathway.

2 Methods

The experimental protocol for this study was approved by the Committee on the Use of Live Animals in Teaching and Research of The University of Hong Kong. Male Sprague-Dawley rats aged 28 days were randomly divided into groups for the treatment of CH or IH and for normoxic (Nx). While the Nx controls were kept in room air with the same housing and maintenance, both CH and IH rats were kept in acrylic chambers for normobaric hypoxia and had free access to water and chow (Tipoe and Fung 2003). The oxygen fraction inside the chamber was kept at 10 ± 0.5%, 24 h per day for the CH group and was cyclic from 21 to 5 ± 0.5% per min, 60 cycles/h, 8 h per day diurnally for the IH group in which the inspired oxygen level fell to 4–5% (nadir arterial oxygen saturation ca. 70%) for about 15 sec per min, which mimicks the recurrent episodic hypoxemia in patients with obstructive sleep apnoea (Fletcher 2001). The desired oxygen level was established by a mixture of room air and nitrogen that was regulated and monitored by an oxygen analyzer (Vacumetrics Inc., CA, USA). Carbon dioxide was absorbed by soda lime granules and excess humidity was removed by a desiccator. The chamber was opened twice a week for an hour to clean the cages and replenish food and water. The rats were exposed to hypoxia for 3, 14 and 28 days and were immediately used in experiments after taken out of the chambers.

Following deep anesthesia with halothane, the rat was decapitated and the carotid bifurcation was excised rapidly. The CB was carefully dissected free from the bifurcation and was fixed in neutral buffered formalin for 72 hour. Tissues were processed routinely for histology and embedded in paraffin blocks. Serial sections of 5 μm thickness were cut and mounted on silanized slides (DAKO, Denmark). Sections were kept in the oven overnight at 56°C. Consequently, sections were deparaffinized, rehydrated and immunostained with antiserum to the following proteins: EPO (rabbit IgG antibody, 1:500 dilution, Cat # E0271, Sigma, USA); EPOr (goat IgG antibody, 1:500 dilution, Cat # E4644, Sigma, USA), using LSAB kit (K0690, DAKO). In brief, sections were immersed in antigen retrieval solution (0.1 M citric acid buffer, pH 6.0) for 10 min at 98°C. To block endogenous peroxidase activity, the sections were immersed in 3% hydrogen peroxide for 5 min at room temperature. All sections were immersed in a solution containing 0.01% trypsin and calcium chloride for 5 min at room temperature and were pre-incubated with 20% normal serum for 2 hour to reduce non-specific binding for the antiserum. Then sections were incubated with the corresponding primary antibodies in 0.05 M Tris-HCl buffer, respectively, containing 2% bovine serum albumin overnight at 4°C. Sections were washed three times in PBS, and then incubated with biotinylated link agent and streptavidin peroxidase for 30 min at room temperature. Finally, sections were washed and the peroxidase was visualized by immersing in 0.05% diaminobenzidine containing 0.03% H_2O_2 in Tris-HCl buffer (pH 7.5). Sections were counterstained mildly with hematoxylin. Control sections were prepared by substitution of primary antibodies with buffer or normal serum, and liquid phase pre-adsorption of the used antibodies with excess blocking peptides (Sigma, USA).

The immunoreactivities of EPO and EPOr were measured using the Leica QWIN Imager Analyzer (Cambridge, UK). Immunostained sections were captured with a CCD JVC camera using a Zeiss Axiophot microscope at 100X objective. The luminance incident light passing through each section was calibrated using the setup menu where the grey pixel values were set to 0 and 1.00. Once the setup was done, five fields per section from one CB of one animal were measured. The percent area of positive stain for both EPO and EPOr protein was measured by detecting the positive brown cytoplasmic stain divided by the sum areas of the reference field.

A total of 20 fields for 4 CBs from four different animals at each time-point namely day 3, 14 and 28 were determined. The mean value of the 20 fields was calculated to represent each time-point.

GraphPad Prism® software (GraphPad Software, Inc., San Diego, USA) was used to analyze the data. A non-parametric Mann-Whitney U-test was used to compare differences between time-points. A p-value < 0.05 was considered statistically significant.

3 Results

Immunohistochemical studies revealed that the EPO and EPOr proteins were positively stained in most of cells throughout the CBs of rats exposed to CH and IH treatment ($n=5$). The positive staining of EPO and EPOr were localized in CB glomic clusters (Figs. 1A and 2A). Moreover, the immunostaining was not observed

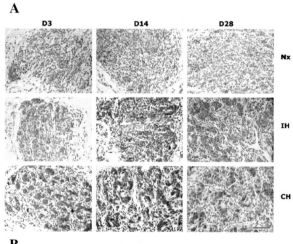

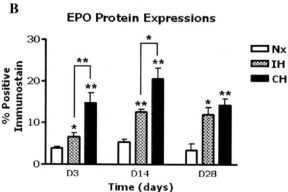

Fig. 1 Immunohistochemical localization of EPO in 3-, 14-, and 28-day treatment of IH, CH and their corresponding Nx in rat CB. Bar = 40 μm. (**A**) Protein expressions of EPO in carotid body of rat in Nx, IH and CH for 3 (D3), 14 (D14) and 28 days (D28). Data are presented in % area with positive staining of EPO immunoreactivity and are expressed as mean ± SEM (n=5 for each group)(**B**). $^*p<0.05$; $^{**}p<0.005$

Fig. 2 Immunohistochemical localization of EPOr in 3-, 14- and 28-day treatment of IH, CH and their corresponding Nx in rat CB. Bar = 40 μm (**A**). Protein expressions of EPOr in carotid body of rat in Nx, IH and CH for 3 (D3), 14 (D14) and 28 days (D28). Data are presented in % area with positive staining of EPOr immunoreactivity and are expressed as mean ± SEM ($n = 5$ for each group) (**B**). $^*p<0.05$; $^{**}p<0.005$

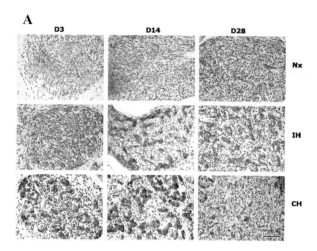

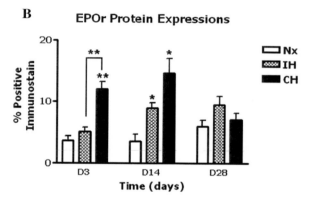

in the CB of the corresponding control sections incubated with normal serum instead of the primary antibodies (data not shown).

The proportional amount of cells with positive staining of EPO immunoreactivities (IR) was significantly increased in both IH and CH groups when compared with the Nx control. The EPO expression was markedly increased in the 3-day group of both IH and CH rats, while the elevation was less in IH than CH. A sustained significant increase of EPO expression was found throughout the time course and reached a peak level at day 14 in both IH and CH rats, but augmented in a greater extent in CH than IH (Fig. 1B).

The positive staining of EPOr expression was markedly increased in the 3-day groups of CH rats. By day 14, the EPOr expression elevated considerably and reached a peak level in both the IH and CH rats, whereas the elevation was greater

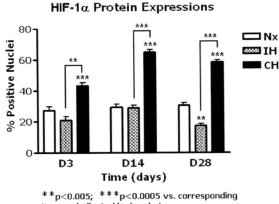

Fig. 3 Protein expressions of HIF-1α in the carotid body of rat in Nx, IH and CH for 3 (D3), 14 (D14) and 28 days (D28). Data are presented in % nuclei with positive staining of HIF-1α immunoreactivity and are expressed as mean ± SEM (n=5 for each group). Adapted and modified from Lam et al., *Histol Histopathol* (2008) 23:271–280.
** $p<0.005$; *** $p<0.0005$

in CH than IH. Moreover, EPOr expression reduced to levels above the Nx control in the CH group but remained elevated in IH by day 28 (Fig. 2B).

As reported previously, positive staining of HIF-1α in the rat CB is significantly more in the CH, but not in the IH group (Fig. 3 modified from (Lam et al. 2008). Collectively, these results suggest an upregulation of the expression of EPO and its receptor in the rat CB under IH and CH conditions, presumably mediated by the activation of HIF-1 pathway. These observations are consistent with the hypothesis that EPO plays a functional role in the enhancement of CB excitability under hypoxic conditions.

4 Discussion

EPO acts as a key regulator of erythropoiesis, by promoting the survival and proliferation of erythroid precursor cells (Jelkmann 1992). The kidney produces EPO in response to hypoxia, and thereby increases the number of red blood cells in order to increase the tissue oxygen supply (Pagel et al. 1992). Evidence suggest that the CB is an endocrine gland playing a part in the control of erythropoiesis despite its classical chemoreceptive function (Tramezzani et al. 1971). The present study demonstrated that (1) EPO and its receptors, EPOr were expressed and localized in CB glomic clusters; (2) The proportional amount of cells with positive staining of EPO and EPOr immunoreactivities were significantly increased in both IH and CH groups when compared with the Nx control. These findings unequivocally support the hypothesis that CH and IH upregulate the EPO and its receptor expression in the rat CB.

Physiological responses to hypoxia involve changes in gene expression that are mediated by the transcriptional activator HIF-1. As cellular oxygen concentration decreases, expression of HIF-1α increases exponentially with the upregulation of various proteins such as EPO to exert the physiological effects (Wang et al. 1995).

Along with the presence and upregulation of HIF-1α in the nuclei of the CB in hypoxia, it has been demonstrated that EPO is indeed expressed in the CB glomus cells and its expression is increased in CH and IH. Our study showed that the expression was more markedly increased in the CH than that of the IH groups throughout the time course, reaching a peak level at day 14. The differential expression of HIF-1α in CH and IH might account for the expression profile of EPO under two different hypoxic conditions. Nonetheless, elevated EPO expression observed in the CB during CH and IH might contribute to the activation of the peripheral chemoreceptors in CB and the modulation of the CB chemotransduction in hypoxia.

EPO has been shown to modulate the ventilatory response to reduced oxygen supply despite its novel erythropoietic function in enhancing red blood cells production. Recent study has showed that exogenous EPO activates the peripheral chemoreceptor in CH. The expression of EPOr in the medullary respiratory centre as well as in the CB chemosensitive type-I cells suggests that EPO could modulate the hypoxic ventilatory response at both the central and peripheral levels (Soliz et al. 2005, 2007). In agreement, our results demonstrated that EPOr was expressed in the CB glomus cells and the expression was upregulated by both CH and IH. Thus, the positive EPOr immunostaining was increased significantly in the 3-day CH group and was elevated considerably at peak levels by day 14 in both the IH and CH rats with a greater extent of elevation in the CH group. This expression profile is in conjunction to the HIF-1α expression in the CB, although the EPOr expression was partially returned to the Nx level at day 28.

In summary, these results suggest an upregulation of the expression of EPO and its receptor in the CB glomic clusters in the CH and IH rats. The increased EPO expression in the CB may be mediated by the activation of HIF-1 pathway under hypoxic conditions. It is speculated that the increased expression of EPO and its receptor might be functionally related to the augmented CB hypoxic chemosensitivity and ventilatory chemoreflex under CH and IH conditions, which are relevant to the physiological acclimation to high altitude and the pathophysiological changes in patients with sleep-disordered breathing.

Acknowledgments We thank Mr. W.B. Wong and Ms. K.M. Leung for their technical assistance. This work was supported by research grant (HKU 7510/06M) from the Research Grants Council, HKSAR and the University of Hong Kong.

References

Bee, D., and Howard, P., 1993, The carotid body: a review of its anatomy, physiology and clinical importance, *Monaldi Arch Chest Dis* 48(1):48–53.

Chen, Y., Tipoe, G. L., Liong, E., Leung, S., Lam, S. Y., Iwase, R., Tjong, Y. W., and Fung, M. L., 2002, Chronic hypoxia enhances endothelin-1-induced intracellular calcium elevation in rat carotid body chemoreceptors and up-regulates ETA receptor expression, *Pflugers Arch* 443(4):565–573.

Fletcher, E. C., 2001, Invited review: Physiological consequences of intermittent hypoxia: systemic blood pressure, *J Appl Physiol* 90(4):1600–1605.

Fung, M. L., 2003, Hypoxia-inducible factor-1: a molecular hint of physiological changes in the carotid body during long-term hypoxemia? *Curr Drug Targets Cardiovasc Haematol Disord* 3(3):254–259.

Gonzalez, C., Almaraz, L., Obeso, A., and Rigual, R., 1994, Carotid body chemoreceptors: from natural stimuli to sensory discharges, *Physiol Rev* 74(4):829–98.

Heath, D., Edwards, C., and Harris, P., 1970, Post-mortem size and structure of the human carotid body, *Thorax* 25(2):129–140.

Jelkmann, W., 1992, Erythropoietin: structure, control of production, and function, *Physiol Rev* 72(2):449–489.

Kim, D. K., Natarajan, N., Prabhakar, N. R., and Kumar, G. K., 2004, Facilitation of dopamine and acetylcholine release by intermittent hypoxia in PC12 cells: involvement of calcium and reactive oxygen species, *J Appl Physiol* 96(3):1206–1215; discussion 1196.

Lam, S. Y., Tipoe, G. L., Liong, E. C., and Fung, M. L., 2008, Differential expressions and roles of hypoxia-inducible factor-1alpha, -2alpha and -3alpha in the rat carotid body during chronic and intermittent hypoxia, *Histol Histopathol* 23(3):271–280.

Pagel, H., Engel, A., and Jelkmann, W., 1992, Erythropoietin induction by hypoxia. A comparison of in vitro and in vivo experiments, *Adv Exp Med Biol* 317:515–519.

Peng, Y. J., Rennison, J., and Prabhakar, N. R., 2004, Intermittent hypoxia augments carotid body and ventilatory response to hypoxia in neonatal rat pups, *J Appl Physiol* 97(5):2020–2025.

Prabhakar, N. R., 2001, Oxygen sensing during intermittent hypoxia: cellular and molecular mechanisms, *J Appl Physiol* 90(5):1986–1994.

Soliz, J., Joseph, V., Soulage, C., Becskei, C., Vogel, J., Pequignot, J. M., Ogunshola, O., and Gassmann, M., 2005, Erythropoietin regulates hypoxic ventilation in mice by interacting with brainstem and carotid bodies, *J Physiol* 568(Pt 2):559–571.

Soliz, J., Soulage, C., Hermann, D. M., and Gassmann, M., 2007, Acute and chronic exposure to hypoxia alters ventilatory pattern but not minute ventilation of mice overexpressing erythropoietin, Am J Physiol Regul Integr Comp Physiol 293(4):R1702–R1710.

Tipoe, G. L., and Fung, M. L., 2003, Expression of HIF-1alpha, VEGF and VEGF receptors in the carotid body of chronically hypoxic rat, *Respir Physiol Neurobiol* 138(2–3):143–154.

Tramezzani, J. H., Morita, E., and Chiocchio, S. R., 1971, The carotid body as a neuroendocrine organ involved in control of erythropoiesis, *Proc Natl Acad Sci USA* 68(1):52–55.

Wang, G. L., Jiang, B. H., Rue, E. A., and Semenza, G. L., 1995, Hypoxia-inducible factor 1 is a basic-helix-loop-helix-PAS heterodimer regulated by cellular O_2 tension, *Proc Natl Acad Sci USA* 92(12):5510–5514.

Iron Chelation and the Ventilatory Response to Hypoxia

Mieczyslaw Pokorski, Justyna Antosiewicz, Camillo Di Giulio and Sukhamay Lahiri

Abstract Chelation of iron in in vitro carotid body emulates the effects of hypoxia. The role iron plays in in vivo ventilatory responses is unclear. In the current study we addressed this issue by examining the effects of chronic iron chelation on the hypoxic ventilatory response in 9 conscious Wistar rats. Acute responses to 14 and 9% O_2 in N_2 were recorded in the same rat before and then after 7 and 14 days of continuous iron chelation. Iron chelation was carried out with ciclopirox olamine (CPX) in a dose of 20 mg/kg daily, i.p. Ventilation was recorded with whole body plethysmography. We found that the peak hypoxic ventilation (V_E) achieved during 14 and 9% hypoxia was lower by 239.6±55.4(SE) and 269.6.2±69.2 ml·min^{-1}·kg^{-1}, respectively, in the rats treated with CPX for 7 days. The decreases were not intensified by a longer duration of iron chelation. CPX failed to alter hypoxic sensitivity, assessed from the gain of peak V_E with increasing strength of the hypoxic stimulus. In conclusion, we believe we have shown that iron is operational in shaping the hypoxic ventilatory response, but is not liable to be the underlying determinant of the hypoxic chemoreflex.

Keywords Carotid body · Ciclopirox olamine · Hypoxic sensitivity · Hypoxic ventilatory response · Iron · Iron chelation

1 Introduction

Iron metabolism is essential for hypoxia-sensing in the carotid body. The hypoxia-inducible factor (HIF)-1α, a protein in control of O_2 homeostasis, is degraded under the normoxic condition by hydroxylation requiring iron ions (Semenza 1999, Linden et al. 2003). Hypoxia reduces hydroxylase activity, which stabilizes the HIF-1α protein. HIF-1α is present in carotid chemoreceptor cells and is activated by

M. Pokorski (✉)
Department of Respiratory Research, Medical Research Center, Polish Academy of Science, Warsaw, Poland
e-mail: m_pokorski@hotmail.com

hypoxia (Baby et al. 2003, Roy et al. 2004a, b). Therefore, it is a reasonable assumption that doing away with iron would have a profound effect on the hypoxia-sensing process. Indeed, chelation of iron in in vitro glomus cells emulates the effects of hypoxia in that the chemosensory discharge, the intracellular level of Ca^{2+}, and the HIF-1α expression all increase (Daudu et al. 2002, Roy et al. 2004a, b). Moreover, iron chelation-induced increase in chemosensory discharge of an in vitro rat carotid body replaces the sensory excitatory effect of hypoxia (Daudu et al. 2002). Iron chelation also induces HIF-1α target gene expression (Linden et al. 2003). However, functional studies of the role of iron in the in vivo hypoxic chemoreflex are missing. Here, we analyzed the effects of chronic iron chelation on the hypoxic ventilatory response in awake rats. The results of the present study lend support for the regulatory role of iron in hypoxia-sensing.

2 Methods

The study was approved by a local Ethics Committee. Experiments were performed in 9 adult male Wistar rats aged 10–11 weeks and weighing 294–340 g. The rats received food and water ad libitum and were housed on a 12 h light/12 h dark cycle throughout the study time.

Ventilatory measurements were made in a whole body plethysmograph (Buxco Electronics, Wilmington, NC). The system consisted of a chamber equipped with two pneumotachographs. Pressure difference between the experimental and reference chambers was measured with a differential pressure transducer. The pressure signal was amplified and integrated by data analysis software (Biosystem XA for Windows, Buxco Electronics). Minute ventilation (V_E ml·min^{-1}, BTPS), tidal volume (V_T), and respiratory frequency (f) were computed breath-by-breath and stored for off-line analysis. Ten second averages of each variable were taken every 30 sec during hypoxic exposure.

After the rat acclimated to the chamber, basal ventilation was recorded. Next, the rat was randomly exposed to two levels of decreasing inspired O_2: 14 and 11% O_2 in N_2. Hypoxic tests took 3 min and were separated by 15 min normoxic recovery intervals. The switch of inspired gas took approximately 40 sec to achieve equilibrium in the chamber, and then the time count of a hypoxic test started. During hypoxia, the expired CO_2 was allowed to run free.

Ventilatory studies were repeated thrice in each animal in like manner in the three consecutive experimental conditions: basal control, then after 7 and 14 days of iron chelation; the study covered a 2-week period. Iron chelation was carried out with ciclopirox olamine (CPX), a cell penetrating, lipophilic chelator, in a dose of 20 mg/kg daily, i.p., in a volume of 0.75 ml.

Data were expressed as means ±SE. Ventilatory variables were normalized to weight in kg. Differences in resting ventilation and in peak hypoxic stimulation at the corresponding time points across the three experimental conditions were estimated with a Kruskal-Wallis test followed by a Mann-Whitney U test.

Significance of the hypoxic V_E increases from baseline to peak at 30 sec and decreases from peak to depressant nadir at 180 sec was evaluated with a paired t-test. Criterion for statistical significance was $P<0.05$.

3 Results

3.1 Baseline Ventilation

Baseline animal characteristics and ventilatory data during air breathing before the start of hypoxic tests are shown in Table 1. There was a slight, not exceeding 5–10%, gain in the animals' weight during the 2-week study period. Resting control V_E amounted to 1092.8 ± 97.2 ml·min^{-1}·g^{-1}. Iron chelation caused a modest decrease in resting V_E, which amounted to about 13% after 7 days, with no further decline during continued chelation for up to 2 weeks; the decreases failed to reach statistical significance compared with the control baseline level. Both frequency and tidal breathing contributed to those declines in resting V_E during iron chelation (Table 1).

Table 1 Animals' data and resting ventilation across the three experimental conditions: control, 7 and 14 days of iron chelation

	Control	CPX – 7 Days	CPX – 14 Days
n	9	9	9
Age (wk)	10–11	11–12	12–13
Weight (g)	294–340	250–328	280–360
V_E (ml·min^{-1}·kg^{-1})	1092.8 ±97.2	948.7 ±94.1	967.6 ±61.4
f (breaths·min^{-1})	99.2 ±6.4	97.2 ±3.7	96.1±3.2
V_T (ml·kg^{-1})	11.2 ±0.5	10.0 ±1.1	10.2 ±0.6

Values are means ±SE; n – number of animals studied.

3.2 Hypoxic Ventilatory Responses

In this article, attention was focused on the three key points in the course of hypoxic ventilatory responses: baseline control, hypoxic stimulatory peak, and hypoxic depressant nadir. The mean V_E values at the above-mentioned time points during 14 and 9% hypoxia in the control condition and after 7 and 14 days of iron chelation are displayed in Table 2. The control ventilatory responses showed a typical biphasic pattern, with the peak ventilation noted at 30 sec from the beginning of a test and depressant nadir at 180 sec; the ending time point of the test. The stronger hypoxic stimulus, expectedly, caused a 2-fold greater increase in peak V_E than the milder level of hypoxia.

Iron chelation changed the profile of the hypoxic ventilatory response. Rats treated with CPX for 7 days still responded to both 14 and 9% hypoxia with a rapid increase in V_E ($P<0.03$ for 14% and $P<0.0001$ for 9% hypoxia, t-test) (Table 2).

Table 2 Minute ventilation (V_E) at the three time points of interest during the hypoxic runs: baseline – 0 sec, peak – 30 sec, and nadir – 180 sec during 14 and 9% hypoxia across the three experimental conditions studied: control – untreated rats, CPX-treated for 7 days, and CPX-treated for 14 days

Hypoxic test	14% Hypoxia V_E (ml·min^{-1}·kg^{-1})			9% Hypoxia V_E (ml·min^{-1}·kg^{-1})		
	Baseline	Peak	Nadir	Baseline	Peak	Nadir
Time of test (sec)	0	30	180	0	30	180
Control- untreated	1081.9	1499.0	1225.4	1078.4	2142.1	1974.0
	±95.6	±116.1*	±97.4	±74.9	±82.7*	±186.4
CPX-treated for 7 days	913.3	1079.2	935.2	885.1	1731.5	1810.9
	±74.3	±108.0	±76.4	±88.4	±149.3	±161.9
CPX-treated for 14 days	881.1	1088.9	1020.7	911.2	1702.0	1709.3
	±62.6	±53.3	±67.6	±62.3	±157.5	±161.8

Values are means ±SE from nine rats; CPX - ciclopirox olamine, 20 mg/kg given i.p. daily; *P<0.03 for differences in the peak hypoxic V_E in the untreated *vs.* both CPX-treated for 7 and 14 days rats for the two levels of hypoxia.

However, the peak hypoxic V_E in the CPX-treated rats failed short of that reached by the untreated rats (p<0.03, Table 2). The dampening effect of CPX was evident during the responses to both levels of hypoxia recorded after 7 days. A second week of CPX chelation remained without any further appreciable effect on the V_E. Since the depressant nadir of the response decreased disproportionately less than the magnitude of peak V_E, the profile of the hypoxic response was flattened out, with the distinction between the peak and nadir levels being blurred (Table 2).

Both frequency and volume components contributed to the hypoxic ventilatory responses in both control and CPX-treated conditions. The proportion of this contribution did not change significantly after CPX. However, the dampening of peak hypoxic V_E after 7 days' CPX treatment tended to be due mostly to inhibition of V_T, with smaller changes in breath frequency. On average, peak V_T decreased by about 18 and 13% and f by 7 and 1% during the responses to 14 and 9% hypoxia, respectively. These changes persisted throughout the second week of CPX treatment.

3.3 Hypoxic Ventilatory Gain

The hypoxic ventilatory gain may be evaluated from the increment in peak hypoxic V_E with increasing strength of the hypoxic stimulus. The peak V_E achieved at the two levels of hypoxia in the three experimental conditions are graphically illustrated in Fig. 1. CPX given for 7 days decreased the peak V_E by about the same magnitude at both levels of hypoxia and the decreases did not progress after additional 7 days of iron chelation. Therefore, the response lines, describing hypoxic sensitivity were shifted downward in a parallel manner. The inset in Fig. 1 depicts the mean increments in peak V_E, calculated from individual differences between the peak V_E achieved in response to 9 and 14% hypoxia in each condition. In the

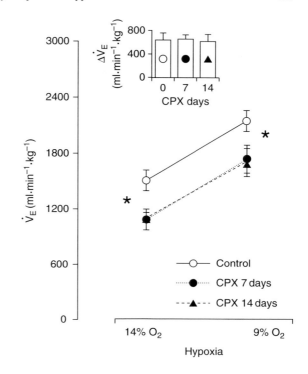

Fig. 1 Hypoxic ventilatory gain assessed from the increment in peak hypoxic V_E achieved between 14 and 9% hypoxia in the three experimental conditions: control, CPX-7 days, and CPX-14 days. *P<0.03 for differences between peak V_E in the untreated vs. both CPX-7 and CPX-14 days' treated rats for both levels of hypoxia (Kruskal-Wallis followed by Mann-Whitney U test). *Inset* - the mean increments in peak V_E with the stronger hypoxic stimulus in each condition. There were no significant differences among the V_E increments

control condition, the V_E increment amounted to 643.1 ± 101.2 ml·min^{-1}·kg^{-1} and remained grossly unchanged after 7 and 14 days of iron chelation.

4 Discussion

The present study demonstrates that chronic iron chelation decreased, but did not abolish, the ventilatory response to acute hypoxia in conscious rats. The decrease was fairly modest, albeit statistically significant, and concerned mostly the fast stimulatory phase of the response. No major effects of iron chelation on the subsequent ventilatory roll-off could be discerned. Since the rapid hypoxic stimulation is mainly, if not solely, mediated by the carotid body (Lahiri et al. 2001), we may well reason that iron ions are operational in shaping the nascent excitatory reaction generated by carotid chemoreceptor cells in response to the institution of hypoxia. The role of iron, however, does not seem to have extended to the innate mechanisms controlling the hypoxic reactivity, as we could not substantiate any influence of iron chelation on the hypoxic ventilatory gain with increasing strength of the hypoxic stimulus. The latter finding should be interpreted with caution, as in the present study we used just two levels of hypoxic stimulation, which may not entirely forejudge the effect of iron on hypoxic sensitivity.

CPX is a lipophilic iron chelator that permeates through the plasma membrane and binds intracellular free iron (Linden et al. 2003). The corollary is that it is free iron in the chemoreceptor cells that is liable to interact with the stimulatory phase of the hypoxic ventilatory response. In the present study, the effect of CPX on hypoxic ventilation was evident after 7 days and an extension of its administration failed to make a difference. It seems, therefore, that whatever iron there was to be accessed by the chelator was bound during the first week.

The present findings seem at variance with previous studies that showed quite a dedicated effect of iron chelation on the carotid body or glomus cells in vitro (Daudu et al. 2002). In those studies, CPX emulated the effect of hypoxia, leading to increased chemosensory discharge and Ca^{2+} influx in the glomus cells. These effects were instantaneous, requiring seconds to be expressed, and could be specifically ascribed to the lack of active iron ions, as they were reversed by the addition of ferrous iron (Daudu et al. 2002). It is unknown how the in vitro findings translate into modifications of the hypoxic ventilatory response in conscious animals. The dampening of the ventilatory response observed in the present study may be reminiscent of chronic hypoxic blunting (Pokorski and Lahiri 1991). Both chronic hypoxia in vivo (Tipoe and Fung 2003, Di Giulio et al. 2005) and chronic iron chelation (Linden et al. 2003, Wang and Semenza 1993) induce and stabilize HIF-1α. CPX mimics hypoxia in that it activates HIF-1α due most likely to inhibition of the oxygen-dependent protein hydroxylases (Hofer et al. 2001). Chronic iron chelation and overexpression of HIF-1α may thus limit the availability of functional protein hydroxylases that instantaneously induce HIF-1α in response to acute hypoxia (Jewell et al. 2001), which, in turn, may result in lower ventilatory responses, a brainstem-mediated effect of hypoxic carotid body excitation. The present findings of CPX-blunted hypoxic hyperventilation are, in a way, in accord with the in vitro observations in which CPX-enhanced carotid chemosensory discharge substituted, at least in part, for that generated by the overlapping acute hypoxia (Daudu et al. 2002); a smaller ventilatory effect of hypoxia is then explicable.

There are other possible mechanisms by which iron chelation could affect the magnitude of the hypoxic ventilatory response either way. Iron catalyzes hydrogen peroxide in the Fenton reactions to form highly reactive hydroxyl radicals. Iron chelation is thus bound to decrease hydroxyl radical formation. To the extent that intracellular free radicals are released in hypoxia (Paddenberg et al. 2003), participate in hypoxic activation of HIF-1α (Bel Aiba and Görlach 2003), and thus are at play in the initiation of the hypoxia sensing process, iron chelation would dampen chemoreceptor hypoxic excitation, and, consequently, the ensuing hyperventilation. On the other side, there are compensatory mechanisms, at both central and peripheral levels, which, particularly in conscious animals, could supersede the failing, iron-chelated carotid body and maintain a level of hypoxic ventilation (Martin-Body et al. 1985, Olson et al. 1988). The hampering effect of iron chelation on carotid body function would then be underestimated. These mechanisms were uncontrolled for in the present study, as they require alternative study designs.

In conclusion, the findings of the present study show that iron is germane to the hypoxic ventilatory response, but is not a prior condition of this response.

Acknowledgments Supported in parts by the statutory budget of the Polish Academy of Sciences Medical Research Center in Warsaw, Poland, LLP-Erasmus Program for teaching staff mobility - academic year 2007/2008, Convenzione tra l'Università degli Studi "G.d'Annunzio" di Chieti e Pescara in Italy, and Accademia Natzionale dei Lincei in Rome, Italy.
The authors had no conflicts of interest to declare in relation to this article.

References

Baby, S.M., Roy, A., Mokashi, A.M., & Lahiri, S. 2003, Effects of hypoxia and intracellular iron chelation on hypoxia-inducible factor-1α and -1β in the rat carotid body and glomus cells. *Histochem Cell Biol*, 120: 343–352.

Bel Aiba, R.S. & Görlach, A. 2003, Regulation of the hypoxia-inducible transcription factor HIF-1 by reactive oxygen species in smooth muscle cells. *Adv Exp Med Biol*, 536: 171–178.

Daudu, P.A., Roy, A., Rozanov, C., Mokashi, A., & Lahiri S. 2002, Extra- and intracellular free iron and the carotid body responses. *Respir Physiol Neurobiol*, 130: 21–31.

Di Giulio, C., Bianchi, G., Cacchio, M., Artese, L., Rapino, C., Macri, M.A., & Di Ilio, C. 2005, Oxygen and life span: chronic hypoxia as to model for studying HIF-1alpha, VEGF and NOS during aging. *Respir Physiol Neurobiol*, 147: 31–38.

Hofer, T., Desbaillets, I., Höpfl, G., Gassmann, M., & Wenger, R.H. 2001, Dissecting hypoxia-dependent and hypoxia-independent steps in the HIF-1α activation cascade: implications for HIF-1α gene therapy. *FASEB J*, 15: 2715–2717.

Jewell. U.R., Kvietikova, I., Scheid, A., Bauer, C., Wenger, R.H. & Gassmann, M. 2001, Induction of HIF-1α in response to hypoxia is instantaneous. *FASEB J*, 15: 1312–1314.

Lahiri, S., Rozanov, C., Roy, A., Storey, B., & Buerk, D.G. 2001, Regulation of oxygen sensing in peripheral arterial chemoreceptors. *Int J Biochem Cell Biol*, 33: 755–774.

Linden, T., Katschinski, D.M., Eckhardt, K., Scheid, A., Pagel, H., & Wenger, R.H. 2003, The antimycotic ciclopirox olamine induces HIF-1α stability, VEGF expression, and angiogenesis. *FASEB J*, 17: 761–763.

Martin-Body, R.L., Robson, G.J., & Sinclair J.D. 1985, Respiratory effects of sectioning the carotid sinus glossopharyngeal and abdominal vagal nerves in the awake rat. *J Physiol*, 361: 35–45.

Olson, E.B. Jr., Vidruk, E.H., & Dempsey, J.A. 1988, Carotid body excision significantly changes ventilatory control in awake rats. *J Appl Physiol*, 64: 666–671.

Paddenberg, R., Goldenberg, A., Faulhammer, P., Braun-Dullaeus, R.C., & Kummer, W. 2003, Mitochondrial complex II is essential for hypoxia-induced ROS generation and vasoconstriction in the pulmonary vasculature. *Adv Exp Med Biol*, 536: 163–169.

Pokorski, M. & Lahiri, S. 1991, Endogenous opiates and ventilatory acclimatization to chronic hypoxia in the cat. *Respir Physiol*, 83: 211–222.

Roy, A., Li, J., Baby, S.M., Mokashi, A., Burek, D.G., & Lahiri, S. 2004a, Effects of iron-chelators on ion-channels and HIF-1α in the carotid body. *Respir Physiol Neurobiol*, 141: 115–123.

Roy, A., Volgin, D.V., Baby, S.M., Mokashi, A., Kubin, L. & Lahiri, S. 2004b, Activation of HIF-1α by hypoxia and iron chelation in isolated rat carotid body. *Neurosci Lett*, 363: 229–232.

Semenza, G.L. 1999, Regulation of mammalian O_2 homeostasis by hypoxia-inducible factor 1. *Ann Rev Cell Dev Biol*, 15: 551–578.

Tipoe, G.L. & Fung, M.L. 2003, Expression of HIF-1α, VEGF and VEGF receptors in the carotid body of chronically hypoxic rat. *Respir Physiol Neurobiol*, 138: 143–154.

Wang, G.L. & Semenza, G.L. 1993, Desferrioxamine induces erythropoietin gene expression and hypoxia-inducible factor 1 DNA-binding activity: Implications for models of hypoxia signal transduction. *Blood*, 82: 3610–3615.

Systemic Effects Resulting from Carotid Body Stimulation – *Invited Article*

Prem Kumar

Abstract The carotid body is stimulated by a number of blood-borne stimuli, ranging from increasing intensities of hypoxia, hypercapnia or acidosis to less studied stimuli, including hyperthermia, hyperosmolarity and hyperkalaemia. Although there exists heterogeneity in type I cell structure and function, there is no evidence to demonstrate that individual afferent fibres of the carotid sinus nerve subserve separate stimulus modalities. Thus, afferent information appears graded only for intensity (although this may be time-dependent) and not stimulus type. The response to carotid body stimulation, therefore, is to produce a graded, stereotypic response that can broadly be defined as the primary respiratory and cardiovascular reflexes of hyperventilation, bradycardia and peripheral vasoconstriction, originating directly as a consequence of increased afferent traffic in the carotid sinus nerve leading to augmented discharge in phrenic and intercostals nerves and altered activity in sympathetic and parasympathetic efferents. Where ventilation is not controlled, the secondary cardiovascular responses of tachycardia and peripheral vasodilation arise subsequent to the hyperventilation as a result of increased respiratory drive in the brainstem and increased activity of slowly adapting pulmonary stretch receptors in the airways. A role for the carotid body in mediating the hyperpnea of exercise, the aetiology of certain cardiovascular diseases, including sleep-apnoea derived hypertension and heart failure is supported by evidence but these complex situations generate reflex responses that can be mediated via a number of sensory and effector systems and the precise role of the carotid body is yet to be defined.

Keywords Carotid body · Chemoreceptor · cardiorespiratory reflexes · Chemoreflex · Hypoxia · Adequate stimuli · Sleep apnoea · Hypertension · Heart failure

P. Kumar (✉)
School of Experimental Medicine, College of Medical and Dental Sciences,
University of Birmingham, B15 2TT, UK
e-mail: p.kumar@bham.ac.uk

1 Introduction

The carotid bodies are peripheral chemoreceptors that were long believed to be unique sensors of acute reductions in Po_2 (hypoxic hypoxia), acting to translate this stimulus into afferent neural discharge in the carotid sinus branch of the glossopharyngeal nerve. It is, however, now recognised that many, if not all, tissues share this ability of the carotid bodies to detect hypoxia, as demonstrated by an array of alterations in cellular biochemistry and function that are observable when tissue Po_2 falls below critical levels. However, whilst most tissues appear to sense hypoxia for the preservation of their own integrity, usually by reducing energy expenditure, the carotid bodies couple their sensory function with an active, energy consuming response aimed at an increased delivery of systemic oxygen rather than its local conservation. In this respect, they form part of a small but important group of tissues that also contains erythropoietin-secreting cells of the renal cortex and pulmonary artery smooth muscle cells. Their uniqueness, therefore, now must lie with their initiation of appropriate and corrective, systemic cardiorespiratory reflexes in response to the transduction of arterial hypoxia rather than to O_2 transduction per se. In addition, it is now also recognised that the carotid body is adapted to sample arterial blood not just for hypoxia, but also for a variety of other physico-chemical stimuli and may thus be considered as a polymodal receptor. Irrespective of the stimulus, or stimuli, being sensed, the carotid body transduction process is extremely rapid, occurring within seconds and without need for altered protein expression, although a longer-term alteration of function is a recognised, characteristic feature of these organs that must occur via stimulus-induced changes in transcription. The adaptation of the carotid body to its function appears to depend crucially upon the relation between its size and its blood supply, as a 4-fold increase in its volume between the fetal and adult forms of a single species, the cat, occurs with a concomitant and proportional increase in extravascular volume and small vessel volume, such that the endothelial surface area of these small vessels remains essentially unchanged over a lifetime (Clarke et al., 1990). In addition, the small vessel volume, when expressed as a percentage of total carotid body volume, remains unchanged at around 5–6%, irrespective of carotid body size across a number of species, ranging from primates to rat (Clarke et al., 1993). Postnatal changes in chemosensitivity, that have often been described, must therefore occur as a consequence of changes in cellular protein sensors or downstream transducing elements in the carotid body rather than via any age-dependent, alteration in the delivery of these stimuli. What these cellular protein sensors may be remains a key question in carotid body research and although progress in recent years has identified a number of likely candidates, including AMP-activated protein kinase (Wyatt et al., 2007, Williams et al., 2004) and haemoygenase 2 (Williams et al., 2004), amongst others, further definitive experiments are required and there is nothing to preclude a number of sensors acting together to provide the full response over a wide range of stimulus intensities. A second, related key question is: do peripheral chemoreceptors have any role in the aetiology of human disease? This very important question is linked to the former by the question of what are the precise reflex responses to carotid body stimulation? For an answer to that, it is necessary to appreciate the stimuli of the carotid body and their interactions as well as the interaction between the various reflexes that they may initiate. A more complete account can be found elsewhere (Daly, 1997).

2 Natural Stimuli of the Carotid Body

Hypoxia remains the defining natural stimulus of the carotid body. Hypoxia is, of course, a lack of something – namely oxygen – rather than an applied stimulus in its own right and hypoxic hypoxia, leading to reductions in both P_{O_2} and O_2 content, raises chemoafferent discharge and ventilation in a similar hyperbolic, or single exponential, fashion such that the P_{O_2}-afferent and P_{O_2} ventilation curves can essentially be superimposed. A consideration of the differences between O_2 partial pressure and content as stimuli has been detailed elsewhere (Kumar and Bin-Jaliah, 2007). The response is rapid in onset and non-adapting, at least if all stimulus levels are maintained constant. Discharge rises significantly only from values of P_{O_2} a little below arterial normoxia (ca. 90 mmHg) and peak discharge occurs at arterial P_{O_2} values between ca. 20–30 mmHg after which it may fall, concomitantly, in vivo, with the blood pressure. Conversely, if arterial P_{O_2} is raised from normoxia to hyperoxia by addition of O_2 to inspired gas, then chemoafferent discharge falls to around half to one third of its normoxic level. This suggests that there exists an elevated chemoafferent tone in normoxia and its importance may be ascertained by noting that sympathetic tone to skeletal muscle in humans is reduced by up to 25% by 3–4 minutes of breathing pure O_2 (Seals et al., 1991). In addition, the sensitivity of the carotid body to P_{O_2} can vary with steady or intermittent changes in the arterial blood levels of this gas and this plasticity occurs whether such changes are brought about naturally, such as in the early postnatal period or on ascent to altitude, or pathologically as seen in various disease aetiologies such as sleep apnoea or heart failure (Prabhakar et al., 2007).

The carotid body is also particularly sensitive to changes in CO_2 and pH. As all physiological responses to hypoxia occur in a background of steady, or changing, PCO_2, the interaction between these stimuli is an important factor in determining the magnitude of the chemoafferent response and thus the extent and degree of the reflex response. The carotid body responds to rising CO_2 with a linearly increasing discharge between PCO_2 levels of ca. 20 to 80 mmHg, being abolished at even lower levels and failing to be maintained at even higher levels. Thus, at low levels of PCO_2, even the effect of severe hypoxia upon chemodischarge is significantly blunted and this negative interaction explains the blunted ventilatory response to so-called 'aviator's hypoxia' observed when CO_2 is allowed to fall with increasing ventilation (poikilocapnia). On the other hand, when hypoxia increases with hypercapnia, the resulting chemoafferent discharge frequency is greater than the sum of the two responses when applied independently (Lahiri and DeLaney, 1975). This 'multiplicative interaction' is a characteristic feature of carotid body function and can be observable in vitro and, importantly, translates to the level of ventilation in human (Nielsen and Smith, 1952, Honda and Hashizume, 1991). Although, in vivo, the interaction between stimuli may have some additional input from an interaction with central chemoreceptors, this appears to be a small component of the overall response. Thus, in one of the relatively few experiments performed with carotid body resected patients (Honda and Hashizume, 1991), whilst the resected patients failed, as expected, to show an elevated ventilation when hypoxia was added, the ventilatory response to CO_2 was not sharpened by hypoxia and was even often reduced at PCO_2 levels in excess of 40mmHg. The mechanism underlying multiplicative interaction in the carotid body is not known, but it does develop with increasing postnatal age (Pepper et al., 1995) and is correlated with an increasing density of maxi-K ion channels in type I cells (Hatton et al., 1997). It is tempting to speculate that this development may underlie some of the increasing ventilatory sensitivity to hypoxia known to occur during this time (Calder et al., 1994, Williams et al., 1991).

In contrast to the response to hypoxia, the carotid body response to CO_2, when applied at a sufficiently abrupt rate of change, shows an over/undershooting response with adaptation to increases and decreases in stimulus intensity, respectively. As blood gas tensions are known to vary with a respiratory periodicity, this dynamic response to CO_2 provides a potential, oscillating signal in chemodischarge that is independent of the mean level of blood gas tensions and thus could provide feedforward information to the CNS regarding, for example, exercise intensity (Band and Wolff, 1978, Kumar et al., 1988).

In addition to the well-characterised blood gas and pH stimuli, many other blood-borne stimuli can excite chemoafferent discharge with a high sensitivity and fidelity. These stimuli include hyperkalaemia, hyperthermia, hypo-osmolarity, various hormones, including certain catecholamines and steroid hormones and possibly even hypoglycaemia (Kumar and Bin-Jaliah, 2007). These stimuli all appear to fulfil the various definitions presently used to characterize an adequate stimulus with the carotid body appearing to be duly adapted for the detection of the specific energy of each stimulus. However, it still remains to be established whether these stimuli are being sensed in their own right or whether they act, instead, to modulate the hypoxia/hypercapnia response. It is not clear, presently, how this conundrum may be resolved as all stimuli appear to have a potential for non-specific actions upon the blood gas transduction process. Thus, each of the non-blood gas stimuli can interact with hypoxia and hypercapnia, most usually in a simple additive way, to further increase chemoafferent discharge and this action leads to an augmented reflex response. To date, there has been no evidence to suggest that there exists modality-specific afferent discharge, i.e. that specific chemoafferent fibres respond only to specific stimuli, or that specific stimuli give rise to specific reflexes and thus the carotid body can be described as demonstrating univariant behaviour. The carotid body simply senses an alteration in blood that threatens homeostasis by initiating a stereotypic reflex response that appears only to be threshold/intensity dependent and not stimulus dependent. However, these non-blood gas stimuli differ in one significant way from the blood gas and pH stimuli in that they do not form part of a ventilatory reflex feedback loop. Instead, their action upon chemoafferent discharge is maintained in a feedforward manner, as cardiorespiratory reflexes do not alter their intensity.

Thus, arterial Po_2, Pco_2 and pH are just three of a number of physico-chemical stimuli of the carotid body that are able to act synergistically to provide afferent, feedback and feedforward information, via the nucleus tractus solitarius (NTS) of the brainstem, to the central pattern generator(s) that drive the respiratory motorneurones that control ventilation and cardiovascular reflexes. In the case of blood gases, this simple feedback loop acts to increase ventilation during asphyxia and so counteracts, or at least limits, the blood gas changes. For the other, non-blood gas, stimuli it remains to be seen whether they are, despite their fulfilling the criteria for adequate stimuli, of functional importance in homeostasis or whether they might simply act to augment the blood gas stimuli as the greater the stimulus intensity, the greater the afferent discharge and thus the magnitude of the cardioventilatory response over a large range of blood gas tensions.

3 Cardiorespiratory Integration

Maintaining an optimal, systemic oxygen delivery requires the integration of corrective cardiovascular and respiratory reflexes following perturbations in blood gas tensions. Thus, in addition to the elevated ventilation observable upon carotid body stimulation, a variety of sympathetic and parasympathetic autonomic effectors mediate a range of cardiovascular reflexes that act upon heart rate, arterial blood pressure, cardiac output and vascular resistance as well as catecholamine release from the adrenal medulla (Fig. 1). The close association between respiratory and cardiovascular reflexes can be most clearly seen in the respiratory-related rhythm that can be found in the discharge pattern of both sympathetic and vagal motor neurones and the reflex modulation of this pattern by stimulation of cardiorespiratory, baro- and chemoreceptors (Daly, 1997). These various reflexes demonstrate a high degree of plasticity and whilst some of the plasticity resides in alterations in carotid body sensitivity, some also resides in the CNS in a distinct form of Hebbian learning. Thus, a significantly reduced synaptic branching was observable in a number of cardiorespiratory-related areas of the hindbrain, including the posterior hypothalamus, periaqueductal grey and NTS, in rats that had undergone 3–4 months of spontaneous exercise in a running wheel as compared to non-exercising animals (Nelson et al., 2005). These anatomical changes were correlated with decreases in resting heart rate.

The stimulus-independent, excitation of carotid bodies initiates a range of cardiovascular reflexes with the precise response depending principally upon whether ventilation was, or was not, elevated concomitantly. Thus, if ventilation is elevated, a tachycardia and decreased vascular resistance are most often observed. However, if ventilation is not elevated either by external, experimental control, or if an apnoea occurs as it might in obstructive respiratory disorders, then the reflex response to

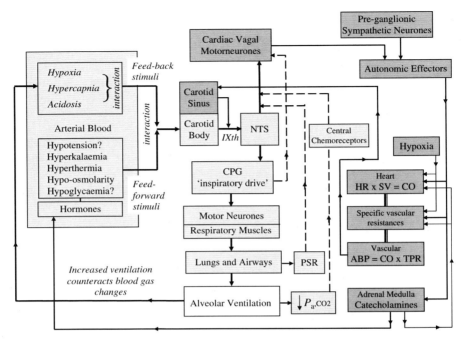

Fig. 1 **Cardiorespiratory reflex responses to carotid body stimulation.** Schematic representation of the various routes for cardiovascular and respiratory reflex integration following stimulation of carotid body chemoreceptors. Respiratory related components are shaded lighter than cardiovascular related components. Dashed lines indicated inhibitory influences. Stimulation of the carotid body by blood gas or pH alterations induces a feedback loop that acts, via augmented ventilation, to correct this disturbance. The magnitude of the carotid body response is augmented by a multiplicative interaction between these stimuli and by a, largely, additive, interaction with other, non-blood gas stimuli. These latter stimuli act in a feedforward fashion upon ventilation, as they are not themselves altered in magnitude by the reflex increase in ventilation. The increase in ventilatory drive, if not due to an increased metabolism, induces a fall in $PaCO_2$ and an increased pulmonary stretch receptor activation with all three acting to decrease vagal motorneurone tone. The consequent effect upon cardiovascular parameters of heart rate, blood pressure and peripheral resistance depends upon the interplay between autonomic reflexes and is also dependent upon the magnitude of the ventilatory response which itself is determined by a number of factors including species and type of anaesthetic. Catecholamine release can augment the cardiovascular and respiratory reflex responses

carotid body stimulation is most often one of bradycardia and increased vascular resistance in specific vascular beds, most notably skeletal muscle, sphlancnic and renal. Thus, the direct consequence of carotid body stimulation of bradycardia and increased vascular resistance is often masked by the indirect consequence of an elevated ventilatory response and, as such, an accelerated heart rate can be observed if carotid body stimulation is removed transiently, for example by application of a few breaths of 100% O_2 or following the first breath after immersion under water (De Burgh Daly et al., 1977). The precise pattern of cardiovascular response will depend upon the degree of ventilation. For example, if the bradycardia is profound enough to reduce the cardiac output, then arterial blood pressure can fall despite increases in total peripheral resistance. However, if the bradycardia is less pronounced, as may occur if ventilation is increased, then cardiac output is maintained and arterial blood pressure may rise. If vagotomised, then carotid body stimulation induces a rise both in total peripheral resistance and blood pressure. The reflex response also shows some species dependency, with the consequences of elevated ventilation being more pronounced in dog than in rat (Marshall, 1994).

The effect of increased ventilation upon cardiovascular reflexes occurs as a consequence of the carotid body-mediated, increase in central respiratory drive, the augmented pulmonary stretch receptor activation and, to a lesser degree, the fall in $PaCO_2$ that occurs when ventilation is stimulated independently of an elevated systemic metabolism. In addition, if hypoxia is used systemically as a stimulus, then additional direct effects of the lowered O_2 tension upon the heart and vasculature, as well as temporal effects, need also to be considered. Thus, if a lowered inspired O_2 is applied to a Saffan- anaesthetised rat (Neylon and Marshall, 1991, Marshall and Metcalfe, 1988), an increase in ventilation is observed at the onset of the hypoxaemia that is associated with a tachycardia, an increase in vascular conductance in most circulations, most notably in skeletal muscle, and a fall in arterial blood pressure. If the hypoxia is maintained for more than 2–3 minutes, ventilation gradually wanes (a biphasic response commonly observed in neonates and/or small mammals in response to hypoxia, but present to a lesser extent in large adults (Mortola, 2004)) and heart rate returns to, or below, the pre-hypoxia control levels. Arterial blood pressure, however, continues to fall or remains lowered, due now to the dominant, local effects of hypoxia upon the cardiovascular system. In these studies, the use of a steroid anaesthetic enabled components of a carotid body dependent, hypothalamic mediated, alerting response to be observed. This response, also seen in conscious animals where it is accompanied with behavioural arousal, includes the cardiorespiratory reflexes, mentioned above, and also papillary dilatation, piloerection, urination and daefecation (Hilton and Marshall, 1982).

To further complicate the answer to the question of what is the cardiovascular consequence of carotid body stimulation, we need also to consider the phenomenon of laterality of response and the interaction of baroreceptors and chemoreceptors. In an artificially ventilated, anaesthetised cat preparation, when the carotid body was stimulated unilaterally with lobeline, a bradycardia was observed and sympathetic discharge increased bilaterally in sympathetic fibres innervating the kidney and skeletal muscle vasoconstrictors, as an expected consequence of an elevated blood pressure (Kollai et al., 1978). However, cardiac sympathetic activity was inhibited only on the ipsilateral side to the stimulation and was markedly increased on the contralateral side. The ipsilateral inhibition of cardiac sympathetic activity was converted to a mild excitation when baroreceptor stimulation was prevented by maintaining blood pressure and the elevated discharge on the contralateral side was further augmented. The bradycardia was reduced in this circumstance. Thus, carotid body stimulation induces a differential effect upon left and right cardiac sympathetic nerves that is amplified by an interaction with baroreceptors. This action in mediated presumably through the CNS and its specific, reciprocal nature may be related to the more complex, autonomic innervation of the heart relative to other organs.

4 Role of the Carotid Body in Cardiorespiratory-Related Pathologies

The stimulation of carotid body chemoreceptors, without a consequent increase in ventilation, or with apnoea, leads to the direct effects of bradycardia and vasoconstriction as can be seen for example during diving, or even simply by activating trigeminal receptors in the skin of the face. Apnoea may also occur, in the intact animal, consequent to laryngeal stimulation. If apnoea occurs, the reflex inhibition of respiration occurs despite the increasing severity of the arterial asphyxial stimulus, at least until a breakpoint is reached. The strength of this reflex inhibition of respiration and its unmasking of the direct effects of chemoreceptor stimulation, may underlie some cases of sudden death due to a vagal-mediated, cardiac arrest. Persistent apnoea may lead to the augmented carotid body sensitivity seen in models of intermittent hypoxia (Prabhakar et al., 2007) and this may itself induce a chronically elevated arterial blood pressure through augmented vasoconstrictor influences (Fletcher et al., 1992). Add to this an impaired baroreflex and the blood pressure elevation may lead to the pathological consequences, including stroke and heart failure, attributed to hypertension (Kiely and McNicholas, 2000). Evidence for such an outcome comes from a variety of sources, including direct recordings of sympathetic neural activity by muscle microneurography (Narkiewicz et al., 1998), where patients with obstructive sleep apnoea (OSA), but not obese, non-sleep disordered control subjects, showed a significant decrease in muscle sympathetic activity and blood pressure when given 100% inspired O_2, suggesting that tonic chemoexcitation might underlie the sympathetic hypertension in this group of patients. In addition, as an habituation of the alerting response might act to reduce the adverse cardiovascular impact of repeated bouts of hypoxia, it has been speculated that a reduced habituation may underlie the development of hypertension in susceptible individuals (Marshall, 1994).

It is worth speculating that the recurring apnoeic episodes that lead to hypertension via elevated sympathetic activity, impaired baroreflex activity and vasoconstriction may be a foreunner of heart failure (Fig. 2). In a rabbit model of pacing-induced heart failure (Sun et al., 1999), basal renal sympathetic nerve activity in conscious animals was greatly enhanced relative to sham controls and this activity was elevated to a greater extent at a number of graded levels of hypoxia in the heart failure animals. As the ventilatory sensitivity to hypoxia was also augmented in the heart failure animals, this points to an increased chemoreflex gain. Central chemosensitivity was unaffected, thus indicating a peripheral chemoreceptor origin for the pathological increased in chemoreflex mediated, sympathetic nerve activity. As with the human OSA studies described above, hyperoxia decreased sympathetic activity only in the heart failure animals, although only to a level that was still greater

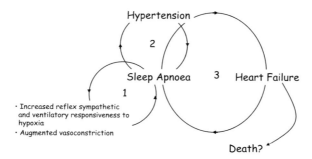

Fig. 2 Carotid body dependent, pathological consequences of sleep apnoea. In the absence of the 'braking' effect of hyperventilation, the asphxial drive to carotid bodies during sleep apnoea can induce adverse cardiovascular consequences that arise from plastic changes in chemosensitivity and baroreceptor sensitivity. Thus, three interacting cycles can be established with sleep apnoea predisposing to hypertension and hypertension predisposing to heart failure with possible fatal outcome

than that of the sham control, suggesting that other factors, that may include altered baroreceptor sensitivity, may also contribute to the increased sympathetic drive in heart failure. Although not consistent with some human studies on heart failure patients (Narkiewicz et al., 1999), this may, in part, be due to variation in the specific sympathetic nerve being recorded.

5 Concluding Remarks

Taken together, it can be seen that the reflex consequence of carotid body stimulation depends upon a number of interacting factors, that include, as well as the magnitude of the ventilatory response, the degree of pressor response evoked, the integrity of the vagi and CNS structures involved in cardiorespiratory interaction and even the side of the body that is stimulated. In addition, although stimulus independent, the reflex consequence of carotid body excitation will be graded with stimulus strength and may be species and/or age dependent. Although not discussed here, the reflex will depend also upon the type and depth of anaesthesia, if used. The phase of respiration can also affect the reflex, as can interaction with baroreceptor influences. Finally, cardiorespiratory reflexes exhibit plasticity, habituation and/or conditioning and are selectively augmented by apnoea, which may predispose to the severity of a number of cardiorespiratory disease states with their associated morbidity and mortality. A more complete understanding of the interactions of the various reflexes that can be elicited by carotid body stimulation may therefore be of benefit in establishing novel treatments for the betterment of human health.

Acknowledgments I gratefully acknowledge the generous support of the British Heart Foundation.

References

Band, D. M. & Wolff, C. B. 1978, Respiratory oscillations in discharge frequency of chemoreceptor afferents in sinus nerve and anaesthetized cats at normal and low arterial oxygen tensions, *J Physiol*, 282: 1–6.

Calder, N. A., Williams, B. A., Kumar, P., & Hanson, M. A. 1994, The respiratory response of healthy term infants to breath-by-breath alternations in inspired oxygen at 2 postnatal ages, *Pediatric Res*, 35: 321–324.

Clarke, J. A., Daly, M. D., & Ead, H. W. 1990, Comparison of the size of the vascular compartment of the carotid-body of the fetal, neonatal and adult cat, *Acta Anatomica*, 138: 166–174.

Clarke, J. A., Daly, M. D., Ead, H. W., & Kreclovic, G. 1993, A morphological-study of the size of the vascular compartment of the carotid-body in a nonhuman primate (cercopithecus-ethiopus), and a comparison with the cat and rat, *Acta Anatomica*, 147: 240–247.

Daly, M. D. B. 1997, *Peripheral Arterial Chemoreceptors and Respiratory-Cardiovascular Integration*, Oxford Medical Publications.

De Burgh Daly, M., Elsner, R., & Angell-James, J. E. 1977, Cardiorespiratory control by carotid chemoreceptors during experimental dives in the seal, *Am J Physiol*, 232: H508–H516.

Fletcher, E. C., Lesske, J., Behm, R., Miller, C. C., Stauss, H., & Unger, T. 1992, Carotid chemoreceptors, systemic blood-pressure, and chronic episodic hypoxia mimicking sleep-apnea, *J Appl Physiol*, 72: 1978–1984.

Hatton, C. J., Carpenter, E., Pepper, D. R., Kumar, P., & Peers, C. 1997, Developmental changes in isolated rat type I carotid body cell K^+ currents and their modulation by hypoxia, *J Physiol*, 501(Pt 1): 49–58.

Hilton, S. M. & Marshall, J. M. 1982, The pattern of cardiovascular-response to carotid chemoreceptor stimulation in the cat, *J Physiol Lond*, 326: 495–513.

Honda, Y. & Hashizume, I. 1991, Evidence for hypoxic depression of CO_2-ventilation response in carotid body-resected humans, *J Appl Physiol*, 70: 590–593.

Kiely, J. L. & McNicholas, W. T. 2000, Cardiovascular risk factors in patients with obstructive sleep apnoea syndrome, *Eur Respir J*, 16: 128–133.

Kollai, M., Koizumi, K. & Brooks, C. M. 1978, Nature of differential sympathetic discharges in chemoreceptor reflexes, *Proc Natl Acad Sci USA*, 75: 5239–5243.

Kumar, P. & Bin-Jaliah, I. 2007, Adequate stimuli of the carotid body: more than an oxygen sensor? *Respir Physiol Neurobiol*, 157: 12–21.

Kumar, P., Nye, P. C., & Torrance, R. W. 1988, Do oxygen tension variations contribute to the respiratory oscillations of chemoreceptor discharge in the cat? *J Physiol*, 395: 531–552.

Lahiri, S. & DeLaney, R. G. 1975, Stimulus interaction in the responses of carotid body chemoreceptor single afferent fibers, *Respir Physiol*, 24: 249–266.

Marshall, J. M. 1994, Peripheral chemoreceptors and cardiovascular regulation, *Physiol Rev*, 74: 543–594.

Marshall, J. M. & Metcalfe, J. D. 1988, Analysis of the cardiovascular changes induced in the rat by graded levels of systemic hypoxia, *J Physiol*, 407: 385–403.

Mortola, J. P. 2004, Implications of hypoxic hypometabolism during mammalian ontogenesis, *Respir Physiol Neurobiol*, 141: 345–356.

Narkiewicz, K., Pesek, C. A., Vandeborne, P., Kato, M., & Somers, V. K. 1999, Enhanced sympathetic and ventilatory responses to central chemoreflex activation in heart failure, *Circulation*, 100: 262–267.

Narkiewicz, K., Van De Borne, P. J., Montano, N., Dyken, M. E., Phillips, B. G., & Somers, V. K. 1998, Contribution of tonic chemoreflex activation to sympathetic activity and blood pressure in patients with obstructive sleep apnea, *Circulation*, 97: 943–945.

Nelson, A. J., Juraska, J. M., Musch, T. I., & Iwamoto, G. A. 2005, Neuroplastic adaptations to exercise: neuronal remodeling in cardiorespiratory and locomotor areas, *J Appl Physiol*, 99: 2312–2322.

Neylon, M. & Marshall, J. M. 1991, The role of adenosine in the respiratory and cardiovascular response to systemic hypoxia in the rat, *J Physiol*, 440: 529–545.

Nielsen, M. & Smith, H. 1952, Studies on the regulation of respiration in acute hypoxia, *Acta Physiologica Scandanavica*, 24: 293–313.

Pepper, D. R., Landauer, R. C., & Kumar, P. 1995, Postnatal development of CO_2-O_2 interaction in the rat carotid body in vitro, *J Physiol*, 485(Pt 2): 531–541.

Prabhakar, N. R., Peng, Y. J., Kumar, G. K., & Pawar, A. 2007, Altered carotid body function by intermittent hypoxia in neonates and adults: relevance to recurrent apneas, *Respir Physiol Neurobiol*, 157: 148–153.

Seals, D. R., Johnson, D. G., & Fregosi, R. F. 1991, Hyperoxia lowers sympathetic activity at rest but not during exercise in humans, *Am J Physiol*, 260: R873–R878.

Sun, S. Y., Wang, W., Zucker, I. H., & Schultz, H. D. 1999, Enhanced activity of carotid body chemoreceptors in rabbits with heart failure: role of nitric oxide, *J Appl Physiol*, 86: 1273–1282.

Williams, B. A., Smyth, J., Boon, A. W., Hanson, M. A., Kumar, P., & Blanco, C. E. 1991, Development of respiratory chemoreflexes in response to alternations of fractional inspired oxygen in the newborn-infant, *J Physiol Lond*, 442: 81–90.

Williams, S. E., Wootton, P., Mason, H. S., Bould, J., Iles, D. E., Riccardi, D., Peers, C., & Kemp, P. J. 2004, Hemoxygenase-2 is an oxygen sensor for a calcium-sensitive potassium channel, *Science*, 306: 2093–2097.

Wyatt, C. N., Mustard, K. J., Pearson, S. A., Dallas, M. L., Atkinson, L., Kumar, P., Peers, C., Hardie, D. G., & Evans, A. M. 2007, AMP-activated protein kinase mediates carotid body excitation by hypoxia, *J Biol Chem*, 282: 8092–8098.

Bicarbonate-Regulated Soluble Adenylyl Cyclase (sAC) mRNA Expression and Activity in Peripheral Chemoreceptors

A.R. Nunes, E.C. Monteiro, S.M. Johnson and E.B. Gauda

Abstract Peripheral arterial chemoreceptors in the carotid body (CB) are modulated by pH/CO_2. Soluble adenylyl cyclase (sAC) is directly stimulated by bicarbonate ions (HCO_3^-). Because CO_2/HCO_3^- mediates depolarization in chemoreceptors, we hypothesized that sAC mRNA would be expressed in the CB, and its expression and function would be regulated by CO_2/HCO_3^-.

Sprague-Dawley rats at postnatal days 16–17 were used to compare sAC mRNA gene expression between CB and non-chemosensitive tissues: superior cervical (SCG), petrosal (PG) and nodose ganglia (NG) by quantitative real time-PCR. Rat sAC gene expression was standardized to the expression of GAPDH (housekeeping gene) and the data were analyzed with the Pfaffl method. Gene and protein expression, and sAC regulation in the testis was used as a positive control. To determine the regulation of sAC mRNA expression and activity, all tissues were exposed to increasing concentrations of bicarbonate (0, 24, 44 mM, titrated with CO_2 and maintained a constant pH of 7.40). RESULTS: sAC mRNA expression was between 2–11% of CB expression in the SCG, PG and NG. Furthermore, only in the CB did HCO_3^- upregulate sAC gene expression and increase cAMP levels. *Conclusion*: sAC mRNA and protein expression is present in peripheral arterial chemoreceptors and non-chemoreceptors. In the CB, CO_2/HCO_3^- not only activated sAC but also regulated its expression, suggesting that sAC may be involved in the regulation of cAMP levels in response to hyper/hypocapnia.

Keywords cAMP · Carotid body (CB) · soluble adenylyl cyclase (sAC) · Peripheral arterial chemoreceptors · CO_2/HCO_3^- · qRT-PCR

1 Introduction

Breathing is modulated by peripheral arterial chemoreceptors distributed in Type I cells in the carotid body (CB), which are sensitive to changes in pCO_2/pH (for a review see (Gonzalez et al. 1994)). cAMP is an important mediator of

E.B. Gauda (✉)
Pediatrics, Johns Hopkins Medical Institutions, Baltimore, MD, USA
e-mail: egauda@mail.jhmi.edu

O_2-chemotransduction (for a review see (Gonzalez et al. 1994)) through activation of transmembrane adenylyl cyclase (tmAC), and hypercapnia increases cAMP levels in the CB (Perez-Garcia et al. 1990); however, the specific cellular mechanisms responsible for CO_2 sensitivity in this organ are not well understood. Recently, a soluble adenylyl cyclase (sAC) that is activated and regulated by changes in HCO_3^-/CO_2 has been characterized in testis and somatic tissues (Buck et al. 1999; Schmid et al. 2007; Sinclair et al. 2000; Sun et al. 2003; Wang et al. 2005). Unlike tmAC, sAC is insensitive to G proteins or to foskolin, and is unique in its activation by HCO_3^- (reviewed in (Kamenetsky et al. 2006)). In this study, we examined the role of sAC in chemotransduction and hypothesized that sAC mRNA will be expressed in the CB and that changes in CO_2/HCO_3^- levels will regulate sAC mRNA expression and activity as evidenced by changes in cAMP levels.

2 Methods

Sprague-Dawley rats at postnatal days 16–17 were used. Animals were anesthetized with isoflurane and decapitated. Carotid body (CB), superior cervical (SCG), petrosal (PG), nodose (NG) ganglia and testis were removed, and prepared for quantitative real-time polymerase chain reaction (qRT-PCR), immunoblotting or cAMP measurements. Experiments were approved by the Animal Care and Use Committee at the Johns Hopkins University School of Medicine.

2.1 sAC mRNA Gene Expression and Regulation

We characterized sAC mRNA gene expression in the CB and non-chemoreceptor tissues pooled from six rats from the same litter. The experiment was done in triplicate using a total of 18 animals from 3 different litters.

In order to study the effect of HCO_3^- on sAC mRNA levels in the CB, SCG and testis, the tissues were pre-incubated in Krebs modified solution containing (in mM) NaCl 116; KCl 5; $CaCl_2$ 2; $MgCl_2$ 1.1; HEPES 10; glucose 5.5, pH 7.40 (Perez-Garcia et al. 1990) with 0 mM HCO_3^- / 0%CO_2/60%O_2, for 30 minutes. The tissues were then incubated in Krebs modified solution containing one of the following conditions: (1) 0 mM HCO_3^-/0% CO_2/60% O_2, (2) 24 mM HCO_3^-/5% CO_2/60% O_2, and (3) 44 mM HCO_3^-/10% CO_2/60% O_2 at 37°C, pH 7.4 (4 rats per condition, n=3 independent experiments). After 1 hour, the tissues were stored at −80°C and then processed for qRT-PCR.

The primer sequences used for the target gene and house keeping gene were : sAC, sense 5′-catgagtaaggaatggtggtactca-3′ (complementary to 4961–4986 base pairs of rat sAC gene (AF081941)) and anti-sense 5′-agggttacgttgcctgatacaatt-3 (complementary to the 5049-5072 base pairs) and G6PDH, sense 5′-gaagcctggcgtatcttcac-3′ and anti-sense 5′-gtgagggttcacccacttgt-3′ (Pastor-Soler et al. 2003). Relative expression of sAC between the tissues was analysed by the excel Pfaff method (Pfaffl 2001). qRT-PCR was performed with a MyiQ iCycler qRT-PCR system (Bio-Rad Laboratories Inc., Hercules, CA) with Syber Green detection.

Immunoblotting was performed using 20 μL of protein from crude homogenates of pooled CBs, SCGs and testes, and sAC protein was detected with monoclonal anti-sAC R21 (1:500, (Zippin et al. 2004), provided as a generous gift from Drs. Jochen Buck and Lonny R. Levin). Protein expression levels were normalized with β-actin detected with a goat anti-mouse monoclonal β-actin antibody (1:10000; Sigma Chemicals Company, St. Louis, MO).

2.2 Bicarbonate-Regulated cAMP Levels

sAC activity was detected indirectly by measuring changes in cAMP levels in response to increasing concentrations of HCO_3^-. Tissues (CB, SCG, PG and testes) were removed and pooled from 2 pups per litter, from a total of 3–8 litters. Tissues were incubated in Krebs modified solution without HCO_3^-/CO_2 and in the presence of MDL-12,330A (500 μM; a specific tmAC inhibitor, Sigma) for 30 minutes. Then the tissues were transferred to Krebs modified solution in the presence of MDL-12,330A (500 μM), IBMX (500 μM; a non-specific phosphodiesterase inhibitor; Sigma), plus HCO_3^-/CO_2 in one of three concentrations: 0 mM HCO_3^-/0 % CO_2, 24 mM HCO_3^-/5% CO_2, or 44 mM HCO_3^-/10 % CO_2, pH 7.4, and incubated for 30 minutes at 37°C.

Cyclic nucleotides were extracted from tissues as described by Batuca and co-workers (Batuca et al. 2003) with cAMP quantification by enzyme immunoassay, using an EIA commercial kit (RPN 2255, GE Healthcare Bio-Sciences AB, Piscataway, NJ). Protein levels were measured via standard Bradford Protein assay (Bio-Rad) and fluorescent detection using a NanoOrange Protein Quantification Kit (Invitrogen, Eugene, OR). cAMP levels were expressed in fentomoles per microgram of protein (fmol/μg protein).

2.3 Data Analysis and Statistical Procedures

The data are represented as mean ± SEM, and differences ($P<0.05$) between the experimental groups were calculated by non-parametric Spearman's Rank-Order Correlation and by Kruskal-Wallis one-way ANOVA with the Dunnett's post hoc test using GraphPad Prism (GraphPad Software, Inc., version 4, San Diego, CA) or SPSS (SPSS Inc., version 12, Chicago, IL).

3 Results

3.1 sAC mRNA Gene Expression and Regulation

sAC mRNA expression was lower in SCG, PG and NG than in CB and testis removed from rats at postnatal days 16–17 (Table 1).

Table 1 sAC mRNA expression in peripheral chemoreceptors and non-chemoreceptors relative to CB

	Chemo-sensitive	Non- chemosensitive			
Tissues	CB	SCG	NG	PG	testis
Percentage of CB. sAC expression (m±SEM)	100	11.3±1.33	3.05±0.22	3.78±0.54	254±47.9

Increasing concentrations of HCO_3^- upregulated sAC gene expression only in the CB ($p < 0.01$; Spearman's correlation), but did not affect sAC mRNA gene expression in other tissues, including testis (Table 2).

Western immunoblot was used to determine the presence of sAC protein detected with the monoclonal anti-sAC antibody, identified as R21 (Fig. 1). We detected an intense band at 50 kDa in the CB, SCG and testis corresponding to the sACt isoform. In the testis we also detected bands at 130 kDa and 31 kDa.

Table 2 Effect of HCO_3^-/CO_2 on sAC mRNA relative gene expression (relative expression to 0mM HCO_3^- for each tissue)

	HCO_3^- (mM) / CO_2 (%)		
Tissues	0 mM / 0 %	24 mM / 5%	44 mM / 10%
CB	1.00 ± 0.00	1.58 ± 0.15	2.12 ± 0.30
SCG	1.00 ± 0.00	1.00 ± 0.14	0.90 ± 0.04
Testis	1.00 ± 0.00	0.67 ± 0.47	0.97 ± 0.23

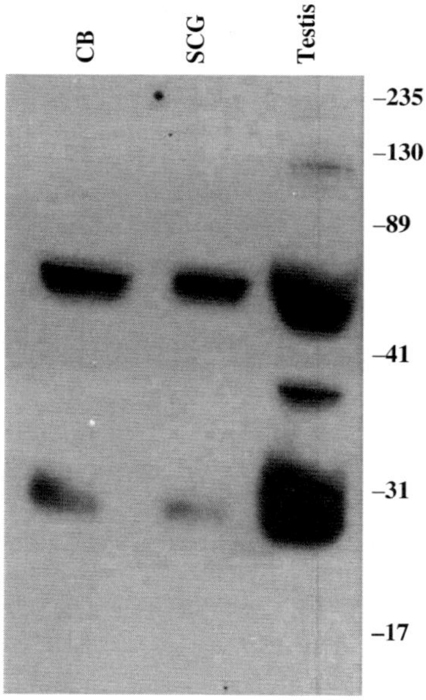

Fig. 1 Immunoblot showing sAC protein expression in CB, SCG and testis. *CB*, carotid body, *SCG*, superior cervical ganglia

3.2 Bicarbonate-Regulated cAMP Levels

The effect produced by increasing the concentrations of HCO_3^- on cAMP levels in CB and non-chemoreceptor tissues (SCG, PG, testis) in the presence of IBMX (500 μM) and MDL-12,330A (500 μM) is shown in Table 3. Previous experiments have demonstrated that MDL-12,330A (500 μM) was effective in blocking tmAC. In the CB, increasing HCO_3^- concentrations from 0 to 24 mM significantly increased cAMP levels from 15.27 ± 3.22 to 35.27 ± 6.04 fmol/μg protein ($p<0.01$). Exposing the CB to higher concentrations of HCO_3^-/CO_2 (44 mM HCO_3^- with 10%CO_2) did not produce any additional increase in cAMP levels in this tissue. Increasing HCO_3^- concentrations did not increase cAMP levels in non-chemoreceptor tissues (SCG, PG and testis). These results suggest that sAC could be involved in chemoreception of HCO_3^-/CO_2 in peripheral arterial chemoreceptors.

Table 3 Effects of changes in HCO_3^-/CO_2 on cAMP levels (fentomoles /μg of tissue)

Tissues	HCO_3^- (mM) / CO_2 (%)		
	0 mM / 0%	24 mM / 5%	44 mM / 10%
CB	15.3 ± 3.22 (n=16)	35.3 ± 6.04 (n=14)**	26.3 ± 2.83 (n=14)*
SCG	31.6 ± 5.41 (n=9)	43.5 ± 14.8 (n=7)	36.06 ± 6.96 (n=8)
PG	16.8 ± 6.125 (n=6)	15.9 ± 6.69 (n=6)	17.9 ± 8.09 (n=5)
Testis	9.21 ± 1.75 (n=10)	8.99 ± 1.49 (n=10)	9.32 ± 1.44 (n=10)

*$p<0.05$ **$p<0.01$ by one-way Kruskal-Wallis test with the Dunnett's post-test corresponding to differences between cAMP levels in response to increasing concentration of concentrations of HCO_3^-.

4 Discussion

We describe the novel finding of increased sAC gene expression and regulation in the carotid body, which is sensitive to changes in pH/CO_2. These results demonstrate that sAC gene expression occurs in both the CB and in non-chemoreceptor tissues: SCG, PG, NG. Unlike non-chemoreceptor tissues, sAC gene expression is greatest in the CB, where it is modulated by HCO_3^- in a concentration-dependent manner.

We identified several protein isoforms of sAC using homogenates of testicular tissue, CB, and the SCG that were consistent with the previously described truncated sAC isoform (sAC_t: 48 kDA) (Buck et al. 1999) reported for testicular tissue. Our data also show that other novel isoforms of sAC are expressed in the homogenates of the testis but the functions of these isoforms have not been characterized.

We tested the role of HCO_3^-/CO_2 on sAC gene expression and activity in CB and SCG, PG and testis, and showed that increasing concentrations of CO_2/HCO_3^- increased sAC mRNA gene expression in a concentration-dependent manner and induced an increased in cAMP levels but *only* in the CB. We excluded the effect of external pH variations by maintaining the extracellular pH at 7.4, but the intracellular pH was not measured. However, it has been shown that when the extracellular pH maintains constant and pCO_2 increases, there is little change in intracellular pH in Type I cells (Buckler et al. 1991) of the carotid body. Additionally, sAC activity is also independent of changes in intracellular pH (Chen et al. 2000). Taken together, sAC gene expression in the CB is most likely regulated by changes in HCO_3^- concentration. Yet, we can not be certain whether these findings are specific to Type I cells, since homogenates of the CB were used. Similar to other reports, we detected sAC gene expression in testicular tissue. However, in response to CO_2/HCO_3^-, we did not observe an increase in cAMP levels in this tissue, which could be explained by the fact that we used whole testis homogenates from younger animals than has been previously reported (Jaiswal and Conti 2001).

In this study, we describe for the first time that sAC gene expression and its activity differs between CB chemosensitive tissues and non- chemosensitive tissues (PG, NG, and SCG). The regulation of sAC gene expression and activity suggests different cellular roles for sAC in these tissues. Our results also suggest that in peripheral arterial chemoreceptors, sAC may mediate cAMP-dependent pathways, responding to changes in CO_2, while in response to hypoxia these pathways are activated by tmAC.

Acknowledgments We would like to thank Dr. Jochen Buck and Dr. Lonny R. Levin from Cornell University Medical College for the generous gift of the sAC antibody (R21), Mr. Devin Mack for assistance with Western blotting, Dr. Raul Chavez for helpful suggestions and Dr Gabrielle Mclemore for the review of the paper. This work was supported by grant R01 EBG, HL 072748; AR Nunes was supported by FLAD, 155/2007.

References

Batuca, J. R., Monteiro, T. C., & Monteiro, E. C. 2003, Contribution of dopamine D_2 receptors for the cAMP levels at the carotid body, *Adv. Exp. Med. Biol.*, 536: 367–373.

Buck, J., Sinclair, M. L., Schapal, L., Cann, M. J., & Levin, L. R. 1999, Cytosolic adenylyl cyclase defines a unique signaling molecule in mammals, *Proc. Natl. Acad. Sci. U.S.A.*, 96: 79–84.

Buckler, K. J., Vaughan-Jones, R. D., Peers, C., Lagadic-Gossmann, D., & Nye, P. C. 1991, Effects of extracellular pH, PCO_2 and HCO_3^- on intracellular pH in isolated type-I cells of the neonatal rat carotid body, *J. Physiol.*, 444: 703–721.

Chen, Y., Cann, M. J., Litvin, T. N., Iourgenko, V., Sinclair, M. L., Levin, L. R., & Buck, J. 2000, Soluble adenylyl cyclase as an evolutionarily conserved bicarbonate sensor, *Science*, 289: 625–628.

Gonzalez, C., Almaraz, L., Obeso, A., & Rigual, R. 1994, Carotid body chemoreceptors: from natural stimuli to sensory discharges, *Physiol Rev.*, 74: 829–898.

Jaiswal, B. S. & Conti, M. 2001, Identification and functional analysis of splice variants of the germ cell soluble adenylyl cyclase, *J. Biol. Chem.*, 276: 31698–31708.

Kamenetsky, M., Middelhaufe, S., Bank, E. M., Levin, L. R., Buck, J., & Steegborn, C. 2006, Molecular details of cAMP generation in mammalian cells: a tale of two systems, *J. Mol. Biol.*, 362: 623–639.

Pastor-Soler, N., Beaulieu, V., Litvin, T. N., Da Silva, N., Chen, Y., Brown, D., Buck, J., Levin, L. R., & Breton, S. 2003, Bicarbonate-regulated adenylyl cyclase (sAC) is a sensor that regulates pH-dependent V-ATPase recycling, *J. Biol. Chem.*, 278: 49523–49529.

Perez-Garcia, M. T., Almaraz, L., & Gonzalez, C. 1990, Effects of different types of stimulation on cyclic AMP content in the rabbit carotid body: functional significance, *J. Neurochem.*, 55: 1287–1293.

Pfaffl, M. W. 2001, A new mathematical model for relative quantification in real-time RT-PCR, *Nucleic Acids Res.*, 29: e45.

Schmid, A., Sutto, Z., Nlend, M. C., Horvath, G., Schmid, N., Buck, J., Levin, L. R., Conner, G. E., Fregien, N., & Salathe, M. 2007, Soluble adenylyl cyclase is localized to cilia and contributes to ciliary beat frequency regulation via production of cAMP, *J. Gen. Physiol*, 130: 99–109.

Sinclair, M. L., Wang, X. Y., Mattia, M., Conti, M., Buck, J., Wolgemuth, D. J., & Levin, L. R. 2000, Specific expression of soluble adenylyl cyclase in male germ cells, *Mol. Reprod. Dev.*, 56: 6–11.

Sun, X. C., Zhai, C. B., Cui, M., Chen, Y., Levin, L. R., Buck, J., & Bonanno, J. A. 2003, HCO_3^--dependent soluble adenylyl cyclase activates cystic fibrosis transmembrane conductance regulator in corneal endothelium, *Am. J. Physiol. Cell Physiol.*, 284: C1114–C1122.

Wang, Y., Lam, C. S., Wu, F., Wang, W., Duan, Y., & Huang, P. 2005, Regulation of CFTR channels by HCO_3^--sensitive soluble adenylyl cyclase in human airway epithelial cells, *Am. J. Physiol. Cell Physiol.*, 289: C1145–C1151.

Zippin, J. H., Farrell, J., Huron, D., Kamenetsky, M., Hess, K. C., Fischman, D. A., Levin, L. R., & Buck, J. 2004, Bicarbonate-responsive "soluble" adenylyl cyclase defines a nuclear cAMP microdomain, *J. Cell Biol.*, 164: 527–534.

Developmental Maturation of Chemosensitivity to Hypoxia of Peripheral Arterial Chemoreceptors – *Invited Article*

Estelle B. Gauda, John L. Carroll and David F. Donnelly

Abstract Peripheral arterial chemoreceptors, particularly the carotid body chemoreceptors, are the primary sites for the detection of hypoxia and reflexly increase ventilatory drive and behavioral arousal during hypoxic or asphyxial events. Newborn infants are at risk for hypoxic and asphyxial events during sleep, yet, the strength of the chemoreceptor responses is low or absent at birth and then progressively increases with early postnatal development. This review summarizes the available data showing that even though the "oxygen sensor" in the glomus cells has not been unequivocally identified, it is clear that development affects many of the other properties of the chemoreceptor unit (glomus cell, afferent nerve fibers and neurotransmitter profile at the synapse) that are necessary and essential for the propagation of the "sensing" response, and exposure to hypoxia, hyperoxia and nicotine can modify normal development of each of the components leading to altered peripheral chemoreceptor responses.

Keywords Carotid body · Development · Chemoreception · Hypoxia · New born · Sleep · Reflexes · Environment · PO_2 · Glomus cells · Afferent fibers · Calcium responses · Dopamine · Dopaminergic · Tyrosine hydroxylase · Adenosine · Acetylcholine · Cholinergic

1 Introduction

Human infants, particularly preterm infants, often present with apnea, with and without upper airway obstruction, and may have reduced arousal responses when exposed to hypoxia or an asphyxial event during sleep. Peripheral arterial chemoreceptors, particularly the carotid body chemoreceptors, are the primary sites for the detection of hypoxia and reflexly increase ventilatory drive and behavioral arousal during hypoxic or asphyxial events. The strength of this reflex increases

E.B. Gauda (✉)
Johns Hopkins Medical Institutions, Department of Pediatrics, Division of Neonatology, Baltimore, MD 21287, USA
e-mail: egauda@mail.jhmi.edu

with maturation. Near term, fetal chemoreceptor activity is generally absent, and when present, responds poorly to a given level of hypoxia (Blanco et al., 1984). An increase in the organ sensitivity to hypoxia appears to be initiated with birth associated with the large increase in PaO$_2$ that takes place at birth along with the increase in sympathetic activity. The post-natal developmental process takes approximately 1–2 weeks, and, during that period, spontaneous action potential (AP) activity is lower at all oxygen levels compared to the adult response (Mulligan, 1991)

Carotid body chemoreceptors consist of (1) type I or glomus cells, similar to presynatpic neurons, that are chemosensitive and contain neurotransmitters and autoreceptors (2) postsynaptic afferent, nerve fibers from the carotid sinus nerve which appose glomus cells. These fibers contain neurotransmitters and postsynaptic receptors and have cell bodies in the petrosal ganglion, (3) type II cells similar to glial cells, that are not chemosensitive, (4) microganglion cells, that express cholinergic traits, and (5) blood vessels and sympathetic fibers innervating these vessels. Outlined below is a short review of the molecular and cellular events associated with maturation of hypoxic chemosensitivity focusing on the changes in the intrinsic properties of glomus cells and chemoreceptor afferent neurons, and changes in the neurotransmitter profiles at the synapse. The effect of environmental exposures in early development leading to aberrations of peripheral chemoreceptor maturation will be discussed.

2 Developmental Changes and Environmental Perturbations to the Properties of the Glomus Cell

The glomus cell, a secretory cell apposed to the afferent nerve endings, is widely regarded as the O$_2$-sensing element in the carotid body. Although the precise nature of the O$_2$ sensor remains unknown, it is clear that exposure to acute hypoxia induces glomus cell depolarization, activation of voltage-dependent calcium currents, a rise in intracellular calcium and a resulting increase in neurotransmitter release. All of these steps undergo developmental changes in the postnatal period.

2.1 Glomus Cell Membrane Depolarization with Hypoxia

Glomus cell membrane potential is sensitive to changes in local PO$_2$. Although glomus cell depolarization by hypoxia is likely a critically important step in transduction, the mechanism remains unresolved. According to current theories, hypoxic inhibition of one or more K$^+$ currents leads to cell depolarization. Indeed, glomus cells express several types of O$_2$-sensitive K$^+$ currents which can be broadly categorized as: (i) transient (Lopez-Lopez et al., 1993), (ii) calcium dependent (Peers, 1990) and (iii) background or "leak" (Buckler, 1997). However, the sensitivity of the transient K$^+$ current to hypoxia appears to be too high, with complete inhibition at PO$_2$ levels higher than that for the [Ca^{2+}]$_i$ response or CB neural response to hypoxia (Ganfornina and Lopez-Barneo, 1991). This current also appears to be present only in rabbit glomus cells and not in those from other species; thus, it is not a critical element in the transduction process. In the case of the calcium-dependent current or other large conductance K^+ currents, inhibition with TEA/4AP actually enhances the glomus cell response to hypoxia (Wasicko et al., 2006), suggesting that these currents normally serve to damp depolarization but not initiate depolarization. The PO_2-current inhibition relationship of the background K^+ current (Buckler, 1997) most closely matches the PO_2-$[Ca^{2+}]_i$ relationship of dissociated glomus cells (Buckler, 1997; Wasicko et al., 1999) and the PO_2-nerve activity relationship of the carotid body (Roumy, 1994). However, it remains unclear whether the small-conductance background K^+ current is capable of initiating and driving hypoxia-induced depolarization in glomus cells.

Developmental studies have been reported for the calcium-dependent current and the background K^+ current and both currents mature during early postnatal development. The magnitude of the calcium-dependent current increases with age over the first 10 days of life in rats and is further increased in adults (Hatton et al., 1997). In addition, the proportion of glomus cell K^+ current inhibited by hypoxia increases with age over the same time course (Hatton et al., 1997). Similarly, a recent study from one of our laboratories demonstrated that the magnitude of hypoxia-induced inhibition of the background or "leak" K^+ conductance (Buckler, 1997) is small in glomus cells from newborn rats and increases considerably over the first 2 weeks of life in rats (Wasicko et al., 2006). Although the maturational time course for both currents matches the known time course of glomus cell $[Ca^{2+}]_i$ response maturation (Wasicko et al., 1999) and nerve activity hypoxia response maturation (Kholwadwala and Donnelly, 1992), a causal nature to the relationship has not been unequivocally established for either oxygen-sensitive current.

2.2 Calcium Responses of Glomus Cells During Development

Depolarization of the glomus cell leads to activation of voltage-dependent calcium currents and a rise in intracellular calcium. Recordings from a large number of glomus cells demonstrate that the magnitude of the calcium response is several fold greater in glomus cells harvested from older rats compared to newborns (Wasicko et al., 1999). Wasicko et al., characterized the $[Ca^{2+}]_i$ response to graded hypoxia in glomus cells isolated from term-fetal, P1, P3, P7, P11, P14, and P21 rats (Wasicko et al., 1999). The response was hyperbolic at all ages, the maximal $[Ca^{2+}]_i$ response to hypoxia and anoxia increased with age and there was an apparent rightward shift in the response curve with age. The age-related increase in the glomus cell $[Ca^{2+}]_i$ response to graded hypoxia occurred mainly between P3 and P11-14, which is in agreement with the reported time course for maturation of rat CB neural responses to anoxia (Kholwadwala and Donnelly, 1992).

The smaller calcium response of the immature cell cannot be accounted for by a lower density of voltage-gated calcium channels, since the calcium channel density is similar between P4 and P10 (Hatton et al., 1997). In addition, the glomus cell $[Ca^{2+}]_i$ response to elevated extracellular K^+ is unchanged between P3 and P14, which is the same time frame during which all of the developmental increase in the $[Ca^{2+}]_i$ response to hypoxia occurs (Wasicko et al., 1999). Considering elevated extracellular K^+ as a non-specific depolarizing stimulus, these results demonstrate that the immature glomus cell can mobilize large amounts of $[Ca^{2+}]_i$ with depolarization, suggesting that maturation of the $[Ca^{2+}]_i$ response is not due to changes in voltage-dependent calcium currents.

In addition to excitatory mechanisms that may enhance glomus cell depolarization, another possibility involves postnatal changes in expression of ion channels that damp, limit or suppress depolarization in the newborn. One of our laboratories reported that HERG-like K^+ channels play a major role in limiting the hypoxia-induced depolarization and rise in glomus cell $[Ca^{2+}]_i$ of newborns but not P14 rats (Kim et al., 2005). These findings support the hypothesis that postnatal glomus cell maturation may involve high expression levels, in the newborn, of non-O_2-sensitive K^+ currents that damp excitability; subsequent decline in expression of these currents would result in increasing excitability (removal of damping) with age. Other candidate K^+ channels that may damp or inhibit the newborn glomus cell depolarization response to hypoxia are currently under investigation.

Another potential site of maturational change is in the number of hypoxia-sensitive glomus cells. Some glomus cells fail to depolarize with hypoxia (Pang and Eyzaguirre, 1992) and some fail to increase their intracellular calcium levels with hypoxia (Bright et al., 1996; Donnelly and Kholwadwala, 1992; Roumy, 1994; Sterni et al., 1995). Why some cells fail to respond is unclear, although cell preparation methods and stimulus intensity may be important factors. For example, Bright et al used 35 mmHg superfusate PO_2 as the "hypoxia" challenge and found that only 20% of isolated glomus cells responded with an increase in $[Ca^{2+}]_I$ (Bright et al., 1996) . However, 35 mmHg PO_2 is in the normoxic range of carotid body microvascular (tissue) PO_2 (Rumsey et al., 1991), which may explain the low proportion of glomus cell $[Ca^{2+}]_i$ responses reported by these authors. In a study using superfusate PO_2's of 0, 2, 7, and 14 mmHg, the great majority of glomus cells significantly raised $[Ca^{2+}]_i$ in response to hypoxia (Wasicko et al., 1999). Nevertheless, patch clamp recordings of intact glomus cells demonstrate a suppression of calcium-dependent K^+ current in many cells under normoxic conditions (Donnelly, 1995), suggesting that calcium currents may be inhibited, in situ. If so, then the number of responding cells may (potentially) increase post-natally, providing an enhanced response with development.

2.3 *Environmental Perturbations to Glomus Cell Maturation*

As with nerve activities, environmental oxygen levels in the post-natal period exert major effects on the glomus cell developmental pattern. Birthing into a low oxygen environment results in an impaired response to hypoxia when tested at points where a mature response is anticipated (Sterni et al., 1999). Similarly, exposure to a high oxygen environment (e.g., $FiO_2 = 0.6$) in the post-natal period results in a profound reduction in the CB single unit neural response to acute hypoxia challenge

(Donnelly et al., 2005). In parallel, the same exposure to hyperoxia results in a similarly profound reduction in the glomus cell $[Ca^{2+}]_i$ response to hypoxia (Gonzalez et al., 2009). This hyperoxia alteration may be due to altered channel expression. Real-time PCR of whole carotid bodies from hyperoxia-exposed developing rats revealed an ~70% reduction, compared to normoxia-reared controls, in the expression (RNA) of TASK1, TASK3 and TASK5 background K^+ channels (Kim et al., 2006). Taken together, these findings suggest that the profound reduction in glomus cell background K^+ channel expression, the $[Ca^{2+}]_i$ response to hypoxia and the whole organ neural response to hypoxia may be related, although this remains to be proven. If so, then postnatal maturation of glomus cell hypoxia sensitivity may depend on developmental changes in background K^+ current expression, O_2 sensitivity of background K^+ channels, sensor-to-ion channel signaling or other factors involved in hypoxia-mediated glomus cell depolarization.

3 Developmental Changes in Afferent Nerve Activity and Environmental Modifications to Afferent Nerve Function

3.1 Developmental Changes in Afferent Nerve Activity

The post-natal development of hypoxic chemosensitivity takes approximately 1–2 weeks, and, during that period, spontaneous action potential (AP) activity is lower at all oxygen levels compared to the adult response (Mulligan, 1991). At least part of the enhanced response to hypoxia with maturation is an alteration in the interaction between CO_2 and hypoxia, both of which stimulate nerve activity and which interact in a multiplicative fashion (i.e., greater than the sum of individuals). Kumar et al apportioned the nerve response into three terms reflecting the sensitivity to hypoxia alone, to hypercapnia alone and to an interaction term, hypoxia × hypercapnia (Pepper et al., 1995). In the postnatal period, only the interaction term increased in magnitude, suggesting that the maturational site lies downstream of the sensing sites for hypoxia and hypercapnia.

Along with the higher level of AP activity, the mature chemoreceptor is better able to sustain a higher level of activity. In piglets, a strong hypoxic stimulus ($PaO_2 < 30$ mmHg) resulted in a decrease, rather than an increase in AP activity (Mulligan, 1991). Similarly, in P1-P3 kittens, half of the chemoreceptor fibers accommodated to a hypoxic stimulus which was not observed in any chemoreceptor fibers from P15 kittens.

3.2 Electrophysiologic and Morphologic Changes in Chemoreceptor Nerve Fibers During Development

Although the primary site for maturational change is likely to be at elements presynaptic to the nerve endings (see below), some changes in the afferent nerve morphology and channel expression contribute to organ maturation. A single afferent nerve fiber undergoes branching within the carotid body and establishes synaptic relations with multiple glomus cells (Kondo, 1975). The number of synapses per 600 μm increases approximately 4-fold in the postnatal period, consistent with an increase in the number of synaptic interactions per afferent axon. In addition, the number of afferent axons may be increasing over this period. For instance, the number of labeled somas in the petrosal ganglion following antegrade staining from the carotid body doubled between P3 and P7 and then decreased by half by P14, suggesting an active period of neuron growth and pruning (Tolosa et al., 2005).

The excitability of the afferent axon may also be increased due to an increase in Na^+ channel expression. The voltage activated fast-Na^+ current recorded at the soma of chemoreceptor neurons approximately doubles from P3 to P18 in the rat (Cummins et al., 2002). This is primarily due to an increase in somal size since the density of Na^+ current is not changed over the period. Similarly, no change in the voltage dependence of activation and inactivation is observed, suggesting isoform expression and modulatory subunits do not change over the period.

During the early postnatal period, the phenotype of the afferent neuron becomes established. Approximately half the neurons projecting to the carotid body are tyrosine-hydroxylase (TH) positive (Finley et al., 1992). Establishment of the TH phenotype likely reflects AP activity in the newborn period, because exogenous depolarizing stimuli such as high K^+ can induce the TH positive phenotype in all petrosal sensory neurons if applied in early development. Once the TH phenotype is established, catecholamine release from the neuronal soma may occur during afferent stimulation and potentially modulate the excitability of neighboring neurons.

3.3 Environmental Factors Altering Post-Natal Development of Chemoreceptor Afferent Nerve Fibers

As mentioned above, the rise in PaO_2 at the time of birth is likely to be the initiating factor for chemoreceptor maturation. If the rise in PaO_2 is limited, for instance by being born into a low oxygen atmosphere or clinical entities such as cyanotic heart disease, then the chemoreceptor and respiratory response to acute hypoxia remains blunted, i.e., the immature response persists. Thus, rats or kittens born into FiO_2=8–10% have a profoundly depressed ventilatory and chemoreceptor response to acute hypoxia (Hanson et al., 1989).

Unlike chronic hypoxia, intermittent hypoxia (e.g., episodes of breathing 5% O_2, lasting 15s followed by 5 min of 21% O_2) results in an augmentation of the acute response to hypoxia (Peng et al., 2004). The enhanced response is also maintained despite a return to continuous normoxic conditions. This suggests that intermittent hypoxia and factors produced by intermittent hypoxia, but not by chronic hypoxia, enhance chemoreceptor maturation.

Perhaps surprisingly, being born into a *high* oxygen atmosphere produces a similar phenotype as being born in an hypoxic environment. Levels of O_2 as low as 30%, lasting 2 weeks to 1 month, produce impairment in hypoxic sensitivity (Hanson et al., 1989) lasting months to years despite a return to a normoxic environment. At least part of this reduced response is due to axonal loss of chemoreceptor axons which decrease by >40% following one month of 60% FiO_2 (Erickson et al., 1998). For those axons remaining, the discharge frequencies during exposure to normoxia or hypoxia are profoundly depressed compared to control chemoreceptor axons and the conduction time for the axons is longer, suggesting a change in axonal morphology or ion channel expression (Donnelly et al., 2005).

Nicotine, as an environmental toxin, also alters the maturational process. For instance, lambs exposed to prenatal nicotine exhibit a reduced ventilatory response to 10% O_2 along with depressed arousal responses (Hafstrom et al., 2000, 2002). However, this nicotine effect may be species dependent since rat pups exposed to prenatal nicotine had similar ventilatory responses as unexposed pups.

4 Developmental Changes in Carotid Body Systems of Neurotransmitters/ Neuromodulators

Chemical transmission is essential for propagation of electrical activity from the glomus cells to afferent nerve fibers that then synapse on to 2nd order neurons in the nucleus tractus soltarii. Neurotransmitters released from glomus cells (1) bind to autoreceptors, modulating further release of the neurotransmitters/modulators, and (2) bind to postsynaptic receptors, modifying propagation of the impulse. Summation of these events determines the output from the carotid body in response to the depolarizing stimuli, i.e. hypoxia and hypercapnia.

4.1 Dopaminergic Mechanisms Mediating Hypoxic Chemosensitivity During Development

Multiple putative inhibitory and excitatory transmitters, of which dopamine is the most abundant, have been identified in CBs from a variety of mammalian species and has been recently reviewed (Prabhakar, 2006). Experiments in adult and newborn animals consistently demonstrate that dopamine inhibits hypoxic chemosensitivity likely through binding to inhibitory D_2-dopamine autoreceptors and postsynaptic receptors (Dinger et al., 1981). Recent experiments, suggest that D_2-autoreceptors have a greater role on dopamine release in the carotid body than D_2-postsynaptic receptors on inhibitory afferent activity in mice who do not express D_2-dopamine receptors (Prieto-Lloret et al., 2007). Direct examination of the postsynaptic maturation of functional D_2 receptors has not been done. However, indirect assessment of the summation of autoreceptor and postsynatptic effects suggest that D_2-dopamine receptors mediate a greater inhibition of hypoxic chemosensitivity in kittens than adult cats (Tomares et al., 1994).

Tyrosine hydroxylase, the rate-limiting enzyme for catecholamine synthesis, gene and protein expression and mRNAs for D_2-dopamine autoreceptors have been localized to glomus cells from newborn animals, and the functional role of autoreceptors in mediating neurotransmitter release has been recently reviewed (Bairam and Carroll, 2005). D_2-dopamine autoreceptor mRNA expression increases during the first 3 days of life in rats and rabbits with little change thereafter. However, D_2 autoregulatory function in newborn rabbits is less than in adults resulting in a greater release of dopamine from the carotid body in response to hypoxia with increasing maturation. Maturation of dopamine secretion from glomus cells also parallels the maturational increase in hypoxic-induced carotid sinus nerve (CSN) activity in newborn rats (Kholwadwala and Donnelly, 1992). Thus, paradoxically, the maturational increase in hypoxic sensitivity is associated with a greater release of the inhibitory neurotransmitter, dopamine.

4.2 Cholinergic Mechanisms Mediating Hypoxic Chemonsensitivity During Development

Co-release of an excitatory neurotransmitter with dopamine in response to hypoxia may be the key to resolving this paradox. The classical neurotransmitter, acetylcholine (ACh), mediates hypoxia-induced chemo excitation in adult cats and in co-cultures of glomus and petrosal ganglion neurons from newborn rat pups (Nurse and Zhang, 1999). Excitatory nicotinic receptors are present on glomus cells and afferent nerve fibers (Gauda, 2002). It is possible that ACh may be co-released with dopamine from glomus cells. However, controversy exist as to the source of ACh within the rat carotid body (Gauda et al., 2004) and its role in hypoxic chemosensitivity (Zapata, 2007) in general. ACh markers are present in microganglion cells interspersed throughout the carotid body and within the carotid sinus nerve (Gauda et al., 2004) in the rat. With maturation, choline acetyltransferase (ChAT, the rate-limiting enzyme for aetylcholine synthesis), increases by 2.5 fold in the cat carotid body (Bairam et al., 2007). However, even larger increases in ChAT protein expression are found in the petrosal ganglion and superior cervical ganglion. The source of cholinergic markers in the sympathetic ganglion is from preganglionic fibers. Thus, whether ACh is released from microganglion cells or glomus cells within the carotid body, in response to hypoxia, it will bind to excitatory nicotinic receptors on glomus cells and afferent nerve fibers, thereby leading to augmentation of carotid sinus nerve activity in response to hypoxia. With maturation the mRNAs for different nicotinic receptors also change (Bairam et al., 2007). Whether the sum of these changes translates to an increase in hypoxic chemosensitivity related to ACh has not been determined.

4.3 Purinergic Mechanisms Mediating Hypoxic Chemosensitivity During Development

ATP and adenosine (ADO) modulate hypoxic chemosensitivity. ATP is synthesized, stored and co-released with classical neurotransmitters, including ACh, and catecholamines. Experiments using knock-out mice confirm that ATP is an excitatory neurotransmitter mediating hypoxic chemosensitivity by binding to postsynaptic P2X$_2$ excitatory inotropic receptors.

In response to hypoxia, ATP is released from glomus cells or carotid body slices from newborn rat pups (Buttigieg and Nurse, 2004). Exposure to acidic hypercapnia also induces ATP release, measured with a novel microbiosensor, from the superfused carotid bodies of rats between 16 and 20 postnatal days (Masson et al., 2008). In this age group, ATP release in response to hypoxia is 2.5 fold greater compared that evoked by acidic hypercapna (unpublished data). With maturation, ATP content, measured with the luciferin-luciferase assay, decreased by 50% from the isolated cat carotid from day 1 to 2 months postnatal age (Bairam et al., 2007). Although the measuring tools exist, at the current time, the role of ATP release in mediating maturation of hypoxic chemosensitivity has not been reported. P2X$_2$ – immunoreactivity (-ir) is present in nerve fibers innervating the carotid body of rat pups at 5 and 27 postnatal days, and a 2 fold increase in P2X$_2$ – mRNAs in the petrosal ganglion during this developmental period also occurs. Thus, co-release of ATP with dopamine from glomus cells binding to excitatory P2X$_2$ postsynaptic receptors could explain why high levels of dopamine (an inhibitory neurotransmitter) are associated with high levels of CSN activity in response to hypoxia with maturation.

ADO is released from metabolically active cells by facilitated diffusion and extracellular ATP is rapidly converted to ADO. ADO modifies cellular function by binding to specific G-protein-coupled receptors (GPCRs) cell-surface receptors (A1, A2a, A2b and A3). A1- and A3-Rs inhibit adenylyl cyclase (AC), and hyperpolarize cells by G-protein-coupled K$^+$ channels, while A2a- and A2b-Rs activate AC leading to depolarization. ADO is released from CBs of adult rats in response to hypoxia (Conde and Monteiro, 2004), and many pharmacological experiments suggest that ADO plays an excitatory role in augmenting hypoxic chemosensitivity through binding to A2a-Rs in adult models. Centrally, ADO is a potent inhibitor of the hypoxic ventilatory response and largely accounts for the hypoxic ventilatory decline observed in immature animals. In newborn rabbits and human infants, indirect evidence suggests that ADO may also inhibit peripheral arterial responses to hypoxia and baseline ventilation, respectively.

Both A$_{2a}$ (Gauda et al., 2000) and A$_{2b}$ autoreceptors (Conde et al., 2006) are present on glomus cells. In different model systems, ADO binding to A$_{2a}$ autoreceptors depolarizes or hyperpolarizes the glomus cell. No data are currently available in immature animals demonstrating the functional role of A$_{2a}$ autoreceptors on glomus cell function. Significant interaction occurs between A$_{2a}$ and D$_2$ autoreceptors in central neurons that modifies transmitter release, specifically, dopamine. This interaction may also occur in the carotid body of mature animals as suggested by Conde et al (this volume). The interaction of these receptors on dopamine release during maturation has not been determined.

Only indirect measures of the role of postsynaptic A$_{2a}$ receptors exist. A$_{2a}$ receptors are present in petrosal ganglion neurons that are positive for TH phenotype. While carotid body A$_{2a}$ gene and protein expression does not change during the first two weeks of postnatal development, they are increased 2-fold in the petrosal ganglion. A$_1$-adenosine (highest-affinity) receptor mRNAs are abundantly expressed throughout the petrosal ganglion and appear to have an inhibitory role on hypoxic chemosensitivity (Gauda et al., 2006). The balance of less inhibitory postsynaptic A_1 and more excitatory A_{2a} receptors on chemoafferents may contribute to the maturation of hypoxic chemosensitivity.

4.4 Environmental Exposures Modify Neurotransmitter Profiles in the Carotid Body During Development

Exposures to hypoxia/hyperoxia and nicotine during prenatal and early postnatal development induces histological changes in the carotid body and petrosal ganglion and either ablates or accentuates hypoxic chemosensitivity depending on the level and paradigm of exposure. Newborn rat pups born in an hypoxic environment have higher levels of dopamine (DA) content (Hertzberg et al., 1992), have increased TH mRNA levels in the carotid body and have reduced peripheral chemoreceptor responses. On the other hand, neonatal rat pups who are exposed to intermittent hypoxia are more susceptible to sensitization of hypoxic chemosensitivity than adult rats (Pawar et al., 2008), and this response is associated with increased TH-ir in carotid bodies. Chronic hyperoxic exposure during the first 28 days of postnatal life is associated with atrophy of the carotid bodies, a twofold increase in DA concentration, but a 50% reduction in the rat of DA synthesis (Prieto-Lloret et al., 2004), and a reduction in TH –ir in chemoafferents (Erickson et al., 1998).

Prenatal nicotine exposure alters cardiorespiratory responses, reviewed in (Hafstrom et al., 2005). Acute exposure to nicotine increases TH mRNA levels and dopamine release from carotid bodies of newborn rats, while prenatal and early postnatal exposure to nicotine increases TH and dopamine-β-hydroxylase, norepinephrine synthesizing enzyme, gene expression but does not affect D_2 dopamine autoreceptor or postsynaptic receptor gene expression (Gauda et al., 2001) in the carotid body. Decreased chemosensitivity associated with prenatal nicotine exposure in newborn lambs may be mediated through the D_2 dopamine receptor (Hafstrom et al., 2002). The role of dopamine in mediating plasticity induced as a result of environmental exposures has been most frequently reported, however, there are many other neurotransmitter systems, that may be adversely affected during critical developmental periods that will ultimately modify the function of peripheral arterial chemoreceptors throughout life.

5 Conclusion

In summary, reflexes involving peripheral arterial chemoreceptors in the carotid body are important in health and disease, and hypoxic chemosensitivity increases within the first 3 weeks after birth in essentially all mammalian models studied. During this maturational period, the peripheral arterial chemoreceptors are particularly vulnerable to adverse environmental exposures that can adversely affect chemoreceptor function. The infant that is born prematurely is especially at risk of the adverse effects of these environmental exposures. We have briefly reviewed what is known about how normal development and environmental exposures affect the structure, molecular and cellular responses of the "chemoreceptor unit" (glomus cell, afferent fibers and the synapse). It is concluded that even though the "oxygen sensor" in the glomus cell has not been unequivocally identified, it is clear that development affects many of the other properties of the chemoreceptor unit that are necessary and essential for the propagation of the "sensing" response, and exposure to hypoxia, hyperoxia and nicotine can modify normal development of each of the components leading to altered peripheral chemoreceptor responses.

Acknowledgments This work was supported by the following NIH GRANTS R01HL80735; R21HL082860-02.

References

Bairam, A. and Carroll, J.L. 2005 Neurotransmitters in carotid body development. *Respir. Physiol. Neurobiol.* 149: 217–232.
Bairam, A., Joseph, V., Lajeunesse, Y., and Kinkead, R. 2007 Developmental profile of cholinergic and purinergic traits and receptors in peripheral chemoreflex pathway in cats. *Neuroscience* 146: 1841–1853.
Blanco, C.E., Dawes, G.S., Hanson, M.A., and McCooke, H.B. 1984 The response to hypoxia of arterial chemoreceptors in fetal sheep and new-born lambs. *J. Physiol.* 351: 25–37.
Bright, G.R., Agani, F.H., Haque, U., Overholt, J.L., and Prabhakar, N.R. 1996 Heterogeneity in cytosolic calcium responses to hypoxia in carotid body cells. *Brain Res.* 706: 297–302.
Buckler, K.J. 1997 A novel oxygen-sensitive potassium current in rat carotid body type I cells. *J. Physiol. (Lond.)* 498(Pt 3): 649–662.
Buttigieg, J. and Nurse, C.A. 2004 Detection of hypoxia-evoked ATP release from chemoreceptor cells of the rat carotid body. *Biochem. Biophys. Res. Commun.* 322: 82–87.
Conde, S.V. and Monteiro, E.C. 2004 Hypoxia induces adenosine release from the rat carotid body. *J. Neurochem.* 89: 1148–1156.
Conde, S.V., Obeso, A., Vicario, I., Rigual, R., Rocher, A., and Gonzalez, C. 2006 Caffeine inhibition of rat carotid body chemoreceptors is mediated by A2A and A2B adenosine receptors. *J. Neurochem.* 98: 616–628.
Cummins, T.R., Dib-Hajj, S.D., Waxman, S.G., and Donnelly, D.F. 2002 Characterization and developmental changes of Na+ currents of petrosal neurons with projections to the carotid body. *J. Neurophysiol.* 88: 2993–3002.
Dinger, B., Gonzalez, C., Yoshizaki, K., and Fidone, S.J. 1981 [3H] Spiroperidol binding in normal and denervated carotid bodies. *Neurosci. Lett.* 21: 51–55.
Donnelly, D.F. 1995 Modulation of glomus cell membrane currents of intact rat carotid body. *J. Physiol.* 489(Pt 3): 677–688.
Donnelly, D.F. and Kholwadwala, D. 1992 Hypoxia decreases intracellular calcium in adult rat carotid body glomus cells. *J. Neurophysiol.* 67: 1543–1551.
Donnelly, D.F., Kim, I., Carle, C., and Carroll, J.L. 2005 Perinatal hyperoxia for 14 days increases nerve conduction time and the acute unitary response to hypoxia of rat carotid body chemoreceptors. *J. Appl. Physiol.* 99: 114–119.
Erickson, J.T., Mayer, C., Jawa, A., Ling, L., Olson, E.B., Jr., Vidruk, E.H. et al. 1998 Chemoafferent degeneration and carotid body hypoplasia following chronic hyperoxia in newborn rats. *J. Physiol.* 509(Pt 2): 519–526.
Finley, J.C.W., Polak, J., and Katz, D.M. 1992 Transmitter diversity in carotid body afferent neurons: Dopaminergic and peptidergic phenotypes. *Neuroscience* 51: 973–987.

Ganfornina, M.D. and Lopez-Barneo, J. 1991 Single K$^+$ channels in membrane patches of arterial chemoreceptor cells are modulated by O$_2$ tension. *Proc. Natl. Acad. Sci. U.S.A.* 88: 2927–2930.

Gauda, E.B. 2002 Gene expression in peripheral arterial chemoreceptors. *Microsc. Res. Tech.* 59: 153–167.

Gauda, E.B., Cooper, R., Akins, P.K., and Wu, G. 2001 Prenatal nicotine affects catecholamine gene expression in newborn rat carotid body and petrosal ganglion. *J. Appl. Physiol.* 91: 2157–2165.

Gauda, E.B., Cooper, R., Johnson, S.M., McLemore, G.L., and Marshall, C. 2004 Autonomic microganglion cells: a source of acetylcholine in the rat carotid body. *J. Appl. Physiol.* 96: 384–391.

Gauda, E.B., Cooper, R.Z., Donnelly, D.F., Mason, A., and McLemore, G.L. 2006 The effect of development on the pattern of A1 and A2a-adenosine receptor gene and protein expression in rat peripheral arterial chemoreceptors. *Adv. Exp. Med. Biol.* 580: 121–129.

Gauda, E.B., Northington, F.J., Linden, J., and Rosin, D.L. 2000 Differential expression of A$_{2A}$, A$_1$-adenosine and D$_2$-dopamine receptor genes in rat peripheral arterial chemoreceptors during postnatal development. *Brain Res.* 872: 1–10.

Gonzalez, C., Almaraz, L., Obeso, A., Rigual, R. 1994 Carotid body Chemoreceptors: from natural stimuli to sensory discharges. *Physiol. Rev.* 74: 829–98.

Hafstrom, O., Milerad, J., Asokan, N., Poole, S.D., and Sundell, H.W. 2000 Nicotine delays arousal during hypoxemia in lambs. *Pediatr. Res.* 47: 646–652.

Hafstrom, O., Milerad, J., Sandberg, K.L., and Sundell, H.W. 2005 Cardiorespiratory effects of nicotine exposure during development. *Respir. Physiol. Neurobiol.* 149: 325–341.

Hafstrom, O., Milerad, J., and Sundell, H.W. 2002 Prenatal nicotine exposure blunts the cardiorespiratory response to hypoxia in lambs. *Am. J. Respir. Crit. Care Med.* 166: 1544–1549.

Hanson, M.A., Eden, G.J., Nijhuis, J.G., and Moore, P.J. 1989 Perpherial chemoreceptors and other O$_2$ sensors in the fetus and newborn . In *Chemoreceptors and Reflexes in Breathing.* Pack AI (ed.), New York: Oxford University Press, pp. 113–120.

Hatton, C.J., Carpenter, E., Pepper, D.R., Kumar, P., and Peers, C. 1997 Developmental changes in isolated rat type I carotid body cell K+ currents and their modulation by hypoxia. *J. Physiol.* 501(Pt 1): 49–58.

Hertzberg, T., Hellstrom, S., Holgert, H., Lagercrantz, H., and Pequignot, J.M. 1992 Ventilatory response to hyperoxia in newborn rats born in hypoxia – possible relationship to carotid body dopamine. *J. Physiol. (Lond.)* 456: 645–654.

Kholwadwala, D. and Donnelly, D.F. 1992 Maturation of carotid chemoreceptor sensitivity to hypoxia: In vitro studies in the newborn rat. *J. Physiol.* 453: 461–473.

Kim, I., Boyle, K.M., and Carroll, J.L. 2005 Postnatal development of E-4031-sensitive potassium current in rat carotid chemoreceptor cells. *J. Appl. Physiol.* 98: 1469–1477.

Kim, I., Donnelly, D.F., and Carroll, J.L. 2006 Modulation of gene expression in subfamilies of TASK K+ channels by chronic hyperoxia exposure in rat carotid body. *Adv. Exp. Med. Biol.* 580: 37–41.

Kondo, H. 1975 A light and electron microscopic study on the embryonic development of the rat carotid body. *Am. J. Anat.* 144: 275–293.

Lopez-Lopez, J.R., De Luis, D.A., and Gonzalez, C. 1993 Properties of a transient K+ current in chemoreceptor cells of rabbit carotid body. *J. Physiol.* 460: 15–32.

Masson, J.F., Kranz, C., Mizaikoff, B., and Gauda, E.B. 2008 Amperometric ATP microbiosensors for the analysis of chemosensitivity at rat carotid bodies. *Anal. Chem.* 80: 3991–3998.

Mulligan, E.M. 1991 Discharge properties of carotid bodies: developmental aspects. In *Developmental Neurobiology of Breathing.* Haddad, G.G. and Faber, J.P. (eds.), New York, NY: Marcel Dekker, Inc, pp. 321–340.

Nurse, C.A. and Zhang, M. 1999 Acetylcholine contributes to hypoxic chemotransmission in co-cultures of rat type 1 cells and petrosal neurons. *Respir. Physiol.* 115: 189–199.

Pang, L. and Eyzaguirre, C. 1992 Different effects of hypoxia on the membrane potential and input resistance of isolated and clustered carotid body glomus cells. *Brain Res.* 575: 167–173.

Pawar, A., Peng, Y.J., Jacono, F.J., and Prabhakar, N.R. 2008 Comparative analysis of neonatal and adult rat carotid body responses to chronic intermittent hypoxia. *J. Appl. Physiol.* 104: 1287–1294.

Peers, C. 1990 Hypoxic suppression of K^+ currents in type I carotid body cells: selective effect on the Ca2(+)-activated K^+ current. *Neurosci. Lett.* 119: 253–256.

Peng, Y.J., Rennison, J., and Prabhakar, N.R. 2004 Intermittent hypoxia augments carotid body and ventilatory response to hypoxia in neonatal rat pups. *J. Appl. Physiol.* 97: 2020–2025.

Pepper, D.R., Landauer, R.C., and Kumar, P. 1995 Postnatal development of CO_2-O_2 interaction in the rat carotid body in vitro. *J. Physiol.* 485(Pt 2): 531–541.

Prabhakar, N.R. 2006 O_2 sensing at the mammalian carotid body: why multiple O_2 sensors and multiple transmitters? *Exp. Physiol.* 91: 17–23.

Prieto-Lloret, J., Caceres, A.I., Obeso, A., Rocher, A., Rigual, R., Agapito, M.T. et al. 2004 Ventilatory responses and carotid body function in adult rats perinatally exposed to hyperoxia. *J. Physiol.* 554: 126–144.

Prieto-Lloret, J., Donnelly, D.F., Rico, A.J., Moratalla, R., Gonzalez, C., and Rigual, R.J. 2007 Hypoxia transduction by carotid body chemoreceptors in mice lacking dopamine D(2) receptors. *J. Appl. Physiol.* 103: 1269–1275.

Roumy, M. 1994 Cytosolic calcium in isolated type I cells of the adult rabbit carotid body: effects of hypoxia, cyanide and changes in intracellular pH. *Adv. Exp. Med. Biol.* 360: 175–177.

Rumsey, W.L., Iturriaga, R., Spergel, D., Lahiri, S., and Wilson, D.F. 1991 Optical measurements of the dependence of chemoreception on oxygen pressure in the cat carotid body. *Am. J. Physiol.* 261: C614–C622.

Sterni, L.M., Bamford, O.S., Tomares, S.M., Montrose, M.H., and Carroll, J.L. 1995 Developmental changes in intracellular Ca^{2+} response of carotid chemoreceptor cells to hypoxia. *Am. J. Physiol.* 268: L801–L808.

Sterni, L.M., Bamford, O.S., Wasicko, M.J., and Carroll, J.L. 1999 Chronic hypoxia abolished the postnatal increase in carotid body type I cell sensitivity to hypoxia. *Am. J. Physiol.* 277: L645–L652.

Tolosa, J.N., Cooper, R., Myers, A.C., McLemore, G.L., Northington, F., and Gauda, E.B. 2005 Ontogeny of retrograde labeled chemoafferent neurons in the newborn rat nodose-petrosal ganglion complex: an ex vivo preparation. *Neurosci. Lett.* 384: 48–53.

Tomares, S.M., Bamford, O.S., Sterni, L.M., Fitzgerald, R.S., and Carroll, J.L. 1994 Effects of domperidone on neonatal and adult carotid chemoreceptors in the cat. *J. Appl. Physiol.* 77: 1274–1280.

Wasicko, M.J., Breitwieser, G.E., Kim, I., and Carroll, J.L. 2006 Postnatal development of carotid body glomus cell response to hypoxia. *Respir. Physiol. Neurobiol.* 154: 356–371.

Wasicko, M.J., Sterni, L.M., Bamford, O.S., Montrose, M.H., and Carroll, J.L. 1999 Resetting and postnatal maturation of oxygen chemosensitivity in rat carotid chemoreceptor cells. *J. Physiol. (Lond.)* 514(Pt 2): 493–503.

Zapata, P. 2007 Is ATP a suitable co-transmitter in carotid body arterial chemoreceptors? *Respir. Physiol. Neurobiol.* 157: 106–115.

Physiological Carotid Body Denervation During Aging

C. Di Giulio, J. Antosiewicz, M. Walski, G. Petruccelli, V. Verratti, G. Bianchi and M. Pokorski

Abstract Aging is characterized by a lower homeostatic capacity and the carotid body (CB) plays an important role during aging. Here, we sought to elucidate whether the aging effects on the oxygen-sensitive mechanisms in CB cells occur through a reduction of the contact surfaces in the synaptic junctions. The hypothesis was that the CB would undergo a "physiological denervation" in old age. Two groups of male Wistar rats, young (2–3 months old) and senescent (22 months old) were used. CBs were rapidly dissected and the specimens were subjected to a routine transmission electron microscopic procedure. Expressions of HIF-1α, VEGF and NOS-1 were evaluated by immunohistochemical analysis. Our results show that in the old CB, HIF-1α, VEGF and NOS-1 expressions decrease. The cell volume, the number of mitochondria and that of dense-cored vesicles were reduced, and the nucleus shrank. There also was an accumulation of lipofuscin and a proliferation of extracellular matrix. Most importantly, there were fewer synaptic connections between chemoreceptor cells. The total number of synapses observed in all electronograms decreased from 125 in the young to 28 in the old CB. These results suggest the aging CB undergoes a "physiological denervation" leading to a reduction in homeostatic capacity. The age-related reduction of synaptic junctions may be a self-protective mechanism through which cells buffer themselves against reactive oxygen species accumulation during aging.

Keywords Carotid body · Aging · Denervation · HIF-1α · VEGF · NOS-1 · Dense core vesicles · Mitochondria · Synaptic junctions

1 Introduction

To protect against the consequences of oxygen consumption throughout an organism lifespan, aging tissues must protect themselves "against oxidative damage" by decreasing the local oxygen supply. This occurs through a reduction of vascular

C. Di Giulio (✉)
Department of Basic and Applied Medical Sciences, University of Chieti, Chieti, Italy
e-mail: digiulio@unich.it

compliance and an increase in the thickness of the extracellular matrix, increasing the distance the oxygen must diffuse across to reach tissues. The consequent overall reduction in tissue PO$_2$, termed the "aging effect," is a cellular self-protection mechanism induced by long-term oxygen consumption. Oxygen consumption is related to the activity of several enzymes and is correlated with the stability of the "milieu intern" through the speed of various homeostatic survival mechanisms. Thus, changes in oxygen lead to changes in metabolic activity, and the aging effect may act as a secondary response, compensating for alterations in the metabolic machinery.

During aging there is a general decline in the arterial partial pressure of oxygen (PaO$_2$) (Cerveri et al. 1995, Guenard 1998, Gunnarsson et al. 1996). A characteristic reduction in nerve conductivity is also seen; the maximum number of nerve impulses per minute decreases and the sensitivity of peripheral receptors is reduced, leading to a lower homeostatic capacity with higher latency of adaptation responses. In this context, the CB plays an important role during aging. The CB is an important organ for body homeostasis with lobular structure. We previously showed that the number of cells in the CB decreases with age, concomitant with increases in local levels of extracellular matrix and fibronectin, and with reductions in the number of dense-cored vesicles and granules (Dymecka et al. 2006). We found that the type II CB cells were not altered, but the observed age-related changes in type I cells of CB were associated with decreased mitochondrial volume and an increased cytoplasm/mitochondria volume ratio. We noted that the hyperplasic response of CB cells during chronic hypoxia was less evident in aged CB samples than in younger tissues, probably related to age-related decreases in the release of growth factors. Decreased cellular PO$_2$ induces Nitric Oxide Synthase-1 (NOS-1) release and angiogenesis through Vascular Endothelial Growth Factor (VEGF) accumulation and, with Hypoxic Inducible Factor-1α (HIF-1α) playing a physiological role in aged tissues (Di Giulio et al. 2003).

The correlation between arterial and tissue PO$_2$ levels warrants further study in the context of aging, since the noted "PO$_2$ drop" would be expected to be a key regulator of gene expression during aging. Such intercellular communication is important, with each cell influencing the surrounding milieu, and vice versa. Here, we sought to elucidate whether the aging effects on the oxygen-sensitive mechanisms in CB cells occur through a reduction of the contact surfaces in the synaptic junctions. Synaptic junctions were measured and counted from the glomus cells of rat CB taken from young and senescent animals. The hypothesis was that the CB would undergo a "physiological denervation" with increased age, leading to a reduction in homeostatic capacity and a decreased ability to handle stress.

2 Methods

Two groups of male Wistar rats (250 g), young (2–3 months old) and senescent ($\geq$22 months old), were used. The animals were anesthetized with α-chloralose and urethane (15 mg and 75 mg/100 g for young and 10 mg and 50 mg/100 g for senescent, i.p.) and euthanized by perfusion through the heart with a mixture of aldehydes (2% paraformaldehyde/2.5% glutaraldehyde). The CBs were rapidly dissected, post-fixed in the same mixture and then routinely processed for post-embedding electron microscopy. Synapses were defined as plasma membrane thickenings with perimembrane densities, comprising two cells flanking each other across a clearly apparent synaptic gap. Electronograms were taken from 7 young and 4 old carotid bodies. Synapses were counted on electronograms showing from 3 to 13 glomus cells, under primary magnifications ranging from 2.5 to 7.5k for young and 6–15k for old CBs.

Parallel quantification of the results was made by stereological measurements using a Bioquant system interfaced through a digitizing tablet to a microcomputer. Photomicrographs were taken and subjected to visual examination. Expressions of HIF-1α, VEGF and NOS-1 were evaluated by immunohistochemical analysis. Slides were preincubated in PBS for 5 min and then with the appropriate antibodies against rat HIF-1α, VEGF and NOS-1. The experimental data for each group were analyzed by one-way ANOVA. Values of $p < 0.05$ were considered significant.

3 Results

The electronograms revealed that the average numbers of glomus cells in young and old carotid bodies were 6.6 and 5.6, respectively (magnification, 6k). Synapses were confirmed by two independent observers prior to inclusion in the study count, in order to minimize intraobserver variability (Table 1). In old animals, there were significantly fewer synaptic junctions between glomus cells (Fig. 1). In addition, the synapses in older tissues had perimembrane densities of the two adjacent cells that were clearly thinner than those in samples from young animals (Fig. 2).

Both the distribution patterns of the synapses between young and old carotid bodies and the differences in distribution patterns were reproducible, suggesting that these patterns have functional significance. Finally, carotid body expression of HIF-1α, VEGF and NOS-1 was less evident in old versus young tissues (Fig. 3). Our results show that during aging CB shows:

- HIF-1α, VEGF and NOS-1 expressions decrease
- Reduction in cell volume
- Shrinking nucleus

Table 1 Synaptic junctions among carotid body glomus cells of young and old rats

	Young	Old
Age (mo)	2–3	≥ 22
Total number of electronograms	37	36
Number of electronograms with visible synapses	29	7
Total number of synapses on all electronograms	125	28
Average number of synapses per electronogram	3.37	0.77
Range of the number of synapses found on a single glomus cells	1–13	1–7

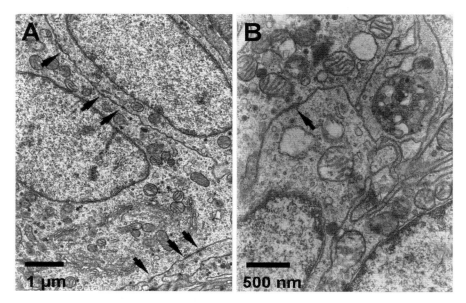

Fig. 1 Electron microscopy of rat carotid body. Young rat carotid body (**A**) and senescent rat carotid body (**B**). Primary magnifications: A – 6k and B – 15k (Part A is reproduced from Pokorski et al. 2004.)

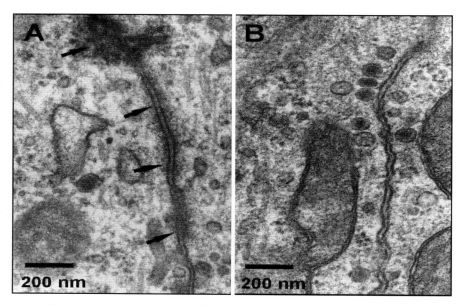

Fig. 2 Electron microscopy of rat carotid body. Young rat carotid body (**A**) and senescent rat carotid body (**B**). Primary magnifications: 40k for both

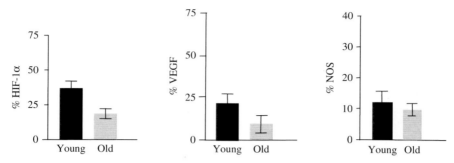

Fig. 3 Carotid body expression of HIF-1α, VEGF and NOS-1 in young and old carotid bodies. Quantitative analyses are expressed as changes in integrated optical intensity (I.O.I. %).*p < 0.05

- Reduction in mitochondria number and volume
- Increase of extracellular matrix
- Lipofuscin accumulation
- Reduction of dense vesicles
- Reduction of synaptic connections

4 Discussion

The increased extracellular matrix deposition, the reduction in number and volume of mitochondria, the reduction of contact surface between glomus cells, and the decrease in number of neurotransmitter-containing vesicles that occur with age suggest that the CB becomes less responsive with age, potentially explaining the observed decreases in chemosensory responses during aging. This suggests that the aging CB undergoes a "physiological denervation." Consistent with this, Dejours described the "Bert effect" wherein CB exposed to acute hyperoxia show a physiological denervation, leading to silencing of the CB (Dejours and Dejours 1992). In the context of more chronic stimuli, there are many examples of decreases in oxygen-sensitive mechanisms, such as hypoxia-induced erythropoietin release in the kidney (Bunn and Poyton 1996, Ratcliffe et al. 1996). Another example is the observation that chronic hyperoxia (100% O_2) attenuated the CB response to hypoxia without altering the CO_2 response (Lahiri et al. 1987). We observed a similar phenomenon during aging (personal observation), prompting us to consider whether hyperoxia might have some link with CB denervation and aging. We hypothesized that the aged CB resembles a situation of CB denervation, and the age-related reduction in the hypoxic response would arise from the same site that is responsible for the reduced responsiveness to hypoxia revealed after chronic hyperoxia (Lahiri et al. 1987).

We herein propose that the point at which mitochondrial activity intersects with neural activity might be the site of action. The mitochondria/cytoplasm ratio determines the oxygen consumption of the cell, with a lower mitochondrial volume and density associated with a decreased cellular capacity for work. A loss of mitochondrial function is one of the prime factors of the aging process. However, it is not yet clear whether the total mitochondria volume for single organs be taken as a lifespan index for the cells. Conde et al. found that hypoxic stimulation elicited a smaller catecholamine release in aged versus young rats, and that chemosensory activity elicited by hypoxia but not hypercapnic acidosis decreased with aging, distinguishing the two pathways (Conde et al., 2006). Decreases in cellular PO_2 induce angiogenesis through VEGF release and accumulation of HIF-1α, the latter of which plays a decreasing role with age (Di Giulio et al. 2003). The correlation between arterial PO_2 and tissue PO_2 warrants additional study in the context of aging, especially since decreases in PO_2 regulate gene expression, and should be tightly correlated with aging. During aging, we found decreases in the expression of HIF-1α, VEGF, and NOS-1 in the CB, suggesting a smaller basal responsiveness of aging carotid body. The age-related increases in fibronectin and reductions of mitochondria are consistent with the idea that senescent tissues have a smaller need for increasing the oxygen consumption. Arteriosclerosis, for example, induces a state of hypoxia in the CB, stimulating deposition of additional extracellular matrix. The observation that both hypoxia and hyperoxia are associated with accumulation of lipofuscin, which is a general reaction to stress (Martinelli et al., 1990, Amicarelli et al. 1999), is consistent with the accumulation of lipofuscin seen during aging. Oxygen sensitivity decreases as tissues age, yet free radicals continue to form, resulting in need for additional homeostatic capacities to buffer the tissue's metabolic needs. A strong increase in ventilation during aging would result in dangerous consequences, as increased oxygen supply to the cells would lead to greater damage through increased reactive oxygen species (ROS) production.

In conclusion, aging leads to CB adaptation through reductions in the sensitivity to stimuli. Changes in the set-point sensitivity for the chemosensory peripheral drive include reduced release of several substances, including HIF-1α, VEGF, and NOS-1. Decreases of these levels in the CB during aging are important for the adaptation processes that allow long-term life. Similarly, the observed age-related reduction of synaptic junctions is a self-protective mechanism through which cells buffer themselves against Reactive Oxygen Species (ROS) accumulation during aging.

Acknowledgments Supported in parts by the statutory budget of the Polish Academy of Sciences. Medical Research Center in Warsaw, Poland, LLP-Erasmus Program for teaching staff mobility – academic year 2007/2008. Convenzione tra l'Università degli Studi "G.d'Annunzio" di Chieti e Pescara in Italy, and Accademia Nazionale dei Lincei in Rome, Italy. The authors had no conflicts of interest to declare in relation to this article.

References

Amicarelli, F., Ragnelli, A.M., Aimola, P., Bonfigli, A., Colafarina, S., Di Ilio, C., Miranda, M. 1999, Age-dependent ultrastructural alterations and biochemical response of rat skeletal muscle after hypoxic or hyperoxic treatments, *Biochim Biophys Acta,* 1453: 105–14.

Bunn, A.F., Poyton, R.O., 1996. Oxygen sensing and molecular adaptation to hypoxia, *Physiol Rev,* 76: 839–85.

Cerveri, I., Zoia, M.C., Fanfulla, F., Spagnolatti, L., Berrayah, L., Grassi, M., Tinelli, C. 1995, Reference values of arterial oxygen tension in the middle-aged and elderly, *Am J Respir Crit Care Med*, 152(3): 934–41.

Conde, S., Obeso, A., Rigual, R., Monteiro, E.C., Gonzales, C. 2006, Function of the rat carotid body chemoreceptors in ageing, *J Neurochemistry*, 99: 711–23.

Dejours, P., Dejours, S. 1992, The effects of barometric pressure according to Paul Bert: The question today, *Int J Sports Med*, 13: 1–5.

Di Giulio, C., Bianchi, G., Cacchio, M., Macrì, M.A., Ferrero, G., Rapino, C., Verratti, V., Piccirilli, M., Artese, L. 2003, Carotid body HIF-1alpha, VEGF and NOS expression during aging and hypoxia, *Adv Exp Med Biol*, 536: 603–10.

Dymecka, A., Walski, M., Pokorski, M. 2006, Ultrastructural degradation of the carotid body in the aged rat: Is there a role for atherosclerosis in the main carotid arteries?, *J Physiol Pharmacol*, 57 Supp 4: 85–90.

Guenard, H. 1998, Respiration and aging, *Rev Mal Respir*, 15: 713–21. Review.

Gunnarsson, L., Tokics, L., Brismar, B., Hedenstierna, G. 1996, Influence of age on circulation and arterial blood gases in man, *Acta Anaesthesiol Scand*, 40(2): 237–43.

Lahiri, S., Mulligan, E., Andronikou, S., Shirahata, M., Mokashi, A. 1987, Carotid body chemosensory function in prolonged normobaric hyperoxia in the cat, *J Appl Physiol*, 62(5): 1924–31.

Martinelli, M., Winterhalder, R., Cerretelli, P., Howald, H., Hoppeler, H. 1990, Muscle lipofuscin content and satellite cell volume is increased after high altitude exposure in humans, *Experientia*, 46: 672–76.

Pokorski, M., Walski, M., Dymecka, A., Marczak, M. 2004, The aging carotid body, *J Physiol Pharmacol*, 55: 107–13.

Ratcliffe, P.F., Eckardt, K.-U., Bauer, C., 1996. Hypoxia erythropoietin gene expression, and erythropoiesis. In: *Handbook of Physiology*, ed. M.J. Fregly and C.M. Blatteis, C.M., Section 4: Environmental Physiology, vol. II. American Physiological Society, Oxford University Press, New York, pp. 1125–53.

Does Ageing Modify Ventilatory Responses to Dopamine in Anaesthetised Rats Breathing Spontaneously?

T.C. Monteiro, A. Obeso, C. Gonzalez and E.C. Monteiro

Abstract Dopamine has been widely used in humans in the management of cardiocirculatory shock, and its inhibitory effect on ventilation has received particular attention in clinical situations more prevalent in the elderly. Dopamine has been extensively studied at the carotid body in adult animals but little is known in aged animals. We investigated the ventilatory responses caused by dopamine in 3 and 24 months old rats.

Cumulative intracarotid bolus injections of dopamine were performed in anaesthetised and vagotomised rats, in the absence and in the presence of i.v. infusions of domperidone (23.5–1175 nmol Kg^{-1} min^{-1}). Airflow (V), tidal volume (V_T), respiratory rate (f), arterial blood pressure and heart rate were monitored and respiratory minute volume (V_E) calculated. Basal values of V_E were lower in 24 months rats (322.9±18.8 mL Kg^{-1} min^{-1}) than in 3 months old rats (442.5±24.2 mL Kg^{-1} min^{-1}), mainly due to reductions in V_T. The dose-dependent decreases caused by dopamine (3–100 nmol) in V_T, f and V_E, were totally prevented by section of the carotid sinus nerve and were not modified by ageing. The maximal % antagonism of the inhibitory effect of dopamine on V_E caused by domperidone was similar in both 3 (74.6±2.7) and 24 (70.7±0.8) months old rats. Domperidone alone, increased basal V_E by 59.6±16.6 mL min^{-1} Kg^{-1}, and by 11.8±1.2 mL min^{-1} Kg^{-1}, respectively in 3 and 24 months old rats ($p<0.01$).

The inhibitory basal tonus caused by dopamine in ventilation was reduced in aged rats, although the decrease in V_E caused by its exogenous administration remained unchanged.

Keywords Ageing · Dopamine breathing · Domperidone · Respiratory frequency · Tidal volume · Respiratory minute volume

E.C. Monteiro (✉)
Departamento de Farmacologia, Faculdade de Ciências Médicas, Universidade Nova de Lisboa, Campo Mártires da Pátria, 130, 1169-056 Lisboa, Portugal
e-mail: ecmonteiro.farm@fcm.unl.pt

1 Introduction

Ventilatory responses to hypoxia are attenuated with ageing (Fukuda, 1992; Guenard, 1998; Peterson et al., 1981; Chapman and Cherniack, 1987; Janssens et al., 1999) and the measurement in humans of the respiratory drive suggests that there is an age-related decline in the ability to integrate information received from peripheral and central chemoreceptors (Peterson et al., 1981). It is also known that ageing attenuates the increase in carotid sinus nerve afferent activity in response to hypoxia (Conde et al., 2006).

Carotid body chemosensitivity has been extensively studied in animals assessing its dopaminergic function (for a review see Gonzalez et al., 1994). It is known that in aged rats carotid body exhibits a high catecholamine content and turnover time, no changes on the release of dopamine under normoxic conditions but a diminished responsiveness to hypoxic stimuli to evoke the release of this biogenic amine (Conde et al., 2006).

Exogenous dopamine has been widely used in humans to improve renal and cardiovascular function in the management of cardiocirculatory shock, but its inhibitory effect on ventilation could be deleterious especially when dopamine is used in hypoxic patients, in patients with tenuous cardiorespiratory status, or in patients being weaned off ventilatory support (Van de Borne et al., 1998). The prevalence of these limitations is higher in the elderly but the effects of ageing on the ventilatory responses to exogenous administration of dopamine have never been investigated.

In the present work we compared the ventilatory responses induced by exogenous dopamine and by its non-specific D_2 antagonist (Laduron and Leysen, 1979) domperidone, in 3 and 24 months old rats breathing spontaneously.

2 Methods

2.1 Animals and Surgical Procedures

Experiments were performed on male Wistar rats aged 3 months (360–480 g, body weight), and 24 months (500–900 g, body weight). All along their lifespan, the animals were kept in the vivarium of our university, in an air-conditioned room at $21\pm1°C$, $55\pm10\%$ humity, with a 12:12 h light-dark cycle, and with food and water available ad libitum and experiments were carried out in accordance with the Portuguese regulations for the protection of the animals.

The animals were anaesthetised by a single intraperitoneal injection of sodium pentobarbitone (60 mg Kg^{-1}), supplemented intravenously with 10% of the starting dose as necessary. They were placed supine and breathed spontaneously room air. The trachea was exposed in the neck, sectioned below the larynnx and a pneumotachographic sensor was connected to the distal end of the tracheostomy tubing. Four cannulations were performed under a dissection microscope: the right femoral artery for systemic arterial blood pressure measurement; the right femoral vein for pharmacological infusions (domperidone at a rate of 0.5 ml min^{-1} during 3 min); the left femoral vein for anaesthesic supplements; and the right external carotid for pharmacological bolus (dopamine i.c. in a volume of 0.1 ml, washed in with 0.2 ml 0.9% aqueous sodium chloride). Body temperature was maintained close to 37°C using a heated underblanket governed by a rectal thermistor probe.

Respiratory flow was measured by a HSE-pneumotachometer PTM type, a differential pressure transducer (model DP 45-14 Validyne Engineering, North-Northridge, CA) and a pressure amplifier (Plugsys Housings, model 603, HSE-HA GmgH). Blood pressure was measureed with a pressure transducer (model Isotec, HSE-HA GmgH) and a pressure amplifier (Plugsys Housings, model 603, HSE-HA GmgH).

We used HSE-Harvard Pulmodyn® W software for data acquisition (system for respiratory studies): signals such as pulmonary airflow, tidal volume (V_T), arterial pressure (BP) and O_2 partial pressure (pO_2) ; and parameters that can be derived from these signals e.g., respiratory rate (f), V_T, heart rate (HR) and mean value of blood pressure, were recorded continuously during experiments. Respiratory minute volume (V_E) was calculated as the product of V_T and f.

The intervals between dopamine injections were at least 5 min. Two cumulative dose-response curves for dopamine (i.c.) were performed in each rat: one with the simultaneous infusion of 0.9% aqueous sodium chloride, and the other after 3 min of i.v. infusion of domperidone. Only one dose of domperidone was tested per animal. Control values for V_T, f, BP and HR correspond to the mean value measured in a period of 25 s immediately before drug administration. After drug i.c. administration the values of V_T, f, BP and HR, were taken as the maximal effects measured during the period of 25 s that followed the injections, and were compared with dose measured during the control. The maximal effects induced by i.c. injections of dopamine always occurred in the first 25 s that followed the end of the injections.

2.2 Drugs

Doses of all drugs were calculated on the basis of salt weight. Dopamine was prepared in 0.9% w/v aqueous sodium chloride solution, and domperidone was prepared in 0.1 N HCl. The drugs used were: sodium pentobarbitone (Eutasil-Sanofi Veterinária), sodium heparin and saline (B. Braun), Dopamine (Medopa-Medinfar), and domperidone (Sigma-Aldrich).

2.3 Statistics

Data are expressed as mean ±S.E.M. and compared for statistical significance using a two tails Student t-test for unpaired data. Significance level was established at $P < 0.05$. Models for analysis were developed using GraphPad Prism software (Version 4.03).

3 Results

Basal values of V_E were lower in 24 months old rats than in 3 months old rats, mainly due to reductions in V_T (Table 1). The dose-dependent decreases caused by dopamine (3–100 nmol) in V_T, f and V_E, were totally prevented by section of the carotid sinus nerve (Fig. 1) and were not modified by ageing (Fig. 2).

Table 1 Comparison of basal ventilatory parameters in 3 and 24 months old rats

	3 months old	24 months old	p value
V_T (ml·Kg^{-1})	8.78 ± 0.34	7.02 ± 0.39	0.0029
f (breaths·min^{-1})	49.75 ± 1.20	46.57 ± 1.54	0.0213
V_E (ml·min·Kg^{-1})	442.50 ± 24.23	322.90 ± 18.83	0.0008

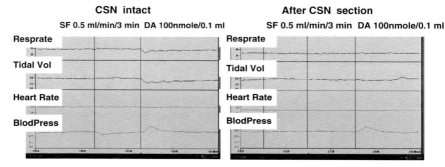

Fig. 1 Effect of i.c. injections of dopamine (100 nmol) on respiratory rate, tidal volume (V_T), heart rate (HR) and mean blood pressure (BP) in a 3 months old rat, anaesthetised and vagotomised, before and after bilateral section of the carotid sinus nerves (CSN)

The maximal % antagonism of the inhibitory effect of dopamine on V_E caused by domperidone was similar in both 3 (74.6±2.7), and 24 (70.7±0.8) months old rats. Domperidone alone (1175 nmol Kg^{-1} min^{-1}), increased basal V_E by 59.6±16.6 mL

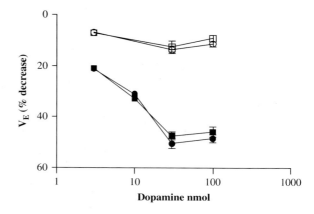

Fig. 2 Comparison between the effects of i.c. injections of dopamine on respiratory minute volume in 3 (■□) and 24 (●○) months old rats in the absence (filled symbols) and during (open symbols) i.v. infusion of domperidone (1175 nmol Kg^{-1} min^{-1})

min^{-1} Kg^{-1}, and by 11.8±1.2 mL min^{-1} Kg^{-1}, respectively in 3 and 24 months old rats (p<0.01).

4 Discussion

In this model of vagotomised anaesthetised Wistar rats, ageing decreased ventilation mainly due to a decrease in V_T. The inhibitory basal dopaminergic tonus in ventilation was attenuated in aged animals, but ageing did not modify the effect of exogenous dopamine.

The reduction in resting breathing with ageing in Wistar rats is consensual although some differences between awake and unrestrained rats and vagotomised anaesthetised animals are apparent. In the former case only changes in frequency are evident (Soulage et al., 2004) in contrast with more pronounced reductions in V_T observed in the present work. These results could be linked to the fact that age-related changes in mechanical properties of the lung and thorax become more visible after removing the control of the vagus nerve upon the respiratory frequency. We have used vagotomised animals in the present work because we were particularly interested in the ventilatory effects of dopamine mediated by the carotid body chemoreceptors.

The inhibitory effect of exogenous dopamine on ventilation is mediated by D_2 receptors present at the carotid body chemoreceptors (Mir et al., 1984) and is known since the seventies in both animals (Cardenas and Zapata, 1981) and humans (Welsh et al., 1978). The effects of domperidone by itself on ventilation were attributed to the antagonism of dopamine receptors at the carotid body activated by the endogenous amine, because domperidone does not cross the blood brain barrier and in the doses used in the present work almost totally block the effects of exogenously applied dopamine. These results agree with those previously found by others (Zapata and Torrealba, 1984). The efficacy of domperidone blocking the effects of dopamine on ventilation also indicate that in the range of doses herein tested, the ventilatory effects of dopamine were due exclusively to activation of D_2 receptors and that the involvement of adrenoceptors can be excluded.

The decrease in the excitatory effects of domperidone observed in 24 months old rats could be caused by both a smaller release of dopamine in normoxic conditions and/or a reduction in the number of active dopamine receptors. The absence of age-related changes in the release of dopamine in normoxic conditions in the carotid body in vitro (Conde et al., 2006) is more consistent with an age-related change in dopaminergic D_2 receptors at the carotid body. However, further experiments are needed to exclude any interpretation because the O_2 conditions of this in vivo animal model, can not be exactly extrapolated to those in vitro (Conde et al., 2006) where ageing reduces catecholamine release induced by hypoxia (2 and 5%O_2) in the carotid body. The finding that venous blood dopamine concentrations are higher in healthy sexagenarian than in young alpinists at rest (Serebrovskaya et al., 2000), corroborates the hypothesis that down-regulation of dopamine receptors at the carotid body could occur in the elderly. Age-related impairment of dopamine receptors have been extensively described in the striatum (for a review sees e.g. Roth and Joseph, 1994).

Although the inhibitory effects of dopamine on ventilation were exclusively mediated by the peripheral chemoreceptors (were abolished by bilateral section of the CSN and dopamine does not cross the blood brain barrier), the absence of differences between the effects of exogenous dopamine in young and aged rats can not necessarily mean that carotid body function is well preserved in ageing. The results of morphological studies show degenerative changes developing with age in the ultrastructure of carotid bodies (Di Giulio et al., 2003; Conde et al., 2006) and carotid sinus nerve activity in response to hypoxia is clearly attenuated in old animals (Conde et al., 2006). The reduction in the excitatory effects of domperidone by itself, found in the present work, also suggests a decrease in carotid body dopaminergic function probably attributed to dopamine receptors desensitization, an effect that can be surpassed by the high doses of dopamine administered intravenously. Since dopamine has mainly inhibitory actions in the peripheral control of breathing, a decrease in endogenous dopaminergic activity at the carotid body with ageing could be considered as a compensatory mechanism to preserve O_2 supply. In addition, ageing by itself did not seem to increase the risk of ventilatory depression caused by the clinical use of dopamine.

Acknowledgments Supported by CEDOC/FCT (Portugal), by BFU2007-61848, CIBER CB06/06/0050 (FISS-ICiii) and by JCYL GR 242 (Spain) and CRUP AI-E99/04.

References

Cardenas, H. & Zapata, P. 1981, Dopamine-induced ventilatory depression in the rat, mediated by carotid nerve afferents, *Neurosci Lett*, 24: 29–33.

Conde, S.V., Obeso, A., Rigual, R., Monteiro, E.C. & Gonzalez, C. 2006, Function of the rat carotid body in ageing, *J Neurochem*, 99: 711–723.

Chapman, K.R. & Cherniack, N.S. 1987, Aging effects on the interaction of hypercapnia and hypoxia as ventilatory stimuli, *J Gerontol*, Mar; 42: 202–9.

Di Giulio, C., Cacchio, M., Bianchi, G., Rapino, C. & Di Ilio, C. 2003, Selected contribution: Carotid body as a model for aging studies: is there a link between oxygen and aging?, *J Appl Physiol*, 95: 1755–8.

Fukuda, Y. 1992, Changes in ventilatory response to hypoxia in the rat during growth and aging, *Pflugers Arch*, 421: 200–3.

Gonzalez, C., Almaraz, L., Obeso, A., & Rigual, R. 1994, Carotid body chemoreceptors: from natural stimuli to sensory discharges, *Physiol Rev*, 74: 829–98.

Guenard, H. 1998, Respiration and aging, *Rev Mal Respir*, 15: 713–21.

Janssens, J.P., Pac, J.C., & Nicod, L.P., 1999, Physiological changes in respiratory function associated with ageing, *Eur Respir J*, 13: 197–205.

Laduron, P.M. & Leysen, J.E. 1979, Domperidone, a specific in vitro dopamine antagonist, devoid of in vivo central dopaminergic activity, *Biochem Pharmacol*, 28: 2161–5.

Mir, A.K., McQueen, D.S., Pallot, D.J. & Nahorski, S.R. 1984, Direct biochemical and neuropharmacological identification of dopamine D_2-receptors in the rabbit carotid body, *Brain Res*, 291: 273–83.

Peterson, D.D., Pack, A.I., Silage, D.A. & Fishman, A.P. 1981, Effects of aging on ventilatory and occlusion pressure responses to hypoxia and hypercapnia, *Am Rev Respir Dis*, 124: 387–91.

Roth, G.S., & Joseph, J.A., 1994, Cellular and molecular mechanisms of impaired dopaminergic function during aging, *Ann N Y Acad Sci*, 719: 129–35.

Serebrovskaya, T.V., Karaban, I.N., Kolesnikova, E.E., Mishunina, T.M., Swanson, R.J., Beloshitsky, P.V., Ilyin, V.N., Krasuk, A.N., Safronova, O.S., & Kuzminskaya, L.A. 2000, Geriatric men at altitude: hypoxic ventilatory sensitivity and blood dopamine changes, *Respiration*, 67: 253–60.

Soulage, C., Pequignot, J.M. & Perrin, D. 2004, Breathing pattern and hypoxic sensitivity during ageing in a new model of obesity-resistant rat, *Respir Physiol Neurobiol*, 144: 45–57.

van de Borne, P., Oren, R., & Somers, V.K. 1998, Dopamine depresses minute ventilation in patients with heart failure, *Circulation*, 98:126–31.

Welsh, M.J., Heistad, D.D. & Abboud, F.M. 1978, Depression of ventilation by dopamine in man. Evidence for an effect on the chemoreceptor reflex, *J Clin Invest*, 61: 708–13.

Zapata, P. & Torrealba, F. 1984, Blockade of dopamine-induced chemosensory inhibition by domperidone, *Neurosci Lett*, 51: 359–64.

The Role of the Carotid Bodies in the Counter-Regulatory Response to Hypoglycemia

Denham S. Ward, William A. Voter and Suzanne Karan

Abstract Although controversial, animal and tissue studies indicate that carotid bodies are sensitive to changes in glucose as well as in oxygen, thereby functioning as metabolic sensors. This study was designed to test the hypothesis that carotid bodies in humans participate in the counter-regulatory response to insulin-induced hypoglycemia.

Dopamine and hyperoxia were used to suppress the carotid bodies' responsiveness in 16 normal subjects. Insulin and glucose infusions were used to clamp the plasma glucose in a step-wise decrease to 2.5 mmol/l over 4 hours while counter-regulatory hormones were measured.

The hypoglycemic trajectories were similar under all three interventions (dopamine, hyperoxia and control), but the total glucose infused was significantly larger for hyperoxia than for dopamine. Cortisol and epinephrine both showed the expected increase with hypoglycemia, but there was no difference among interventions. Glucagon and norepinephrine levels were increased by dopamine, but only the normalized increase in glucagon was lower with dopamine and hyperoxia than control.

The decrease in total glucose required for the dopamine experiments was most likely due to the higher baseline glucagon and norepinephrine levels. Hyperoxia did require more infused glucose, indicating some increased insulin sensitivity, but it was not clearly due to a decrease in cortisol or epinephrine responses. Thus, we did not find direct evidence of the carotid bodies' role in glucose homeostasis in humans.

Keywords Hypoglycemia · Hyperoxia · Dopamine · Carotid bodies · Humans · Counter regulatory response

---

D.S. Ward (✉)
Department of Anesthesiology, University of Rochester School of Medicine and Dentistry, Rochester, NY 14642, USA
e-mail: Denham_Ward@URMC.Rochester.edu

1 Introduction

The physiological response to hypoglycemia is important for normal regulation of blood glucose and also for the protective counter-regulatory response to insulin excess. Besides direct glucose sensing in the alpha and beta cells of the pancreas, controlling secretion of glucagon and insulin respectively, glucose-sensing cells that could particularly affect the autonomic response have been found in several regions of the brain as well as in the portal vein and liver (Levin et al. 2004).

The carotid bodies play a fundamental role in the physiological control of ventilation. In humans, they are the sole stimulatory sensors for hypoxemia. Accumulating evidence indicates the carotid bodies are also glucose sensors (Alvarez-Buylla and Alvarez-Buylla 1988; Lopez-Barneo 2003; Pardal and Lopez-Barneo 2002; Zhang et al. 2007) and could participate in the counter-regulatory response to hypoglycemia (Koyama et al. 2000). This is an attractive hypothesis because of the obvious need to coordinate supply and delivery of the two most important metabolic substrates to the cells. However, the intracellular sensing mechanisms of oxygen and glucose may be different (Garcia-Fernandez et al. 2007) and other studies both in vitro (Conde et al. 2007) and in vivo (Bin-Jaliah et al. 2004, 2005; Kumar 2007; Ward et al. 2007) have called into question an actual direct glucose sensing role for the carotid bodies.

This protocol aimed to determine whether hyperoxia and low-dose dopamine interventions in humans, having been shown to blunt the carotid bodies' response to other stimuli (hypercapnia and hypoxia) (Sabol and Ward 1987), would also cause a decrease in the counter-regulatory response to hypoglycemia (Mitrakou et al. 1991).

2 Methods

The study was completed at the University of Rochester General Clinical Research Center (GCRC) and approved by the local institutional review board. The studies conformed to the standards set by the Declaration of Helsinki and written, informed consent was obtained from all subjects. Subjects were free of significant cardiopulmonary disease and on no prescription medications except for birth control pills.

Each subject was scheduled for three experimental days. The three arms of the protocol were designed to test the counter-regulatory response to hypoglycemia (via glucose clamp, described below) during carotid body inhibition versus control. Carotid body inhibition was accomplished in two settings: either with a continuous dopamine infusion (in normoxia), or in hyperoxia (no dopamine).

At 7:00 a.m. the morning of the experiment, two intravenous lines were placed, one for insulin and glucose infusion and the other for blood sampling. Starting at 7:40 a.m., either dopamine ($3.0\,\mu g \cdot kg^{-1} \cdot min^{-1}$) or placebo infusion was started and the subjects wore a light-weight, non-rebreathing face mask supplied with either air (for the dopamine or control experiments) or 50% O_2 (hyperoxia experiment) at 7 l/min.

Glucose levels were lowered in a step-wise progression starting at 8:00 a.m. via insulin infusion (1.5 mU · kg^{-1} · min^{-1}). A 20% glucose infusion at 1.5 ml · kg^{-1} · min^{-1} was started approximately 15 min later. The target levels for blood glucose levels were 4.16, 3.61, 3.05, and 2.50 mmol/dl to be attained at 20 min before the hour (e.g., 8:40, 9:40 a.m., etc.) by titrating the glucose infusion. The insulin infusion was increased only if the target hypoglycemia was not reached even after the glucose infusion had been titrated to zero. Glucose levels were measured bedside every five minutes (Beckman Glucose Analyzer 2; Beckman Coulter Inc., Brea, CA).

Blood samples for hormone measurements were taken prior to the start of the clamp and every 60 min thereafter (just after each breathing run) with one additional sample taken 30 min before the last breathing run.

Measurements of insulin, glucagon, cortisol, adrenaline and norepinephrine were made on each set of six plasma samples. Glucagon and insulin were measured using the appropriate double antibody radio-immunoassay kit (Linco Research Inc., St. Charles, MO). Cortisol was measured with a coated-tube radioimmunoassay kit (Diagnostic Systems Laboratories Inc., Webster, TX). Epinephrine and norepinephrine were determined by Kat-Combi radioimmunoassay kit, performed by KMI Diagnostics (KMI Diagnostics, Minneapolis, MN). All blood samples were immediately centrifuged and the plasma frozen until analysis. STATA software (Stata Corp., College Station, TX) was used for statistical analysis. Analysis of variance (ANOVA) with repeated measures was used on all measurements with time, intervention (control, hyperoxia and dopamine) and intervention by time interaction as independent factors. When overall significance was found, differences among the three interventions at each time point were isolated using the false discovery rate procedure with P < 0.05 as the significance level (Curran-Everett 2000). Data is reported as mean ± s.d. except for the figures where mean ± s.e.m. is given for clarity.

3 Results

Seventeen subjects enrolled and sixteen (6 female, 10 male) completed the protocol. Average age was 25 ± 2.9 years; average body mass index was 24.3 ± 4.4 kg/m^2.

The hypoglycemic trajectories were similar under all three interventions (Fig. 1), although dopamine did cause a small increase in glucose at t = 0. Although there was considerable variation in the insulin levels, there was no significant difference. Six subjects each in the dopamine infusion and the hyperoxic tests required an increase in insulin infusion to achieve the lowest target glucose, and seven subjects under control interventions required a higher insulin infusion (NS). The total glucose infused for the three interventions was 931 ± 128 mg, 782 ± 123 mg and 1027 ± 126 mg for control, dopamine, and hyperoxia respectively (P = 0.016, with hyperoxia significantly different from the dopamine experiments).

Cortisol and epinephrine both showed the expected increases with the hypoglycemia (Fig. 2), but there was no significant difference among the interventions.

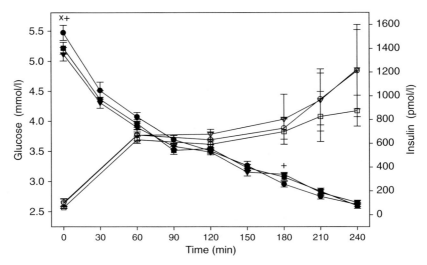

Fig. 1 Glucose (*filled symbols*) **and insulin** (*open*) **levels (mean ± s.e.m.).** There was no differences either overall or at the individual time points for the insulin levels. For glucose ANOVA showed an intervention effect at p=0.04 and an intervention by time effect at p=0.06. For the individual time points, significance indicated by: * hypoxia vs. control; + dopamine vs. hyperoxia; x dopamine vs. control. ○ dopamine, □ control, ∇ hyperoxia

For glucagon, the increase with hypoglycemia was significant and there was an overall intervention effect with the levels during the dopamine infusion being generally higher throughout the hypoglycemia. When the glucagon levels were normalized by the initial glucagon level measured at $t = 0$, an overall intervention effect remained ($p = 0.007$) and while both the hyperoxia and dopamine interventions tended to be less than the control values, this was only significant for dopamine vs. control at $t = 210$ min (Fig. 3). For the norepinephrine response (Fig. 2), there was a significant increase with hypoglycemia and an overall intervention effect with dopamine greater than hyperoxia and control. Normalizing the response showed an overall intervention effect that was still significant but no individual differences at specific time points were significant (Fig. 3).

4 Discussion

Since dopamine and oxygen have different metabolic effects and act at different sites within the carotid bodies, we hypothesized that a unique role for the carotid bodies in the counter-regulatory response to hypoglycemia could be discerned. We found that hyperoxia, but not dopamine, required more exogenous glucose to be infused during the clamp. While this would indicate a degree of increased insulin sensitivity, it was not due to blunted glucagon and cortisol responses as Wasserman's laboratory (Koyama et al. 2000) found in dogs without carotid bodies. For dopamine, there was

Carotid Bodies and Hypoglycemia

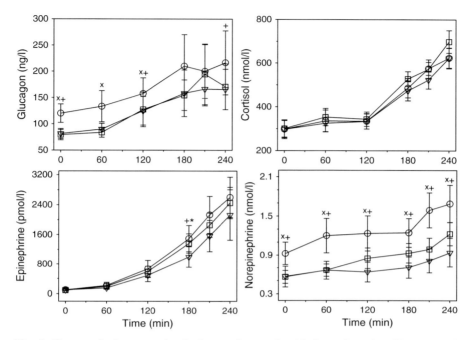

Fig. 2 Changes in hormone levels (mean ± s.e.m.) with hypoglycemia. Glucagon and norepinephrine showed a significant intervention effect (p=0.002 and p<0.001, respectively). Epinephrine showed an intervention effect at the p=0.057 level and the intervention effect on cortisol was NS. No significant intervention by time interaction was found for any of the hormones. For the individual time points, significance indicated by: ∗ hypoxia vs. control; + dopamine vs. hyperoxia; × dopamine vs. control. ○ dopamine, □ control, ▽ hyperoxia

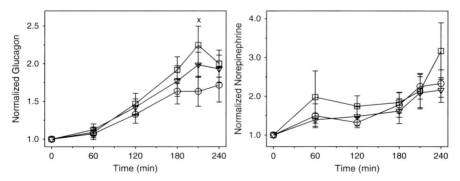

Fig. 3 Glucagon and norepinephrine measurements (mean ± s.e.m.) are plotted after nor-malizing by the initial (t = 0) values. Both showed a significant intervention effect (p=0.007 and p=0.05, for glucagon and norepinephrine respectively) but no intervention by time interaction. For the individual time points significance indicated by: ∗ hypoxia vs. control; +dopamine vs. hyperoxia; × dopamine vs. control. ○ dopamine, □ control, ▽ hyperoxia

a decrease in insulin sensitivity due primarily to an increase in baseline glucagon and norepinephrine. However, consistent with the results of Koyama et al. (Koyama et al. 2000) there did seem to be a smaller increase in the normalized glucagon.

There are several possibilities that could explain the lack of an unequivocal finding in this study: insufficient suppression of the carotid bodies; an effect of glucose on the carotid bodies via a different pathway than those affected by dopamine and hyperoxia; the redundancy of glucose sensitivity in vivo; or the other metabolic effects of our interventions.

In addition the results for in vitro and animal studies are not unequivocal in proving a role for the carotid bodies in the in vivo regulation of glucose. Alvarez-Buylla and colleagues were probably the first to suggest that the carotid chemoreceptors might participate in glucose homeostasis (Alvarez-Buylla and Alvarez-Buylla 1988), although Obeso et al. had found previously that, in vitro, glucose was required for the carotid bodies to respond to hypoxia (Obeso et al. 1986). More recently, additional evidence for a glucose sensitivity in the carotid bodies has come from the laboratories of Nurse and Lopez-Barneo (Lopez-Barneo 2003; Pardal and Lopez-Barneo 2002; Zhang et al. 2007) using tissue and cell culture techniques.

However, recently Bin-Jaliah et al. could not find a change in the carotid sinus nerve firing rate with hypoglycemia in the rat even though the ventilatory response to hypoxia was augmented (Bin-Jaliah et al. 2004, 2005). Ward et al. also found no augmentation of the hypoxic response that could be clearly ascribed to a hypoglycemic sensitivity in the carotid bodies in humans, but instead found an augmentation of the hypoxic ventilatory response with hyperglycemia (Ward et al. 2007). Interestingly, Conde et al. (Conde et al. 2007), using freshly isolated rat carotid bodies, could not confirm any glucose sensitivity in either normoxia or hypoxia.

The measure of carotid body suppression by the levels of dopamine and hyperoxia used in this study were not tested directly, but previous studies would indicate over a 50% reduction in the carbon dioxide sensitivity (Sabol and Ward 1987). While substantial, it is obviously less than from resection of the carotid bodies, and it is not known how closely the reduction in carbon dioxide sensitivity relates to a possible reduction in glucose sensitivity. Further, recent in vitro work (Garcia-Fernandez et al. 2007), using rat carotid body cells, supported the hypothesis that the carotid bodies respond to glucose within the physiologic range. These investigators found a response apparently due to hypoglycemic activation of background Na^+ channels; it was at least additive to the oxygen response. Since the carotid body response to oxygen becomes very flat in hyperoxia, this study would predict that our use of 50% oxygen would cause a marked reduction in the response to hypoglycemia.

Both dopamine and hyperoxia have direct effects on metabolism. Since norepinephrine is a precursor in the catecholamine synthesis pathway, it is not surprising that its levels were increased by dopamine infusion. Pernet et al. (Pernet et al. 1984) studied a two-hour infusion of $3.0\,\mu g \cdot kg^{-1} \cdot min^{-1}$ dopamine and found a small increase in norepinephrine and a doubling of basal glucagon levels, resulting in a small transient elevation in plasma glucose. There have been no studies of the effects of dopamine infusion during a hypoglycemic clamp.

Despite considerable literature on the effects of hypoxia (high altitude) on glucose homeostasis and metabolism, there are few studies on hyperoxia. In an anesthetized newborn piglet model, Bandali et al. (2003) found that hyperoxia rapidly increased arterial glucose (within 15 min). Since there were also increases in glucagon and insulin, but without an increase in epinephrine, they speculated that hyperoxia increased glucagon and the hyperinsulinemia was a consequence of the resulting hyperglycemia. The mechanisms by which hyperoxia might cause this increased glucagon in this animal model are unknown.

Thus the main results of this study do not directly support an important role for the carotid bodies in the in vivo human counter-regulatory response to hypoglycemia. While we did find differences in the response under interventions that suppress the responsiveness of the carotid bodies, these differences could result indirectly from the treatments used rather than from a specific direct suppression of the carotid bodies. The use of pharmacological agents (including oxygen) in humans to modify physiological responses presents many obstacles in interpretation of the results. Agents rarely have a single clean site of action and their actions will present interpretation difficulties in such complex, redundant feedback systems as is the counter-regulatory response to insulin-induced hypoglycemia. An understanding of the full role for the carotid bodies in the in vivo regulation of glucose will require further study.

Acknowledgments Assistance from John Gerich, M.D., Linda Palmer, R.N., Suzanne Donahue, R.N. and the nursing (Ann Miller, R.N. and Mark Cloninger, R.N.) and laboratory (David Robson and Noya Rackovsky, M.S.) staffs of the GCRC is gratefully acknowledged. We wish to thank the American Diabetes Association Clinical Research Grant (1-04-CR-37), University of Rochester Clinical Research Center (5 M01 RR00044 from the National Center for Research Resources, NIH), and the Department of Anesthesiology for financial support.

References

Alvarez-Buylla, R. & Alvarez-Buylla, E.R. 1988, Carotid sinus receptors participate in glucose homeostasis, *Respiration Physiology*, 72: 347–59.

Bandali, K.S., Belanger, M.P. & Wittnich, C. 2003, Does hyperoxia affect glucose regulation and transport in the newborn?, *Journal of Thoracic and Cardiovascular Surgery*, 126: 1730–35.

Bin-Jaliah, I., Maskell, P.D. & Kumar, P. 2004, Indirect sensing of insulin-induced hypoglycaemia by the carotid body in the rat, *Journal of Physiology*, 556: 255–66.

Bin-Jaliah, I., Maskell, P.D. & Kumar, P. 2005, Carbon dioxide sensitivity during hypoglycaemia-induced, elevated metabolism in the anaesthetized rat, *The Journal of Physiology Online*, 563: 883–93.

Conde, S.V., Obeso, A. & Gonzalez, C. 2007, Low glucose effects on rat carotid body chemoreceptor cells' secretory responses and action potential frequency in the carotid sinus nerve, *Journal of Physiology*, 585: 721–30.

Curran-Everett, D. 2000, Multiple comparisons: philosophies and illustrations, *AJP –Regulatory, Integrative and Comparative Physiology*, 279: R1–R8.

Garcia-Fernandez, M., Ortega-Saenz, P., Castellano, A. & Lopez-Barneo, J. 2007, Mechanisms of low-glucose sensitivity in carotid body glomus cells, *Diabetes*, 56: 2893–900.

Koyama, Y., Coker, R.H., Stone, E.E., Lacy, D.B., Jabbour, K., Williams, P.E. & Wasserman, D.H. 2000, Evidence that carotid bodies play an important role in glucoregulation *in vivo*, *Diabetes*, 49: 1434–42.

Kumar, P. 2007, How sweet it is: sensing low glucose in the carotid body, *Journal of Physiology*, 578: 627.
Levin, B.E., Routh, V.H., Kang, L., Sanders, N.M. & Dunn-Meynell, A.A. 2004, Neuronal glucosensing: what do we know after 50 years?, *Diabetes*, 53: 2521–28.
Lopez-Barneo, J. 2003, Oxygen and glucose sensing by carotid body glomus cells, *Current Opinion in Neurobiology*, 13: 493–99.
Mitrakou, A., Ryan, C., Veneman, T., Mokan, M., Jenssen, T., Kiss, I., Durrant, J., Cryer, P. & Gerich, J. 1991, Hierarchy of glycemic thresholds for counterregulatory hormone secretion, symptoms, and cerebral dysfunction, *American Journal of Physiology*, 260: E67–E74.
Obeso, A., Almaraz, L. & Gonzalez, C. 1986, Effects of 2-deoxy-D-glucose on *in vitro* cat carotid body, *Brain Research*, 371: 25–36.
Pardal, R. & Lopez-Barneo, J. 2002, Low glucose-sensing cells in the carotid body, *Nature Neuroscience*, 5: 197–98.
Pernet, A., Hammond, V.A., Blesa-Malpica, G., Burrin, J., Orskov, H., Alberti, K.G. & Johnston, D.G. 1984, The metabolic effects of dopamine in man, *European Journal of Clinical Pharmacology*, 26: 23–28.
Sabol, S.J. & Ward, D.S. 1987, Effect of dopamine on hypoxic-hypercapnic interaction in humans, *Anesthesia & Analgesia*, 66: 619–24.
Ward, D.S., Voter, W.A. & Karan, S. 2007, The effects of hypo- and hyperglycaemia on the hypoxic ventilatory response in humans, *Journal of Physiology*, 582: 859–69.
Zhang, M., Buttigieg, J. & Nurse, C.A. 2007, Neurotransmitter mechanisms mediating low-glucose signaling in cocultures and fresh tissue slices of rat carotid body, *Journal of Physiology*, 578: 735–50.

The Respiratory Responses to the Combined Activation of the Muscle Metaboreflex and the Ventilatory Chemoreflex

C.K. Lykidis, P. Kumar and G.M. Balanos

Abstract The excessive hyperventilation seen during exercise in chronic heart failure (CHF) contributes to the limited exercise capacity in this condition. The hyperactivation of reflexes originating, independently, from muscle (ergoreflex) and from chemoreceptors (chemoreflex) has been suggested to play an important part in the mediation of the CHF ventilatory abnormalities. In this study we aimed to assess the ventilatory responses to the combined activation of the muscle ergoreflex and the ventilatory chemoreflex, achieved by post-exercise circulatory occlusion (PECO) and euoxic hypercapnia (end-tidal PCO_2 = 7 mmHg above normal), respectively.

Three healthy women and three healthy men (29.33 ± 1.28 yrs; mean ± SD) undertook four trials, in random order, separated from each other by 30 min of rest: 2 min of isometric handgrip exercise followed by 2 min of PECO with hypercapnia, 2 min of isometric handgrip exercise followed by 2 min of PECO while breathing room air, 4 min of rest with hypercapnia and 4 min of rest while breathing room air.

Ventilation (V_E) was significantly elevated by the ventilatory chemoreflex and it was further elevated by 5.13±0.83 L/min ($P<0.05$) when the muscle ergoreflex was superimposed upon it. The response to the combination of these stimuli was significantly greater than the sum of the responses to the two stimuli when given independently ($P<0.05$).

The results indicate that the interaction between the two reflexes has an additional stimulatory effect on ventilation and consequently could be involved in the limited exercise capacity in CHF.

Keywords Respiratory responses · Muscle metaboreflex · Ventilatory chemoreflex · Ergoreflex human · Postexercise ventilatory occlusion

---

C.K. Lykidis (✉)
School of Sport and Exercise Sciences, University of Birmingham, Edgbaston, Birmingham, B15 2TT, UK
e-mail: CXL584@bham.ac.uk

1 Introduction

The exaggerated ventilatory increase seen in response to exercise in chronic heart failure (CHF) is regarded as a hallmark of this condition and is adversely related to the quality of life and survival of patients (Tumminelo et al., 2007). Whereas the underlying mechanisms have yet to be clarified, recent work has emphasised the potential role of two impaired reflex cardiorespiratory control systems in CHF, namely the muscle ergoreflex and the ventilatory chemoreflex (Coats, 2001). Therefore the term muscle ergoreflex has referred to the activation of metabolically sensitive afferents in the exercising muscle and the reflexly-induced increases in ventilation and in sympathetic drive. The ventilatory chemoreflex, in turn, has denoted the activation of chemoreceptors by carbon dioxide and the ensuing enhancements in ventilation and sympathetic activity (Schmidt et al., 2005). It has been shown that both reflexes are hyperactive in CHF and also that the degree of their overactivation was a strong and independent prognostic marker of ventilatory abnormalities during exercise and disease state (Chua et al., 1996; Ponikowski et al., 2001). Overactivation of the muscle ergoreflex has been attributed to a number of physiological abnormalities of CHF that include, but they are not limited to, the altered muscle structure and biochemistry (Piepoli et al., 1999) as well as the incapacity to enhance cardiac performance and stroke volume (Crisafulli et al., 2008). In contrast, the reasons underlying the overactivity of the ventilatory chemoreflex remain largely obscure.

Ponikowski et al., (2001), who evidenced that muscle ergoreflex activation was an independent predictor of ventilatory sensitivity to hypercapnia in CHF patients, argued for the existence of an interaction between the two reflexes. Furthermore it was proposed that alternation of signals arising from ergoreceptors might feed in the respiratory control centres and augment the input to central chemoreceptors thereby inducing changes in chemosensitivity (Chua et al., 1996; Coats, 2001; Piepoli et al., 1999; Schmidt et al., 2005). Nevertheless, to our knowledge, this proposition has not been experimentally tested. Therefore the aim of this study was to assess the respiratory responses to the combined activation of the muscle ergoreflex and ventilatory chemoreflex, achieved by post-exercise circulatory occlusion (PECO) and hypercapnia, respectively.

2 Methods

Six healthy and physically active subjects (3 women, 3 men; 29.33 ± 1.28 yrs; mean ± SD) took part in this study, after they had given informed written consent. The investigation was performed according to the Declaration of Helsinki and was approved by the Local Ethics Committee.

Prior to the experimental day subjects underwent a preliminary trial during which they became familiarised to the procedures of the study. On the experimental day subjects resumed the seated position and the maximum voluntary contraction

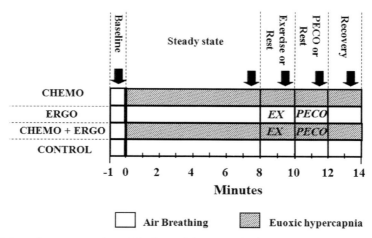

Fig. 1 Schematic representation of the four trials. *Arrows* **indicate where measurements were taken.** EX = exercise

(MVC) force of the forearm muscles at the dominant arm was measured by a handgrip dynamometer. Subjects then rested for 15 min while breathing through a mouthpiece during which their normal end-tidal partial pressure of oxygen ($P_{ET}O_2$) and carbon dioxide ($P_{ET}CO_2$) was determined. Euoxic hypercapnia ($P_{ET}O_2=100$ mmHg, $P_{ET}CO_2=7$ mmHg above normal) was accomplished by a dynamic end-tidal forcing system (Robbins et al., 1982).

Subjects then undertook four trials in random order. The trials were separated by 30 min of rest. A schematic representation of the four trials is shown in Fig. 1. Baseline measurements were obtained before each trial, and all trials were followed by 2 min of recovery. The four trials were as follows:

1. CHEMO, during which the ventilatory chemoreflex was activated via exposing subjects to euoxic hypercapnia whilst they rested for 12 min.
2. ERGO, during which subjects breathed room air whilst they performed 2 min of isometric handgrip exercise at 40% of the MVC, followed by a further 2 min of PECO achieved by inflation of a cuff around the upper exercising arm at 205 mmHg, in order to activate the muscle ergoreflex.
3. CHEMO+ERGO, during which subjects were exposed to euoxic hypercapnia whilst they performed 2 min of isometric handgrip exercise followed by a further 2 min of PECO as in ERGO, so as to investigate the combined activation of the ventilatory chemoreflex and the muscle ergoreflex.
4. CONTROL, during which subjects breathed room air whilst they rested for 12 min.

To ensure that steady state had been achieved in the ventilatory response to euoxic hypercapnia in the CHEMO and CHEMO+ERGO trials, a period of 8 min was allowed before any further intervention was performed. This 8-minute period was applied to all trials for consistency.

Ventilation (V_E) was continuously recorded throughout the duration of the protocol. Minute averages were calculated for baseline, steady state (SS), second minute into exercise, second minute of PECO and finally the last minute of recovery.

Data are expressed as means ± S.E.M. and significance (P<0.05) was tested by means of Friedman's ANOVA followed by Wilcoxon signed rank tests (SPSS 16.0).

3 Results

Changes in V_E taken place during the 4 trials from baseline are illustrated in Fig. 2. Mean V_E at baseline was not different between trials (P>0.05). Thereafter it significantly increased by hypercapnia in the CHEMO and CHEMO+ERGO trials compared to baseline (baseline vs SS; all P<0.001) whereas it remained unchanged in the ERGO and CONTROL trials (baseline vs SS; all P>0.05). Mean ventilation was further increased during exercise compared to SS in both the CHEMO+ERGO and in the ERGO trials, but this did not reach significance (all P>0.05). Subsequently during PECO respiration continued to increase in the CHEMO+ERGO (5.07±0.83 L/min) and it was significantly greater compared to SS (P<0.05). V_E during PECO in the ERGO trial decreased as compared to exercise and it was not significantly different from SS (P>0.05). V_E during recovery in the CHEMO+ERGO and ERGO trials declined towards SS values whereas in the CHEMO and CONTROL it remained at SS levels (SS vs Recovery; all P>0.05).

The ventilatory effect of the interaction between the muscle ergoreflex and ventilatory chemoreflex activation is depicted in Fig. 3. Ventilation increased synergistically by 15.36±1.64 L/min from baseline with the combined activation of CHEMO and ERGO compared with the additive effect of ERGO alone plus CHEMO alone (11.53±1.83 L/min) (P<0.05).

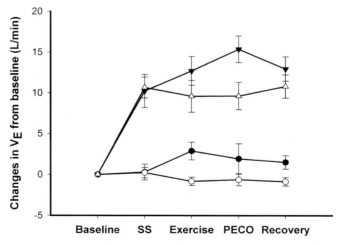

Fig. 2 Mean changes in ventilation from baseline taking place during trials. (Changes taking place during the CHEMO + ERGO, CHEMO, ERGO and NO CHEMO-NO ERGO trials, are shown in lines connected by *filled triangles, open triangles, filled circles* and *open circles*, correspondingly)

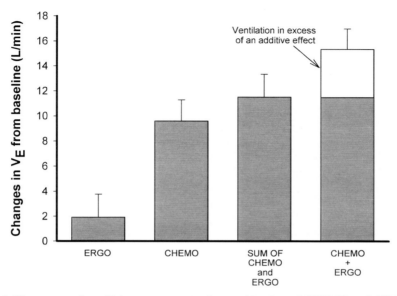

Fig. 3 **The measured ventilatory response to the combination of CHEMO and ERGO is greater than the sum of the ventilatory increases seen during the CHEMO and ERGO trials** suggesting a more complicated interaction between the two mechanisms than simply being additive

4 Discussion

It has been widely reported that the hyperactivation of the ventilatory chemoreflex in CHF is involved in the excessive ventilatory increments seen during exercise in this condition, yet the underlying pathophysiology is not well understood. The present finding that activation of the muscle ergoreflex increases chemo-responsiveness in healthy humans, suggests a potential link between the over-activity of the ventilatory chemoreflex and that of muscle ergoreflex in CHF. Our finding is in line with the work from Jordan et al (2000) who found that the respiratory response to hypercapnia was augmented upon superimposition of the sympathoexcitatory stimulus of head-up tilt (HUT) which, in parallel, caused attenuations in the hypercapnia-induced increases of cerebral blood flow (CBF). Markedly, when the effect of HUT and hypercapnia was studied with complete ganglionic blockade, diminished elevations in ventilation and enhancements in CBF were observed (Jordan et al., 2000).

These authors suggested that sympathoexcitation-induced ventilatory enhancements were possibly mediated via cerebral vasoconstriction and/or increased peripheral chemosensitivity. Interestingly, reduction of brain blood flow (to 70% of normal) in unanesthetized goats provoked increases in the respiratory, HR and BP responses to hypercapnia (Chapman et al., 1979). Whether a diminution in brain blood flow accounted for the PECO-induced elevations in ventilation, HR and BP that we observed during hypercapnia is unknown. When studied under euoxic / eucapnic conditions, PECO failed to affect cerebral vasoreactivity as reflected by the Doppler-derived estimation of middle cerebral artery velocity (Pott et al., 1997). On the other hand, PECO-induced sympathoexcitation and vasoconstriction has been evidenced at extracranial arterial beds, including inactive muscle (Seals, 1989), kidney (Momen et al., 2003) and lung (Lykidis et al., 2008). This raises the possibility for increased carotid chemoresponsiveness to contribute to the increases in ventilation we observed during PECO in hypercapnia, given that sympathetic activation has been shown to increase the carotid body gain to carbon dioxide (Joels and White, 1968; Maskell et al., 2006). Our observations could be also accounted for by a central interaction between the muscle ergoreflex and the ventilatory chemoreflex. As such, neural signals from ergoreceptors that feed into respiratory control centres could cause augmentation of medullary chemoreceptor input and increase central chemoreceptor sensitivity to CO_2 (Clark and Cleland, 2000; Schmidt et al., 2005).

In conclusion, we observed that the superimposition of the muscle ergoreflex activation on that of the ventilatory chemoreflex stimulated ventilation, with the increases being greater than if the effect of the combined activation of the reflexes was simply additive. The ventilatory effects of this multiplicative interaction between the two reflexes could be involved in the exercise dyspnoea seen in CHF.

Acknowledgments We are grateful to Lauro Vianna for his help with the study design and statistical analysis.

References

Chapman, R.W., Santsiago, T.V. and Edelman, N.H. 1979, Effects of graded reduction of brain blood flow on chemical control of breathing. *J Appl Physiol*, 47: 1287–1294

Chua, T.P., Clark, A.I., Amadi, A.A. and Coats, A.J.S. 1996, Relation between chemosensitivity and the ventilatory response to exercise in chronic heart failure. *J Am Coll Cardiol*, 27: 650–657.

Clark, A.L. and Cleland, J.G.F. 2000, The control of adrenergic function in heart failure: therapeutic intervention. *Heart Fail Rev.* 5: 101–114.

Coats, A.J.S. 2001, What causes the symptoms of heart failure? *Heart*, 86: 574–578.

Crisafulli, A., Salis, E., Tocco, F., Melis, F., Milia, R., Pittau, G., Caria, M.A., Solinas, R., Meloni, L., Pagliaro, P. and Concu, A. 2008, Impaired central hemodynamic response and exaggerated vasoconstriction during muscle metaboreflex activation in heart failure patients. *Am J Physiol Heart Circ Physiol*, 292, H2988–H2996.

Joels, N. and White, H. 1968, The contribution of the arterial chemoreceptors to the stimulation of respiration by adrenaline and noradrenaline in the cat. *J Physiol*, 197: 1–23.

Jordan, J., Shannon, J.R., Diedrich, A., Black, B., Costa, F., Robertson, D., Biaggioni, I. 2000, Interaction of carbon dioxide and sympathetic nervous system activity in the regulation of cerebral perfusion in humans. *Hypertension*, 36:383–388.

Lykidis, C.K., White, M.J. and Balanos, G.M. 2008, The pulmonary vascular response to the sustained activation of the muscle metaboreflex in man. *Exp Physiol*, 93: 247–253.

Maskell, P.D., Rusius, C.J., Whitehead, K.J. and Kumar, P. (2006). Adrenaline increases carotid body CO2 sensitivity. *Adv Exp Med Biol*, 580, 245–250.

Momen, A., Leuenberger, U.A., Ray, C.A., Cha, S., Handly, B., and Sinoway, L.I. 2003, Renal vascular responses to static handgrip: role of muscle mechanoreflex. *Am J Physiol Heart Circ Physiol*, 285: H1247–H1253

Piepoli, M., Ponikowski, P., Clark, A.L., Banasiak, W., Capucci, A. and Coats, A.J.S. 1999, A neural link to explain the 'muscle hypothesis' of exercise intolerance in chronic heart failure. *Am Heart J*, 137: 1050–1056.

Ponikowski, P. P., Chua, T.P., Francis, D.P., Capucci, A., Coats, A.J.S. and Piepoli, M.F. 2001, Muscle ergoreceptor overactivity reflects deterioration in clinical status and cardiorespiratory reflex control in chronic heart failure. *Circulation*, 104: 2324–2330.

Pott, F., Ray, C.A., Olesen, H.L., Ide, K and Secher, N.H. 1997, Middle cerebral artery blood velocity, arterial diameter and muscle sympathetic nerve activity during post-exercise muscle iscaemia. *Acta Physiol Scand*, 160: 43–47.

Robbins, P.A., Swanson, G.D., Micco, A.J. and Schubert, W.P. 1982, A fast gas-mixing system for breath-to-breath respiratory control studies. *J Appl Physiol*, 52: 1358–1362.

Seals, D.R. 1989, Sympathetic neural discharge and vascular resistance during exercise in humans. *J Appl Physiol,* 66: 2472–2478.

Schmidt, H., Francis, D.P., Rauchhaus, M., Werdan, K., Piepoli, M.F. 2005, Chemo- and ergoreflexes in health, disease and ageing. *Int J Cardiol*, 98: 369–378.

Tumminelo, G., Guazzi, M., Lancellotti, P. and Piérard, L.A. 2007, Exercise ventilation insufficiency in heart failure: pathophysiological and clinical significance, *Eur Heart J*, 28: 673–678.

Cardiovascular Responses to Hyperoxic Withdrawal of Arterial Chemosensory Drive

Patricio Zapata, Carolina Larrain, Marco-Antonio Rivera and Christian Calderon

Abstract Searching for an arterial chemosensory drive exerted upon the cardiovascular system under eucapnic normoxia, we performed experiments on spontaneously ventilated, pentobarbitone-anesthetized cats, in which ventilatory flow through a pneumo-tachograph, instantaneous respiratory frequency, end-tidal pressure of CO_2, arterial pressure, and instantaneous heart frequency were simultaneously recorded. Repeated exposures to 100% O_2 breathing for 5 to 60 s caused the well-known transient decreases in tidal ventilatory volume and instantaneous respiratory frequency, after which minor decreases in systolic, diastolic and mean arterial pressures, as well as in instantaneous heart frequency were observed. After selective bilateral denervation of carotid sinuses (barodenervation), hyperoxia-induced falls in arterial pressure and heart rate became more evident. Subsequent bilateral section of the carotid nerves (with or without section of the aortic nerves) suppressed these effects. Present results indicate the presence of a chemosensory drive of the cardiovascular system under eucapnic normoxia, although considerably smaller than that exerted upon ventilation. The small magnitude of the decreases in arterial pressure and heart rate observed under control conditions suggests that cardiovascular effects elicited by hyperoxic challenges are normally buffered by carotid baroreflexes.

Keywords Carotid bodies · Carotid sinuses · Circulatory chemosensory drive · Blood pressure · Heart rate

1 Introduction

The peripheral arterial chemoreceptors (carotid and aortic bodies) maintain a tonic rate of afferent discharges when mammals are in a steady-state of normoxic eucapnia (see Eyzaguirre and Zapata, 1984). The transient ventilatory depression upon exposure to breathing pure oxygen (Dejours, 1957) has been shown to result from

P. Zapata (✉)
Laboratorio de Neurobiología, Pontificia Universidad Católica de Chile, Santiago, Chile
e-mail: pzapata@udd.cl

the sudden withdrawal of such chemosensory discharges by hyperoxia (Leitner et al., 1965). Thus, a chemosensory drive of ventilation is commonly accepted (Dejours, 1962).

Since stimulation of carotid chemoreceptors elicits cardio-vascular responses (see Fitzgerald and Shirahata, 1997), one is prompted to ask if brief exposure to hyperoxic breathing -through interruption of the chemosensory discharges observed under normoxic-eucapnic conditions- also results in transient changes in systemic circulation, i.e., to find out if the cardiovascular system is also subjected to a similar chemosensory drive. Since circulatory responses to any given stimuli are truncated and shortened by the rapid intervention of resulting baroreflexes, we decided to study if chemoreflex responses to hyperoxia were modified by selective denervation of both carotid sinuses.

2 Methods

Experiments were performed on 10 adult male cats, weighing 3.3 ± 0.1 kg (mean $\pm$ S.D.), anesthetized with sodium pentobarbitone 40 mg/kg body weight given ip, followed by additional 12 mg doses given iv when required to maintain a light level of surgical anesthesia (stage III, plane 2). At the end of experiments, animals were euthanized by an overdose of pentobarbitone. Procedures here reported had been accepted by the Committee on Bioethics and Biosafety, Faculty of Biological Sciences, P. Catholic University of Chile.

Animals were placed in supine position over a heating blanket, automatically controlled by a thermoregulator, fed from a thermotrode introduced 5 cm into the rectum. Initial rectal temperature was $37.7 \pm 0.5°C$, which was maintained at $37.6°C \pm 0.1°C$ along the experiments.

Cats breathed spontaneously throughout experiments. A tracheal cannula was introduced *per os* and connected to a heated Fleisch pneumotachograph head (N° 00), separately calibrated for air and pure oxygen flows. A Grass volumetric differential pressure transducer allowed recording of ventilatory flow ($\delta V/\delta t$). Full-wave integration of flow signals gave a continuous record of ventilatory tidal volume (V_T). Instantaneous respiratory frequency (f_R) was obtained by tachographic analysis of ventilatory signals. A fine (PE-10) tubing was inserted deep into the tracheal cannula for continuous sampling of a fraction of tracheal air and monitoring of end-tidal CO_2 pressure ($P_{ET}CO_2$) through an infrared gas analyzer (model 703, Infrared Industries, CA).

The right saphenous vein was cannulated for administration of drugs and a tubing (PE-100), filled with heparin 50 IU/ml in saline solution, was inserted into the left femoral artery for arterial pressure (P_a) recording, through a Statham pressure transducer. Mean P_a was obtained by electronic filtering of the upper frequencies of differential arterial pressure. Instantaneous heart frequency (f_H) was obtained by tachographic analysis of pulsatile arterial pressure signal, or derived as reciprocal of R-R intervals from EKG recording obtained from electrodes placed on right shoulder and left groin.

The physiological variables were displayed on a polygraph and a multiple beam oscilloscope. Raw signals were stored on magnetic tapes or video-cassettes for later analyses.

In a series of five cats, 100% O_2 tests of 15 s duration were assayed repeatedly under the following conditions: (i) under control conditions; (ii) after selective bilateral carotid sinus denervation (BCSD), by sectioning and crushing the nerve filaments at the vascular wall between the carotid sinus and carotid body; (iii) after bilateral carotid neurotomy (BCN), by local anesthetic block (lidocaine 2%) followed by section of the carotid nerve, at its emergence from the glossopharyngeal nerve trunk; and (iv) after additional bilateral aortic neurotomy (+BAN), by bilateral section of the aortic nerve at its exit from the angle formed by the superior laryngeal nerve and main trunk of cervical vagus, immediately below the nodose ganglion.

The selectivity of the procedure performed (according to Zapata et al., 1969) for carotid barodenervation had been confirmed in other animals by: (i) carotid sinus nerve recordings showing the disappearance of barosensory discharges, with preservation of chemosensory discharges; (ii) the disappearance of the hypertensive (baroreflex) response to brief bilateral occlusion of common carotids, with preservation of the hyperpneic (chemoreflex) response to brief exposure to 100% N_2. Thus both types of observations had revealed that the chemosensory innervation of the carotid body was unaffected by this surgical procedure performed on its immediate neighborhood.

In another group of five animals, 100% O_2 tests were performed by periods from 5 to 60 s, to search for the time course of hyperoxic responses. In two of these animals, responses to hyperoxia were tested after bilateral section of the aortic nerves, to isolate those responses originating only from carotid receptors. In another animal, hyperoxic responses were tested after section of one carotid nerve and one aortic nerve at the same side.

3 Results

In all cats under control conditions, 100% O_2 breathing for periods from 5 to 60 s resulted in statistically significant reductions in V_T, f_R, P_a and f_H. Their magnitudes were variable but quite reproducible in the same animal. Furthermore, if additional doses of pentobarbitone were required to maintain the adequate level of anesthesia, the depressant effects of this agent on respiratory and cardiovascular parameters dissipated within 3 min and the original strengths of responses to hyperoxia were recovered.

Ventilatory responses to hyperoxic challenges were initiated within two to four ventilatory cycles, preceding by approximately 15 s the beginning of circulatory responses. All responses to 100% O_2 tests were transient, e.g., reductions in V_T lasted no more than 15 s when elicited by 5 s hyperoxic tests. The minimal reductions in mean P_a and f_H elicited by 100% O_2 breathing (Fig. 1) lasted less than 30 s

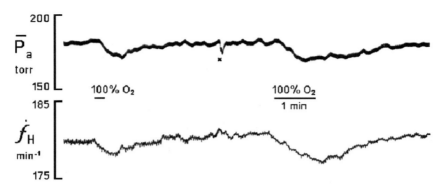

Fig. 1 Effects of breathing 100% O_2 for 15 and 60 s (*horizontal bars*) on mean arterial pressure and instantaneous heart frequency in a cat under control conditions. * Spontaneous gasp

for responses to 5–10 s tests, and they disappeared within 2 min when elicited by hyperoxic tests of 1 min duration.

In the series of five cats in which hyperoxic tests were repeated under different neurological conditions, the surgical procedures performed (BCSD, BCN and BAN) did not result in statistically significant changes in the steady-states of cardio-respiratory parameters studied (Table 1). Carotid barodenervation produced transient hypertension, P_a reaching up to 130% of basal values, and returning to control values within 5–15 min. Carotid neurotomy also provoked transient hypertension, disappearing within 2–5 min, and a transient reduction of V_T, down to 79% of control values, but resolved within 2–3 min.

Table 1 Effects of surgical procedures upon steady-state basal conditions

Parameter	Control	BCSD	BCN	+ BAN	p =
V_T (ml·kg^{-1})	7.7 ± 0.9	8.3 ± 1.2	7.7 ± 0.7	7.3 ± 1.7	0.294
f_R (min^{-1})	25.0 ± 9.4	26.9 ± 7.9	27.6 ± 9.6	27.9 ± 10.0	0.232
$P_{ET}CO_2$ (Torr)	32.3 ± 12.0	31.2 ± 10.8	31.0 ± 10.1	27.9 ± 10.0	0.075
Mean P_a (Torr)	138 ± 20	155 ± 44	154 ± 34	163 ± 35	0.133
f_H (min^{-1})	224 ± 21	224 ± 27	231 ± 20	219 ± 28	0.484

p ascertained by Quade's tests.

3.1 Respiratory Chemoreflex Responses

In intact animals, hyperoxic tests reduced both V_T and f_R (Fig. 2), resulting in increases in $P_{ET}CO_2$. These responses were preserved and not significantly different after BCSD. However, further BCN suppressed the decreases in V_T and f_R and the consequent increase in $P_{ET}CO_2$ evoked by hyperoxic tests.

After complete peripheral chemodenervation by BCN+BAN, ventilatory responses to hyperoxia were absent, including in two animals in which very minor

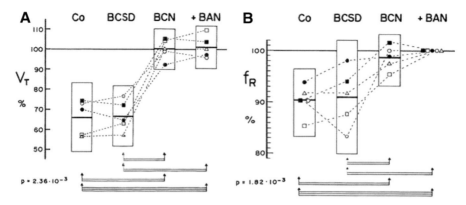

Fig. 2 Effects of breathing 100% O_2 for 15 s periods upon: Ventilatory tidal volume (V_T) and instantaneous respiratory frequency (f_R) under control (Co) conditions, and after bilateral carotid sinus denervation (BCSD), bilateral carotid neurotomy (BCN) and additional bilateral aortic neurotomy (+BAN). Minimal values reached along tests, expressed in percentages of basal values previous to each test. Each symbol represents the mean of two tests performed in each of 5 cats. Central wide horizontal line, arithmetic mean; surrounding rectangle, 95% normal distribution in each condition (arithmetic mean ± 1.96 SD). Comparisons by Quade's tests. Inserts, global **p** values Paired comparisons between *arrows: triple lines*, p < 0.001; *double lines*, p < 0.01; *single lines*, p < 0.05. Paired comparisons

hypopnea or bradypnea were still observed in basal conditions after carotid denervation. On the contrary, in the other series of cats, ventilatory responses to hyperoxic tests observed after BAN were of similar magnitude to those observed under control conditions (not illustrated).

It must be noted that global statistical analyses revealed no significant differences between carotid denervated and carotid plus aortic denervated preparations with regard to hyperoxia-induced respiratory responses (Fig. 2). Ventilatory responses to hyperoxic tests were still seen in one cat which had been subjected to unilateral carotid and aortic neurotomies (not illustrated).

The effectiveness of BCN+BAN to achieve complete peripheral chemodenervation was confirmed by the disappearance of hyperventilatory responses to brief exposure to 100% N_2, tested in all animals.

3.2 Cardiovascular Chemoreflex Responses

In intact animals, breathing 100% O_2 for 15 s periods resulted in statistically significant although small decreases in P_a in control conditions (Figs. 1 and 3). Statistical analysis showed that hyperoxia-induced hypotensive responses were more pronounced after BCSD (p < 0.05) (Fig. 3). Subsequent BCN significantly reduced these responses and no significant changes in P_a were observed after additional BAN (Fig. 3).

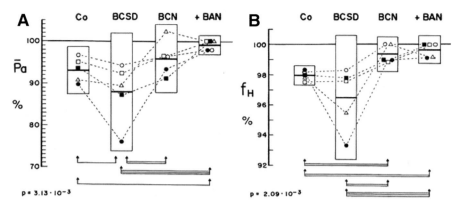

Fig. 3 Effects of breathing 100% O₂ for 15 s periods upon: mean arterial pressure (Pa) and instantaneous heart frequency (f_H) under control (Co) conditions, and after bilateral carotid sinus denervation (BCSD), bilateral carotid neurotomy (BCN) and additional bilateral aortic neurotomy (+BAN). Symbols and statistics as in legend to Fig. 2

Hyperoxia induced small decreases in f_H under control conditions (Fig. 1), which became more pronounced in some cats after BCSD (Fig. 3), but no statistically significant differences were observed between Co and BCSD, as well as between BCN alone and BCN+BAN. However, significant differences were observed between each of the two first conditions vs. each of the two following conditions (Fig. 3).

A study of the bivariation of hyperoxia-induced falls in f_H and P_a revealed that the area covered after BCSD was larger than that recorded under control conditions (not illustrated). Furthermore, linear correlation between the magnitude of falls in f_H and mean P_a provoked by hyperoxic tests was statistically significant when performed after BCSD (r = 0.78; p < 0.025), but no such correlation was observed under control conditions (r = −0.08; p < 0.9). Carotid bodies denervation by BCN reduced the area of bivariation of cardiovascular responses, with disappearance of linear correlation between hyperoxia-induced falls in f_H and P_a (r = 0.54; p < 0.2).

In contrast to the elimination by BAN of the residual hyperoxia-induced hypotensive responses observed in some preparations subjected to BCN, cardiovascular responses to hyperoxic tests were well preserved in the two animals subjected only to bilateral aortic neurotomy, as well as in one cat subjected to unilateral carotid and aortic neurotomies (not illustrated).

4 Discussion

Selective carotid barodenervation provoked immediate hypertension, indicating that normal barosensory activity had been exerting a tonic depressor influence upon Pa. However, this was a transient effect, suggesting that other barostatic mechanisms come into operation to maintain vascular tone within a homeostatic range. This effect has been previously described by other authors (see Abboud and Thames, 1983) upon block or sectioning of carotid (sinus) nerves; an effect that we have repeatedly observed in our preparations.

Selective carotid chemodenervation (as provoked by BCN performed after BCSD) produced immediate but brief hypoventilation, which indicates the interruption of the tonic excitatory influence upon resting ventilation exerted by carotid chemoreceptors (chemosensory drive) in normoxic conditions. We had previously observed a similar transient ventilatory depression upon blocking or sectioning the second of the two carotid nerves (Eugenin et al., 1989), but that was a simultaneous deafferentation of the carotid body and sinus. This is -to our knowledge- the first report on the isolated effect of sudden interruption of carotid chemoafferent fibers. The transient nature of this ventilatory depression upon the withdrawal of carotid chemoreflex activity suggests that the remaining chemoreceptors (central and aortic chemoreceptors) assume a larger control upon resting ventilation, under poikilocapnic hyperoxia. This resetting in ventilatory control has been observed previously (see Dejours, 1962).

Hypertension upon interrupting carotid nerves has been always ascribed to the suppression of barosensory fibers. However, we are here reporting that -after selective barodenervation- carotid neurotomy also induces a discrete and transient increase in Pa, an indication that carotid chemosensory activity is also contributing to the maintenance of a basal vascular tone. One may argue that this minor reflex influence of carotid chemoreceptors upon P_a would appear only after selective carotid sinus denervation has eliminated the major reflex influence of carotid baroreceptors.

Present experiments show hyperoxia-induced depression of pulmonary and alveolar ventilation. Previous studies (Dejours, 1957; Eugenin et al., 1989) indicate that this phenomenon reveals a steady-state of ventilatory drive exerted under eucapnic normoxia, and that such drive results from a tonic reflex from carotid bifurcation receptors, since ventilatory depression as response to hyperoxia disappears after BCN. However, present study has the advantage of discriminating between the influences of carotid sinus baroreceptors and carotid body chemoreceptors, since reflex ventilatory depression induced by hyperoxia was preserved after selective BCSD, but was eliminated by the following BCN, which in this condition only suppress the tonic influence of carotid body chemoreceptors.

Present experiments also show that brief periods of hyperoxia produce transient, very moderate hypotension and bradycardia, an observation -to our knowledge- not previously reported in experimental studies performed on normal mammals. Only the figures illustrating a paper by von Euler and Liljestrand (1943) -reporting on P_a changes recorded in cats chloralose-anesthetized and artificially ventilated with air- show that shifting ventilation to 100% O_2 reduced P_a in some preparations. This lack of information is probably due to the small magnitude of the decreases in arterial pressure and heart rate observed when the carotid nerves are intact. Furthermore, we observed that such hyperoxia-induced vascular effects were enhanced after selective BCSD, suggesting a buffering of chemoreflex-mediated vascular effects when baroreflexes were operating normally.

With regard to vascular effects induced by hyperoxia under pathological conditions, a mild reduction of systolic and diastolic arterial pressures has been reported in young patients with essential hypertension when exposed to 100% O_2 for 20 ventilatory cycles, an effect not observed in age-matched normotensive subjects (Tafil-Klawe et al., 1985; Izdebska et al., 1996). Otherwise, Ciarka et al (2005) reported that breathing 100% O_2 reduces mean P_a and muscle sympathetic nerve activity in hypertensive patients and heart transplant recipients, but not in control human subjects.

Regarding changes in heart frequency elicited by hyperoxia, these have been searched for repeatedly in human subjects. A first report by Parkinson (1912) indicated that heart rate in healthy students was reduced from 68/min to 64/min during 30 min of pure O_2 breathing. Dripps and Comroe (1947) reported a 3.7% decrease of pulse rate during the first 2 min of 100% O_2 inhalation in normal men under resting conditions. Daly and Bondurant (1962) reported a reduction in heart rate from 71/min to 65/min after 7–10 min of breathing 100% O_2 in normal men. Seals et al (1991) reported that hyperoxia reduces heart rate by 6 beats/min, reduces muscle sympathetic activity, but does not modify Pa. However, Drysdale and Petersen (1977) had concluded from their observations that the "primary effect of withdrawal of chemosensory drive in conscious man is tachycardia". One hour oxygen administration in healthy men has been recently reported to slightly reduce heart rate, but increase mean P_a (Waring et al., 2003; Thomson et al., 2006). Another recent report indicates that breathing 100% O_2 for 5 min, although increasing $P_{ET}O_2$ from 102 to 635 Torr and S_aO_2 from 97 to 100%, neither changes mean P_a nor heart rate in normal male subjects (Yamazaki et al., 2007). Thus, contradictory reports can be found in the literature, probably due to differences in procedures and timing of effects.

Summarizing, present experiments show that brief hyperoxic tests reveal a chemosensory drive exerted not only upon ventilation, but also upon arterial pressure and heart rate. The effects here reported are derived from tonic chemoreflex influences initiated from the carotid bodies, since they persist after selective elimination of carotid sinuses innervation, but disappear following interruption of the remaining chemoafferent fibers traveling through the carotid sinus nerves.

Acknowledgments Work financed by FONDECYT (Chilean National Fund for the Development of Science and Technology) and DIUC (Research Division, P. Catholic University of Chile).

References

Abboud, F.M., Thames, M.D. 1983, Interaction of cardiovascular reflexes in circulatory control. In: American Physiological Society: Handbook of Physiology, Sect 2, Vol 3, ed. J.T. Shepherd, F.M. Abboud, Bethesda, MD, pp 675–753.

Ciarka, A., Najem, B., Cuylits, N., Leeman, M., Xhaet, O., Narkiewicz, K., Antoine, M., Degaute, J.P., van de Borne, P. 2005, Effects of peripheral chemoreceptors deactivation on sympathetic activity in heart transplant recipients, Hypertension, 45: 894–900.

Daly, W.J., Bondurant, S. 1962, Effects of oxygen breathing on the heart rate, blood pressure, and cardiac index of normal men – resting, with reactive hyperemia, and after atropine, J Clin Invest, 41: 126–132.

Dejours, P. 1957, Intérêt méthodologique de l'étude d'un organisme vivant à la phase initiale de rupture d'un équilibre physiologique. Compt Rend Acad Sci, Paris, 245: 1946–1948.

Dejours, P. 1962, Chemoreflexes in breathing, Physiol Rev, 42: 335–358.

Dripps, R.D., Comroe, J.H., Jr. 1947, The effect of inhalation of high and low oxygen concentrations on respiration, pulse rate, ballistocardiogram and arterial oxygen saturation (oximeter) of normal individuals, Am J Physiol, 149: 277–291.

Drysdale, D.B., Petersen, E.S. 1977, Arterial chemoreceptors, ventilation and heart rate in man, J Physiol, Lond, 273: 109–120.

Eugenin, J., Larrain, C., Zapata, P. 1989, Correlative contribution of carotid and aortic afferences to the ventilatory chemosensory drive in steady-state normoxia and to the ventilatory chemoreflexes induced by transient hypoxia, Arch Biol Med Exp, 22: 395–408.

Eyzaguirre, C., Zapata, P. 1984, Perspectives in carotid body research, J Appl Physiol 57: 931–957.

Fitzgerald, R.S., Shirahata, M. 1997, Systemic responses elicited by stimulating the carotid body: primary and secondary mechanisms. In: *The Carotid Body Chemoreceptors*, ed. C. Gonzalez, Springer-Verlag, New York, pp 171–191.

Izdebska, E., Izdebski, J., Trzebski, A. 1996, Hemodynamic responses to brief hyperoxia in healthy and in mild hypertensive human subjects in rest and during dynamic exercise. J Physiol Pharmacol 47: 243–256.

Leitner, L.-M., Pagès, B., Puccinelli, R., Dejours, P. 1965, Étude simultanée de la ventilation et des décharges des chémorécepteurs du glomus carotidien chez le chat. I. Au cours d'inhalations brèves d'oxygène pur. Arch Intl Pharmacodyn Thér 154: 421–426

Parkinson, J. 1912, The effect of inhalation of oxygen on the rate of the pulse in health. J Physiol, Lond 44: 54–58.

Seals, D.R., Johnson, D.G., Fregosi, R.F. 1991, Hyperoxia lowers sympathetic activity at rest but not during exercise in humans. Am J Physiol 260: R873–R878.

Tafil-Klawe, M., Trzebski, A., Klawe, J., Pałko, T. 1985, Augmented chemoreceptor reflex tonic drive in early human hypertension and in normotensive subjects with family background of hypertension. Acta Physiol Pol 36: 51–58.

Thomson, A.J., Drummond, G.B., Waring, W.S., Webb, D.J., Maxwell, S.R.J. 2006, Effects of short-term isocapnic hyperoxia and hypoxia on cardiovascular function. J Appl Physiol 101: 809–816.

Waring, W.S., Thomson, A.J., Adwani, S.H., Rosseel, A.J., Potter, J.F., Webb, D.J., Maxwell, S.R.J. 2003, Cardiovascular effects of acute oxygen administration in healthy adults. J Cardiovasc Pharmacol 42: 245–250.

von Euler, U.S., Liljestrand, G. 1943, The rôle of the chemoreceptors of the sinus region for the occlusion test in the cat. Acta Physiol Scand 6: 319–323.

Yamazaki, F., Takahara, K., Sone, R., Johnson, J.M. 2007, Influence of hyperoxia on skin vasomotor control in normothermic and heat-stressed humans. J Appl Physiol 103: 2026–2033.

Zapata, P., Hess, A., Bliss, E.L., Eyzaguirre, C. 1969, Chemical, electron microscopic and physiological observations on the role of catecholamines in the carotid body. Brain Res 14: 473–496.

Time-Dependence of Hyperoxia-Induced Impairment in Peripheral Chemoreceptor Activity and Glomus Cell Calcium Response

J.L. Carroll, I. Kim, H. Dbouk, D.J. Yang, R.W. Bavis and D.F. Donnelly

Abstract In mammals, transient exposure to hyperoxia for a period of weeks during perinatal life leads to impairment of the ventilatory response to acute hypoxia, which may persist long beyond the duration of the hyperoxia exposure. The impairment of the ventilatory response to hypoxia is due to hyperoxia-induced reduction of carotid chemoreceptor sensitivity to hypoxia. We previously demonstrated that hyperoxia exposure in rats, from birth to two weeks of age, profoundly reduced carotid chemoreceptor single axonal responses to acute hypoxia challenge. However, the time course and mechanisms of this impairment are not known. Therefore, we investigated the effect of hyperoxia ($FiO_2 = 0.6$) on neonatal rats after 1, 3, 5, 8, and 14 days of exposure, starting at postnatal day 7. Carotid chemoreceptor single unit activities, nerve conduction time and glomus cell calcium responses to acute hypoxia were recorded in vitro. After 1 day in hyperoxia, single unit spiking rate in response to acute hypoxia was increased compared to controls. After 5 days in hyperoxia, the spiking response to acute hypoxia was significantly reduced compared to controls, nerve conduction time was lengthened and the glomus cell calcium response to acute hypoxia was reduced compared to controls. We conclude that perinatal exposure to hyperoxia, in rats, impairs the glomus cell calcium response (pre-synaptic) and the afferent nerve excitability (post-synaptic). The time course indicates that hyperoxia exerts these effects within days.

Keywords Carotid body · Hyperoxia · Single unit · Perinatal · Hypoxia-induced damage · Intracellular calcium · Isolated glomus cells

1 Introduction

The carotid body chemoreceptors, the main arterial oxygen sensors in mammals, mediate critically important cardiorespiratory responses to hypoxia. In spite of their importance, the carotid bodies are not structurally or functionally mature at birth.

J.L. Carroll (✉)
Department of Pediatrics, University of Arkansas for Medical Sciences, Little Rock, AR, USA
e-mail: carrolljohnl@uams.edu

After birth, important developmental changes occur in their response to low oxygen tension, determination of afferent nerve phenotype and axonal pruning, which are due to local changes in trophic factors such as GDNF and BDNF (Brady et al. 1999). The major determinant of these postnatal changes is the large increase in arterial oxygen tension (PaO$_2$) occurring at birth (Blanco et al. 1984). It is now well-established that birth of mammals into a low PO$_2$ environment blunts postnatal maturation of carotid chemoreceptor and ventilatory responses to hypoxia (Eden and Hanson, 1987; Hanson et al. 1989b; Sterni et al. 1999).

Elevated oxygen levels at birth and during the perinatal period (perinatal hyperoxia) also lead to marked impairment of ventilatory responses to acute hypoxia in rats (Fuller et al. 2001; Ling et al. 1997b) and kittens (Hanson et al. 1989b). This effect persists beyond the perinatal period, may last throughout life and may occur with as little as 30% O$_2$ exposure (Eden and Hanson, 1986; Ling et al. 1996, 1997b).

The marked perinatal hyperoxia-induced impairment of the ventilatory response to hypoxia is primarily due to reduced peripheral chemoreceptor sensitivity to hypoxia (Bisgard et al. 2003; Ling et al. 1997a, b). Consistent with this conclusion, Erikson et al. reported that perinatal hyperoxia exposure in rats resulted in striking structural changes in the carotid bodies and a profound reduction in the number of carotid sinus nerve axons (Erickson et al. 1998), suggesting that impaired function may be due to hyperoxia-related carotid body damage, degeneration and/or impaired structural maturation. However, Fuller et al found that the impaired ventilatory response to acute hypoxia in rats exposed to perinatal hyperoxia could be functionally restored by subsequent exposure to intermittent hypoxia (Fuller et al. 2001). Thus, the hyperoxia-induced functional impairment of carotid chemoreceptor O$_2$ sensitivity is at least partially reversible. This observation raises intriguing questions about the mechanisms involved and suggests hyperoxia is not acting by simple structural degradation; the organs remain sufficiently structurally intact such that full functional recovery can be induced under certain conditions (Fuller et al. 2001).

In order to determine whether the hyperoxia-induced reduction in whole-nerve chemoreceptor activity (Bisgard et al. 2003) could be due entirely to loss afferent axons, we previously studied the activity of carotid single chemoreceptor axons in response to acute hypoxia in rats exposed to hyperoxia from birth to 14 days of age (Donnelly et al. 2005). Two weeks of hyperoxia exposure caused a profound reduction in single unit axonal spiking rates in normoxia and during acute hypoxia and also caused a significant lengthening of nerve conduction time. These results clearly showed that individual carotid chemoreceptor sensory units were profoundly functionally impaired following hyperoxia exposure, suggesting that overall organ functional impairment is not due simply to axon loss. Thus, the functional impairment of single chemosensory units could be due to pre- (glomus cell) and/or post-synaptic (sinus nerve excitability) induced by hyperoxia exposure. The present study was undertaken in order to understand the site of the hyperoxia-induced impairment of carotid body O$_2$-sensing function and to better understand the time course.

2 Methods

2.1 Animal Model

Experiments were undertaken with the approval of the Yale Animal Care and Use Committee and Animal Care and Use Committee of the University of Arkansas for Medical Sciences. Studies were conducted on multiple litters of Sprague-Dawley rat pups, starting at postnatal day 7 (P7). Control litters were maintained in normoxia and test litters were placed in an environmental chamber with an atmosphere of 60% oxygen. The test rats lived in the hyperoxia environment until studied at P8, P10, P12, P15 and P21.

2.2 Tissue Harvest and Recording

The carotid body – ganglion complexes, consisting of the carotid body, carotid sinus nerve, glossopharyngeal nerve and petrosal ganglion, were harvested and prepared as previously described (Donnelly et al. 2005). Recording chamber oxygen tension was measured with a fiberoptic probe coupled to a phosphorescence oxygen probe (Oxy-micro, WPI Instruments).

2.3 Single Unit Recording

Single-unit activity was recorded using a suction electrode advanced into the petrosal ganglion. Electrode tip size was approximately 30 μm in diameter which allowed individual ganglion cells to enter the tip. The pipette potential was amplified with an extracellular amplifier (BAK Instruments, MDA5), filtered (0–5 kHz), displayed on an oscilloscope, digitized (10 kHz sample rate) and stored on computer (Axoscope, Axon Instruments). Unit chemoreceptor activity was discriminated and timed post-hoc using an event detection program which identified the timing and magnitude of action potential events (CLAMPFIT9.0, Axon Instruments). The number of individual spikes per second was calculated and graphed as a function of time and fit to a 3-sec moving average for determining peak firing frequencies (Origin 7.5; Microcal Corporation).

2.4 Nerve Conduction Time

Nerve conduction time was measured using an orthodromic electrical stimulus applied in the center of the carotid body, and measuring the time for the evoked spike to reach the soma of the petrosal neurons being recorded.

2.5 Isolation of Glomus Cells and Measurement of Intracellular Calcium

Rat carotid bodies were isolated from pups at the same ages and treatment period as above, using a protocol as previously specified (Wasicko et al. 1999). Cells were loaded with the calcium-sensitive dye, fura-2 by incubation with 4 mM fura-2 AM ester for 30 min at 37°C in BSS equilibrated with 21% O_2-5% CO_2. Fura-2 fluorescence emission was measured at 510 nm in response to alternating excitation at 340 and 380 nm. Images were acquired and stored using a Nikon TE2000 microscope and (CoolSNAP HQ2) camera under Metafluor software (Molecular Devices).

2.6 Experimental Protocol

Following measurement of nerve conduction time, unit activity was recorded during normoxia perfusion (21% O_2/5% CO_2/balance N_2) and during superfusion with severe hypoxia (0% O_2/5% CO_2/balance N_2) for a period of two minutes, followed by a return to normoxia.

Experiments utilizing isolated glomus cells were preceded by loading cells with fura-2. The coverslip was placed in a closed microscope chamber (0·2 ml total volume) and perfused with a bicarbonate-buffered balanced salt solution (BSS) containing (mM): 118 NaCl, 23 $NaHCO_3$, 3 KCl, 2 KH_2PO_4, 1.2 $CaCl_2$, 1 $MgCl_2$ and 10 glucose. The hypoxic solution consisted of the above BSS equilibrated with 0% O_2 in 5% CO_2 balanced with N_2. While recording fura-2 fluorescence emission, cells were exposed to 20 mM KCl (2 min), allowed to recover (5 min) and exposed to 0% O_2 (2 min). Calcium level was estimated using the method of Grynkiewicz (Grynkiewicz et al. 1985).

2.7 Data Analysis

Single unit baseline activity was quantified as the average spiking rate measured over 60 sec of normoxia perfusion. Peak discharge activity was quantified as the peak spiking rate, averaged over 3 sec, during presentation of the hypoxic stimuli. Spiking rates and conduction times were analyzed using two-way analysis of variance (ANOVA) with age and treatment (normoxia/hyperoxia) as grouping variables. If the critical F value was exceeded following ANOVA then post-hoc testing was performed using Fisher's LSD test with Sidak's correction to the critical t-value in order to maintain the chance of a type 1 error at 0.05. Conduction times were similarly analyzed.

Calcium responses were analyzed as follows: Baseline intracellular calcium $[Ca^{2+}]_i$ level was determined by averaging $[Ca^{2+}]_i$ during the 30 seconds prior to challenge. Peak response was measured as the highest $[Ca^{2+}]_i$ during the challenge and $\Delta[Ca^{2+}]_i$ was calculated by subtracting baseline from peak $[Ca^{2+}]_i$ level.

Calcium responses were analyzed by ANOVA, as above. Significance level for all tests was set at $p < 0.05$. All results are expressed as mean±SEM.

3 Results

3.1 Baseline AP Activity

Baseline spiking activity in normoxia was about 1 Hz in the control group and, although there was a trend towards lower activity in the hyperoxia group, there were no major differences.

3.2 Nerve Responses to Acute Hypoxia

Control carotid bodies responded to 0% O_2 superfusion with a brisk increase in activity (Fig. 1). Peak spiking rates during hypoxia were significantly higher following 1 day of hyperoxia-treatment, but significantly reduced by 5 days of hyperoxia treatment and continued reduced through 14 days of hyperoxia-treatment (Fig. 1).

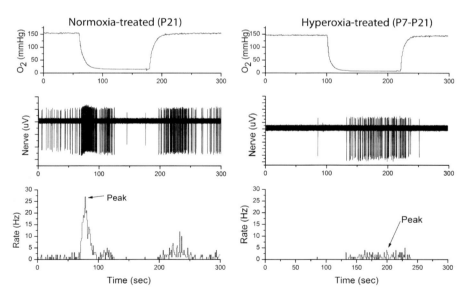

Fig. 1 *Left panel* – single unit recording from carotid body of control rat on day P21 showing active baseline spiking and a brisk increase in activity during hypoxia. *Right panel* – single unit recording from carotid body of a 21 day old rat hyperoxia exposed to hyperoxia between days 7 and 21

3.3 Nerve Conduction Time

Nerve conduction time was not different between controls and hyperoxia-exposed groups at P8–P10. By P12, nerve conduction time was significantly longer in the hyperoxia group.

3.4 Hypoxia-Induced Increase in Intracellular Calcium

Cells exhibiting typical glomus cell morphology (~ 10 m diameter, tendency to form clusters, large nucleus) and rapid rise in intracellular calcium ($[Ca^{2+}]_i$) in response to 20 mM K^+ were presumed to be glomus cells. In cells dissociated from carotid bodies of both groups, superfusion with 20 mM K^+ caused a brisk rise in $[Ca^{2+}]_i$ that did not differ with age. In cells from controls, superfusion of cells with hypoxic solution (equilibrated with 0% O_2 / 5% CO_2) caused a brisk rise in $[Ca^{2+}]_i$ that increased with age, as previously reported. In contrast, in the hyperoxia exposed group acute hypoxia challenge resulted in a larger $[Ca^{2+}]_i$ response, compared to controls, at P8 (not shown) and a smaller response by at P14 (Fig. 2) and P21.

4 Discussion

These preliminary results indicate that exposure of developing rats to hyperoxia profoundly impairs the carotid body single unit spiking response to acute hypoxia challenge, as well as the glomus cell intracellular calcium response to hypoxia. This suggests that both pre- and post-synaptic effects of hyperoxia are likely to contribute to the observed impairment of organ function.

The increase in spiking activity and $[Ca^{2+}]_i$ response to hypoxia after one day of hyperoxia exposure is probably related to the immature, developmental state at P7, when hyperoxia exposure was started. It has been previously reported, for developing mammals, that brief hyperoxia exposure increases the responsiveness of the carotid body to acute hypoxia challenge (Blanco et al. 1988). Thus, the effect of hyperoxia at P8 (1 day of hyperoxia exposure) may have been to accelerate carotid body functional maturation.

After the initial enhancement of the carotid body response to hypoxia at P8, spiking activity and the $[Ca^{2+}]_i$ response to acute hypoxia declined with age, being significantly reduced by 5 days of hyperoxia exposure. This is consistent with effects of chronic hyperoxia exposure on carotid body spiking activity (30% O_2 for weeks) in developing kittens and rats (Eden and Hanson, 1986; Hanson et al. 1989a). However, previous studies used prolonged hyperoxia exposure for many weeks and did not examine the short-term time course of hyperoxia induced impairment of carotid body responsiveness to hypoxia. Our results indicate that hyperoxia exposure in developing mammals can increase carotid body responsiveness in as little as one day and cause a substantial decrease in hypoxia responsiveness by five days of exposure. In our study rats were exposed to 60% O_2 atmosphere; we do not know what the

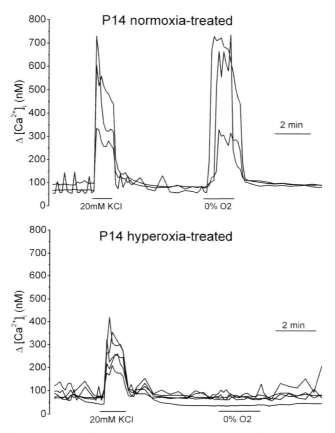

Fig. 2 Carotid body glomus cell responses to acute hypoxia; each line represents an individual glomus cell or cell cluster. *Upper panel* – cells from P14 rats reared in normoxia show a brisk response to extracellular K^+ and to 0% O_2. *Lower panel* – cells from P14 rats exposed to hyperoxia from P7 still show a brisk response to 20 mM K^+ but do not respond to acute hypoxia

short-term time course of carotid body functional impairment would be at other, milder levels of exposure.

The glomus cell is considered to be the presynaptic element in the carotid body. The preliminary results reported here indicate that the effects of hyperoxia exposure on the glomus cell $[Ca^{2+}]_i$ response to acute hypoxia follows a similar pattern to that observed for single unit spiking activity. The hyperoxia-induced lengthening of nerve conduction time occurred earlier than the impairment of the glomus cell $[Ca^{2+}]_i$ response, suggesting that this mechanism may predominate early in the exposure.

Acknowledgments This work was supported by National Institutes of Health grants P20 RR-016463 from the INBRE Program of the National Center for Research Resources (RWB) and HL-084520 (DFD).

References

Bisgard, G. E., Olson, E. B., Jr., Wang, Z. Y., Bavis, R. W., Fuller, D. D., & Mitchell, G. S. 2003, "Adult carotid chemoafferent responses to hypoxia after 1, 2, and 4 wk of postnatal hyperoxia", *J. Appl. Physiol.*, vol. 95, no. 3, pp. 946–952.

Blanco, C. E., Dawes, G. S., Hanson, M. A., & McCooke, H. B. 1984, "The response to hypoxia of arterial chemoreceptors in fetal sheep and new-born lambs", *J. Physiol.*, vol. 351, pp. 25–37.

Blanco, C. E., Hanson, M. A., & McCooke, H. B. 1988, "Effects on carotid chemoreceptor resetting of pulmonary ventilation in the fetal lamb in utero", *J. Dev. Physiol.*, vol. 10, no. 2, pp. 167–174.

Brady, R., Zaidi, S. I., Mayer, C., & Katz, D. M. 1999, "BDNF is a target-derived survival factor for arterial baroreceptor and chemoafferent primary sensory neurons", *J. Neurosci.*, vol. 19, no. 6, pp. 2131–2142.

Donnelly, D. F., Kim, I., Carle, C., & Carroll, J. L. 2005, "Perinatal hyperoxia for 14 days increases nerve conduction time and the acute unitary response to hypoxia of rat carotid body chemoreceptors", *J. Appl. Physiol.*, vol. 99, pp. 114–9.

Eden, G. J. & Hanson, M. A. Effect of hyperoxia from birth on the carotid chemoreceptor and ventilatory responses of rats to acute hypoxia. *J. Physiol.*, vol. 374, 24p. 1986. Ref Type: Abstract

Eden, G. J. & Hanson, M. A. 1987, "Effects of chronic hypoxia from birth on the ventilatory response to acute hypoxia in the newborn rat", *J. Physiol.*, vol. 392, pp. 11–19.

Erickson, J. T., Mayer, C., Jawa, A., Ling, L., Olson, E. B., Vidruk, E. H., Mitchell, G. S., & Katz, D. M. 1998, "Chemoafferent degeneration and carotid body hypoplasia following chronic hyperoxia in newborn rats", *J. Physiol.*, vol. 509 (Pt 2), pp. 519–526.

Fuller, D. D., Wang, Z. Y., Ling, L., Olson, E. B., Bisgard, G. E., & Mitchell, G. S. 2001, "Induced recovery of hypoxic phrenic responses in adult rats exposed to hyperoxia for the first month of life", *J. Physiol.*, vol. 536, no. Pt 3, pp. 917–926.

Grynkiewicz, G., Poenie, M., & Tsien, R. Y. 1985, "A new generation of Ca2+ indicators with greatly improved fluorescence properties", *J. Biol. Chem.*, vol. 260, no. 6, pp. 3440–3450.

Hanson, M. A., Eden, G. J., Nijhuis, J. G., & Moore, P. J. 1989a, "Peripheral chemoreceptors and other oxygen sensors in the fetus and newborn," in *Chemoreceptors and Reflexes in Breathing: Cellular and Molecular Aspects*, S. Lahiri et al., eds., Oxford University Press, New York, pp. 113–120.

Hanson, M. A., Kumar, P., & Williams, B. A. 1989b, "The effect of chronic hypoxia upon the development of respiratory chemoreflexes in the newborn kitten", *J. Physiol.*, vol. 411, pp. 563–574.

Ling, L., Olson, E. B., Jr., Vidruk, E. H., & Mitchell, G. S. 1996, "Attenuation of the hypoxic ventilatory response in adult rats following one month of perinatal hyperoxia", *J. Physiol.*, vol. 495 (Pt 2), pp. 561–571.

Ling, L., Olson, E. B., Vidruk, E. H., & Mitchell, G. S. 1997a, "Integrated phrenic responses to carotid afferent stimulation in adult rats following perinatal hyperoxia", *J. Physiol.*, vol. 500 (Pt 3), pp. 787–796.

Ling, L., Olson, E. B., Vidruk, E. H., & Mitchell, G. S. 1997b, "Phrenic responses to isocapnic hypoxia in adult rats following perinatal hyperoxia", *Respir. Physiol.*, vol. 109, no. 2, pp. 107–116.

Sterni, L. M., Bamford, O. S., Wasicko, M. J., & Carroll, J. L. 1999, "Chronic hypoxia abolished the postnatal increase in carotid body type I cell sensitivity to hypoxia", *Am. J. Physiol.*, vol. 277, no. 3 Pt 1, p. L645–L652.

Wasicko, M. J., Sterni, L. M., Bamford, O. S., Montrose, M. H., & Carroll, J. L. 1999, "Resetting and postnatal maturation of oxygen chemosensitivity in rat carotid chemoreceptor cells", *J. Physiol.*, vol. 514 (Pt 2), pp. 493–503.

Long-Term Regulation of Carotid Body Function: Acclimatization and Adaptation – *Invited Article*

N.R. Prabhakar, Y.-J. Peng, G.K. Kumar, J. Nanduri, C. Di Giulio and Sukhamay Lahiri

Abstract Physiological responses to hypoxia either continuous (CH) or intermittent (IH) depend on the O_2-sensing ability of the peripheral arterial chemoreceptors, especially the carotid bodies, and the ensuing reflexes play important roles in maintaining homeostasis. The purpose of this article is to summarize the effects of CH and IH on carotid body function and the underlying mechanisms. CH increases baseline carotid body activity and sensitizes the response to acute hypoxia. These effects are associated with hyperplasia of glomus cells and neovascularization. Enhanced hypoxic sensitivity is due to alterations in ion current densities as well as changes in neurotransmitter dynamics and recruitment of additional neuromodulators (endothelin-1, ET-1) in glomus cells. Morphological alterations are in part due to up-regulation of growth factors (e.g. VEGF). Hypoxia-inducible factor-1 (HIF-1), a transcriptional activator might underlie the remodeling of carotid body structure and function by CH. Chronic IH, on the other hand, is associated with recurrent apneas in adults and premature infants. Two major effects of chronic IH on the adult carotid body are sensitization of the hypoxic sensory response and long-lasting increase in baseline activity i.e., sensory long-term facilitation (LTF) which involve reactive oxygen species (ROS) and HIF-1. In neonates, chronic IH leads to sensitization of the hypoxic response but does not induce sensory LTF. Chronic IH-induced sensitization of the carotid body response to hypoxia increases the likelihood of unstable breathing perpetuating in more number of apneas, whereas sensory LTF may contribute to increased sympathetic tone and systemic hypertension associated with recurrent apneas.

Keywords Heme proteins · Mitochondria · Ion channels · Transmitters · Reactive oxygen species · Sensory long-term facilitation · Plasticity · HIF-1 transcription factor

---

J. Nanduri (✉)
Center for Systems Biology of O_2 Sensing, Department of Medicine, University of Chicago, IL, USA
e-mail: nanduri@uchicago.edu

1 Introduction

Continuous hypoxia (CH) is experienced during sojourns at high altitude, whereas people residing at sea level encounter intermittent hypoxia (IH) more frequently than continuous hypoxia. Physiological responses to both forms of hypoxia depend on the O_2-sensing ability of the peripheral arterial chemoreceptors, especially the carotid bodies, and the ensuing reflexes play important roles in maintaining homeostasis. The purpose of this article is to summarize the effects of CH and IH on carotid body function and the underlying mechanisms.

2 Effects of Continuous Hypoxia on Carotid Body

Carotid bodies are composed of two cell types: type I and type II. Type I cells (also called glomus cells) express a variety of neurotransmitters, and form synaptic contact with afferent nerve endings whose cell bodies lie in the petrosal ganglion. Type II cells resemble glial cells. Available evidence support the idea that type I cells are the primary site where changes in O_2 are being sensed. Figure 1 illustrates the effects of acute hypoxia on carotid body sensory activity and ventilation. Lowering inspired PO_2 is followed by an immediate increase of carotid chemoreceptor (CC)

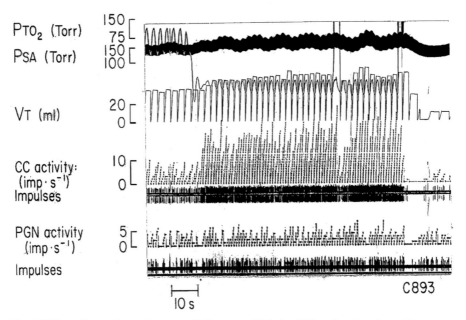

Fig. 1 Effect of acute hypoxia on ventilation, carotid body (CC) and petrosal ganglion nerve (PGN) activities in an anesthetized cat. PTO_2 = Partial pressure of tracheal oxygen; P_{SA} = Arterial blood pressure; VT- tidal volume

and petrosal ganglion nerve (PGN) activities. After a short delay, ventilation starts to increase. Terminating hypoxia results in prompt cessation of CC and PGN activities along with decreases in ventilation.

The transduction of the hypoxic stimulus at the carotid body is complex and involves interactions between heme and/or redox-sensitive proteins and O_2 sensitive K^+ channels that regulate membrane potential (Gonzalez et al., 1994, Lahiri and Cherniack, 2001, Lopez-Barneo et al., 2001, Prabhakar, 2006). The transduction process leads to Ca^{2+}-dependent release of neurotransmitter(s) from the glomus cells, which by acting on the nearby afferent nerve endings produce an increase in sensory nerve activity (sensory transmission).

If hypoxia continues for several days, the carotid chemoreceptor activity continue to increase at the same inspired oxygen levels, contributing to ventilatory acclimatization (Bisgard et al., 1987, Barnard et al., 1987). The following section summarizes the effects of CH on the sensory transduction and transmission at the carotid body.

2.1 Chronic Hypoxia and Ion Channel Expression in the Carotid Body

Ion channels, especially those conducting K^+ and Na^+ ions contribute to the excitability of glomus cells. Three to four weeks of hypoxia decreases the density of K^+ currents in rat glomus cells (Wyatt et al., 1995); whereas 2 weeks of hypoxia increase Na^+ current density via cAMP-dependent mechanisms (Stea et al., 1992). Acute hypoxia increases Ca^{2+} influx in the glomus cells and this effect is mediated entirely by opening of the voltage-dependent Ca^{2+} channels. Four types of voltage dependent Ca^{2+} channels have been identified in glomus cells, including L-, P/Q-, N-, and R- (resistant) type Ca^{2+} currents (Silva and Lewis, 1995; Overholt and Prabhakar, 1997). Hempleman (1996) reported that 5–8 days of hypobaric hypoxia (~0.4 ATM) increases Ca^{2+} channel density in glomus cells from neonatal carotid bodies and markedly attenuated their inactivation properties. Hempleman proposed that increased hypoxic sensitivity is mediated in part by augmented Ca^{2+} influx through voltage-gated Ca^{2+} channels leading to increased stimulus-secretion coupling in glomus cells.

2.2 Effect of Chronic Hypoxia on Heme and/or Redox-Sensitive Proteins

Carotid bodies express a variety of heme-containing proteins including NO synthases (NOS), heme-oxygenase-2 (HO-2), mitochondrial cytochromes, and NADPH-oxidases. Di Giulio et al. (1998) reported increases in nNOS protein in the carotid body after 12 days of hypoxia. Ye et al. (2002) found no changes in nNOS expression in the carotid body after 4 weeks of hypoxia, but there was a robust increase in iNOS expression, which was normally absent in the glomus tissue. These investigators further showed that chronic hypoxia increases NO production in the carotid body. NO being inhibitory to the carotid body activity, it was proposed that elevated NO might contribute to blunting of the hypoxic sensitivity of the carotid body during prolonged hypoxia. Little information is available on the effects of chronic hypoxia on HO-2, mitochondrial cytochromes, and NADPH-oxidases in the carotid body.

2.3 Chronic Hypoxia and Neurotransmitters in the Carotid Body

Carotid body expresses a variety of neurotransmitters/modulators. Some of them excite and others inhibit afferent nerve activity. Because chronic hypoxia sensitizes the carotid body to subsequent hypoxic stimuli, it was thought that prolonged exposure to low oxygen down regulate the "inhibitory" transmitters and up-regulate the "excitatory" ones. This idea has been extensively tested and the following section will briefly summarize these studies.

Carotid bodies express dopamine (DA), which may function as an inhibitory modulator (Gonzalez et al., 1994; Bisgard, 2000; Prabhakar, 2000). Tyrosine hydroxylase (TH), the enzyme responsible for DA synthesis, is expressed in glomus cells as well as in nerve fibers and ganglion cells of the carotid body (Gonzalez et al., 1994). Chronic hypoxia up-regulates TH mRNA, protein, as well as the enzyme activity (Czyzyk-Krzeska et al., 1992), and also increases DA turnover in the glomus tissue (Hanbauer et al., 1981). Dopaminergic D_2 receptors which mediate the inhibitory actions of DA are initially down-regulated, but later up-regulated by chronic hypoxia in the carotid body (Huey and Powell, 2000). These observations indicate that although DA is inhibitory, chronic hypoxia up-regulates the dopaminergic system in the carotid body.

Acetylcholine (ACh) stimulates the carotid body activity and its stimulatory actions are mediated by nicotinic cholinergic receptors localized to glomus cells as well as the afferent fiber terminals (Shirahata et al., 1998). Chronic hypoxia (9–14 days) up-regulates nicotinic receptor expression ($\alpha 3$ and $\alpha 7$ subunits) on petrosal neurons that provide afferent innervation to the carotid body (Dinger et al., 2003). However, mecamylamine, a blocker of nicotinic receptors had little effect on chronic hypoxia-induced hypersensitivity of the carotid body. Endothelin-1 (ET-1) and its receptors are either undetectable or expressed in low abundance in glomus cells from normal carotid bodies. but are up-regulated by chronic hypoxia in glomus cells (He et al., 1996; Chen et al., 2002a, b). ET-1 by itself has no appreciable excitatory effect on carotid body afferent nerve activity but markedly augments the hypoxic sensory response (Chen et al., 2000). Blockade of ET_A receptors prevents chronic hypoxia-induced hypersensitivity of the carotid body to acute low PO_2. Taken together these observations suggest that the enhanced hypoxic sensitivity elicited by chronic hypoxia is mediated by recruiting excitatory neuromodulator(s), which are not normally expressed in the carotid body, rather than tipping the balance between the inhibitory and excitatory transmitters.

3 Molecular Mechanisms Underlying the Effects of Continuous Hypoxia on the Carotid Body

Transcriptional regulation of genes is one of the pivotal mechanisms that underlie long-term adaptations to chronic hypoxia (Bunn and Poyton, 1996). The following section summarizes studies on the transcriptional activator hypoxia-inducible factor-1 in mediating carotid body response to continuous hypoxia.

3.1 Role of HIF-1 on Carotid Body Changes by Continuous Hypoxia

HIF-1 is a heterodimeric protein composed of a constitutively expressed HIF-1β subunit and an O_2-regulated HIF-1α subunit. HIF-1 is a global transcriptional regulator of oxygen homeostasis that controls multiple physiological processes (Semenza, 2000). Complete loss of function in *Hif1a* gene in mice results in embryonic lethality at mid-gestation with major malformations of the heart and vasculature (Iyer et al., 1998). On the other hand, HIF-1α heterozygous mice (i.e., partially deficient in HIF-1α) develop normally and are indistinguishable from wild type controls. However, upon exposure to chronic hypoxia, adult $Hif1a^{+/-}$ mice exhibit impaired physiological adaptations as evidenced by reduced erythropoietic response and pulmonary vascular remodeling (Yu et al., 1999). Kline et al. (2002) reported severely impaired hypoxic sensing of the carotid bodies, and absence of ventilatory acclimatization in $Hif1a^{+/-}$ mice. How might HIF-1 contribute to carotid body adaptations to chronic hypoxia? It is known that HIF-1 regulates several genes including ET-1 (Semenza, 2000). Therefore, HIF-1 might participate in functional changes in the carotid body by regulating ET-1.

Chronic hypoxia increases the number and size of glomus cells and increased blood vessels in the carotid body (McGregor et al., 1984; Wilson et al., 2005). HIF-1 might also contribute to structural alterations of the chemoreceptor organ during chronic hypoxia by up-regulating growth factors. HIF-1 is a potent inducer of vascular endothelial growth factor (VEGF; Semenza, 2000). Carotid body expresses VEGF in type I cells, and PD-ECGF expression is confined to the extra-cellular stroma (Jyung et al., 2000; Chen et al., 2003). Carotid bodies also express VEGF receptors, Flt-1 and Flk-1 (Jyung et al., 2000; Chen et al., 2003). VEGF acting on Flk-1 regulates hyperplasia of the type I cells, and promotes neo-vascularization by acting on Flt-1. PD-ECGF contributes more to angiogenesis as it does elsewhere in the body including the lung (Jyung et al., 2000). Thus, HIF-1 might participate in both functional and structural re-organization of the carotid bodies during chronic hypoxia.

4 Intermittent Hypoxia and Carotid Body Function

Recurrent apneas are characterized by periodic cessations of breathing resulting in cyclical decreases in arterial blood O_2 or intermittent hypoxia (IH). Studies in humans and rodents suggest that the carotid bodies constitute the "frontline" defense system for detecting IH. Recurrent apnea patients exhibit more pronounced ventilatory depression with brief hyperoxic challenge (Dejour's test, a measure of peripheral chemoreceptor sensitivity) than control subjects (Tafil-Klawe et al., 1991; Kara et al., 2003). Furthermore, glomectomized subjects with sleep apneas do not develop hypertension (see discussion in Somers and Abboud, 1993). Likewise, rats exposed to 30d of IH develop hypertension and increased sympathetic nerve activity and chronic bilateral sectioning of sinus nerves prevented these responses (Lesske et al., 1997). The following section summarizes recent studies reporting the effects of IH on carotid body function in experimental animal models.

4.1 Effects of IH on Hypoxic Sensory Response of the Adult Carotid Body

Hypoxic but not hypercapnic sensory response of the carotid body was augmented in adult rats exposed to 10d of IH (Peng and Prabhakar, 2004). Similar augmentation of the hypoxic sensory response was also reported in cats (Rey et al., 2004, 2006) and mice (Peng et al., 2006), suggesting that chronic IH uniformly augments the hypoxic sensory response in three species studied thus far. The augmented hypoxic response was reversed by re-exposing IH rats to 10 d of normoxia (Peng et al., 2003). IH had no significant effect on carotid body morphology (Peng et al., 2003).These observations suggest that IH leads to selective sensitization of the hypoxic sensory response.

4.2 IH –Induces Sensory Long-Term Facilitation of the Carotid Body in Adults

When anesthetized rats were exposed to acute intermittent hypoxia (AIH; 15s of hypoxia followed by 5 min of re-oxygenation, 10 episodes), sensory activity increased with each episode and promptly returned to baseline after terminating AIH. In striking contrast, in chronic IH exposed animals, AIH resulted in long-lasting increase in baseline activity that persisted for 60 min after terminating AIH. This long-lasting increase in baseline sensory activity has been termed sensory LTF (Peng et al., 2003). Thus, in addition to sensitizing the response to hypoxia, IH also induces hitherto uncharacterized form of plasticity manifested as sensory long-term facilitation (LTF; Peng et al., 2003, 2006).

Unlike AIH, acute intermittent hypercapnia (10 episodes of 15 sec of 7% CO_2 +93% O_2 interspersed with 5 min of 100% O_2), another "physiological" stimulus to the carotid body was ineffective in evoking sensory LTF in IH exposed rats (Peng et al., 2003). The induction of chronic IH-induced sensory LTF is a time-dependent phenomenon, in that it was apparent after 3d of IH, and the magnitude further increased following 10d of IH. More interestingly, the magnitude of sensory LTF was not dependent on the severity of hypoxia used for IH conditioning, because it was indistinguishable with either FiO_2 of 5 or 10% O_2 used for chronic

IH conditioning (Peng et al., 2003). Like the sensitization of the hypoxic response, sensory LTF was reversible after re-exposing chronic IH rats to 10d of normoxia.

4.3 Comparison of IH with Comparable Duration of Continuous Hypoxia in Adult Rodents

Studies described thus far suggest that IH exerts two major effects on the carotid body that include (a) selective sensitization of the hypoxic sensory response and (b) induction of sensory LTF. The total duration of hypoxia accumulated over 10d of chronic IH corresponds to 4h of low O_2. A single episode of 4h of hypoxia or 4h of hypoxia/day for 10 days, however, were ineffective in inducing sensitization of the hypoxic sensory response and sensory LTF of the carotid body (Peng et al., 2003; Peng and Prabhakar, 2004). These findings demonstrate that for a given duration of hypoxic exposure, IH is more effective than continuous hypoxia in altering the carotid body function.

4.4 Effects of IH on Neonatal Carotid Body Activity

Carotid bodies are immature at birth and respond poorly to hypoxia compared with adults (Carroll, 2003). Clinical studies have shown that nearly 70–90% of prematurely born infants experience IH as a consequence of recurrent apneas (Poets et al., 1994). Recent studies examined carotid body response to hypoxia in neonatal rat pups (P0) exposed to 8h/day IH for 1-10d (Peng et al., 2004; Pawar et al., 2008a, b). Carotid body sensory response to hypoxia was markedly augmented in IH exposed pups and the magnitude was even greater compared to adults. A systematic comparison with adult carotid bodies showed that: (a) neonatal carotid bodies are more susceptible to IH than adults, (b) unlike adult carotid bodies, the effects of neonatal IH were not reversible following re-exposures to normoxia for ~2 months, and (c) the effects of neonatal IH were associated with increased number of glomus cells (Pawar et al., 2008a).

In striking contrast to adult rats, IH was ineffective in inducing sensory LTF in neonatal carotid bodies (Pawar et al., 2008a). Previous studies reported blunting of hypoxic sensory response in neonatal rats exposed to continuous hypoxia (Carroll, 2003). Thus, there are striking differences between the effects of intermittent and continuous hypoxia, in that the former facilitates, whereas the later reduces hypoxic sensing ability of neonatal carotid bodies.

4.5 Cellular Mechanism(S) Associated with IH

4.5.1 Role of Reactive Oxygen Species (ROS)

IH increases ROS generation in the carotid body as evidenced by decreased activity of the red-ox sensitive enzyme aconitase (Peng et al., 2003). Rats treated with a stable superoxide dismutase mimetic (MnTMPyP), a potent scavenger of ROS, attenuated or abolished sensitization of the hypoxic response as well as the sensory LTF in adult rats (Peng et al., 2003; Peng and Prabhakar, 2004) and the augmented hypoxic response in neonates (Pawar et al., 2008b).

Cellular sources of ROS generation involve inhibition of complex I and III of the mitochondrial electron transport chain (ETC) as well as activation of several oxidases. IH down-regulated mitochondrial complex I but not the complex III activity, whereas IH up-regulated NADPH oxidase activity in adult rat carotid bodies (Peng et al., 2003), suggesting that both mitochondrial electron transport chain at complex I and NADPH oxidase contribute to elevated levels of ROS. These observations demonstrate that chronic IH such as that seen during recurrent apneas facilitate hypoxic sensing in the carotid body by increasing the levels of ROS. In other words, ROS are acting as "amplifiers" of brief hypoxic signals associated with IH.

4.5.2 Role of Neurotransmitters/Modulators in IH-Evoked Functional Changes in the Carotid Body

A study by Rey et al. (2006) on adult cat carotid body suggests that up-regulation of endothelin-1 (ET-1) in glomus cells and ET-1 receptors contribute to IH-induced sensitization of the hypoxic sensory response. Pawar et al. (2008b) showed that ET-1 and ET_A receptors also important for IH-induced sensitization in neonatal carotid bodies. Furthermore, in neonates IH- enhances ET-1 release from the carotid body and up-regulates ET_A receptor mRNA and both these responses to IH involve ROS-mediated signaling (Pawar et al., 2008b).

5-Hydroxytryptamine (5-HT) evokes long-term activation of neuronal activity elsewhere in the nervous system. Carotid bodies express 5-HT. Peng et al., (2006) reported that spaced (3 × 15 s of 100 nM at 5 min intervals) but not mass (300 nM, 45 s) application of 5-HT elicited LTF, whereas both modes of 5-HT application evoked initial sensory excitation of the carotid bodies in rats. Ketanserin, a 5-HT(2) receptor antagonist prevented sensory LTF but not the initial sensory excitation. Spaced application of 5-HT activated protein kinase C (PKC) and these effects were abolished by ketanserin as well as bisindolylmaleimide (Bis-1), an inhibitor of PKC. Bis-1 prevented 5-HT-evoked sensory LTF. Furthermore, 5-HT-evoked sensory LTF of the carotid body involves a novel signaling mechanism coupled to PKC-dependent activation of NADPH oxidase.

Further studies are needed to examine whether 5-HT signaling contributes to chronic IH-evoked sensory LTF.

4.6 Molecular Mechanisms Associated with IH

4.6.1 Hypoxia-Inducible Factor-1 (HIF-1) in IH-Induced Changes in the Carotid Body Function

IH up-regulates HIF-1α protein in glomus cells of the rat carotid body (Lam et al., 2008; and J. Nanduri, 2007 Unpublished observations). Peng et al (2006) examined the effects of chronic IH on carotid body activity in WT and HET mice partially deficient in HIF-1α. In IH exposed WT mice hypoxic sensory response of the carotid body was augmented, and acute intermittent hypoxia (AIH) evoked sensory LTF. In striking contrast, IH-evoked sensitization of the hypoxic sensory response and sensory LTF were absent in mice with heterozygous deficiency of HIF-1α. These observations suggest that HIF-1-mediated transcriptional activation is an important molecular mechanism underpinning the carotid body changes by IH. However, HIF-1 regulated down stream genes contributing to CIH-evoked changes in the carotid body remains to be determined.

Acknowledgments The research is supported by grants from National Institutes of Health (Heart, Lung and Blood Institute) HL-90554 and HL-76537.

References

Barnard P, Andronikou S, Pokorski M, Smatresk N, Mokashi A, Lahiri S (1987) Time-dependent effect of hypoxia on carotid body chemosensory function. J Appl Physiol 63: 685–691

Bisgard GE (2000) Carotid body mechanisms in acclimatization to hypoxia. Respir Physiol 121: 237–246.

Bisgard GE, Kressin NA, Nielsen AM, Daristotle L, Smith CA, Forster HV (1987) Dopamine blockade alters ventilatory acclimatization to hypoxia in goats. Respir Physiol. 69, 245–55.

Bunn HF and Poyton RO (1996) Oxygen Sensing and Molecular Adaptation to Hypoxia. Physiol Rev 76: 839–885.

Carroll JL (2003) Developmental plasticity in respiratory control. J Appl Physiol 94: 375–389.

Chen J, Dinger B, Jyung R, Stensaas L and Fidone S (2003) Altered expression of vascular endothelial growth factor and FLK-1 receptor in chronically hypoxic rat carotid body. Adv Exp Med Biol 536: 583–591.

Chen J, He L, Dinger B and Fidone S (2000) Cellular mechanisms involved in rabbit carotid body excitation elicited by endothelin peptides. Respir Physiol 121: 13–23.

Chen J, He L, Dinger B, Stensaas L and Fidone S (2002a) Role of endothelin and endothelin A-type receptor in adaptation of the carotid body to chronic hypoxia. Am J Physiol Lung Cell Mol Physiol 282: L1314–L1323.

Chen Y, Tipoe GL, Liong E, Leung S, Lam SY, Iwase R, Tjong YW and Fung ML (2002b) Chronic hypoxia enhances endothelin-1-induced intracellular calcium elevation in rat carotid body chemoreceptors and up-regulates ETA receptor expression. Pflugers Arch 443: 565–573.

Czyzyk-Krzeska MF, Bayliss DA, Lawson EE and Millhorn DE (1992) Regulation of tyrosine hydroxylase gene expression in the rat carotid body by hypoxia. J Neurochem 58: 1538–1546.

Di Giulio C, Grilli A, De Lutiis MA, Di Natale F, Sabatino G and Felaco M (1998) Does chronic hypoxia increase rat carotid body nitric oxide? Comp Biochem Physiol A Mol Integr Physiol 120: 243–247.

Dinger B, He L, Chen J, Stensaas L and Fidone S (2003) Mechanisms of morphological and functional plasticity in the chronically hypoxic carotid body, in *Oxygen Sensing: Responses and Adaptation to Hypoxia* (Lahiri S, Semenza G and Prabhakar NR eds) pp 439–465, Marcel Dekker, New York.

Gonzalez C, Almaraz L, Obeso A and Rigual R (1994) Carotid body chemoreceptors: From natural stimuli to sensory discharges. Physiol Rev 74: 829–898.

Hanbauer I, Karoum F, Hellstrom S and Lahiri S (1981) Effects of hypoxia lasting up to one month on the catecholamine content in rat carotid body. Neuroscience 6: 81–86.

He L, Chen J, Dinger B, Stensaas L and Fidone S (1996) Endothelin modulates chemoreceptor cell function in mammalian carotid body. Adv Exp Med Biol 410: 305–311.

Hempleman SC (1996) Increased calcium current in carotid body glomus cells following in vivo acclimatization to chronic hypoxia. J Neurophysiol 76: 1880–1886.

Huey KA and Powell FL (2000) Time-dependent changes in dopamine D(2)-receptor MRNA in the arterial chemoreflex pathway with chronic hypoxia. Brain Res Mol Brain Res 75: 264–270.

Iyer NV, Kotch LE, Agani F, Leung SW, Laughner E, Wenger RH, Gassmann M, Gearhart JD, Lawler AM, Yu AY and Semenza GL (1998) Cellular and developmental control of O2 homeostasis by hypoxia-inducible factor 1 Alpha. Genes Dev 12: 149–162.

Jyung RW, LeClair EE, Bernat RA, Kang TS, Ung F, McKenna MJ and Tuan RS (2000) Expression of angiogenic growth factors in paragangliomas. Laryngoscope 110: 161–167.

Kara T, Narkiewicz K, Somers VK, (2003) Chemoreflexes – physiology and clinical implications. Acta Physiol Scand 177, 377–384.

Kline DD, Peng YJ, Manalo DJ, Semenza GL and Prabhakar NR (2002) Defective carotid body function and impaired ventilatory responses to chronic hypoxia in mice partially deficient for hypoxia-inducible factor 1 alpha. Proc Natl Acad Sci U S A 99: 821–826.

Lam S-Y, Tipoe GL, Liong EC, and Fung ML (2008) Differential expressions and roles of hypoxia-inducible factor-1α,-2α and 3α in the rat carotid body during chronic and intermittent hypoxia. Histol Histopathol 23: 271–280.

Lahiri S, Cherniack NS (2001) Cellular and molecular mechanisms of O2 sensing with special reference to the carotid body. In *High Altitude: An Exploration of Human Adaptation* (Hornbein T and Schoene RB eds), pp 101–130, Marcel Dekker Inc., New York.

Lesske J, Fletcher EC, Bao G, Unger T (1997) Hypertension caused by chronic intermittent hypoxia – influence of chemoreceptors and sympathetic nervous system. J Hypertens 15, 1593–1603.

Lopez-Barneo J, Pardal R, Ortega-Sánez P (2001) Cellular mechanisms of oxygen sensing. Annu Rev Physiol 63: 259–287

McGregor KH, Gil J, Lahiri S (1984) A morphometric study of the carotid body in chronically hypoxia rats. J Appl Physiol 57: 1430–1438.

Overholt JL and Prabhakar NR (1997) Ca2+ current in rabbit carotid body glomus cells is conducted by multiple types of high-voltage-activated Ca2+ channels. J Neurophysiol 78: 2467–2474.

Pawar A, Peng YJ, Jacono FJ, Prabhakar NR (2008a) Comparative analysis of neonatal and adult rat carotid body responses to chronic intermittent hypoxia. J Appl Physiol 104:1287–94.

Pawar A, Nanduri J, Khan SA, Wang N, Prabhakar NR (2008b) Reactive oxygen species-dependent endothelin signaling is required for augmented hypoxic sensory response of the neonatal carotid bodies by intermittent hypoxia. Am J Physiol (Regulatory, Comparative and Integrated). Am J Physiol Regul Integr Comp Physiol. 2008 Dec 24. [Epub ahead of print].

Peng YJ, Overholt JL, Kline D, Kumar GK, Prabhakar NR (2003) Induction of sensory long-term facilitation in the carotid body by intermittent hypoxia: implications for recurrent apneas. Proc Natl Acad Sci U S A 100, 10073–10078.

Peng YJ, Prabhakar NR (2004) Effect of two paradigms of chronic intermittent hypoxia on carotid body sensory activity. J Appl Physiol 96, 1236–1242; discussion 1196.

Peng YJ, Rennison J, Prabhakar NR (2004) Intermittent hypoxia augments carotid body and ventilatory response to hypoxia in neonatal rat pups. J Appl Physiol 97, 2020–2025.

Peng YJ, Yuan G, Jacono FJ, Kumar GK, Prabhakar NR (2006) 5-HT evokes sensory long-term facilitation of rodent carotid body via activation of NADPH oxidase. J Physiol 576, 289–295.

Peng YJ, Yuan G, Ramakrishnan D, Sharma SD, Bosch-Marce M, Kumar GK, Semenza GL, Prabhakar NR (2006) Heterozygous Hif-1 deficiency impairs carotid body-mediated cardio-respiratory responses and Ros generation in mice exposed to chronic intermittent hypoxia. J Physiol 577:705–16.

Poets CF, Samuels MP, Southall DP, 1994. Epidemiology and pathophysiology of apnoea of prematurity. Biol Neonate 65, 211–219.

Prabhakar NR (2000) Oxygen sensing by the carotid body chemoreceptors. J Appl Physiol. 88, 2287–95.

Prabhakar NR (2006) O2 sensing at the mammalian carotid body: why multiple O2 sensors and multiple transmitters? Exp Physiol 91:17–23.

Rey S, Del Rio R, Alcayaga J, Iturriaga R (2004) Chronic intermittent hypoxia enhances cat chemosensory and ventilatory responses to hypoxia. J Physiol 560, 577–586.

Rey S, Del Rio R, Iturriaga R (2006) Contribution of endothelin-1 to the enhanced carotid body chemosensory responses induced by chronic intermittent hypoxia. Brain Res 1086, 152–159.

Semenza GL (2000) HIF-1: Mediator of physiological and pathophysiological responses to hypoxia. J Appl Physiol 88: 1474–1480.

Shirahata M, Ishizawa Y, Rudisill M, Schofield B and Fitzgerald RS (1998) Presence of nicotinic acetylcholine receptors in cat carotid body afferent system. Brain Res 814: 213–217.

Silva MJ and Lewis DL (1995) L- and N-Type Ca2+ Channels in adult rat carotid body chemoreceptor type I cells. J Physiol 489 (Pt 3): 689–699.

Somers VK, Abboud FM (1993) Chemoreflexes–responses, interactions and implications for sleep apnea. Sleep. 16(8 Suppl), S30–3.

Stea A, Jackson A and Nurse CA (1992) Hypoxia and N6,O2'-Dibutyryladenosine 3',5'-Cyclic Monophosphate, but not nerve growth factor, induce Na+ channels and hypertrophy in chromaffin-like arterial chemoreceptors. Proc Natl Acad Sci U S A 89: 9469–9473.

Tafil-Klawe M, Thiele AE, Raschke F, Mayer J, Peter JH, von Wichert W, (1991) Peripheral chemoreceptor reflex in obstructive sleep apnea patients; a relationship between ventilatory response to hypoxia and nocturnal bradycardia during apnea events. Pneumologie 45(Suppl 1), 309–311.

Wilson DF, Roy A and Lahiri S (2005) Immediate and long-term responses of the carotid body to high altitude. High Alt Med Biol 6:97–111.

Wyatt CN, Wright C, Bee D and Peers C (1995) O2-Sensitive K+ currents in carotid body chemoreceptor cells from normoxic and chronically hypoxic rats and their roles in hypoxic chemotransduction. Proc Natl Acad Sci U S A 92: 295–299.

Ye JS, Tipoe GL, Fung PC and Fung ML (2002) Augmentation of hypoxia-induced nitric oxide generation in the rat carotid body adapted to chronic hypoxia: An involvement of constitutive and inducible nitric oxide synthases. Pflugers Arch 444: 178–185.

Yu AY, Shimoda LA, Iyer NV, Huso DL, Sun X, McWilliams R, Beaty T, Sham JS, Wiener CM, Sylvester JT and Semenza GL (1999) Impaired physiological responses to chronic hypoxia in mice partially deficient for hypoxia-inducible factor 1alpha. J Clin Invest 103: 691–696.

Effects of Intermittent Hypoxia on Blood Gases Plasma Catecholamine and Blood Pressure

M.C. Gonzalez-Martin, V. Vega-Agapito, J. Prieto-Lloret, M.T. Agapito, J. Castañeda and C. Gonzalez

Abstract Obstructive sleep apnoea syndrome (OSAS) is a disorder characterized by repetitive episodes of complete (apnoea) or partial (hypopnoea) obstruction of airflow during sleep. The severity of OSAS is defined by the apnoea hypopnoea index (AHI) or number of obstructive episodes. An AHI greater than 30 is considered severe, but it can reach values higher than 100 in some patients. Associated to the OSA there is high incidence of cardiovascular and neuro-psychiatric pathologies including systemic hypertension, stroke, cardiac arrhythmias and atherosclerosis, diurnal somnolence, anxiety and depression. In the present study we have used a model of intermittent hypoxia (IH) of moderately high intensity (30 episodes/h) to evaluate arterial blood gases and plasma catecholamines as main effectors in determining arterial blood pressure. Male rats were exposed to IH with a regime of 80s, 20% O_2 // 40s, 10%O_2, 8 h/day, 8 or 15 days.

Lowering the breathing atmosphere to 10% O_2 reduced arterial blood PO_2 to 56.9 mmHg (nadir HbO_2 86, 3%). Plasma epinephrine (E) and norepinephrine (NE) levels at the end of 8 and 15 days of IH showed a tendency to increase, being significant the increase of norepinephrine (NE) levels in the group exposed to intermittent hypoxia during 15 days. We conclude that IH causes an increase in sympathetic activity and a concomitant increase in NE levels which in turn would generate an increase in vascular tone and arterial blood pressure.

Keywords Obstructive sleep apnoea · Intermittent hypoxia · Blood gases · Plasma norepinephrine · Arterial blood pressure · HbO_2 saturation

---

M.C. Gonzalez-Martin (✉)
Departamento de Bioquímica y Biología Molecular y Fisiología, Instituto de Biología y Genética Molecular, Ciber de Enfermedades Respiratorias, CIBERES, Instituto de Salud Carlos III.
Universidad de Valladolid/CSIC, Valladolid, Spain
e-mail: constanc@ibgm.uva.es

1 Introduction

OSAS is defined as a repeated stop of breathing during sleep. Apnoeas can last a minute or longer, can reach frequencies of 30 times/h or more and may total >240 episodes during a single night. Sleep apnoea can be caused by either complete obstruction of the airway (obstructive apnoea) or partial obstruction (obstructive hypopnoea-hypopnoea is slow, shallow breathing), both of which can wake up the patients. There are three types of sleep apnoea—obstructive, central, and mixed. Of these, obstructive sleep apnoea (OSAS) is the most common and occurs in approximately 2 percent of women and 4 percent of men over the age of 35.

The causes of OSAS are unknown. The site of obstruction in most patients is the soft palate, extending to the region at the base of the tongue. There are no rigid structures, such as cartilage or bone, in this area to hold the airway open. During the day, muscles in the region keep the passage wide open. But as a person with OSA falls asleep, these muscles relax to a point where the airway collapses and becomes obstructed. When the airway closes, breathing stops, and the sleeper awakens to open the airway. The arousal from sleep usually lasts only a few seconds, but brief arousals disrupt continuous sleep and prevent the person from reaching the deep stages of slumber, such as rapid eye movement (REM) sleep, which the body needs in order to rest and replenish its strength. Once normal breathing is restored, the person falls asleep only to repeat the cycle throughout the night. Typically, the frequency of waking episodes is somewhere between 10 and 60. A person with severe OSA may have more than 100 waking episodes in a single night.

The clinical picture of OSA is made out of snoring with discomfort for bed partner, a marked status of somnolence in daytime with poor performance at jobs and very high rate of traffic accidents, a frequent clinical picture of anxiety created by poor sleep that can abut in depression. Associated to these primary symptoms there is a much higher prevalence of systemic hypertension, acute cardiovascular accidents (heart and brain infarction) and diabetes among OSAS patients than in normal age matched subjects (Yaggi et al., 2005). From these perspective it is important to consider in some detail the pathophysiology of OSAS and associated pathology.

Although as stated above the cause of OSAS is unknown there is information enough to state that an inadequate functioning of the muscles openers of the upper airways is at the heart of the obstruction. The muscles of the larynx, pharynx, tongue and soft palate are multifunctional, participating in breathing, vocalization, swallowing and reflexes, such as cough and sneezing (van Lunteren 1997). The pattern of activity of the striated muscles of the upper airways varies during the respiratory cycle. For example the openers of the upper airways are activated during inspiration increasing the airflow to any given level of diaphragm contraction. The activity of these muscles precedes that of the diaphragm and dilate and give strength to the upper airways before the airflow starts; their contraction is maintained during the entire inspiration and is proportional to the airflow. Hypoxia, hypercapnia and any pharmacological agent that activates the carotid body (CB) chemoreceptors activate these muscles in parallel or even at higher strength than diaphragm itself; the counterpart is also true, a decreased drive from the CB chemoreceptors decreases the activity of the openers of the upper airways. During sleep, particularly in phases 3 and 4 of non-REM and during the REM periods there is a decrease drive to air pumping muscles (diaphragm and external intercostal muscles) as well as to the airway opener muscles (Dempsey et al., 1997; Kubin et al., 1998) with the result of a diminished air-pumping power and increased air ways resistance. This behaviour is physiological, but an abnormally decreased drive towards the muscles dilator of the upper airways, or an abnormal response of the muscles, produces obstruction of the respiratory ways (Morrison et al., 1993), existing evidence for diminished electromyographic activity in the genioglossi (an opener of the upper airways) in patients with OSAS: this reduced activity causes the tongue to drop backwards and close the airways (Eisele et al., 2003; Ryan and Bradley, 2005). Each episode of obstruction causes an increase in arterial PCO_2 and a decrease in arterial PO_2 and activation of the CB chemoreceptors. The increased drive of chemoreceptors impinges on the brainstem centres regulators of the respiratory rhythm and to the centres controlling the sleep-wake cycle generating an increase output to the air pumping and airways dilator muscles usually associated to a transitory waking reaction which in itself increases the output towards the muscles and the obstruction resumes. In mouse and in cat it has been demonstrated that CB chemoreceptor drive activates the openers of the upper airways preferentially over the air-pumping muscles (Oku et al., 1993; Gauda et al., 1994). Other alterations in the upper airways muscles have been described (Dempsey et al., 1996). Once blood gases are restored, another cycle of obstruction-apnoea is generated and it repeats to frequencies that might reach >30 apnoeas/h.

According to several authors, the daily activation of the CB chemoreceptors ends up producing a sensitization of the CB chemoreceptors (Gonzalez et al., 1994) with a sustained increased drive from the CB chemoreceptors which produces a permanent activation of the sympathetics which in turn appear to be at the core of the cardiovascular pathology associated to OSAS (Lattimore et al., 2003; Peng et al., 2003; Peng and Prabhakar, 2004; Kumar et al., 2006). The state of anxiety generated by poor sleep can contribute to the increased sympathetic tone, which in addition to cardiovascular diseases seems to be linked to the high prevalence of type II diabetes (Boethel CD, 2002; Chasens et al., 2003). In addition, with each episode of OSA there is fight of inspiratory muscles (a big respiratory effort) aiming to defeat the obstruction. As result, during each apnoeic episode thoracic pressure becomes intensely negative causing many haemodynamic changes (facilitation of venous return, engorgement of pulmonary vessels with blood and a decrease in cardiac output).

Animal models of OSA are limited to the generation of IH mimicking that encountered in patients. Seldom the complete picture of blood gases alterations are mimicked by addition of CO_2 to the hypoxic picture. As a consequence experimentally we produce a hypocapnic hypoxia while the clinical picture is that of hypercapnic hypoxia. In addition, animal models do not mimic the cardiovascular modifications encountered in each apnoeic episode in patients.

2 Material and Methods

2.1 Animals and Anesthesia

Male adult Wistar rats (270–310 g body weight) were housed four per cage under controlled conditions of constant temperature, humidity and a stationary light-dark cycle with free access to standard rat solid diet (A04, Panlab SL, Barcelona, Spain) and drinking water. Rats were weighed at the start and the end of the experiments and no differences were observed in weight gain between control and experimental animals. All experimental protocols, except for the intermittent hypoxia exposure, were performed in anesthetized rats (60 mg/Kg body weight sodium pentobarbital or 133 mg/kg body weight ketamine). Experiments were performed in the morning (from 8:30 to 11 a.m.) and the rats were eating and drinking ad libitum until brought to the laboratory. The Institutional Committee of the University of Valladolid for Animal Care and Use approved the protocols.

2.2 Exposure to Intermittent Hypoxia

Four rats were housed in special transparent methacrylate chambers that shield hermetically. Each chamber of 16 l of volume can hold 4 rats with food and water. Each chamber has an inlet for gas entry and in the front of the inlet, 3 cm separated from the wall of the chamber, there is an incomplete wall that breaks the gas jet and allows a even steady smooth and low velocity flow of gas in the rats room . At the other end of the chamber there are two outlets for the exit of gas of the chamber. In one of them there is an O_2 meter to continuously monitor the gas flowing out of the chamber. The desired gas flow into the chamber from gas tanks connected in cascade with stainless steel tubing and intercalated manometers to assure several days of supply of the desired gases. The entry of the different gas mixtures into the chamber is controled by electrovalves, commanded by a microprocessor that allows the determination of time and duration of entry of the desired gas. The electrovalve system is provided with a battery which assures the functioning of the entire system in case of a failure of the electrical power. In preliminary experiments without rats in the cages the parameters of gas flow were adjusted to achieve at the outlet of the chambers. With our system it is possible to achieve any desired pattern of intermittent hypoxia. In present experiments the pattern of intermittent hypoxia achieved is shown in Fig. 1. We exposed the animals to intermittent hypoxia (10–20% O_2) for 8 and 15 days, 8 h a day from 8 a.m. to 4 p.m (30 episodes/h)

2.3 Blood Gases Measurements

One femoral artery of sodium pentobarbitone anaestetized rats was cannulated and the rats were introduced into the exposure cages one per cage and the catheter was driven out the cages through a hole in the cage cover. This allowed to anaerobically

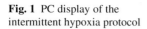
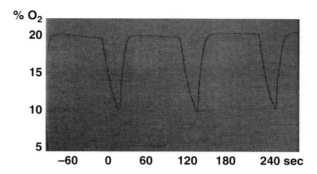

Fig. 1 PC display of the intermittent hypoxia protocol

sampling blood at desired times while the rats remained anaesthetized and spontaneously breathing. Blood gases tensions were immediately measured. Blood sampling started after the animals have been exposed to intermittent hypoxia for 2 h.

2.4 Plasma Catecholamine and Arterial Pressure Measurements

On days 9 and 16 rats were anaesthetized with ketamine, tracheostomized and a carotid artery cannulated to withdraw blood and to measure blood pressure.

Citrated-blood was centrifuged at $1000 \times g$ for 5 min at room temperature (R.T.). Supernatant was transfered to tubes containing 60 mg/ml sodium metabisulfite and frozen at $-80°C$ until the analysis of CA was performed. An aliquot (0.5 ml) was absorbed in Waters Oasis-HLB cartridges (Waters, Milford, MA, USA) and eluted with 0.5 ml of citric acid (26.7 mM) and 2.5% methanol (pH = 2.9) for HPLC-ED analysis, as described by Raggi et al. (1999). 50 µl of the eluates were injected into the HPLC system through a 100 µl loop. The HPLC-ED system was composed of a Milton Roy CM 400 pump (Milton Roy USA; Pennsylvania, USA), a 5 µm Gemini C18 110A column, a Rheodyne 9725 injector (Rheodyne LLC; California, USA) and an electrochemical detector BAS LC-4A (BAS; Indiana, USA). The potential was set at 0.65 V and at a sensitivity of 1–5 nA. The signal was fed to an analogue-to-digital converter controlled by Peak Simple Chromatography System software (Buck Scientific, East Northwalk, CT, USA). The mobile phase was a mixture of NaH_2PO_4 (25 mM) and MeOH (6%) pH (3.6). Identification and quantification of endogenous CA in plasma were performed with external standards (Norepinephrine, Epinephrine, 3, 4-dihydroxyphenylacetic acid (DOPAC), serotonin and Dopamine) previously injected in the HPLC-ED.

Tracheostomized rats were ventilated with a normal mixture of air (20% O_2, 5% CO_2 and 75% N_2) by pump (CL Palmer; London , UK) at a frequency of 40 cycles/min and a positive expiratory pressure of 2 cm H_2O. Arterial pressure was recorded with a catheter connected to a pressure transducer. (Power Lab 16SP; ADI Instruments; Castle Hill, Australia) that sent the signal to the computer for

3 Results

3.1 Arterial Blood Gases

The mean arterial blood gases measured at 20 s intervals during successive episodes of IH are shown in Fig. 2. When the animal cages were fluxed with air in the periods immediately prior to hypoxic episode arterial PO_2 averaged 97 ± 10 mmHg. The lowest value of PO_2 registered during the hypoxic episodes was 56.9 ± 2.5 mmHg. In all likelihood the real level of hypoxia attained was a little more intense because the nadir of PO_2 in the chamber lasted a few seconds and the withdrawal of blood took 20 seconds. PO_2 in blood samples withdrawn immediately after the hypoxic episodes was above 110 mmHg. HBO_2 prior to hypoxic episodes was $96 \pm 0.5\%$, it reached a lowest value of $86.3 \pm 1.6\%$ and rose to 98.5% immediately after the hypoxic episodes. Arterial PCO_2 dropped to 27.1 ± 2.1 at the nadir of the hypoxia and in the period immediately posthypoxia it remained below the control PCO_2 of 38.5 mmHg to recover before the onset of the next episode of hypoxia. The changes in PCO_2 clearly evidence the hyperventilation triggered by hypoxia and are opposite to those encountered in patients with OSA (Chin et al., 1997).

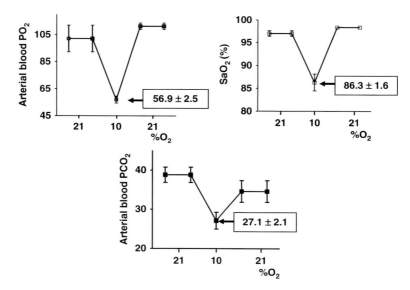

Fig. 2 Arterial blood gases and HBO_2 saturation in episodes of intermittent hypoxia. (10% O_2) Data are means $\pm$ SEM of six individual data obtained from 4 animals

3.2 Plasma Catecholamine Levels and Mean Arterial Blood Pressure

The rationale to measure plasma CA derives from the presumed increased sympathetic tone encountered in OSA patients. Increased sympathetic tone should promote CA secretion from adrenal medulla (norepinephrine and epinephrine) and increase the outflow of NE from the sympathetic endings (Goldstein et al., 2003). Table 1 shows the findings in our group of animals.

From Table 1 it is evident that plasma epinephrine levels increased very significantly in the animals exposed to IH for 8 and 15 days. Norepinephrine levels showed a more gradual time dependent increase with differences between control and experimental animals reaching a statistically significant difference in the group exposed for 15 days. Consistent with this increase in plasma CA levels mean arterial blood pressure increased in both experimental groups.

Table 1 Plasma CA levels in control and intermittent hypoxic animals

Experimental conditions	Control	IH 8 days	IH 15 days
Plasma Epinephrine (pmole/ml)	13.77±2.43 n=17	40.64±6.21 n=8**	37.54±7.81 n=11**
Plasma Norepinephrine (pmole/ml)	15.73±3.59 n=18	18.66±5.27 n=9	32.73±5.77 n=12*
Mean arterial blood pressure (mmHg)	97.30±7.20 n=18	137.70±10.0 n=10*	125.60±11.70 n=11

Means ± SEM of 11 to sixteen individual data. + p = 0.058; *p<0.05; **p<0.01.

4 Discussion

Our findings demonstrate that our home-built intermittent hypoxia system generates adequate patterns of hypoxia allowing to measure the real PO_2 and PCO_2 in blood and to study the animals after different periods of exposure. Findings also evidence that intermittent hypoxia generates a moderate hypertension, as well as concurrent modifications in plasma CA levels. Thus our data validate our model of intermittent hypoxia.

Experimental animals subjected to episodes of intermittent hypoxia represent the most frequently used model of the OSAS. As a model it is imperfect because it does not mimic the hypercapnia associated to each episode of hypoxia in OSAS (although it is easily corrected) and it does not mimic the hemodynamic changes produced by the ventilatory effort directed to overcome the obstructions. Yet, it appears that out of the tree components seen in patients, intermittent hypoxia, hypercapnia and hemodynamic changes, the intermittent hypoxia is pathophysiologically the most important because by itself generates hypertension in experimental animals (Allahdadi et al., 2008; Zoccal et al., 2008; Troncoso Brindeiro et al., 2007; Min Lin et al., 2007) although recently described models of intermittent hypoxia in healthy humans no changes in arterial blood pressure have been noticed (Foster et al., 2005;

Table 2 Some commonly used models of intermittent hypoxia

Author	Lowest O_2(%)	Duration of hypoxia (s)	Cycles of hypoxia/h	Hours/ day	Duration of the IH (days)	Species
Allahdadi et al. (2008)	5*	90	20	7	14	Rat
Zoccal et al. (2008)	6	40	6.6	8	10	Rat
Troncoso et al. (2007)	5*	90	20	7	14	Rat
Lin et al. (2007)	5.7	360	6	12	90	Mice
Prabhakar et al. (2005)	5	15	9	8	10	Rat
Fletcher et al. (2002)	3–5	30	60	7	35	Rat
Sica et al. (2000)	5–7	30	60	8	30	Rat
Kanagy et al. (2001)	5*	90	20	8	11	Rat
Hinojosa-Laborde and Mifflin (2005)	10	180	10	8	7	Rat
Lai et al. (2006)	2–5	30	48	6	30	Rat

Qi Fu et al., 2007). It should be noted however that these human models of IH do not reproduce either the intensity of hypoxia or the frequency of hypoxic episode encountered amongst OSS patients. In this context we want to point out that the dispersion of animal models of intermittent hypoxia is regrettably wide (Table 2), and therefore it can explain differences in results reported by different authors

Acknowledgments We are grateful to Mª Llanos Bravo for technical assistance. This work was supported by grants BFU2007-61848 (MEC, Spain), CIBER CB06/06/0050 (FISS-ICiii) and by JCyL grant: GR242.

References

Allahdadi KJ, Cherng TW, Pai H, Silva AQ, Walker BR, Nelin LD, Kanagy NL. 2008. Endothelin type A receptor antagonist normalizes blood pressure in rats exposed to eucapnic intermittent hypoxia. *Am J Physiol Heart Circ Physiol.* 295:H434–40.
Boethel CD. 2002. Sleep and the endocrine system: new associations to old diseases. *Curr Opin Pulm Med.* 8:502–5.
Chasens ER, Weaver TE, Umlauf MG. 2003. Insulin resistance and obstructive sleep apnea: is increased sympathetic stimulation the link? *Biol Res Nurs.* 5:87–96.
Chin K, Hirai M, Kuriyama T, Fukui M, Kuno K, Sagawa Y, Ohi M. 1997. Changes in the arterial PCO2 during a single night's sleep in patients with obstructive sleep apnea. *Intern Med.* 36:454–60.
Dempsey JA, Smith CA, Harms CA, Chow C, Saupe KW. 1996. Sleep-induced breathing instability. University of Wisconsin-Madison Sleep and respiration Research Group. *Sleep.* 19:236–47.
Dempsey JA. 1997. Sleep apnea causes daytime hypertension. *J Clin Invest.* 99:1–2.
Eisele DW, Schwartz AR, Smith PL. 2003. Tongue neuromuscular and direct hypoglossal nerve stimulation for obstructive sleep apnea. *Otolaryngol Clin North Am.* 36:501–10.
Fletcher EC, Lesske J, Behm R, Miller III, CC, Staus H, Unger T. 1992. Carotid chemoreceptors, systemic blood pressure and chronic episodic hypoxia mimicking sleep apnea. *J. Appl. Physiol.* 72:1978–1984.
Foster GE, McKenzie DC, Milsom WK, Sheel AW. 2005. Effects of two protocols of intermittent hypoxia on human ventilatory, cardiovascular and cerebral responses to hypoxia. *J Physiol* Sep 1;567:689–99.

Fu Q, Townsend NE, Shiller SM, Martini ER, Okazaki K, Shibata S, Truijens MJ, Rodríguez FA, Gore CJ, Stray-Gundersen J, Levine BD. 2007. Intermittent hypobaric hypoxia exposure does not cause sustained alterations in autonomic control of blood pressure in young athletes. *Am J Physiol Regul Integr Comp Physiol.* 292:R1977–84.

Gauda EB, Carroll TP, Schwartz AR, Smith PL, Fitzgerald RS. 1994. Mechano- and chemoreceptor modulation of respiratory muscles in response to upper airway negative pressure. *J Appl Physiol.* 76:2656–62.

Goldstein DS, Eisenhofer G, Kopin IJ. 2003. Sources and significance of plasma levels of catechols and their metabolites in humans. *J Pharmacol Exp Ther.* 305:800–11.

Gonzalez C, Almaraz L, Obeso A, Rigual R. 1994. Carotid body chemoreceptors: from natural stimuli to sensory discharges. *Physiol Rev.*74:829–98.

Hinojosa-Laborde C, Mifflin SW. 2005. Sex differences in blood pressure response to intermittent hypoxia in rats. *Hypertension.* 46:1016–21.

Kanagy NL, Walker BR, Nelin LD. 2001 Role of endothelin in intermittent hypoxia-induced hypertension. *Hypertension.*37:511–5

Kubin L, Davies RO, Pack AI. 1998. Control of upper airway motoneurons during REM Sleep. *News Physiol Sci.* Apr;13:91–97.

Kumar GK, Rai V, Sharma SD, Ramakrishnan DP, Peng YJ, Souvannakitti D, Prabhakar NR. 2006. Chronic intermittent hypoxia induces hypoxia-evoked catecholamine efflux in adult rat adrenal medulla via oxidative stress. *J Physiol.* 575:229–39.

Lai CJ, Yang CC, Hsu YY, Lin YN, Kuo TB. 2006. Enhanced sympathetic outflow and decreased baroreflex sensitivity are associated with intermittent hypoxia-induced systemic hypertension in conscious rats. *J Appl Physiol.* 100:1974–82.

Lattimore JD, Celermajer DS, Wilcox I. 2003. Obstructive sleep apnea and cardiovascular disease. *J Am Coll Cardiol.* 41(9):1429–37.

Lin M, Liu R, Gozal D, Wead WB, Chapleau MW, Wurster R, Cheng ZJ. 2007. Chronic intermittent hypoxia impairs baroreflex control of heart rate but enhances heart rate responses to vagal efferent stimulation in anesthetized mice. *Am J Physiol Heart Circ Physiol.* 293:H997–1006.

Morrison KJ, Gao Y, Vanhoutte PM. 1993. Beta-adrenoceptors and the epithelial layer in airways. *Life Sci.* 52(26):2123–30.

Oku Y, Bruce EN, Richmonds CR, Hudgel DW. 1993. The carotid body in the motorneuron response to protriptyline. *Respir Physiol.* 93:41–9.

Peng YJ, Overholt JL, Kline D, Kumar GK, Prabhakar NR. 2003. Induction of sensory long-term facilitation in the carotid body by intermittent hypoxia: implications for recurrent apneas. *Proc Natl Acad Sci U S A.* 100:10073–8.

Peng YJ, Prabhakar NR. 2004. Effect of two paradigms of chronic intermittent hypoxia on carotid body sensory activity. *J Appl Physiol.* 96:1236–42.

Prabhakar NR, Peng YJ, Jacono FJ, Kumar GK, Dick TE. 2005. Cardiovascular alterations by chronic intermittent hypoxia: importance of carotid body chemoreflexes. *Clin Exp Pharmacol Physiol.* 32, 447–449.

Raggi MA, Sabbioni C, Casamenti G, Gerra G, Calonghi N, Masotti L. 1999. Determination of catecholamines in human plasma by high-performance liquid chromatography with electrochemical detection. *J Chromatogr B Biomed Sci Appl.* 730:201–11.

Ryan CM, Bradley TD. 2005. Pathogenesis of obstructive sleep apnea. *J Appl Physiol.* 99:2440–50.

Sica AL, Greenberg HE, Ruggiero DA, Scharf SM. 2000. Chronic-intermittent hypoxia: a model of sympathetic activation in the rat. *Respir Physiol.* 121:173–84.

Troncoso Brindeiro CM, da Silva AQ, Allahdadi KJ, Youngblood V, Kanagy NL. 2007. Reactive oxygen species contribute to sleep apnea-induced hypertension in rats. *Am J Physiol Heart Circ Physiol.* 293:H2971–6.

Van Lunteren E. 1997. Upper airway effects on breathing. In *The Lung: Scientific Foundations*, Crystal RG, West JB, Weibel ER, Barnes PJ, eds. Philadelphia, Lippincott-Raven. pp. 2073–2084.

Yaggi HK, Concato J, Kernan WN, Lichtman JH, Brass LM, Mohsenin V. 2005. Obstructive sleep apnea as a risk factor for stroke and death. *N Engl J Med.* 353:2034–41.

Zoccal DB, Simms AE, Bonagamba LG, Braga VA, Pickering AE, Paton JF, Machado BH. 2008. Increased sympathetic outflow in juvenile rats submitted to chronic intermittent hypoxia correlates with enhanced expiratory activity. *J Physiol.* 586:3253–65.

Cardioventilatory Acclimatization Induced by Chronic Intermittent Hypoxia

R. Iturriaga, S. Rey, R. Del Rio, E.A. Moya and J. Alcayaga

Abstract It has been proposed that chronic intermittent hypoxia (CIH) contributes to generate hypertension in patients with obstructive sleep apnea syndrome and animal models, due to an enhanced sympathetic outflow. A possible contributing mechanism to the CIH-induced hypertension is a potentiation of carotid body (CB) chemosensory responses to hypoxia, but early changes that precede the CIH-induced hypertension are not completely known. Since the variability of heart rate (HRV) has been used as an index of autonomic influences on cardiovascular system, we studied the effects of short and long-term CIH exposure on HRV in animals with or without hypertension. In cats exposed to CIH (PO_2 ~75 Torr, 10 times/hr during 8 hr) for 4 days, the ventilatory response to acute hypoxia was potentiated, the arterial pressure remained unchanged, but the HRV power spectrum showed a shift towards the low frequency band. Exposure of rats to CIH (PO_2 ~37.5 Torr, 12 times/hr during 8 hr) for 12 days enhanced the ventilatory response to acute hypoxia, but did not increase the arterial pressure. After 21 days of CIH, we found a significant increase of arterial pressure and a shift of the HRV power spectrum towards the low frequency band. Thus, our results support the idea that hypertension induced by long-term CIH was preceded by alterations in the autonomic balance of HRV, associated with an enhance CB chemoreflex sensitivity to hypoxia. Therefore, few days of CIH are enough to enhance the CB reactivity to hypoxia, which contribute to the augmented ventilatory response to hypoxia, and to the early alterations in the autonomic balance of HRV.

Keywords Chronic intermittent hypoxia · Carotid body · Heart rate variability · Cardioventilatory acclimatization

R. Iturriaga (✉)
Laboratorio de Neurobiología, Facultad de Ciencias Biológicas, P. Universidad Católica de Chile, Santiago, Chile
e-mail: riturriaga@bio.puc.cl

1 Introduction

Chronic intermittent hypoxia (CIH), characterized by cyclic hypoxic episodes of short duration followed by normoxia is a feature of obstructive sleep apnea (OSA). The OSA syndrome has been associated with high cardiovascular risk and hypertension (Quan and Gersh 2004). Patients with recently diagnosed OSA present large cardioventilatory responses to acute hypoxia attributed to an enhanced carotid body (CB) sensitivity to hypoxia (Narkiewicz et al. 1998a, 1999). Experiments performed in rats and cats have shown that CIH enhances hypoxic ventilatory responses (Ling et al. 2001, Rey et al. 2004) and produces long-term facilitation of respiratory motor activity (Peng et al. 2003). Moreover, Peng et al. (2001, 2003) found that basal CB discharges and chemosensory responses to acute hypoxia were higher in rats exposed to a CIH pattern applied for 8 hr during 10 days. Using a similar protocol, we studied the early effects of CIH on cat CB chemosensory responses to hypoxia (Rey et al. 2004). We found that cats exposed to CIH ($PO_2 \sim 75$ Torr followed by normoxia during 8 hr) for 4 days showed enhanced CB chemosensory and ventilatory responses to acute hypoxia. Thus, exposure to CIH increases both CB chemosensory and ventilatory responses to acute hypoxia, a phenomenon similar to the ventilatory acclimatization induced by sustained hypoxia.

Patients with OSA and rats exposed to CIH develop systemic hypertension (Fletcher et al. 1992; Somers et al. 1995). The hypertension has been attributed to an increased sympathetic outflow due to the repetitive hypoxic CB stimulation. Fletcher et al. (1992) reported an elevated arterial pressure in rats after 35 days of CIH exposure. In cats exposed to CIH for 4 days, we found that arterial blood pressure did not increase, but the power spectrum of heart rate variability (HRV) was modified (Rey et al. 2004). The spectral analysis of HRV has two major components defined as a low-frequency (LF) band, considered related to combined sympathetic and parasympathetic influences, and a high-frequency (HF) band related to vagal influences and respiratory sinus arrhythmia (Task Force 1996). The LF/HF ratio is believed as an index of the sympathovagal balance on heart rate (Task Force 1996).

Thus, the available data suggest that CIH-induced hypertension is linked to autonomic regulation failure, but it is not well known if changes of HRV may precede the onset of hypertension. Thus, we compared effects of short and long-term intermittent hypoxia on HRV in cats and rats. Cats exposed to short-term CIH showed enhanced reflex ventilatory responses to hypoxia, but not hypertension. Rats exposed to long-term CIH developed enhanced ventilatory reflex responses to acute hypoxia, as well a significant increase in blood arterial pressure.

2 Methods

2.1 Models of CIH

Experiments were performed on male adult cats (2.0–4.5 kg) and male Sprague-Dawley rats (200–250 g). The experimental protocol was approved by the Bioethical Committee of the Facultad de Ciencias Biológicas of the P. Universidad Católica de

Chile. Cats were housed in a cylindrical chamber (35.2 L), which was alternatively flushed with 100% N_2 and compressed air using timed solenoid valves. This gas alternation produced a see-saw pattern of PO_2, which dropped to $\sim$ 75 Torr and then returned to normoxia. The intermittent hypoxic pattern was repeated 10 times/hr during 8 hr/day for 4 days. Rats were housed in individual chambers (2.2L) and exposed a similar CIH pattern (PO_2 $\sim$37.5 Torr, 12 times/hr repeated during 8 hr) for 21 days. As a control group, cats and rats were subjected to a sham pattern of gas alternation by replacing N_2 with compressed air. The PO_2 in the chamber was monitored using an oxygen analyzer. Acute experiments were performed the morning of the day after the end of hypoxic exposures.

2.2 Recordings of Ventilatory and Cardiovascular Responses

Cats were anaesthetized with sodium pentobarbitone (60 mg/kg, i.p.), additional doses (8–12 mg, i.v.) were given when necessary to maintain a level of surgical anaesthesia. Rats were anesthetized with sodium pentobarbitone (40 mg/kg, i.p.). The trachea of the animals was cannulated for recording the airflow signal with appropriate pneumotachographs. Arterial blood pressure was recorded from one femoral (cat) or one carotid (rat) arteries through a cannula filled with 50 IU/ml of heparin in saline solution, and connected to a pressure transducer. The ECG was continuously recorded using the II lead. All the physiological variables were fed to an analog-digital system (PowerLAB/8SP, AD Instruments, Australia), connected to a PC computer. Minute inspiratory volume (V_I), tidal volume and respiratory frequency were digitally obtained from the airflow signal. Heart rate was obtained from the ECG, and mean arterial pressure (Pa) from the arterial pressure signal. The physiological data was analyzed using the LabChart 6.1 Pro software (AD Instruments, Australia). To assess the effects of CIH on the sensitivity and reactivity of peripheral chemoreceptor reflexes, we studied the ventilatory responses elicited by several isocapnic levels of PO_2 ($\sim$ 20 to 740 Torr) maintained during 1–2 min in cats and 30 s in rats. To determine the time-course of the onset of the hypertension in rats, we measured the systolic pressure in the tail of awake rats every 3 days from the beginning of CIH exposure. The ventilatory response to hypoxia (100% N_2 for 15 s) was measured in awake rats using a plethysmograph.

2.3 Spectral Analysis of Heart Rate Variability

Under anaesthesia the ECG was recorded for 5 min at 2 kHz. After ECG acquisition, the R-R intervals were measured. The HRV was analysed with software developed by The Biomedical Signal Analysis Group, Department of Applied Physics, University of Kuopio, Finland (Tarvainen et al. 2002) or with the HRV module of the LabChart 6.1 Pro software (AD Instuments, Australia). To measure HRV, the R-R interval data was analysed with a Fast Fourier Transform (FFT) algorithm. Spectral data obtained from cat R-R intervals was analyzed according to the following frequency bands: very-low frequency (VLF, DC-0.025 Hz), low frequency (LF, 0.025–0.140 Hz), and high frequency (HF, 0.14–0.6 Hz). Rats

R-R intervals variability was analyzed using the following frequency bands: very-low frequency (VLF, DC-0.004 Hz), low frequency (LF, 0.004–0.6 Hz), and high frequency (HF, 0.6–2.4 Hz).

3 Results

3.1 Effects CIH on Basal Blood Arterial Pressure

Exposure of cats to CIH for 4 days did not increase basal arterial pressure (Fig. 1A). In awake rats, we found that CIH produced a progressive increase of systolic blood pressure measured in the tail, but only at 20 days of CIH, arterial pressure was significantly higher (data not shown). This result was confirmed in anesthetized rats (Fig. 1B) exposed to CIH for 21 days (P<0.01, Student t-Test).

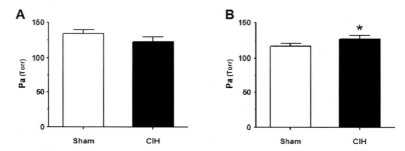

Fig. 1 Effects of CIH on mean arterial pressure (Pa) in cats exposed to CIH for 4 days (A) and rats exposed to CIH for 21 days (B). Data are means from 8 CIH and 12 sham cats and 19 CIH and 21 sham rats. *p<0.01 (Student t-test)

3.2 Effect of CIH on Ventilatory Response to Acute Hypoxia

Cats exposed to CIH for 4 days showed enhanced reflex ventilatory responses (Fig. 2A). In awake rats, we found a significant increase in the ventilatory response to acute hypoxia (100% N_2 for 15 s) after 12 days of CIH exposure. Since arterial blood pressure increased after 20 days of CIH, we measured the ventilatory response (Fig. 2B) at the day 21 and found a significant increase in hypoxic response (P<0.01, Student t-Test).

3.3 Effect of CIH on Heart Rate Variability

Figure 3A illustrates the distribution of the power spectral components of HRV normalized to the total power in CIH and sham cats. In the CIH cats, the power spectral density analysis showed a predominance of the low frequency component over the high frequency, while the opposite situation is observed in sham cats. As

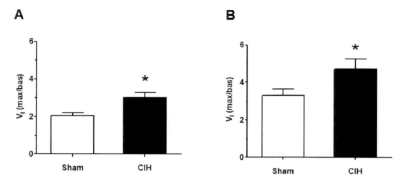

Fig. 2 Ventilatory responses elicited by 5% O_2 in cats exposed to CIH for 4 days (A) and rats exposed to CIH for 21 days. *p<0.01, Student-t-test (4 CIH and 4 sham cats; 13 CIH and 11 sham rats)

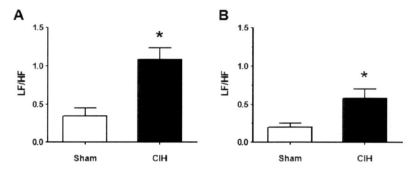

Fig. 3 Effects of CIH on LF/HF ratio in cats exposed to CIH for 4 days (A) and rats exposed for 21 days. *p<0.01, Student-t-test. (8 CIH and 12 sham cats, 11 CIH and 11 Sham rats)

result of the change in the distribution of the relative spectral components of HRV, cats exposed to CIH showed a significantly higher LF/HF ratio than sham cats (P < 0.01, Student t-Test). In rats, exposed to CHI for 21 days, we found a similar shift of the power spectral components of the HRV (Fig. 3B).

4 Discussion

Long-term exposure to CIH in rats induces facilitation of ventilatory responses to hypoxia, hypertension and modified HRV. In cats, short-term CIH potentiates the ventilatory responses to acute hypoxia, and change HRV. However, these early alterations occurred without any increase in arterial blood pressure. Thus, our results suggest that cyclic hypoxic stimulation of the CB by short-term CIH produces selective alterations of HRV and enhances CB chemoreception to hypoxia in normotensive animals.

Chronic intermittent hypoxia has been associated with an enhanced ventilatory response to hypoxia in OSA patients and animal models. Narkiewicz et al. (1998a, 1999) found that ventilatory, sympathetic and cardiovascular responses to acute hypoxia were enhanced in patients with recently diagnosed OSA. Studies performed in rats have shown different effects of CIH on the hypoxic ventilatory response, which appear to be dependent on the pattern and severity of CIH. Ling et al. (2001) reported that the phrenic response to hypoxia was enhanced in rats exposed to 5 min of 10% O_2, followed by 5 min of air, 12 hr/night during 7 nights. On the contrary, Greenberg et al. (1999) did not find any differences in the ventilatory response to hypoxia in rats exposed for 30 s to 7% O_2 followed by 30 s of normoxia, for 8 hr/day during 10 days. Thus, it is likely that effects of CIH on ventilatory response to hypoxia are dependent on the pattern and severity of CIH. Our results show that CIH enhances rat ventilatory responses to hypoxia after 12 days. Thus, our results confirm previous observations of Peng et al. (2001, 2003). They found that CIH enhances CB chemosensory response to acute hypoxia in rats exposed to CIH for 10 days.

Long-term CIH exposure leads to hypertension, which has been attributed to an increased sympathetic outflow due to repetitive CB stimulation (Greenberg et al. 1999; Fletcher et al. 1992). Fletcher et al. (1992) found an elevated arterial pressure in rats after 35 days of CIH exposure, but not after 10 days. On the contrary, Sica et al. (2000) and Peng et al. (2003) found that systolic arterial pressure increases after 7 days of CIH. Our results indicate that the time required for increasing mean arterial pressure in rats was 20 days. The spectral analysis of the R-R interval data indicate that autonomic modulation of HRV is early modified by CIH in cats. Indeed, the LF/HF ratio was significantly higher in CIH cats, although arterial pressure did not increase. The spectral analysis shows a clear effect of CIH that resembles what is observed in OSA patients. In fact, patients with OSA show an increased LF/HF ratio, with a relative predominance of the LF component indicating a predominance of sympathetic modulation (Narkiewicz et al. 1998b).

In summary, our results support the idea that few days of CIH are enough to enhance the cat CB reactivity to hypoxia, which may contribute to the augmented ventilatory response to hypoxia, and to early alterations in the autonomic balance of HRV.

Acknowledgments This work was supported by grant 1070854 from the National Fund for Scientific and Technological Development of Chile (FONDECYT).

References

Fletcher, E.C., Lesske, J., Behm, R., Miller, C.C., Stauss, H., & Unger, T. (1992). Carotid chemoreceptors, systemic blood pressure, and chronic episodic hypoxia mimicking sleep apnoea. *J Appl Physiol* 72, 1978–1984.

Greenberg, H.E., Sica, A., Batson, D., & Scharf S.M. (1999). Chronic intermittent hypoxia increases sympathetic responsiveness to hypoxia and hypercapnia. *J Appl Physiol* 86, 298–305.

Ling, L., Fuller, D.D., Bach, K.B., Kinkead, R., Olson, E.B., & Mitchell, G.S. (2001). Chronic intermittent hypoxia elicits serotonin-dependent plasticity in the central neural control of breathing. *J Neurosci* 21, 5381–5388.

Narkiewicz, K., van de Borne, P. J., Montano, N., Dyken, M. E., Phillips, B. G., & Somers, V. K. (1998a). Contribution of tonic chemoreflex activation to sympathetic activity and blood pressure in patients with obstructive sleep apnea. *Circulation* 97, 943–945.

Narkiewicz, K., Montano, N., Cogliati, C., Van de Borne, P.J., Dyken. M.E., & Somers, V.K. (1998b). Altered cardiovascular variability in obstructive sleep apnea. *Circulation* 98, 1071–1077.

Narkiewicz, K., Van de Borne P.J., Pesek, C.A., Dyken, M.E., Montano, N., & Somers, V.K. (1999). Selective potentiation of peripheral chemoreflex sensitivity in obstructive sleep apnea. *Circulation* 99, 1183–1189.

Peng, Y.J. ,Kline, D.D., Dick, T.E., & Prabhakar, N. R. (2001) Chronic intermittent hypoxia enhances carotid body chemoreceptor responses to low oxygen. *Adv Exp Med Biol* 499: 33–38.

Peng, Y.J., Overholt . l.J., Kline, D., Kumar, G.K., & Prabhakar, NR. (2003) Induction of sensory long term facilitation in the carotid body by intermittent hypoxia: Implications for recurrent apneas. *Proc Natl Acad Sci USA*. 100, 10073–10078

Quan, S. F., & Gersh, B. J. (2004) Cardiovascular consequences of sleep-disordered breathing: past, present and future: report of a workshop from the National Center on Sleep Disorders Research and the National Heart, Lung, and Blood Institute. *Circulation* 109, 951–957.

Rey, S., Del Rio, R., Alcayaga, J., & Iturriaga, R. (2004) Chronic intermittent hypoxia enhances cat chemosensory and ventilatory responses to hypoxia. *J Physiol* 560, 577–586.

Sica, A.L., Greenberg, H.E., Ruggiero, D.A., & Scharf, S.M. (2000). Chronic-intermittent hypoxia: a model of sympathetic activation in the rat. *Respir Physiol* 121, 173–184.

Somers, V.K., Dyken, M.E., Clary, M.P., & Abboud, F.M. (1995). Sympathetic neural mechanisms in obstructive sleep apnea. *J Clin Invest* 96: 1897–1904

Tarvainen, M.P., Ranta-Aho P.O., & Karjalainen, P.A. (2002). An advanced detrending method with application to HRV analysis. *IEEE Trans Biomed Eng* 49, 172–175.

Task Force of the European Society of Cardiology and the North American Society of Pacing and Electrophysiology. Heart rate variability. Standards of measurement, physiological interpretation, and clinical use. *Eur Heart J* 17: 354–381, 1996.

Ventilatory Drive Is Enhanced in Male and Female Rats Following Chronic Intermittent Hypoxia

D. Edge, J.R. Skelly, A. Bradford and K.D. O'Halloran

Abstract Obstructive sleep apnoea is characterized by chronic intermittent hypoxia (CIH) due to recurrent apnoea. We have developed a rat model of CIH, which shows evidence of impaired respiratory muscle function. In this study, we wished to characterize the ventilatory effects of CIH in conscious male and female animals. Adult male (n=14) and female (n=8) Wistar rats were used. Animals were placed in chambers daily for 8 h with free access to food and water. The gas supply to one half of the chambers alternated between air and nitrogen every 90 s, for 8 h per day, reducing ambient oxygen concentration in the chambers to 5% at the nadir (intermittent hypoxia; n=7 male, n=4 female). Air supplying the other chambers was switched every 90 s to air from a separate source, at the same flow rates, and animals in these chambers served as controls (n=7 male, n=4 female). Ventilatory measurements were made in conscious animals (typically sleeping) after 10 days using whole-body plethysmography. Normoxic ventilation was increased in both male and female CIH-treated rats compared to controls but this did not achieve statistical significance. However, ventilatory drive was increased in CIH-treated rats of both sexes as evidenced by significant increases in mean and peak inspiratory flow. Ventilatory responses to acute hypoxia (F_IO_2 = 0.10; 6 min) and hyperoxic hypercapnia (F_ICO_2 = 0.05; 6 min) were unaffected by CIH treatment in male and female rats (P>0.05, ANOVA). We conclude that CIH increases respiratory drive in adult rats. We speculate that this represents a form of neural plasticity that may compensate for respiratory muscle impairment that occurs in this animal model.

Keywords Chronic intermittent hypoxia · Hypoxic ventilatory response · Hypercapnic ventilatory response · Plethysmography · Respiratory plasticity · Sleep apnoea

---

K.D. O'Halloran (✉)
UCD School of Medicine and Medical Science, University College Dublin, Dublin 4, Ireland
e-mail: ken.ohalloran@ucd.ie

1 Introduction

Plasticity, a fundamental property of neural systems, may be defined as a persistent change in a neural control system based on prior experience. Compelling evidence has accumulated in recent years demonstrating that the respiratory control system exhibits diverse and considerable plasticity. In many cases, plasticity in respiratory control may be beneficial for respiratory 'adaptation'. It is also recognized however that potentially pathological forms of plasticity may present depending upon, amongst other things, the nature, intensity, pattern and duration of the stimulus and the timing of the perturbation across the life cycle. There is now evidence to suggest that long-term exposure to intermittent hypoxia elicits considerable 'maladaptive' changes within the nervous system, including the neural control systems that govern respiration. For example, chronic intermittent hypoxia (CIH) has been shown to induce neural injury and spatial learning deficits in rats. Indeed, several neuronal populations show altered function following intermittent hypoxic exposure including chemosensitive afferents involved in the control of breathing (Peng et al. 2003), central respiratory structures involved in integration of sensory inputs (Reeves et al. 2006a), as well as respiratory motor neurones concerned with the regulation of airway patency (O'Halloran et al. 2002; Veasey et al. 2004; Ray et al. 2007). Thus, a picture is emerging that CIH may alter the control of breathing at multiple levels of the reflex loop. There is a growing literature concerning the effects of CIH on breathing and ventilatory responsiveness with evidence of potentially 'adaptive' and 'maladaptive' consequences for respiratory homeostasis. Recently, we reported respiratory muscle dysfunction following a 9-day CIH protocol in adult rats (Skelly et al. 2008). The major aim of the present study was to examine the effects of this experimental paradigm of CIH on respiratory control and, since sex steroid hormones can influence respiratory plasticity (Behan et al. 2003), a second aim was to explore if sex differences exist in the effects of CIH on breathing.

2 Methods

2.1 Chronic Intermittent Hypoxia

Adult male (n=14) and female (n=8) Wistar rats purchased from Harlan (UK) were used in this study. Animals were housed in a temperature- and humidity-controlled facility, operating on a 12 h light:12 h dark cycle with food and water available ad libitum. For the chronic gas exposures, animals were placed in 4 identical custom-built chambers for 8 h (08.00–16.00h) with free access to food and water. The gas supply to one half of the chambers alternated between air and nitrogen every 90 s, reducing ambient oxygen concentration in the chambers to 5% at the nadir (intermittent hypoxia; n=7 male, n=4 female). Air supplying the other chambers was switched every 90 s to air from a separate source, at the same flow rates, and animals in these chambers served as sham controls (n=7 male, n=4 female). Exposures were carried out in this fashion for 9 days over a 10-day treatment period (i.e. 6d CIH or sham, 1d 'Rest', 3d CIH or sham). Ventilatory measurements were made by the barometric method on the day that followed the final treatment day.

2.2 Whole-Body Plethysmography

Respiratory parameters were continuously acquired in freely-behaving animals using the barometric method. Male and female rats were studied on separate occasions. Animals were placed in chambers (Buxco Europe Ltd.) at least 30 min prior to the start of each protocol. Pressure changes in the chamber due to inspiratory and expiratory temperature changes were measured using a differential pressure transducer. Analog signals were continuously digitized and analyzed on-line using commercially available software (Buxco Europe Ltd). A rejection algorithm was included in the breath-by-breath analysis to exclude motion-induced artifacts. Measurements were compensated for temperature and humidity, which were continuously recorded. Two chambers were used in parallel, allowing simultaneous measurement of respiratory parameters in sham and CIH-treated rats, thereby minimizing the influence of external factors on recorded variables.

2.3 Experimental Protocol

Recordings were made 15–26 h after the termination of the sham or CIH protocol. Tidal volume (V_T), respiratory frequency (f), minute ventilation (V_E), inspiratory time (T_I), expiratory time (T_E), and mean inspiratory flow (V_T/T_I) were computed and stored for subsequent analysis off-line. After a 30–60 min settling down period, respiratory parameters were recorded for 10 min of stable normoxic ventilation. Next, chambers were flushed with a pre-mixed hypercapnic gas mixture (5% CO_2, balance O_2) for 6 min. A second 10 min baseline period was recorded after the animals had recovered from the hypercapnic challenge. We then performed a hypoxic challenge by introducing 10% O_2 to the chambers for a period of 6 min. Oxygen concentration was continuously monitored from a side-port in the inlet tubing to the chambers. Finally, after recovery, a third 10 min baseline normoxic period was recorded. The behaviour of the rats was continuously monitored during the experimental protocol. Recordings were made whilst the rats were inactive and most often asleep (behaviourally-defined) and no differences in behaviour were observed between the two groups. Occasional trials where one of the experimental pair of animals displayed a different behaviour to the other (e.g. active, exploring, grooming, feeding etc.) were excluded and repeated (i.e. both animals exposed to the test gas a second time).

2.4 Data Analysis

All values are shown as means±SE. Ventilatory measures were averaged over the 10 min of stable baseline breathing. There was no statistical difference between the 3 baseline periods recorded and therefore values were pooled to generate a single set of baseline normoxic respiratory parameters in sham and CIH-treated animals. For the hypercapnic and hypoxic challenges, ventilatory measures were averaged on a minute-by-minute basis. Peak and mean values were computed for all animals. Baseline parameters for sham and CIH-treated rats were compared using Student's *t* test. Two-way analysis of variance was employed to compare measures during normoxic, hypoxic and hypercapnic exposures. Differences were evaluated via post-hoc analysis using the Neuman Keuls test. Significance was established at $P<0.05$.

3 Results

3.1 Body Mass, Haematocrit and Heart Mass

For male and female rats, there was no significant difference in body mass between the two groups (sham vs. CIH) at the beginning of the study. However, after 10 days of treatment, CIH-treated rats had a lower body mass compared to sham animals (Table 1) which was statistically significant for female animals. Furthermore, haematocrit and right, but not left, cardiac ventricular mass, were significantly elevated in CIH-treated rats (Table 1).

Table 1 Body mass, haematocrit, right and left cardiac ventricular mass in sham and CIH-treated rats

	Body mass (g)	Haematocrit (%)	Right ventricular mass (mg/100 g)	Left ventricular mass (mg/100 g)
Males				
Sham	271 ± 14	44 ± 1	50 ± 2	213 ± 7
CHI	257 ± 16	56 ± 1*	69 ± 3*	222 ± 10
Females				
Sham	206 ± 4	45 ± 1	55 ± 2	203 ± 4
CIH	189 ± 5*	54 ± 1*	76 ± 2*	211 ± 13

Values are mean±SE. BM = body mass. * indicates significant difference from corresponding sham value (Student's t test, $P<0.05$).

3.2 Baseline Normoxic Ventilation

Normoxic ventilation was increased in CIH-treated rats compared to sham controls but this did not achieve statistical significance. Thus, V_E (ml/min/100 g) was 62.0±3.6 vs. 71.2±4.2 (males) and 69.2±3.7 vs. 76.8±3.1 (females), sham vs. CIH-treated respectively. However, ventilatory drive was increased in CIH-treated rats of both sexes as evidenced by significant increases in mean inspiratory flow (Fig. 1).

3.3 Hypoxic and Hypercapnic Challenges

Mean and peak values for V_E during acute hypoxia and hyperoxic hypercapnia were not different ($P>0.05$, ANOVA) in sham and CIH-treated animals for both male and female groups.

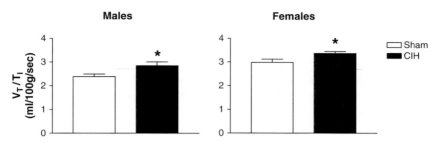

Fig. 1 Mean inspiratory flow (V_T/T_I) was significantly elevated in CIH-treated male (n=7) and female (n=4) rats compared to sham controls (n=7 male; n=4 female). Values are mean±SE. * indicates significant difference from corresponding sham value (Student's *t* test, P<0.05)

4 Discussion

Intermittent hypoxia (IH) is common in respiratory disease and is a central feature of the sleep apnoea syndrome – a prevalent and often debilitating disorder associated with significant cardiovascular, metabolic and neurocognitive dysfunction. Short episodes of IH are known to produce long-term facilitation (LTF) of breathing, a form of central respiratory neuroplasticity defined as a long-lasting increase in breathing that arises following acute exposure to intermittent, but not continuous, hypoxia. The development of animal models of sleep apnoea has allowed an examination of the effects of CIH on respiratory control. We have reported previously that CIH impairs isolated respiratory muscle function, reflex control of upper airway EMG activity and ventilatory responses to moderate hypoxia (O'Halloran et al. 2002; Bradford et al. 2005; O'Halloran et al. 2007; Dunleavy et al. 2008; Skelly et al. 2008). It is also reported that CIH impairs hypoglossal motor neurone excitability (Veasey et al. 2004) and is associated with a decrease in upper airway stability (Ray et al. 2007). Taken together, these findings suggest that CIH – a feature of sleep-disordered breathing due to recurrent apnoea – may be implicated in the on-going pathophysiology of the disorder serving to exacerbate and perpetuate the condition. Thus, characterization of the effects of CIH on ventilatory control, exploration of the underlying mechanisms of hypoxia-induced respiratory plasticity, and the development of intervention strategies designed to ameliorate or prevent the deleterious effects of CIH on breathing may have significant clinical relevance. The main findings of the present study are: (1) CIH elicits plasticity in the respiratory control system with evidence of a significant increase in respiratory 'drive' (V_T/T_I) in conscious animals, (2) Our experimental paradigm of CIH does not affect the acute hypoxic or hypercapnic ventilatory response in adult rats, (3) CIH does not differentially affect breathing in male and female adult rats.

There is a growing interest in the examination of the effects of CIH on breathing because of the relevance to sleep-disordered breathing and other conditions characterized by intermittent hypoxia. CIH elicits 'ventilatory adaptation' (i.e. enhanced normoxic ventilation) in rats (Reeves and Gozal 2006) and mice (Reeves and

Gozal 2004). Furthermore, it enhances LTF in awake rats (McGuire et al. 2003). Conversely, CIH was shown to have no effect on eupnoeic breathing in sleeping dogs (Katayama et al. 2007) and we previously reported no change in baseline normoxic ventilation in rats after 5 weeks of intermittent hypercapnic hypoxia, although the pattern of respiration was different in treated animals (O'Halloran et al. 2007). In the present study, normoxic ventilation was increased in male and female CIH-treated rats compared to sham controls but this was not statistically significant. However, we did observe small, but significant, increases in ventilatory 'drive'. Thus, although there are exceptions, CIH can elicit functional plasticity in the respiratory control system. CIH elicits sensory LTF in the carotid body (Peng et al. 2003), the peripheral sensor of hypoxia, and carotid body chemosensory responses to hypoxia are elevated following CIH (Rey et al. 2004; Pawar et al. 2008). One might therefore expect that the hypoxic ventilatory response (HVR) would be elevated following CIH. Indeed, this is the case in studies of cats (Rey et al. 2004), rats (Peng et al. 2004; Julien et al. 2008) and mice (Reeves and Gozal 2004) following CIH. However, others have reported that the HVR is reduced (Waters and Tinworth 2001; Reeves et al. 2006b; O'Halloran et al. 2007) or unaffected (Katayama et al. 2007; this study) in CIH-treated animals. We observed no difference in ventilation during hypercapnic challenge between sham and CIH rats in this study, whereas an increase in the hypercapnic ventilatory response was observed by others (Katayama et al. 2007). Presumably, differences in the experimental paradigm of CIH are at least partly responsible for the differences in the results of these studies, as well as perhaps age-related and species differences. It is apparent that CIH-induced 'ventilatory adaptation' is an age-dependent phenomenon (Reeves and Gozal 2006). There is increased capacity for IH-induced respiratory plasticity in early life (Reeves and Gozal 2006; Pawar et al. 2008). Early postnatal CIH alters ventilatory control (Julien et al. 2008), with effects that persist for several weeks after treatment (Reeves et al. 2006b). Thus early neonatal life may represent a vulnerable period of development for IH-induced 'maladaptation'. This is of considerable clinical significance as apnoea of prematurity is a well-recognized phenomenon. Adult animals would appear to have reduced capacity for CIH-induced respiratory plasticity and the duration, intensity and pattern of CIH are likely important determinants of alterations in the respiratory phenotype.

There are a number of methodological considerations that deserve some discussion. First, we assessed respiratory parameters in rats whilst the animals were sleeping (behaviourally-defined); it is possible that respiratory behaviour shows state-dependent differences following CIH and that 'ventilatory adaptation' described by Reeves and Gozal (2006) is present during wakefulness but not during sleep. Because of the relevance of CIH to sleep-disordered breathing, we wished to characterize breathing during sleep. It would be very interesting to examine the respiratory effects of CIH in animals across the sleep-wake cycle. Second, we assessed breathing several hours following the termination of the CIH protocol. It is conceivable that respiratory behaviour may have altered and 'recovered' within this timeframe, and as such we may have underestimated the impact of CIH on breathing in this study. Third, the duration of the experimental paradigm may have been too short to elicit major persistent changes in breathing. Of note, 'ventilatory adaptation' was reported following CIH for 30 days (Reeves and Gozal 2006) and long-term ventilatory facilitation was shown to be time-dependent in mice, present after 14, but not 7, days of treatment (Reeves and Gozal 2004). However, arguing against the latter two points is the fact that our paradigm impairs respiratory muscle function (Skelly et al. 2008) demonstrating a sufficient stimulus to elicit muscle plasticity in our model. Fourth, we did not measure metabolism and it is possible that CIH significantly affects oxygen consumption and carbon dioxide production. However, other studies have demonstrated that CIH does not alter these parameters (Peng et al. 2004; Reeves and Gozal 2006). Fifth, preliminary results from this study suggest that there are no sex differences in the effects of CIH on breathing, but the sample size of female rats was small and we did not determine the phase of the oestrus cycle in our animals on the study day. Finally, a sinusoidal-like pattern of IH commonly employed in other studies (Reeves et al. 2003; Reeves and Gozal 2006; Reeves et al. 2006b; Katayama et al. 2007; O'Halloran et al. 2007) was also used in this study. Discussion following our presentation at the ISAC meeting in Valladolid raised a concern that these paradigms may result in persistent background hypoxaemia (i.e. no recovery to normoxia during air breathing) during CIH exposures. We suspect that 90 s of air breathing is sufficient to re-establish normoxia in conscious rats, but measures of arterial PO_2 or O_2 saturation during CIH in our animals are necessary to demonstrate that normoxic conditions are re-established during the protocol. Of note, studies employing the cyclical equal duration 'on-off' paradigm of IH highlighted above are known to have widely different effects on breathing suggesting that other features of the experimental protocol are more important.

We conclude that the paradigm of CIH employed in this study does not affect eupnoeic ventilation or ventilation during hypoxic and hypercapnic challenges but increases respiratory 'drive' in sleeping adult rats. We speculate that the latter represents a form of neural plasticity that may compensate for respiratory muscle impairment that occurs in this animal model (Skelly et al. 2008). Interestingly, chronic antioxidant treatment completely prevents CIH-induced respiratory muscle dysfunction (Skelly et al. 2008) suggesting a role for reactive oxygen species, which been implicated in CIH-induced plasticity in other studies. The mechanism for respiratory plasticity following CIH in our animal model remains unknown but clearly warrants further investigation.

Acknowledgments Supported by the Health Research Board, Ireland (RP/2006/140), University College Dublin and the Royal College of Surgeons in Ireland. JR Skelly is a UCD *Ad Astra* Research Scholar.

References

Behan, M., Zabka, A.G., Thomas, C.F., Mitchell, G.S. 2003, Sex steroid hormones and the neural control of breathing, *Respir Physiol Neurobiol* 136: 249–63.
Bradford, A., McGuire, M., O'Halloran, K.D. 2005, Does episodic hypoxia affect upper airway dilator muscle function? Implications for the pathophysiology of obstructive sleep apnoea, *Respir Physiol Neurobiol* 147: 223–34.

Dunleavy, M., Bradford, A., O'Halloran, K.D. 2008, Oxidative stress impairs upper airway muscle endurance in an animal model of sleep-disordered breathing, *Adv Exp Med Biol* 605: 458–62.

Julien, C., Bairam, A., Joseph, V. 2008, Chronic intermittent hypoxia reduces ventilatory long-term facilitation and enhances apnea frequency in newborn rats, *Am J Physiol Regul Integr Comp Physiol* 294: R1356–66.

Katayama, K., Smith, C.A., Henderson, K.S., Dempsey, J.A. 2007, Chronic intermittent hypoxia increases the CO_2 reserve in sleeping dogs, *J Appl Physiol* 103: 1942–9.

McGuire, M., Zhang, Y., White, D.P., Ling, L. 2003, Chronic intermittent hypoxia enhances ventilatory long-term facilitation in awake rats, *J Appl Physiol* 95: 1499–508.

O'Halloran, K.D., McGuire, M., O'Hare, T., Bradford, A. 2002, Chronic intermittent asphyxia impairs rat upper airway muscle responses to acute hypoxia and asphyxia, *Chest* 122: 269–75.

O'Halloran, K.D., McGuire M, Bradford, A. 2007, Respiratory plasticity following chronic intermittent hypercapnic hypoxia in conscious rats. In *Proceedings of the Joint Meeting of the Slovak Physiological Society, the Physiological Society and the Federation of European Physiological Societies*, ed. V Strbak, Medimond S. r. l., Italy, pp. 99–103.

Pawar, A., Peng, Y.J., Jacono, F.J., Prabhakar, N.R. 2008, Comparative analysis of neonatal and adult rat carotid body responses to chronic intermittent hypoxia, *J Appl Physiol* 104: 1287–94.

Peng, Y.J., Overholt, J.L., Kline, D., Kumar, G.K., Prabhakar, N.R. 2003, Induction of sensory long-term facilitation in the carotid body by intermittent hypoxia: implications for recurrent apneas, *Proc Natl Acad Sci USA* 100: 10073–8.

Peng, Y.J., Rennison, J., Prabhakar, N.R. 2004, Intermittent hypoxia augments carotid body and ventilatory response to hypoxia in neonatal rat pups, *J Appl Physiol* 97: 2020–5.

Ray, A.D., Magalang, U.J., Michlin, C.P., Ogasa, T., Krasney, J.A., Gosselin, L.E., Farkas, G.A. 2007, Intermittent hypoxia reduces upper airway stability in lean but not obese Zucker rats, *Am J Physiol Regul Integr Comp Physiol* 293: R372–8.

Rey, S., Del Rio, R., Alcayaga, J., Iturriaga, R. 2004, Chronic intermittent hypoxia enhances cat chemosensory and ventilatory responses to hypoxia, *J Physiol* 560: 577–86.

Reeves, S.R., Gozal, D. 2004, Platelet-activating factor receptor modulates respiratory adaptation to long-term intermittent hypoxia in mice, *Am J Physiol Regul Integr Comp Physiol* 287: R369–74.

Reeves S.R., Gozal, D. 2006, Changes in ventilatory adaptations associated with long-term intermittent hypoxia across the age spectrum in the rat, *Respir Physiol Neurobiol* 150:135–43.

Reeves S.R., Gozal E., Guo S.Z., Sachleben, L.R. Jr, Brittian K.R., Lipton A.J., Gozal D. 2003, Effect of long-term intermittent and sustained hypoxia on hypoxic ventilatory and metabolic responses in the adult rat. *J Appl Physiol.* 95: 1767–74.

Reeves, S.R., Guo, S.Z., Brittain K.R., Row, B.W., Gozal, D. 2006a, Anatomical changes in selected cardio-respiratory brainstem nuclei following early post-natal chronic intermittent hypoxia, *Neurosci Lett* 402: 233–7.

Reeves, S.R., Mitchell, G.S., Gozal, D. 2006b, Early postnatal chronic intermittent hypoxia modifies hypoxic respiratory responses and long-term phrenic facilitation in adult rats, *Am J Physiol Regul Integr Comp Physiol* 290: R1664–71.

Skelly, J.R., Bradford A., O'Halloran, K.D. 2008, Tempol, a SOD-mimetic, improves muscle function in a rat model of sleep apnoea, *Proc Physiol Soc*, in press.

Veasey, S.C., Zhan, G., Fenik, P., Pratico, D. 2004, Long-term intermittent hypoxia: reduced excitatory hypoglossal nerve output, *Am J Respir Crit Care Med* 170: 665–72.

Waters, K.A. and Tinworth, K.D. 2001, Depression of ventilatory responses after daily, cyclic hypercapnic hypoxia in piglets, *J Appl Physiol* 90: 1065–73.

Contrasting Effects of Intermittent and Continuous Hypoxia on Low O_2 Evoked Catecholamine Secretion from Neonatal Rat Chromaffin Cells

Dangjai Souvannakitti, Ganesh K. Kumar, Aaron Fox and Nanduri R. Prabhakar

Abstract In the present study we examined the effects of intermittent (IH) and sustained hypoxia (SH) on low PO_2-evoked catecholamine (CA) secretion from neonatal rat chromaffin cells. Experiments were performed on chromaffin cells isolated from rat pups exposed to either IH (P0–P5; 15 s hypoxia-5 min normoxia; 8 h/day) or SH (hypobaric hypoxia; 0.4ATM). CA secretion from chromaffin cells was monitored by amperometry. Control chromaffin cells, from P5 rat pups, exhibited robust CA secretion in response to acute hypoxia. IH *facilitated* whereas SH *attenuated* hypoxia-evoked CA secretion. IH increased the epinephrine and norepinephrine content of the adrenal medulla whereas SH had no effect. These results demonstrate that neonatal exposures IH and SH exert diametrically opposed effects on acute hypoxia-evoked CA secretion from chromaffin cells and CA contents.

Keywords Intermittent hypoxia · Chromaffin cells · Neonates · Recurrent apneas

1 Introduction

In neonates, sympathetic nervous system is not well developed and as a consequence catecholamine (CA) secretion from adrenal medulla becomes critical for maintaining homeostasis under a variety of stress conditions including hypoxia. Unlike adults, hypoxia-evoked catecholamine secretion from neonatal adrenal medulla is non-neurogenic and is due to direct actions of low O_2 on chromaffin cells (Seidler & Slotkin, 1985; Thompson et al., 1997). Several lines of evidence suggest that chronic

D. Souvannakitti (✉)
The Center for Systems Biology, Departments of Medicine, University of Chicago, Chicago, IL 60637, USA
e-mail: dangjai@uchicago.edu

changes in environmental O_2 in the neonatal period markedly influence cellular response to acute hypoxia. For instance, exposure of neonatal rats to several days of sustained hypoxia (SH) or hyperoxia attenuate sensory response of the carotid body chemoreceptors to subsequent exposure to acute hypoxia (Donnelly & Doyle, 1994). On the other hand, neonates, especially pre-mature infants (70–90%) experience intermittent hypoxia (IH) often as a consequence of recurrent apneas characterized by periodic cessations of breathing leading to cyclical decreases in arterial blood O_2 saturations (Poets et al., 1994; Stokowski, 2005). Recent studies have shown that exposing neonatal rats to IH markedly augment carotid body chemoreceptor responses to acute hypoxia (Pawar et al., 2008), an effect that is opposite to that reported with SH (Donnelly & Doyle, 1994). Whether SH and IH also affect hypoxia evoked catecholamine secretion from chromaffin cells from neonatal rats has not yet been examined. Therefore, in the present study, we examined the effects of neonatal exposures to SH and IH on hypoxia-evoked CA secretion from chromaffin cells.

2 Materials and Methods

2.1 General Methods

Experiments were performed on neonatal Sprague-Dawley rats pups (P0–P35). Rat pups (P0) along with their mothers were exposed to IH for 5 days (between 9:00 am and 5:00 pm) as described previously (Pawar et al., 2008). To determine the effects of SH, rat pups along with their mother were exposed to hypobaric hypoxia (0.4 ATM) for 24 h or 5 days. Acute experiments were performed on anesthetized pups (Urethane $1.2\,g\,kg^{-1}$; I.P.) 6–10 h following either IH or normoxia.

2.2 Preparation of Chromaffin Cells and Cell Culture

Adrenal glands were harvested from anesthetized rat pups exposed either IH or SH or normoxia (controls). Chromaffin cells were enzymatically dissociated as described previously (Grabner & Fox, 2006). Cells were plated on collagen (type VII; Sigma) coated coverslips and maintained at 37°C in a 5% CO_2 incubator for 12–24 h. The growth medium consisted of F-12 K medium (Invitrogen) supplemented with 10% horse serum, 5% fetal bovine serum and 1% Penicillin/streptomycin/glutamine cocktail (Invitrogen).

2.3 Amperometry

Catecholamine secretion from chromaffin cells was monitored by amperometry using carbon fiber electrodes as described previously (Grabner et al., 2006).

2.4 Recording Solutions and Stimulation Protocols

Amperometric recordings were made from adherent cells that were under constant perfusion (flow rate of about 1.0 mL/min: chamber volume ~80 μL). All experiments were performed at ambient temperature (23±2°C), and the solutions had the following composition (in mM): 1.26 $CaCl_2$, 0.49 $MgCl_2$-$6H_2O$, 0.4 $MgSO_4$-$7H_2O$, 5.33 KCl, 0.441; KH_2PO_4, 137.93, NaCl, 0.34 Na_2HPO_4-$7H_2O$, 5.56 Dextrose and 20 mM Hepes at pH 7.35 and 300mOsm. Normoxic solutions were equilibrated with room air (PO_2 ~146 mmHg). For challenging with hypoxia, solutions were degassed and equilibrated with appropriate gas mixtures that resulted in final PO_2 of ~ 30 mmHg.

2.5 Measurement of Catecholamine Content

CA content of adrenal medulla were determined by high pressure liquid chromatography coupled with electrochemical detection (HPLC-ECD) as previously described (Kumar et al., 2006). Norepinephrine and epinephrine contents were expressed as nmoles per mg of protein.

3 Results

3.1 Effects of IH on Hypoxia-Evoked CA Secretion

Rat pups were exposed either to IH (P0–P5) or normoxia (controls). CA secretion from chromaffin cells was compared between both groups (i.e., IH vs Controls). The magnitude of hypoxia-evoked CA secretion (PO_2 =~30 mmHg) was higher in the IH-treated than in control chromaffin cells. The number of secretory events increased 4 fold and the amount of CA release per event increased ~6 fold in IH-treated cells compared to controls (Controls, n=17 and IH, n=18 cells; P<0.01).

3.2 Effects of SH on Hypoxia-Evoked CA Secretion

Exposing rat pups to 24 h of SH had no significant effect on hypoxia (PO_2 = ~30 mmHg)-evoked CA secretion in terms of either number of events or the amount of CA secreted per event (Control and SH n=12 cells each; P>0.05). However, after 5d of SH, the magnitude of hypoxia-evoked CA secretion was significantly decreased primarily due to reduction in the amount of CA released per secretory event (5±0.8 to 2.7±0.6 10^5 molecules/event), whereas the number of secretory events was unchanged (P>0.05).

3.3 IH but Not SH Increases the Catecholamine Content of Adrenal Medullae

The effects of IH and SH on the norepinephrine and epinephrine contents of adrenal medullae were examined. Both the epinephrine (E) and norepinephrine (NE) contents of adrenal medullae were significantly elevated in IH (E; 373±11, NE; 49±3.7 nmol/mg protein and P < 0.01), but not from 5d SH-treated pups (E; 187±7.3, NE; 21±1.3 nmole/mg protein P>0.05) comparing with control (E; 170±14, NE; 19±1.4 nmole/mg protein).

4 Discussion

A novel finding of the present study was that neonatal exposures IH and SH exert diametrically opposed effects on acute hypoxia-evoked CA secretion from chromaffin cells. Neonatal IH facilitated whereas SH attenuated CA secretion by hypoxia. These findings are similar to previously reported effects of IH and SH on neonatal carotid body sensory response to low O_2 (Donnelly & Doyle, 1994; Pawar et al., 2008). Elevated CA content might account in part for facilitated CA secretion in IH pups, whereas the attenuated transmitter release seen in SH pups was independent of alterations in CA content. A previous study reported that SH facilitates hypoxia-evoked CA secretion from rat PC12 cells (Taylor & Peers, 1999), which is opposite to that seen in neonatal chromaffin cells (this study). The differences could either be due to the effects of SH on cell culture versus intact animals and/or due to the propagating phenotype of PC12 cells versus the non-propagating nature of native chromaffin cells.

The magnitude of CA secretion from chromaffin cells is determined by the number of secretory events and the amount of transmitter secreted per event. SH selectively attenuated the amount of neurotransmitter released per hypoxia-evoked event but not the number of vesicles released, indicating reduced emptying of vesicles and/or fusion of the vesicles to the membrane. On the other hand, IH-evoked facilitation of CA secretion was associated with increases in both the number of vesicles released and the amount of neurotransmitter released per event. This facilitated secretion by IH might be due to larger depolarization in response to acute hypoxia leading to robust activation of voltage-gated Ca^{2+} channels and greater influx of Ca^{2+} promoting greater emptying of each vesicle. Further studies are needed to test this possibility.

Acknowledgments The research is supported by grants from National Institutes of Health (HL-76537, HL-90554, HL-86493 (NRP)); Philip Morris USA Inc. and Philip Morris International grant (AF).

References

Donnelly DF & Doyle TP. (1994). Hypoxia-induced catecholamine release from rat carotid body, in vitro, during maturation and following chronic hypoxia. *Adv Exp Med Biol* **360**, 197–199.

Grabner CP & Fox AP. (2006). Stimulus-dependent alterations in quantal neurotransmitter release. *J Neurophysiol* **96**, 3082–3087.

Grabner CP, Price SD, Lysakowski A, Cahill AL & Fox AP. (2006). Regulation of large dense-core vesicle volume and neurotransmitter content mediated by adaptor protein 3. *Proc Natl Acad Sci U S A* **103**, 10035–10040.

Kumar GK, Rai V, Sharma SD, Ramakrishnan DP, Peng YJ, Souvannakitti D & Prabhakar NR. (2006). Chronic intermittent hypoxia induces hypoxia-evoked catecholamine efflux in adult rat adrenal medulla via oxidative stress. *J Physiol* **575**, 229–239.

Pawar A, Peng YJ, Jacono FJ & Prabhakar NR. (2008). Comparative analysis of neonatal and adult rat carotid body responses to chronic intermittent hypoxia. *J Appl Physiol* **104**, 1287–1294.

Poets CF, Samuels MP & Southall DP. (1994). Epidemiology and pathophysiology of apnoea of prematurity. *Biol Neonate* **65**, 211–219.

Seidler FJ & Slotkin TA. (1985). Adrenomedullary function in the neonatal rat: responses to acute hypoxia. *J Physiol* **358**, 1–16.

Stokowski LA. (2005). A primer on Apnea of prematurity. *Adv Neonatal Care* **5**, 155–170; quiz 171–154.

Taylor SC & Peers C. (1999). Chronic hypoxia enhances the secretory response of rat phaeochromocytoma cells to acute hypoxia. *J Physiol* **514 (Pt 2)**, 483–491.

Thompson RJ, Jackson A & Nurse CA. (1997). Developmental loss of hypoxic chemosensitivity in rat adrenomedullary chromaffin cells. *J Physiol* **498 (Pt 2)**, 503–510.

Hypoxic Pulmonary Vasoconstriction – *Invited Article*

A. Mark Evans and Jeremy P.T. Ward

Abstract Hypoxic pulmonary vasoconstriction (HPV) is an adaptive mechanism that in the face of localised alveolar hypoxia diverts blood away from poorly ventilated regions of the lung, thereby preserving ventilation/perfusion matching. HPV has been recognised for many years, but although the underlying mechanisms are known to reside within the arteries themselves, their precise nature remains unclear. There is a growing consensus that mitochondria act as the oxygen sensor, and that Ca^{2+} release from ryanodine-sensitive stores and Rho kinase-mediated Ca^{2+} sensitisation are critical for sustained vasoconstriction, though Ca^{2+} entry via both voltage-dependent and/or -independent pathways has also been implicated. There is, however, controversy regarding the signalling pathways that link the oxygen sensor to its effectors, with three main hypotheses. The AMP-activated protein kinase (AMPK) hypothesis proposes that hypoxic inhibition of mitochondrial function increases the AMP/ATP ratio and thus activates AMPK, which in turn mediates cADPR-dependent mobilisation of ryanodine-sensitive sarcoplasmic reticulum Ca^{2+} stores. In contrast the two other hypotheses invoke redox signalling, albeit in mutually incompatible ways. The Redox hypothesis proposes that hypoxia suppresses mitochondrial generation of reactive oxygen species (ROS) and causes the cytosol to become more reduced, with subsequent inhibition of K_V channels, depolarisation and voltage-dependent Ca^{2+} entry. In direct contrast the ROS hypothesis proposes that hypoxia causes an apparently paradoxical increase in mitochondrial ROS generation, and it is this increase in ROS that acts as the signalling moiety. In this article we describe our current understanding of HPV, and evidence in support of these models of oxygen-sensing.

Keywords Hypoxic pulmonary vasoconstriction · Alveolar hypoxia · Ventilation · Mitochondria · Oxygen-sensor · Calcium · AMP-activated protein kinase · AMPK · Sarcoplasmic reticulum · Kv chennel

---

A.M. Evans (✉)
Centre for Integrative Physiology, College of Medicine and Veterinary Medicine, Hugh Robson Building University of Edinburgh, Edinburgh EH8 9XD, UK
e-mail: Mark.Evans@ed.ac.uk

1 Introduction

Whereas most systemic arteries dilate in response to hypoxia, those of the pulmonary circulation constrict. This is known as hypoxic pulmonary vasoconstriction (HPV), an important adaptive mechanism that optimises pulmonary ventilation-perfusion matching in the face of localised alveolar hypoxia. HPV also contributes to the high pulmonary vascular resistance (PVR) of the fetus. Whilst normally beneficial, HPV is responsible for the detrimental increase in PVR and consequent pulmonary hypertension in conditions associated with global hypoxia, including COPD, respiratory failure and ascent to altitude. The mechanisms of HPV remain incompletely understood, though it has been known for many years that they reside within the pulmonary artery, and that the primary site of HPV is at the level of the small muscular pulmonary arteries of 100–500 μm internal diameter. There is no doubt that key mechanisms of HPV are contained with the pulmonary artery smooth muscle (PASM), as endothelium-denuded PA and isolated PASM cells respond to hypoxia, but there is also strong evidence that the endothelium is required for full development of sustained HPV. There remains, however, considerable controversy concerning the sub-cellular mechanisms of HPV, specifically those relating to the O_2 sensor and its distal signalling pathways. In this article we briefly describe our current understanding of those responsible for the contraction itself, including Ca^{2+} mobilisation and Ca^{2+} sensitisation, and of the signalling pathways by which these are activated by hypoxia.

2 Characteristics of HPV

In the whole animal and in perfused lungs, a reduction in alveolar PO_2 below ∼50–60 mmHg results in a rapid increase in PVR and pulmonary artery pressure (PAP) proportional to the fall in PO_2, which is then maintained for as long as hypoxia is present. This is usually but not always monophasic in profile, whereas the response in isolated small pulmonary arteries (PA) is commonly biphasic, with a rapid transient constriction (phase 1) superimposed on a more slowly developing sustained constriction that is largely dependent on an intact endothelium (phase 2; Fig. 1) (Leach et al., 1994; Dipp et al., 2001). This is associated with a biphasic elevation in PASM intracellular $[Ca^{2+}]$ (Robertson et al., 1995). The physiological relevance of the transient phase 1 has been questioned, and it has become clear that its underlying mechanisms differ from those of phase 2 (Dipp & Evans, 2001; Aaronson et al., 2006).

2.1 Ca^{2+} Mobilization by Hypoxia in PASM

Early studies implicated activation of voltage–dependent Ca^{2+} entry in HPV, as it was shown that hypoxia causes inhibition of K^+ channels and depolarisation in PASM cells, with consequent Ca^{2+} entry via L-type Ca^{2+} channels (Post et al., 1993).

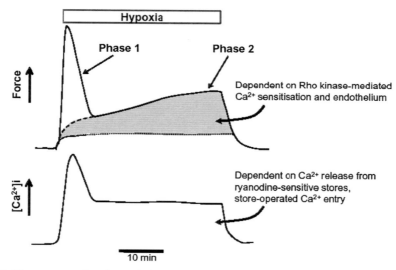

Fig. 1 Diagram showing the typical response to moderate hypoxia (15–30 mmHg) in small intrapulmonary arteries, the *upper trace* showing developed force, the lower intracellular [Ca^{2+}] (as measured with Fura-2). Both force and [Ca^{2+}] show a biphasic response, with a transient (phase 1) superimposed on a more slowly developing sustained constriction (phase 2). The *shaded area* is an approximation of the component of phase 2 that is related to Rho kinase-mediated Ca^{2+} sensitisation, and the influence of the endothelium. Phase 1 is independent of both the endothelium and Rho kinase-mediated Ca^{2+} sensitisation

This became a key plank of the Redox hypothesis of HPV (Weir & Archer, 1995) (see below). However, many studies do not report ablation of HPV with L-type Ca^{2+} channel blockers, and it has been demonstrated that HPV can be fully elicited in isolated PA under conditions in which depolarisation and voltage-dependent Ca^{2+} entry cannot occur, implying that the elevation in intracellular [Ca^{2+}] must involve voltage–independent Ca^{2+} entry and Ca^{2+} release from stores (Robertson et al., 2000b; Dipp et al., 2001). It was subsequently shown that Ca^{2+} release from ryanodine-sensitive stores is a critical event for HPV, mediated by an increase in cyclic ADP ribose (cADPR) accumulation, an endogenous activator of ryanodine receptors (Wilson et al., 2001). Consequent activation of voltage-independent store-operated Ca^{2+} entry (SOCE) then contributes to the elevation of PASM intracellular [Ca^{2+}] (Ng et al., 2005). There is still debate, however, about the relative importance of voltage-dependent and -independent Ca^{2+} entry in HPV.

2.2 Endothelium and Ca^{2+} Sensitisation

Although hypoxia can elevate intracellular [Ca^{2+}] in isolated PASM cells and IPA denuded of endothelium, there is strong evidence that the endothelium is required for full development of sustained HPV (Leach et al., 1994; Dipp et al., 2001).

However, removal of the endothelium does not affect the Ca^{2+} response, and in intact IPA the elevation in intracellular $[Ca^{2+}]$ remains stable whilst tension continues to develop, both of which imply HPV also involves an increase in PASM myofilament Ca^{2+} sensitivity, mediated in part by the endothelium (Robertson et al., 1995). Myofilament Ca^{2+} sensitivity in smooth muscle is largely dependent on the activity of myosin phosphatase, which is regulated by Rho kinase through its targeting sub-unit MYPT-1, and by protein kinase C via the inhibitor CPI-17. An important finding therefore is that HPV in both isolated IPA and perfused lungs is strongly inhibited by the Rho kinase inhibitor Y-27632 and with an IC_{50} in the range (nM) of its Kd for Rho kinase (Robertson et al., 2000a). Although hypoxia also increases Rho kinase activity in cultured PASM cells and denuded IPA, the increase is dramatically enhanced in the presence of the endothelium (Wang et al., 2001; Robertson et al., 2005). The means by which the latter occurs is uncertain, though a small, as yet unidentified vasoconstrictor that increases myofilament Ca^{2+} sensitivity but not intracellular $[Ca^{2+}]$ has been shown to be released from the lung during hypoxia (Robertson et al., 2001).

Whilst there is now a general (but not universal) consensus that sustained HPV is critically dependent on both Ca^{2+} release from stores and Ca^{2+} sensitisation, the mechanisms by which hypoxia leads to activation of these processes is fraught with controversy. In the remainder of this article we will therefore focus on this aspect.

3 The Mitochondrial O_2 Sensor

Over the years, a number of O_2 sensitive processes have been hypothesised to act as sensors for acute changes in PO_2 in a number of O_2 sensitive tissues, including haem-containing proteins, p450 cytochromes, haemoxygenases, NAD(P)H oxidases and the mitochondria (Ward, 2008), but for HPV at least there is now wide agreement that the O_2 sensor resides in the mitochondria. Mitochondria are the major consumers of O_2 in the cell, and in the presence of substrate such as glucose slowing of electron transport due to O_2 limitation will reduce the electron transport chain (ETC) proximal to cytochrome aa3 oxidase, and changes in factors that could potentially be used for signalling, including β-NADH and redox state, ATP and energy state, and reactive oxygen species (ROS). Although cytochrome aa3 has a very high affinity for O_2 in most tissues, with a P50 <1 mmHg, moderate hypoxia has been shown to affect mitochondrial function in isolated PA as reflected by increased NAD(P)H autofluorescence (Leach et al., 2001), which progressively increases as PO_2 falls over a similar range as for HPV (Ward, 2008). This implies that the effective P50 for PA mitochondria is much higher than for most other tissues, as previously suggested for glomus cells of the carotid body (Duchen & Biscoe, 1992). Various mechanisms have been proposed to account for this, including specific cytochrome oxidases, altered coupling between the proximal and distal ETC, competition for the O_2 binding by NO or H_2S, and cytochrome oxidase-independent mechanisms (discussed in Ward, 2008).

Evidence that mitochondria act as the O_2 sensor in PA has been gained from numerous studies showing that inhibitors of the proximal electron transport chain (ETC) selectively block the effects of hypoxia on PA, though the data concerning distal ETC inhibitors such as cyanide are less clear cut (e.g. Leach et al., 2001; Waypa et al., 2001; Michelakis et al., 2002), and loss of O_2 sensing in PASM cells lacking functional mitochondria (Waypa et al., 2001). A key and unresolved disparity, however, concerns the way in which proximal ETC inhibitors cause suppression of HPV: whilst some report that they mimic hypoxia and therefore prevent further activation (Michelakis et al., 2002; Moudgil et al., 2005), others report that they suppress HPV without mimicking hypoxia (Leach et al., 2001; Waypa et al., 2001; Waypa & Schumacker, 2005). Such differences underscore the controversy regarding both the signalling pathways linking the mitochondrial O_2 sensor to the effector mechanisms and the selectivity / relative potency of pharmacological inhibitors of the ETC.

Of the three main hypotheses of signal transduction in HPV that are currently most favoured, two invoke changes in mitochondrial ROS generation and/or redox state (albeit in opposite directions), and one invokes changes in energy state as reflected by β-NADH levels and the AMP/ATP ratio (Fig. 2). Whilst an increase in both β-NADH levels and the AMP/ATP ratio have been measured in response to hypoxia and can be directly inferred from reduced ATP production during O_2 limitation, alterations in mitochondrial ROS generation are more difficult to predict or indeed measure.

3.1 The Redox Hypothesis

The Redox hypothesis of HPV proposes that during hypoxia ROS generation falls and the PASM cell cytosol becomes more reduced, leading to diminished oxidation of sulphydryl groups on K_V channels which are therefore inhibited, with consequent depolarisation and Ca^{2+} entry via L-type Ca^{2+} channels (Weir & Archer, 1995; Michelakis et al., 2002; Moudgil et al., 2005). Even if measurements of ROS are excluded, there is still a fair amount of circumstantial evidence to support this hypothesis, including the effects of exogenous oxidants and reducing agents on K_V channel function (discussed in Moudgil et al., 2005), and the elegant simplicity of the model ensured its acceptance as the mechanism of HPV for more than a decade. However, results from later studies have thrown doubt on the Redox hypothesis, as, for example, blockade of L-type channels does not necessarily abolish HPV (see above), antioxidants do not (as would be predicted) mimic hypoxia, and the model cannot account for the mobilisation of sarcoplasmic reticulum Ca^{2+} stores, release of an endothelium-derived vasoconstrictor and / or activation of Rho kinase-mediated myofilament Ca^{2+} sensitisation.

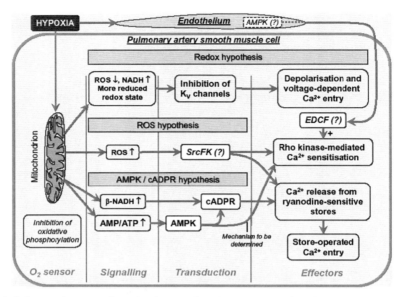

Fig. 2 Schema of proposed mechanisms and pathways underlying the three hypotheses of HPV discussed in the text. Note that little is known about the pathways activated by hypoxia in endothelium, although a rise in intracellular Ca^{2+} may be required (Dipp et al., 2001), and AMPK induces an endothelium-dependent component of constriction (Evans et al., 2005), similar to that induced by hypoxia (EDCF = endothelium derived constricting factor)

3.2 The ROS Hypothesis

A few percent of electrons travelling through the ETC are constitutively lost by single electron donation to O_2 to form superoxide (O_2^-). This occurs in complex I due to auto-oxidation of flavins, where superoxide enters the matrix, and in complex III due to electron donation by the Q cycle intermediate ubisemiquinone. Superoxide formed by ubisemiquinone immediately distal to the Rieske Fe-S group enters the mitochondrial intermembrane space, whereas that formed by ubisemiquinone following its reincarnation by cytochrome b_H enters the matrix (Turrens, 2003). Superoxide is dismuted to H_2O_2 by superoxide dismutases in the matrix, intermembrane space and cytosol.

Whilst O_2 limitation might be expected to reduce mitochondrial ROS generation simply because there is less substrate (i.e. O_2) to form superoxide, reduction of the proximal ETC promotes superoxide generation (Turrens, 2003). The relative balance between these two effects will determine whether ROS increases or decreases during hypoxia, and indeed may be affected by coupling between the proximal and distal components of the ETC (discussed in Ward, 2008), and potentially by heterogeneity between mitochondria from O_2 sensing and non-sensing tissues (Michelakis et al., 2002). Estimations of ROS generation in PA during hypoxia have, however, proved to be highly contentious, primarily due to limitations of the methodologies (discussed in Moudgil et al., 2005; Waypa & Schumacker, 2005), though there is some agreement that the major source of ROS is from complex III, whether it increases or decreases in hypoxia (Leach et al., 2001; Waypa et al., 2001; Michelakis et al., 2002). The question is whether this is of any relevance to HPV.

The hypothesis that O_2 sensing and detection of hypoxia by the mitochondria is transduced by an increase in ROS generation is not limited to PA and HPV, but includes the response of several other tissues such as cardiomyocytes (see Chandel & Schumacker, 2000); thus one must ask how this process may provide for the pulmonary selective effects of hypoxia. Apart from direct measurements of ROS, the evidence for this hypothesis includes differential effects of proximal and distal inhibitors of the ETC on Ca^{2+} mobilisation in PASMCs and Ca^{2+} and force production in isolated PA, with proximal inhibitors (e.g. rotenone) abolishing but not mimicking hypoxia, and distal inhibitors (e.g. antimycin A, cyanide) mimicking hypoxia and/or promoting HPV (e.g. Leach et al., 2001; Waypa et al., 2001). In addition, antioxidants tend to suppress HPV, and overexpression of glutathione peroxidise and catalase has been shown to depress the response to hypoxia in PASMCs (Weissmann et al., 2000; Waypa et al., 2006; Wang et al., 2007). As yet an increase in ROS has not been shown to precisely mimic HPV, but ROS are known to be capable of activating mechanisms that have been implicated in this process, including Ca^{2+} release from ryanodine sensitive stores, voltage-independent Ca^{2+} entry pathways, HIF-1 stabilisation and Rho kinase activation (discussed in Ward et al., 2004). Furthermore, recent work has implicated a role for ROS-sensitive Src family kinases (SrcFK) in these responses (unpublished observations). An argument against this hypothesis concerns the fact that acute hyperoxia, which is generally accepted to increase ROS formation, fails to induce an increase in pulmonary vascular resistance or alter the distribution of blood flow in the lung (Hambraeus-Jonzon et al., 1997).

3.3 The AMPK Hypothesis

Most recently, an alternative hypothesis has been proposed that neatly links changes in energy state to HPV, and is independent of any changes in general redox status or ROS. Cytosolic ATP is maintained relatively constant by adenylate kinase, which converts two ADP to AMP + ATP, with the consequence that a fall in ADP/ATP ratio results in a much greater rise in AMP/ATP ratio. AMP-activated protein kinase (AMPK) is a metabolic fuel gauge which is exquisitively sensitive to small changes in AMP/ATP ratio, due to the fact that AMP regulates AMPK via a triple mechanism which: (1) confers allosteric regulation via the γ subunit (2) permits phosphorylation of the α subunit at Thr-172 by an upstream kinase that is a complex between the tumour suppressor kinase LKB1 and two accessory proteins STRAD and MO25 and (3) inhibits dephosphorylation of AMPK. In the absence of metabolic stress, each of these processes is antagonized by high concentrations of ATP. AMPK modulates a wide variety of processes that regulate ATP supply, but also regulates non-metabolic processes. This led to the proposal that AMPK, possibly in a manner dependent on specific isozymes, could act as a mediator between the mitochondria and effector mechanisms in O_2 sensitive tissues (Evans, 2006). Consistent with this hypothesis, AMPK expression in PA is four-fold that in systemic arteries and AMP/ATP ratio doubles in PA during moderate hypoxia. Most significantly, activation of AMPK represents the only mechanism that has been shown to induce PA constriction with the same characteristics as HPV (Evans et al., 2005), and inhibition of AMPK has been reported to reverse HPV (Evans, 2006). This hypothesis therefore provides an elegant link between changes in energy state caused by hypoxic inhibition of oxidative phosphorylation and a key component of HPV. It should be noted, however, that cADPR-dependent SR calcium release in the PASM cells triggered by hypoxia may be consequent to both AMPK activation and an increase β-NADH levels, which may modulate this process in discrete ways (Evans, 2006).

4 Conclusion

HPV is a complex, multifactorial mechanism, with evidence for Ca^{2+} entry and release, Rho kinase-mediated Ca^{2+} sensitisation, and an important role for the endothelium. However, the relative importance of these components may differ between preparations and in particular depending on the degree of hypoxia, and could potentially be over- or under-emphasised in cultured PASM cells, where ion channel expression and Ca^{2+} release mechanisms are known to be altered. Such aspects have made the search for the key mechanisms of HPV all the more difficult. The area of greatest controversy, however, lies with the mechanisms of O_2-sensing in PA. Whilst there is finally some consensus that mitochondria act as the O_2 sensor, there is still strong support for each of the three current hypotheses of how inhibition of mitochondrial function leads to activation of the contractile mechanisms (Fig. 2). However, we believe that the evidence for the Redox hypothesis is least strong, not least because it can only easily account for one mechanism (voltage-dependent Ca^{2+} entry), which has been shown not to be essential for HPV. In contrast, there is a substantial amount of evidence for both the ROS hypothesis and the AMPK hypothesis, and both can be invoked to account for the majority of mechanisms activated during HPV. Whilst the authors of this article have rather differing opinions as to the role of ROS in HPV, it is interesting to note that there is a possible link between the ROS and AMPK hypotheses, as AMPK has been reported to be activated by ROS (e.g. Quintero et al., 2006). Further studies will no doubt eventually determine the underlying processes of this important and intriguing mechanism.

Acknowledgments Work in the Authors' laboratories is supported by the Wellcome Trust and British Heart Foundation

References

Aaronson PI, Robertson TP, Knock GA, Becker S, Lewis TH, Snetkov V & Ward JP. (2006). Hypoxic pulmonary vasoconstriction: mechanisms and controversies. *J Physiol* **570,** 53–58.
Chandel NS & Schumacker PT. (2000). Cellular oxygen sensing by mitochondria: old questions, new insight. *J Appl Physiol* **88,** 1880–1889.

Dipp M & Evans AM. (2001). Cyclic ADP-ribose is the primary trigger for hypoxic pulmonary vasoconstriction in the rat lung in situ. *Circ Res* **89,** 77–83.

Dipp M, Nye PC & Evans AM. (2001). Hypoxic release of calcium from the sarcoplasmic reticulum of pulmonary artery smooth muscle. *Am J Physiol Lung Cell Mol Physiol* **281,** L318–325.

Duchen MR & Biscoe TJ. (1992). Mitochondrial function in type I cells isolated from rabbit arterial chemoreceptors. *J Physiol* **450,** 13–31.

Evans AM. (2006). AMP-activated protein kinase and the regulation of Ca2+ signalling in O2-sensing cells. *J Physiol* **574,** 113–123.

Evans AM, Mustard KJ, Wyatt CN, Peers C, Dipp M, Kumar P, Kinnear NP & Hardie DG. (2005). Does AMP-activated protein kinase couple inhibition of mitochondrial oxidative phosphorylation by hypoxia to calcium signaling in O2-sensing cells? *J Biol Chem* **280,** 41504–41511.

Hambraeus-Jonzon K, Bindslev L, Mellgard AJ & Hedenstierna G. (1997). Hypoxic pulmonary vasoconstriction in human lungs. A stimulus-response study. *Anesthesiology* **86,** 308–315.

Leach RM, Hill HM, Snetkov VA, Robertson TP & Ward JP. (2001). Divergent roles of glycolysis and the mitochondrial electron transport chain in hypoxic pulmonary vasoconstriction of the rat: identity of the hypoxic sensor. *J Physiol* **536,** 211–224.

Leach RM, Robertson TP, Twort CH & Ward JP. (1994). Hypoxic vasoconstriction in rat pulmonary and mesenteric arteries. *Am J Physiol* **266,** L223–231.

Michelakis ED, Hampl V, Nsair A, Wu X, Harry G, Haromy A, Gurtu R & Archer SL. (2002). Diversity in mitochondrial function explains differences in vascular oxygen sensing. *Circ Res* **90,** 1307–1315.

Moudgil R, Michelakis ED & Archer SL. (2005). Hypoxic pulmonary vasoconstriction. *J Appl Physiol* **98,** 390–403.

Ng LC, Wilson SM & Hume JR. (2005). Mobilization of SR stores by hypoxia leads to consequent activation of capacitative Ca^{2+} entry in isolated canine pulmonary arterial smooth muscle cells. *J Physiol* **563,** 409–419.

Post J, Weir EK, Archer SL & Hume J. (1993). Redox regulation of K+ channels and hypoxic pulmonary vasoconstriction. In *Ion Flux in Pulmonary Vascular Control*, ed. Weir EK, Hume J & Reeves J. Futura Publishing Company, New York.

Quintero M, Colombo SL, Godfrey A & Moncada S. (2006). Mitochondria as signaling organelles in the vascular endothelium. *Proc Natl Acad Sci U S A* **103,** 5379–5384.

Robertson T, Hashmi-Hill M, Vandenplas ML & Lewis SJ. (2005). Endothelium-dependent activation of Rho-kinase during hypoxic pulmonary vasoconstriction in rat intrapulmonary arteries. *FASEB J* **19,** A1277.

Robertson TP, Aaronson PI & Ward JP. (1995). Hypoxic vasoconstriction and intracellular Ca^{2+} in pulmonary arteries: evidence for PKC-independent Ca^{2+} sensitization. *Am J Physiol* **268,** H301–307.

Robertson TP, Dipp M, Ward JP, Aaronson PI & Evans AM. (2000a). Inhibition of sustained hypoxic vasoconstriction by Y-27632 in isolated intrapulmonary arteries and perfused lung of the rat. *Br J Pharmacol* **131,** 5–9.

Robertson TP, Hague D, Aaronson PI & Ward JP. (2000b). Voltage-independent calcium entry in hypoxic pulmonary vasoconstriction of intrapulmonary arteries of the rat. *J Physiol* **525 Pt 3,** 669–680.

Robertson TP, Ward JP & Aaronson PI. (2001). Hypoxia induces the release of a pulmonary-selective, Ca^{2+}-sensitising, vasoconstrictor from the perfused rat lung. *Cardiovasc Res* **50,** 145–150.

Turrens JF. (2003). Mitochondrial formation of reactive oxygen species. *J Physiol* **552,** 335–344.

Wang QS, Zheng YM, Dong L, Ho YS, Guo Z & Wang YX. (2007). Role of mitochondrial reactive oxygen species in hypoxia-dependent increase in intracellular calcium in pulmonary artery myocytes. *Free Radical Biology & Medicine* **42,** 642–653.

Wang Z, Jin N, Ganguli S, Swartz DR, Li L & Rhoades RA. (2001). Rho-kinase activation is involved in hypoxia-induced pulmonary vasoconstriction. *Am J Respir Cell Mol Biol* **25,** 628–635.

Ward JP. (2008). Oxygen sensors in context. *Biochim Biophys Acta* **1777,** 1–14.

Ward JP, Snetkov VA & Aaronson PI. (2004). Calcium, mitochondria and oxygen sensing in the pulmonary circulation. *Cell Calcium* **36,** 209–220.

Waypa GB, Chandel NS & Schumacker PT. (2001). Model for hypoxic pulmonary vasoconstriction involving mitochondrial oxygen sensing. *Circ Res* **88,** 1259–1266.

Waypa GB, Guzy R, Mungai PT, Mack MM, Marks JD, Roe MW & Schumacker PT. (2006). Increases in mitochondrial reactive oxygen species trigger hypoxia-induced calcium responses in pulmonary artery smooth muscle cells. *Circ Res* **99,** 970–978.

Waypa GB & Schumacker PT. (2005). Hypoxic pulmonary vasoconstriction: redox events in oxygen sensing. *J Appl Physiol* **98,** 404–414.

Weir EK & Archer SL. (1995). The mechanism of acute hypoxic pulmonary vasoconstriction: the tale of two channels. *FASEB J* **9,** 183–189.

Weissmann N, Tadic A, Hanze J, Rose F, Winterhalder S, Nollen M, Schermuly RT, Ghofrani HA, Seeger W & Grimminger F. (2000). Hypoxic vasoconstriction in intact lungs: a role for NADPH oxidase- derived H_2O_2? *Am J Physiol Lung Cell Mol Physiol* **279,** L683–690.

Wilson HL, Dipp M, Thomas JM, Lad C, Galione A & Evans AM. (2001). ADP-ribosyl cyclase and cyclic ADP-ribose hydrolase act as a redox sensor. A primary role for cyclic ADP-ribose in hypoxic pulmonary vasoconstriction. *J Biol Chem* **276,** 11180–11188.

Impact of Modulators of Mitochondrial ATP-Sensitive Potassium Channel (mitoK$_{ATP}$) on Hypoxic Pulmonary Vasoconstriction

R. Paddenberg, P. Faulhammer, A. Goldenberg, B. Gries, J. Heinl and W. Kummer

Abstract Previously, we demonstrated that hypoxic pulmonary vasoconstriction (HPV) of intra-acinar arteries (IAA) requires mitochondrial complex II (= succinate dehydrogenase, SDH) activity (Paddenberg et al., Respir Res, 7:93, 2006). Interestingly, SDH subunits A and B have recently been described as components of a multiprotein mitochondrial ATP-sensitive potassium channel (mitoK$_{ATP}$), together with mitochondrial ATP-binding cassette protein-1, adenine nucleotide translocator (ANT), ATP synthase, and phosphate carrier (Ardehali et al., Proc Natl Acad Sci USA, 101(32):11880–5, 2004). Hence, we tested the hypothesis that such an SDH-containing mitoK$_{ATP}$ is involved in HPV. For this purpose, the impact of modulators of mitoK$_{ATP}$ on HPV of IAA was studied videomorphometrically in precision cut murine lung slices. Inhibitors of mitoK$_{ATP}$ (glibenclamide, 5-hydroxydecanoate) completely suppressed HPV, mitoK$_{ATP}$ activators (pinacidil, diazoxide) even induced vasodilatation, and ANT inhibitors (bongkrekic acid, atractyloside) attenuated HPV. This pharmacological profile differs clearly from that described for mitoK$_{ATP}$. Accordingly, co-immunoprecipitation experiments provided no evidence for association of complex II subunits SDH-A, -B and -C with ANT, ATP synthase or cytochrome c oxidase in murine heart mitochondria. Hence, it is likely that the inhibitory effects on HPV that we observed in our experiments result from modulation of several mitochondrial protein complexes independently involved in the signalling cascade such as ROS-producing complex II and ANT-regulated mitochondrial permeability transition pore.

Keywords Complex II · Hypoxia · Hypoxic pulmonary vasoconstriction HPV · Intra-acinar arteries · Mitochondrial ATP-sensitive potassium channel mitoK$_{ATP}$ · Respiratory chain · Succinate dehydrogenase SDH · Oxygen sensing

---

R. Paddenberg (✉)
Institute of Anatomy and Cell Biology, Justus-Liebig-University, ECCPS, Giessen, Germany
e-mail: Renate.Paddenberg@anatomie.med.uni-giessen.de

1 Introduction

Hypoxic pulmonary vasoconstriction (HPV) is a local reflex directing the blood flow from poorly to well oxygenated regions of the lung thereby matching perfusion to ventilation. In search of the underlying molecular mechanisms of oxygen sensing we observed an essential role of complex II (= succinate dehydrogenase, SDH) of the mitochondrial respiratory chain (Paddenberg et al. 2006). According to work published by Ardehali et al. (2004), SDH is part of a macromolecular supercomplex located in the inner mitochondrial membrane which possesses potassium channel activity. Based on its pharmacological profile, it has been identified as mitochondrial ATP-sensitive potassium channel (mitoK$_{ATP}$) (Ardehali et al. 2004). Interestingly, this channel also participates in another oxygen-regulated process, i.e. hypoxic preconditioning (Gross and Fryer 1999). Ardehali et al. co-immunoprecipitated at least 5 proteins from highly purified fractions of inner membranes of rat liver mitochondria: Adenine nucleotide translocator (ANT), ATP synthase, phosphate carrier, ATP-binding cassette protein-1 and SDH subunits A and B. In proteoliposomes containing this multiprotein complex, potassium transport was increased by mitoK$_{ATP}$ activators (diazoxide, pinacidil), and reduced by its blockers (5-hydroxydecanoate, glibenclamide), SDH-inhibitors (malonate, 3-nitropropionic acid) and ATP. Inhibitors of ANT (atractyloside, bongkrekic acid) had no impact on channel activity.

Here, we analysed whether SDH is also a component of such a multiprotein mitoK$_{ATP}$ complex in cardiovascular mitochondria, and whether its pharmacological modulation interferes with HPV.

2 Methods

2.1 Videomorphometry of IAA

HPV of IAA was estimated by videomorphometric analysis of murine precision cut lung slices (PCLS) as described earlier (Paddenberg et al. 2006). The hypoxic response of individual arteries with inner diameters between 20 and 30 μm was recorded as changes in the luminal area. The values obtained at the beginning of the experiments were set as 100%, and vasoreactivity was expressed as relative decrease or increase of these areas. Data are presented as means ± standard error of the mean (SEM). A single vessel per PCLS was analysed.

For statistical analysis of the impact of various drugs on HPV, the values obtained immediately before exposure to hypoxic gassed medium ± drug were set as 100% (not shown) and the values of corresponding time points were analyzed using SPSS 15.0. The Kruskal-Wallis- followed by the Mann-Whitney-test was performed to compare the means of the different experimental groups. The threshold for significance was set at $p \leq 0.05$.

2.2 Isolation of Mitochondria from Murine Heart

Isolation of mitochondria from murine hearts was performed by a combination of differential centrifugation and sucrose density gradient centrifugation as described in "Isolation of Mitochondria Manual" from MitoSciences (www.mitosciences.com/PDF/mitos.pdf).

2.3 Co-immunoprecipitation of Mitochondrial Proteins and Western Blotting

Co-immunoprecipitation experiments employing extracts of murine heart mitochondria were performed with the Complex II Immunocapture Kit (MitoSciences, Oregon, USA) consisting of protein G-agarose beads conjugated with complex II specific or unspecific (control) antibodies. Eluates of the beads (= pellets) and corresponding supernatants were analysed by Western blotting. As primary antibodies we used mouse anti-OxPhos Complex II 70 kDa (= SDH-A; Molecular Probes/Invitrogen, Karlsruhe, Germany), mouse anti-OxPhos Complex II 30 kDa (= SDH-B; Molecular Probes/Invitrogen), mouse anti-complex V subunit α (= ATP synthase; Molecular Probes/Invitrogen), mouse anti-cytochrome C oxidase subunit IV (Abcam, Cambridge, UK), goat anti-ANT (Santa Cruz, Heidelberg, Germany), as well as an own rabbit anti-SDH-C antibody obtained by immunization against a peptide consisting of amino acids 29-52 of SDH-C. As secondary antibodies HRP-conjugated anti-mouse IgG, anti-goat IgG or anti-rabbit IgG were used (all from Pierce/Perbio Science Deutschland GmbH, Bonn, Germany). Bound antibodies were visualized by enhanced chemiluminescence.

3 Results

3.1 Impact of mitoK$_{ATP}$ Modulators on HPV

The use of PCLS facilitates the investigation of HPV of small IAA. Exposure of the vessels to hypoxic gassed medium induced a distinct reduction of the luminal area (Fig. 1) whereas in normoxic gassed medium the areas were unchanged (not shown). Application of the SDH inhibitor malonate completely suppressed HPV whereas the U46619-induced contraction was unaffected (Fig. 1A). For analysis of the impact of classical modulators of ATP-sensitive potassium channels on HPV, we applied blockers and openers which were either specific for the mitochondrial channel or non-selective in that they modulate both surface and mitochondrial channels. The mitoK$_{ATP}$-specific blocker 5-hydroxy – decanoate (Fig. 1B) and the non-selective inhibitor glibenclamide (not shown) completely suppressed HPV whereas the response to U46619 was unchanged. Application of the mitochondrial channel

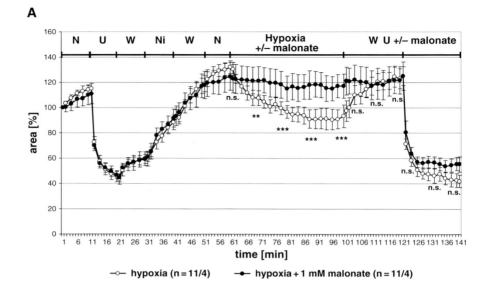
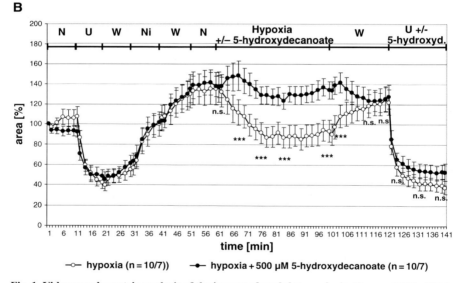

Fig. 1 Videomorphometric analysis of the impact of modulators of mitoK$_{ATP}$ on HPV of IAA. For adaptation to the chamber, PCLS were incubated with normoxic medium (N). The viability of the arteries was assessed by the successive application of the vasoconstrictor U46619 (U) and the vasodilatator nipruss (Ni). After washing out of the drugs (W), PCLS were incubated with hypoxic medium (pregassed with 1% O$_2$) alone or supplemented with modulators of mitoK$_{ATP}$. At the end of each experiment the specificity of the impact of the drug for HPV was investigated by its simultaneous application with U46619. Drugs were added at the concentrations given in the legends of the individual graphs. At the given time points the differences between both groups were tested for significance. n.s.: not significant, *: $p \leq 0.05$, **: $p \leq 0.01$, ***: $p \leq 0.001$

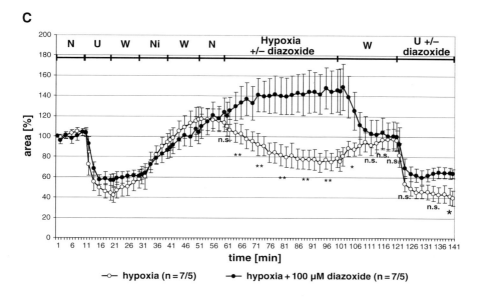

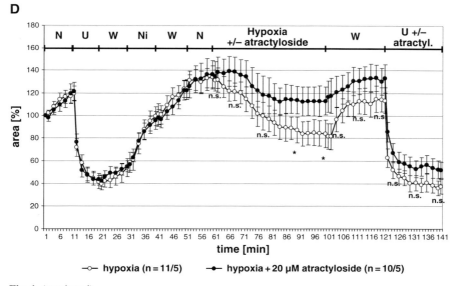

Fig. 1 (continued)

specific activators diazoxide (Fig. 1C) or the non-selective opener pinacidil (not shown) resulted not only in suppression of HPV but even induced vasodilatation. Again, the effects were specific for HPV in that U46619-induced vasoconstriction was unchanged. Finally, we tested the impact of ANT inhibitors on HPV.

Atractyloside (Fig. 1D) and bongkrekic acid (not shown) induced a significant reduction and inhibition of HPV, respectively.

3.2 Co-immunoprecipitation of Mitochondrial Proteins

To clarify whether in the murine cardiopulmonary system SDH is part of a multiprotein complex we performed co-immunoprecipitation experiments on extracts of isolated murine heart mitochondria. Both the precipitates (= pellet) and the corresponding supernatants were analysed by Western blotting (Fig. 2). SDH-A-immunoreactivity was clearly present in the pellet obtained with the beads conjugated with complex II specific antibodies whereas it was absent when beads coupled with unspecific antibodies were used. In addition, both supernatants contained SDH-A (Fig. 2A). Comparable results were obtained for SDH-C (Fig. 2C). Distinct SDH-B-immunoreactivity was precipitated employing specific beads whereas a very weak band was detectable when control beads were used (Fig. 2B). In the corresponding supernatants SDH-B was detectable only after longer exposure of the x-ray films (not shown). Immunoreactivity for cytochrome C oxidase – which is not expected to be part of the multiprotein complex – was detectable exclusively in the supernatants but not in the pellets (Fig. 2D). To evaluate whether the postulated multiprotein complex exists in mouse heart mitochondria we analysed the immunoprecipitates for the presence of ATP synthase and ANT. In both cases we were able to detect the proteins in the supernatants, but not in the pellets (Fig. 2E, F).

4 Discussion

Here, we demonstrate that HPV of IAA can be inhibited by several modulators of the postulated mitoK$_{ATP}$. The pharmacological profile of HPV inhibition, however, differs markedly from that of mitoK$_{ATP}$ modulation. For instance, both 5-hydroxydecanoate and diazoxide potently inhibited HPV whereas they have contrasting effects on mitoK$_{ATP}$ (Ardehali et al. 2004). Thus, HPV is not triggered by simple opening or closure of a multiprotein mitoK$_{ATP}$ in the composition suggested by Ardehali et al. (2004). According to these functional data, our co-immunoprecipitation experiments provide no evidence for interaction of SDH-subunits with ANT and ATP synthase in mitochondria isolated from the cardiovascular system. Hence, it is likely that the inhibitory effects on HPV that we observed in our experiments are not caused by targeting one and the same multiprotein complex by all inhibitors. Instead, they may result from modulation of several, independently involved mitochondrial protein complexes such as ROS-producing complex II (Paddenberg et al. 2006; Guzy et al. 2008) and ANT-regulated mitochondrial permeability transition pore (Leung and Halestrap 2008).

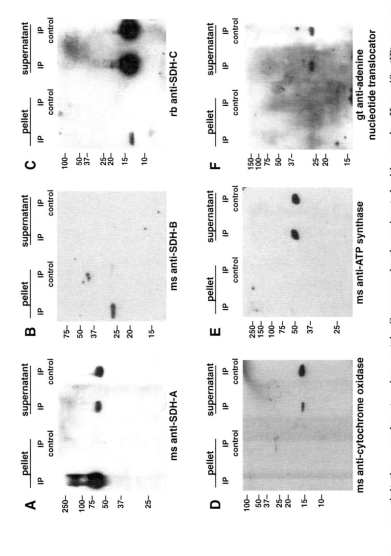

Fig. 2 Co-immunoprecipitation experiments using protein G agarose beads conjugated with complex II specific (IP) or unspecific (control IP) antibodies on mitochondrial extracts. The precipitates (= pellet) and the corresponding supernatants were analysed by Western blotting using antibodies as given in the legends of the individual blots. Molecular weight markers in kDa are given on the left of the blots. ms: mouse, rb: rabbit, gt: goat. Comparable results were obtained in at least 3 independent co-immunoprecipitation experiments

Acknowledgments The financial support of the Deutsche Forschungsgemeinschaft (Sonderforschungsbereich 547, project C1) is gratefully acknowledged.

References

Ardehali H., Chen Z., Ko Y., Mejía-Alvarez R. & Marbán E. 2004, Multiprotein complex containing succinate dehydrogenase confers mitochondrial ATP-sensitive K+ channel activity, Proc Natl Acad Sci U S A, 101(32):11880–5.

Gross G.J. & Fryer R.M. 1999, Sarcolemmal versus mitochondrial ATP-sensitive K+ channels and myocardial preconditioning, Circ Res, 84(9):973–9.

Guzy RD, Sharma B, Bell E, Chandel NS & Schumacker PT. (2008), Loss of the SdhB, but Not the SdhA, subunit of complex II triggers reactive oxygen species-dependent hypoxia-inducible factor activation and tumorigenesis, Mol Cell Biol, 28(2):718–31.

Leung A.W. & Halestrap A.P. 2008, Recent progress in elucidating the molecular mechanism of the mitochondrial permeability transition pore, Biochim Biophys Acta, 1777(7–8):946–52.

Paddenberg R., König P., Faulhammer P., Goldenberg A., Pfeil U. & Kummer W. 2006, Hypoxic vasoconstriction of partial muscular intra-acinar pulmonary arteries in murine precision cut lung slices, Respir Res, 7:93.

Oxygen Sensing in the Brain – *Invited Article*

Frank L. Powell, Cindy B. Kim, Randall S. Johnson and Zhenxing Fu

Abstract Carotid body arterial chemoreceptors are essential for a normal hypoxic ventilatory response (HVR) and ventilatory acclimatization to hypoxia (VAH). However, recent results show that O_2-sensing in the brain is involved in these responses also. O_2-sensing in the rostral ventrolateral medulla, the posterior hypothalamus, the pre-Bötzinger complex and the nucleus tractus solitarius contribute to the acute HVR. Chronic hypoxia causes plasticity in the brain that contributes to VAH and represents another time domain of central O_2-sensing. The cellular and molecular mechanisms of acute O_2-sensing in the brain remain to be determined but they appear to involve O_2-sensitive ion channels and heme oxygenase-2, which acts by a different mechanism than has been described for the carotid body. It is not known if plasticity in such mechanisms of acute central O_2-sensitivity contributes to VAH. However, O_2-sensitive changes in gene expression in the brain do contribute to VAH and demonstrate another mechanism of O_2-sensing that is important for ventilatory control. This time domain of O_2-sensing in the brain involves gene expression under the control of hypoxia inducible factor-1α (HIF-1α) and potentially several HIF-1α targets, such as erythropoietin, endothelin-1, heme oxygenase and tyrosine hydroxylase.

Keywords Central chemoreceptor · Heme oxygenase · Hypoxia inducible factor-1 · Pre-Bötzinger complex · Rostral ventrolateral medulla

1 Background

The effect of hypoxia to stimulate breathing and the cardiovascular system has been well known for more than a century (reviewed by Neubauer 2004). In the 1920s and 1930s, Heymans and his colleagues did critical experiments showing the importance of carotid body chemoreceptors for the hypoxic ventilatory response (HVR)

F.L. Powell (✉)
Department of Medicine, White Mountain Research Station, University of California San Diego, La Jolla, CA 92093, USA
e-mail: fpowell@ucsd.edu

and this work was rewarded with the Nobel Prize for Physiology or Medicine in 1938. However, it was known that hypoxia continued to stimulate ventilation and the cardiovascular system in the absence of carotid body chemoreceptors and this lead Comroe to propose that other arterial chemoreceptors or sites in the central nervous system (CNS) were sensitive to hypoxia. Subsequently, Comroe showed the importance of the aortic body chemoreceptors for hypoxic responses, which established the idea that arterial chemoreceptors are the dominant O_2-sensitive chemoreceptors for cardiopulmonary control. Indeed, when the carotid and aortic body chemoreceptors are eliminated, hypoventilation ensues and the acute HVR is essentially obliterated. Arterial chemoreceptors are also necessary for other time domains of the HVR, such as ventilatory acclimatization to hypoxia (VAH), which is the time-dependent increase in ventilation and O_2-sensitivity that occurs with sustained hypoxia.

Modern studies of different time domains of the HVR have implicated the CNS as an important site of O_2-senstivity also. These include developmental studies of the biphasic HVR in neonates (i.e. hyperventilation followed by ventilatory decline during sustained hypoxia) and studies of respiratory central pattern generators that exhibit intrinsic chemosensitivity. However, the seminal observation of central O_2-sensitivity in the modern era was arguably that of Sun et al. (1992) that showed focal hypoxia produced by cyanide in the medulla excites sympathoexcitatory neurons. This work was quickly recognized by the community of scientists studying respiratory control and has lead to considerable advances in our understanding of the sites, mechanisms and physiological importance of O_2-sensing in the brain.

While the depressing effects of hypoxia on CNS neurons may function as sensing mechanisms instead of representing failing metabolism or energetics, they will not be considered further in this article. Similarly, the activation of inhibitory networks by hypoxia that contribute to ventilatory decline will not be considered because that story is complicated by developmental changes as this phenomenon is observed most prominently in neonatal animals (cf. Neubauer and Sunderram 2004).

2 Sites of Central O_2-Chemosensitivity

Several methods have been used to demonstrate nuclei in the CNS that are directly activated by hypoxia and this was recently reviewed by Neubauer and Sunderram (2004). These include mapping c-fos expression in animals exposed to hypoxia with and without arterial chemoreceptor denervation, neural recordings in a region with focal hypoxia from cyanide microinjections, and neural recordings from cultured cells isolated from putative O_2-sensitive regions. The effects of focal hypoxia from cyanide is reversible, dose dependent and is not replicated by hypercapnia or acidosis. Evidence for O_2-sensitivity is published for the C1 sympathoexcitatory region of the rostral ventrolateral medulla, the posterior hypothalamus, the pre-Bötzinger complex and the nucleus tractus solitarius.

Neurons in the C1 region of the brainstem were among the first to be described as excited by hypoxia and having cardiovascular or respiratory effects, as mentioned above (Sun et al. 1992). Neurons with intrinsic O_2-sensitivity in the posterior hypothalamus also increase sympathetic activity, heart rate and blood pressure and it is interesting that these project to the C1 sympathoexcitatory region (Neubauer and Sunderram 2004). However, experiments using focal hypoxia have demonstrated intrinsic O_2-sensitivity in C1 neurons. The pre-Bötzinger complex is adjacent to the C1 region and also exhibits O_2-sensitivity, which suggests common O_2-sensing mechanisms for both sites. Focal hypoxia in the pre-Bötzinger complex induces gasping, which is a form of autoresuscitation and thought to be physiologically significant O_2-sensitivity in the central rhythm generator for respiration (Solomon and Edelman 2003). O_2-sensitivity in the nucleus tractus solitarius is interesting because this is the site of the primary synapse from carotid body chemoreceptors. This suggests possible interactions between central and peripheral O_2-sensitivity.

3 Mechanisms of Central O_2-Chemosensitivity

Studies of the carotid bodies have revealed several O_2-sensitive ion channels that could play a role in the CNS also (reviewed by Neubauer and Sunderram 2004). Hypoxic inhibition of potassium (K^+) channels contributes to depolarization in glomus cells in the carotid body. For example, voltage-independent leak K^+ currents (TASK-1) channels contribute to O_2-sensing in carotid bodies and are present in the CNS in O_2-sensitive regions. Large-conductance calcium-activated K^+ channels are also inhibited by hypoxia in glomus cells. Finally, hypoxia could decrease K^+ channel conductance by decreasing ATP levels, which could affect ATP-activated K^+ channels, or by producing neuromodulators to act on G-protein coupled inward-rectifying K^+ channels.

Other ion channels proposed to contribute to O_2-sensitivity in carotid bodies include calcium (Ca^{++}) and sodium (Na^+) channels, which can increase conductance in hypoxia (reviewed by Neubauer 2004). L-type Ca^{++} currents are increased by hypoxia in glomus cells and brainstem neurons, and there is indirect evidence for hypoxic activation of Ca^{++} channels in dissociated neurons from the RVLM. Hypoxia generally decreases Na^+ conductivity, presumably to protect neurons during decreased energy supply. However, hypoxia increases the persistent Na^+ current in glomus cells (Donnelly 2008) and this channel is important for the pacemaker activity in the pre-Bötzinger complex (Del Negro et al. 2001). Hence, O_2-sensitivity of persistent Na^+ currents might explain the O_2-sensitivity of the pre-Bötzinger complex described above.

Beside ion channels, there are two forms of heme oxygenase (HO) that could function as an O_2-sensor in the CNS. HO depends on O_2 to catalyze heme into CO, biliverdin and iron, all of which can affect ion channels (reviewed by Powell and Fu 2008). HO-2 is expressed constitutively in O_2-sensitive regions of the RVLM and pre-Bötzinger complex while HO-1 is only observed in these regions after exposure to chronic hypoxia. Hence, HO-2 is a more likely O_2-sensor in the CNS for acute hypoxia. It is interesting that HO is necessary for hypoxic excitation of neurons in the RVLM while it is inhibitory in the carotid body (Neubauer and Sunderram 2004).

A novel mechanism of O_2-sensitivity described for the nucleus tractus solitarius (NTS) involves of S-nitrosothiols (SNOs), which are released from hemoglobin when it is deoxygenated (Lipton et al. 2001). SNOs are complexes of NO bound to the sulfhydryl groups in cysteine. The specific SNO molecule that stimulates ventilation in the NTS appears to be a metabolite of S-nitrosoglutathione (GSNO) that is cleaved by γ-glutamyl transpeptidase (γ-GT). GSNO could be formed when nitric oxide synthase is activated in neurons in the NTS, where it is known to occur. It is suggested that γ-GT is present in neurons but not the vasculature to explain the positive effect of GSNO in the NTS on ventilation versus in its lack of effect in the vasculature, where SNOs transfer NO from hemoglobin and cause vasodilation.

4 Sensitivity to Chronic Hypoxia: O_2-Sensitive Gene Expression in the Brain

Chronic sustained hypoxia increases (a) ventilation in hypoxia, (b) ventilation when normoxia is restored, and (c) ventilatory sensitivity to hypoxia, as demonstrated by an increased slope of the isocapnic HVR (Powell et al. 1998). These changes are collectively called ventilatory acclimatization to hypoxia (VAH), which involves both increased O_2-sensitivity of arterial chemoreceptors (Powell 2007) and plasticity in the CNS (Powell et al. 2000). Hence, the brain has multiple time domains of O_2-sensitivity that can elicit different ventilatory responses to acute versus chronic hypoxia.

The physiological mechanisms of the CNS component of VAH are not known but they may include plasticity in (a) the CNS mechanisms of acute O_2-sensing, (b) the CNS integration of peripheral O_2-sensitive reflexes (e.g. carotid body chemoreflex), and/or (c) the CNS integration of other non-chemoreflex ventilatory control pathways such as reflexes from pulmonary vagal chemoreceptors, or respiratory rhythm generators. VAH does not occur in animals without functional carotid bodies (Dempsey and Forster 1982) so generally it has been assumed that central O_2-sensitivity is not important for VAH. However, modern experimental designs used to study central O_2-sensitivity have not been applied systematically in chronically hypoxic preparations to rigorously test the hypothesis that VAH involves enhancement of CNS mechanisms of acute O_2-sensing. Similar comments apply to the possibilities for plasticity in CNS integration of non-chemoreflex pathways or central pattern generation of the respiratory rhythm. However, there is strong experimental evidence for enhanced CNS integration of arterial chemoreceptor afferent input. In an experimental preparation that isolates the reflex effects of the CNS, the phrenic nerve response to fixed levels of carotid sinus nerve stimulation is increased after chronic hypoxia (Dwinell and Powell 1999).

Any of these mechanisms for the CNS component of VAH are likely to involve changes in gene expression. Therefore, O_2-sensitive gene expression represents another mechanism of O_2-sensing in the brain that can affect cardiopulmonary function. Hypoxia increases immediate early gene expression in the CNS, which in turn produce transcription factors for other genes (Banasiak et al. 2000). Hypoxia also induces transcription factors such as nuclear factor-κB and activator protein-1 but their effects on cardiopulmonary responses to hypoxia are largely unknown (reviewed by Powell and Fu 2008). To date, the most important regulator of O_2-sensitive gene expression in the brain shown to have a role in ventilatory control is hypoxia inducible factor-1 (HIF-1). HIF-1α, which is the oxygen regulated part of the functional HIF-1 heterodimer, increases in respiratory centers of the CNS after as little as one hour and HIF-1α expression in the brain is necessary for normal VAH (reviewed by Powell and Fu 2008). Hence, O_2-sensing by HIF-1α could be involved in VAH by increasing the expression of several genes with products known to modulate the hypoxic ventilatory response. These are considered next.

HIF-1 was originally discovered as an important transcription factor for erythropoietin (EPO), which is released from the kidneys in response to hypoxia and increases the oxygen carrying capacity of blood by stimulating erythropoiesis in bone marrow (Semenza 1999). However, EPO and its receptor (EPO-R) have also been localized to neurons and glia in the CNS, including the pre-Bötzinger complex and nucleus tractus solitarius (Soliz et al. 2005), where it must have different effects. Transgenic mice that overexpress human EPO in the brain can sustain a virtually normal HVR following carotid body denervation, which ablates the HVR in wild type mice, and they show enhanced VAH compared to wild type mice (Soliz et al. 2005). Soluble EPO receptor (sEPO-R) may be the primary mediator for EPO effects on VAH in the brain. Chronic hypoxia downregulates sEPO-R in the brain and intracerebroventricular infusion of sEPO-R decreases EPO and reverses VAH in chronically hypoxic mice (Soliz et al. 2007). Experiments have not been done yet to determine if EPO and its receptor are involved directly in the acute mechanisms of central O_2-sensitivity or modulation of other pathways, e.g. arterial chemoreflexes or respiratory rhythm generators.

Endothelin- (ET-1) is a vasoactive peptide that is regulated by HIF-1 in hypoxia. ET-1 is widely distributed in the brain, where it plays an important role in stress responses, and in other O_2-sensitive tissues such as the carotid body and pulmonary artery (reviewed by Powell and Fu 2008). ET-1 causes enhanced O_2-sensitivity in the carotid bodies with chronic hypoxia (Chen et al. 2002) but it remains to be determined if ET-1 affects acute O_2-sensitivity in the brain or other ventilatory control pathways.

Heme oxygenase-1 (HO-1) is also induced by HIF-1 (reviewed by Powell and Fu 2008). Chronic hypoxia increases HO-1 in the rostral ventrolateral medulla and this has been hypothesized to contribute to ventilatory acclimatization to hypoxia (Mazza et al. 2001). However as discussed above, HO-1 is not expressed in the ventrolateral medulla of normoxic control animals, in contrast to HO-2. It remains to be determined if increased HO-1 with chronic hypoxia acts by the same putative mechanism for HO-2 in acute O_2-sensing in the CNS, and adds to it, or if it even has an effect on ventilation.

Dopamine is another factor that can modulate the HVR by central mechanisms and these effects change during exposure chronic hypoxia (Huey et al. 2003). The rate limiting enzyme for dopamine synthesis is tyrosine hydroxylase, which is generally thought to be under control of HIF-1 (reviewed by Powell and Fu 2008). HIF-1α is increased by hypoxia selectively in cardiorespiratory centers of the brainstem and is co-localized with tyrosine hydroxylase in selected catecholaminergic cell groups in the brainstem, e.g. A1C1 and A2C2 (Pascual et al. 2001). Note these are the same cell groups that were observed to have increased catecholamines in mice that over-express EPO in the brain (Soliz et al. 2005). The precise roles of tyrosine hydroxylase and EPO in such changes with chronic hypoxia remain to be determined.

Summarizing, there are several potential neurochemical mechanisms whereby O_2-sensitive changes in gene expression in the brain, mediated by HIF-1, could contribute to plasticity in the acute hypoxic ventilatory response. However, no direct relationships have been demonstrated yet between acute mechanisms of O_2-sensing in the brain and this longer time domain of O_2-sensing.

5 Lessons from Central CO_2-Sensing in the Brain

The importance of central CO_2 chemoreceptors has been recognized much longer than O_2-sensing in the brain, so it may be instructive to compare and contrast the two phenomena. For CO_2, Guyenet points out a distinction between (a) *chemosensitivity* of CNS neurons and (b) central *chemoreceptors* (Guyenet et al. 2008). This is similar to the distinction that Dawes and Comroe (1954) drew in their review of peripheral chemoreflexes from the heart and lungs for (a) *chemoreflexes*, i.e. "reflexes initiated by a chemical substance whether they act upon true chemoreceptors, other types of sensory receptors or upon nerve endings themselves," versus (b) *chemoreceptors*, i.e. "sensory nerve endings which normally respond to changes in their natural chemical environment in health and disease." While many different experimental methods and preparations have revealed multiple sites and mechanisms of CO_2-sensitivity, Guyenet and colleagues (2008) argue that the evidence for multiple CO_2 chemoreceptors is not as strong. Applying this analysis to the problem of O_2-sensing in the brain, leads to the conclusion that there are multiple sites of O_2-sensitivity in the brain but central O_2-chemoreceptors have not been identified to date.

For central CO_2-sensitivity, there are also competing theories for (a) *distributed chemoreception* versus (b) *specialized chemoreceptors*. Again, Guyenet and colleagues (2008) argue that the evidence for distributed chemoreception is not as strong as that for specialized chemoreceptors that are primarily (if not exclusively) in the retrotrapezoid nucleus. However, there are several studies using whole animal awake or anesthetized preparations showing physiological ventilatory responses to focal hypercapnia or acidosis in sites of the brainstem besides the retrotrapezoid nucleus (e.g. Coates et al. 1993). Further study will be necessary to determine if such distributed responses can be explained by universal properties of H^+/CO_2-sensitivity in neurons that just happen to be in a respiratory pathway or if some neurons in these pathways are true chemoreceptor. The same questions can be asked and need to be answered for O_2-sensing in the brain.

Finally, there are more teleological questions about central CO_2-sensitivity (Guyenet et al. 2008): Why are there so many CO_2-sensitive molecules in CNS neurons and why are there so many CO_2-sensitive sites in the brain? Of course, these questions are impossible to answer experimentally. However, it would increase our understanding of ventilatory control if physiologically significant roles could be described for multiple mechanisms of chemosenstivity or multiple chemoreceptor sites. Multiple CO_2-sensing mechanisms are hypothesized to increase the range and sensitivity of responsiveness (Jiang et al. 2005) and there is evidence that different central CO_2 chemoreceptor sites have different effects depending on sleep state (Nattie and Li 2002). Similar hypotheses should be explored for O_2-sensing in the brain and cardiorespiratory reflexes.

Acknowledgments Supported by NIH R01-HL081823 and the University of California White Mountain Research Station.

References

Banasiak, K. J. Xia, Y. and Haddad, G. G. 2000, Mechanisms underlying hypoxia-induced neuronal apoptosis, *Prog. Neurobiol.* vol. 62, no. 3, pp. 215–249.
Chen, J. He, L. Dinger, B. Stensaas, L. and Fidone, S. 2002, Role of endothelin and endothelin A-type receptor in adaptation of the carotid body to chronic hypoxia, *Am. J. Physiol. Lung Cell. Mol. Physiol.*, vol. 282, no. 6, p. L1314–L1323.
Coates, E. L. Li, A. and Nattie, E. E. 1993, Widespread sites of brain stem ventilatory chemoreceptors, *J. Appl. Physiol.*, vol. 75, pp. 5–14.
Dawes, G. S. and Comroe, J. H. Jr. 1954, Chemoreflexes from the heart and lungs, *Physiol. Rev.* vol. 34, no. 2, pp. 167–201.
Del Negro, C.A., Johsnon, S.M., Butera, J.R. and Smith, J.C. 2001, Models of respiratory rhythm generation in the pre-Bötzinger complex. III. Experimental tests of model predictions. *J. Neurophsiol.* vol. 86, pp. 59–74.
Dempsey, J. A. and Forster, H. V. 1982, Mediation of ventilatory adaptations, *Physiol. Rev.* vol. 62(1), pp. 262–346.
Donnelly, D. F. 2008, Spontaneous action potential generation due to persistent sodium channel currents in simulated carotid body afferent fibers, *J. Appl. Physiol.* vol. 104, no. 5, pp. 1394–1401.
Dwinell, M. R. and Powell, F. L. 1999, Chronic hypoxia enhances the phrenic nerve response to arterial chemoreceptor stimulation in anesthetized rats, *J. Appl. Physiol.* vol. 87, pp. 817–823.
Guyenet, P. G. Stornetta, R. L. and Bayliss, D. A. 2008, Retrotrapezoid nucleus and central chemoreception, *J. Physiol.* vol. 586, no. 8, pp. 2043–2048.
Huey, K. A. Szewczak, J. M. and Powell, F. L. 2003, Dopaminergic mechanisms of neural plasticity in respiratory control: transgenic approaches, *Respir. Physiol. Neurobiol.* vol. 135, no. 2–3, pp. 133–144.
Jiang, C. Rojas, A. Wang, R. and Wang, X. 2005, CO_2 central chemosensitivity: why are there so many sensing molecules?, *Respir. Physiol. Neurobiol.* vol. 145, no. 2–3, pp. 115–126.

Lipton, A. J. Johnson, M. A. Macdonald, T. Lieberman, M. W. Gozal, D. and Gaston, B. 2001, S-Nitrosothiols signal the ventilatory response to hypoxia, *Nature* vol. 413, pp. 171–174.

Mazza, E. Thakkar-Varia, S. Tozzi, C. A. and Neubauer, J. A. 2001, Expression of heme oxygenase in the oxygen-sensing regions of the rostral ventrolateral medulla, *J. Appl. Physiol.* vol. 91, no. 1, pp. 379–385.

Nattie, E. E. and Li, A. 2002, CO_2 dialysis in nucleus tractus solitarius region of rat increases ventilation in sleep and wakefulness, *J. Appl. Physiol.* vol. 92, no. 5, pp. 2119–2130.

Neubauer, J. A. 2004, Comroe's study of aortic chemoreceptors: a path well chosen, *J. Appl. Physiol.*, vol. 97, pp. 1595–1596.

Neubauer, J. A. and Sunderram, J. 2004, Oxygen-sensing neurons in the central nervous system, *J. Appl. Physiol.* vol. 96, no. 1, pp. 367–374.

Pascual, O. Denavit-Saubie, M. Dumas, S. Kietzmann, T. Ghilini, G. Mallet, J. and Pequignot, J. M. 2001, Selective cardiorespiratory and catecholaminergic areas express the hypoxia-inducible factor-1α (HIF-1α) under *in vivo* hypoxia in rat brainstem, *Europ. J. Neurosci.* vol. 14, pp. 1981–1991.

Powell, F. L. 2007, The influence of chronic hypoxia upon chemoreception, *Respir. Physiol. Neurobiol.* vol. 157, no. 1, pp. 154–161.

Powell, F. L. and Fu, Z. 2008, HIF-1 and ventilatory acclimatization to chronic hypoxia, *Respir. Physiol. Neurobiol.* vol. 164, pp. 282–7.

Powell, F. L. Huey, K. A. and Dwinell, M. R. 2000, Central nervous system mechanisms of ventilatory acclimatization to hypoxia, *Respir. Physiol.* vol. 121, pp. 223–236.

Powell, F. L. Milsom, W. K. and Mitchell, G. S. 1998, Time domains of the hypoxic ventilatory response, *Respir. Physiol.* vol. 112, pp. 123–134.

Semenza, G. L. 1999, Regulation of mammalian O_2 homeostasis by hypoxia-inducible factor 1, *Annu. Rev. Cell Dev. Biol.* vol. 15, pp. 551–578.

Soliz, J. Gassmann, M. and Joseph, V. 2007, Soluble erythropoietin receptor is present in the mouse brain and is required for the ventilatory acclimatization to hypoxia, *J. Physiol.* vol. 583, no. Pt 1, pp. 329–336.

Soliz, J. Joseph, V. Soulage, C. Becskei, C. Vogel, J. Pequignot, J. M. Ogunshola, O. and Gassmann, M. 2005, Erythropoietin regulates hypoxic ventilation in mice by interacting with brainstem and carotid bodies, *J. Physiol.* vol. 568, pp. 559–571.

Solomon, I. C. and Edelman, N. H. 2003, Oxygen sensing by the brainstem and respiatory control, in *Oxygen Sensing: Responses and adaptation to hypxoia*, S. Lahiri, G. Semenza, and N. R. Prabhakar, eds. Marcel Dekker, New York, pp. 651–670.

Sun, M. K. Jeske, I. T. and Reis, D. J. 1992, Cyanide excites medullary sympathoexcitatory neurons in rats, *Am. J. Physiol.* vol. 262, no. 2 Pt 2, p. R182–R189.

The Central Respiratory Chemoreceptor: Where Is It Located? – *Invited Article*

Y. Okada, S. Kuwana, Z. Chen, M. Ishiguro and Y. Oku

Abstract We review previous reports on the localization of the central chemoreceptor focusing on our studies that used various experimental techniques including lesioning (brainstem transection and removal of pia mater), analyses of neuronal responses to CO_2 by electrophysiological and optical recording, mapping of CO_2-excitable neurons by c-fos immunohistochemistry and local acidic stimulation. Among these experimental techniques, voltage imaging with calculation of cross correlation coefficients between the respiratory output activity and each pixel, i.e., correlation coefficient imaging technique, enabled us to effectively analyze imaging data without empirical signal processing. The reviewed studies have indicated that the most superficial layer of the rostral ventral medulla, i.e., the surface portions of the nucleus retrotrapezoideus/parafacial respiratory group, nucleus parapyramidal superficialis and nucleus raphe pallidus, is important in central chemoreception. We suggest that one of the major respiratory rhythm generators, i.e., the preBötzinger complex, is not chemosensitive in itself or rather inhibited by CO_2. Based on our detailed analysis of c-fos immunohistochemistry, we propose a cell-vessel architecture model for the central respiratory chemoreceptor. Primary chemoreceptor cells are mainly located beneath large surface vessels within the marginal glial layer of the ventral medulla, and surround fine penetrating vessels that branch from a large surface vessel. Respiratory neurons in the rostral portion of the ventral respiratory group could be intrinsically chemosensitive, but their role in chemoreception might be secondary. Definitive identification of chemosensitive sites and chemoreceptor cells needs further studies.

Keywords Respiratory control · Ventral medulla · Neonatal rat · Isolated brainstem-spinal cord preparation · Marginal glial layer · Voltage-sensitive dye · Voltage imaging · Correlation coefficient imaging · Brainstem lesioning · c-fos

---

Y. Okada (✉)
Department of Medicine, Keio University Tsukigase Rehabilitation Center, Izu, Japan
e-mail: yasumasaokada@1979.jukuin.keio.ac.jp

1 Introduction

The central respiratory chemoreceptor plays an important role in respiratory control and is considered to exist mainly in the lower brainstem. However, the exact localization of the central chemoreceptor in the brain has not been determined, any specific anatomical structure of the central chemoreceptor has not been found, and therefore the chemoreceptor cell has not been identified. In the present article, we review previous reports on the localization of the central chemoreceptor placing emphasis on our own studies that were conducted using various experimental techniques.

2 Lesioning of Brainstem

To analyze the localization of the central chemoreceptor, we conducted lesioning experiments in the isolated brainstem-spinal cord preparation of the neonatal rat (2–4 days old, n=21). Viability of this preparation is maintained without blood circulation, and lesioning procedures, which otherwise should disturb blood circulation, could be effectively applied.

2.1 Transverse Sectioning

By monitoring the respiratory output from the C4 ventral root, we changed the superfusate CO_2 fraction from 2 to 8% to evaluate CO_2 responsiveness as described previously (Okada et al. 1993). We then transversely sectioned the brainstem (n=12) serially in the rostro-caudal direction using fine ophthalmologic scissors as our previous study (Okada et al. 1998). We measured the C4 burst frequency (fR) with 2 and 8% CO_2, and calculated the increase in fR accompanied by elevation of CO_2 as an index of CO_2 responsiveness (ΔfR) at each of 7 sectioning levels. As shown in Fig. 1, fR was increased in response to elevation of CO_2 when the brainstem was sectioned at levels from 1st to 6th, but was decreased when sectioned at the 7th level. These results indicate that the central chemoreceptor is located rostral to the 7th transection level. The 7th level is slightly caudal to the most caudal root of the vagal nerve and corresponds to the most rostral level of the preBötzinger complex (Feldman and Del Negro 2006; Paxinos et al. 1991).

Next, we examined the effect of transverse splitting of the brainstem at the 7th transection level (n=3) by monitoring inspiratory output activities simultaneously from the glossopharyngeal (IX) and C4 ventral roots. Before splitting, the IX and C4 activities were synchronized, and fR of both activities was increased in response to elevation of CO_2 to 8%. After splitting, the IX activity disappeared although the C4 activity was maintained in the control (2% CO_2) condition. By elevation of CO_2, however, the IX activity was restored and the C4 activity disappeared (Fig. 2). These results indicate that there are respiratory rhythm generators both above and below the 7th transection level and they are coupled in the intact brainstem. This observation is compatible with the idea that the parafacial respiratory

The Central Respiratory Chemoreceptor

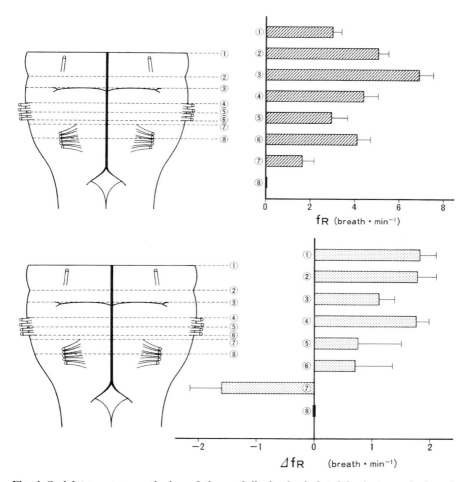

Fig. 1 Serial transverse sectioning of the medulla in the isolated brainstem-spinal cord preparation. *Upper panel*: Effects of sectioning on respiratory frequency (fR) with 2% CO_2. *Lower panel*: Effects of sectioning on fR responses to CO_2, shown as increases in fR (ΔfR). The medullary structure rostral to the 7th splitting level is necessary for chemoresponsiveness

group and preBötzinger complex are coupled and each can generate respiratory rhythm (Onimaru et al. 2003; Oku et al. 2007). These results also suggest that the central chemoreceptor is located mainly rostral to the preBötzinger complex and the preBötzinger complex in itself is inhibited by CO_2.

2.2 Removal of Pia Mater

To evaluate the role of the most superficial layer of the ventral medulla in central chemoreception, we conducted experiments to selectively detach pia mater from the ventral medullary tissue in the isolated brainstem-spinal cord preparation. After CO_2

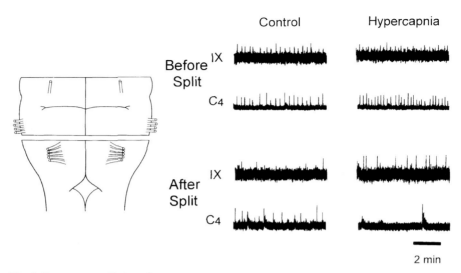

Fig. 2 **Transverse splitting of the medulla in the isolated brainstem-spinal cord preparation.** Before splitting, glossopharyngeal (IX) and C4 ventral root (C4) inspiratory output activities were synchronized and respiratory frequency was increased when CO_2 fraction was elevated from 2% (control) to 8% (hypercapnia). After splitting, IX activity was abolished in the control condition. However, hypercapnia restored IX activity and made C4 activity disappear. These results indicate that the central chemoreceptor is mainly located in the rostral medulla and the respiratory rhythm generator in the caudal portion of the split medulla, i.e., presumably the preBötzinger complex, is rather inhibited by hypercapnia

responsiveness was tested by elevating CO_2 from 2 to 8%, pia mater was removed using fine forceps extremely carefully so that medullary tissue beneath pia mater is minimally damaged in the range from the level of the rostral cut end of the brainstem to the level caudal to the most caudal root of the vagal nerve (n=6). Histological examination showed that this procedure removed medullary tissue with pia mater at most less than 50 μm from the ventral surface. By selectively removing pia mater, fR was remarkably decreased. However, the C4 output pattern was not affected and chemoresponsiveness was not fully abolished. These results indicate that the surface tissue in and/or just beneath pia mater is important in the source of CO_2-sensitive neural drive to the respiratory rhythm generator but is not involved in shaping the inspiratory output pattern (Fig. 3).

3 Neuron Recording

3.1 Electrophysiological Recording

We have applied the isolated brainstem-spinal cord preparation to map CO_2-excitable neurons in the ventral medulla by extracellular recording, and found that chemosensitive tonic (non-respiratory) neurons were densely located in the

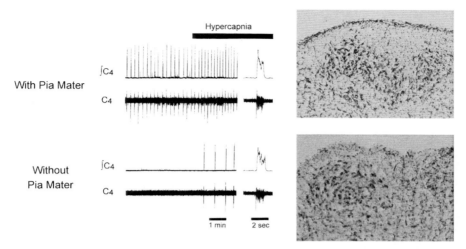

Fig. 3 Removal of pia mater from the ventral medulla in the isolated brainstem-spinal cord preparation. With intact pia mater respiratory frequency was increased in response to hypercapnia (CO_2 8%). Without pia mater respiration stopped in a control condition (CO_2 2%). Although respiration was restored by hypercapnia even after pia was removed, respiratory frequency was extremely low. Histological examination (facial nucleus level) showed that medullary tissue beneath the marginal glial layer was kept intact after pia removal procedure

superficial rostral ventrolateral medulla that corresponds to the nucleus retrotrapezoideus (Okada et al. 1993). We also conducted whole cell recording in the isolated brainstem-spinal cord preparation (Kuwana et al. 1998) and in medullary transverse slice preparations of the neonatal rat (Okada et al. 2002), and found that neurons showing CO_2-induced depolarization were located in close contact with subsurface fine vessels in the rostral ventral medulla. To analyze intrinsic chemosensitivity of medullary neurons, Kawai et al. (1996, 2006) conducted whole cell recording of medullary neurons in the isolated brainstem-spinal cord preparation and tested CO_2 responses under synaptic blockade. They reported that a large portion of tonic and pre-inspiratory neurons in the superficial rostral ventrolateral medulla are intrinsically chemosensitive. However, synaptic blockers (low calcium and high magnesium, or TTX) they used by superfusion will not completely block synaptic transmission, because the isolated brainstem-spinal cord preparation is thick and synaptic transmission could be partially maintained inside the medullary tissue (Kuwana et al. 1998). Although superfusion with synaptic blockers would be sufficient to manipulate the whole tissue environment and to completely block synaptic transmission in slice preparations, intra-arterial perfusion with synaptic blockers would be necessary to fully block synaptic transmission in the isolated brainstem-spinal cord preparation (Okada et al. 2002).

3.2 Optical Recording

Although there are a number of optical recording techniques, optical signals detected by these methods are affected by information other than electrical signals (e.g., by cell swelling). However, among various optical recording methods, a technique of voltage imaging using a voltage-sensitive dye detects changes in fluorescence intensity, which linearly and inversely correlate with changes in membrane potential (Oku et al. 2007). We have successfully applied a voltage imaging technique to analyze the localization of the chemosensitive region and the spatiotemporal response pattern of respiratory neurons to CO_2 (Ito et al. 2004; Okada et al. 2001a, 2007a, b, 2008). When we applied a voltage imaging technique to pontine slice preparations of the neonatal rat under synaptic blockade, we found that A5 area is intrinsically chemosensitive (Ito et al. 2004).

When evaluating voltage imaging signals, various signal processing is often conducted to improve visual impression of images, e.g., by empirically changing the noise-cutting threshold level. However, such empirical processing may greatly influence the result of physiological interpretation. Therefore, when evaluating voltage imaging data obtained from the isolated brainstem spinal cord preparation, we have been calculating cross correlation coefficients between the C4 output activity and each pixel (Okada et al. 2007b; Oku et al. 2007). Using this correlation coefficient imaging technique, we could analyze responses of the respiratory neuronal networks in the medulla to acidosis without ad hoc preprocessing, and revealed that respiratory and metabolic acidosis differentially affect the respiratory neuronal network (Okada et al. 2007a, b).

4 Histological Mapping

Sato et al (1992) first applied a mapping technique using c-fos to the study of chemosensitive cell distribution in the brainstem. The c-fos is an immediate early gene and can be used as a marker of increased neuronal activity. Since their pioneering work, several studies in rats and cats followed and revealed that the nucleus retrotrapezoideus and the nucleus raphe pallidus are the important sites for central chemoreception (Teppema et al. 1994; Okada et al. 2001a). Ribas-Salgueiro et al. (2006) conducted expeiments of Na+/H+ exchanger type 3 inhibition, which induces intracellular acidification. They reported that the inhibition-induced c-fos expression was found in the rostral ventral respiratory column and rostral parapyramidal region, suggesting that these regions could be important in central chemoreception. We also conducted c-fos mapping experiments in the in situ intra-arterially perfused rat, and observed that c-fos positivity disappeared under synaptic blockade in the deeper region but not in the surface region of the rostral ventral medulla, suggesting that surface cells are intrinsically chemosensitive but deeper neurons are not (Okada et al. 2002). We and others also showed that intrinsically chemosensitive surface cells are anatomically arranged in close contact with surface large vessels or with fine penetraing vessels (Bradley et al. 2002; Okada et al. 2002, 2006).

5 Local Acidic Stimulation

Local application of acidic/hypercapnic solution to the brainstem would indicate the site where the respiratory chemoreceptor is located. Nattie and his colleagues conducted in vivo mapping experiments by microinjection of acetazolamide that induces local acidosis in the injected site (Coates et al. 1993; Bernard et al. 1996; Nattie and Li 1996) and by CO_2 microdialysis using a CO_2 diffusion pipette (Li and Nattie 1997). They reported that the nucleus retrotrapezoideus, nucleus raphe pallidus as well as the ventral respiratory group are the important sites for central chemoreception in adult rats. Among their conclusions, we argue that chemosensitivity in the ventral respiratory group, which is supported by Kawai et al. (1996), needs further studies because data of ours (Figs. 1 and 2), Mitchell and Herbert (1974) and Takeda and Haji (1991) are contradictory. Also, in their in vivo experiments, locally applied acidic stimulus could be transported to distant sites by blood flow, and may cause difficulty in interpretation of the experimental results. If the isolated brainstem-spinal cord preparation were used, blood transport of acidic stimulation should not occur because it lacks blood circulation, and experimental results could be straightforwardly interpreted. Then, Issa and Remmers (1992) conducted microinjection of CO_2-enriched saline into the ventral medulla in the isolated brainstem-spinal cord preparation, and reported that rostro-caudally distributing columnar region in the superficial ventral medulla is the site of central chemoreception. We conducted extensive mapping experiments by microinjection of CO_2-enriched saline and reported that rostral ventrolateral, rostral parapyramidal, and rostral midline regions in the superficial ventral medulla are the main sites of chemoreception in the neonatal rat. In this study respiratory augmentation induced by local acidic stimulation was stronger when CO_2-enriched saline was microinjected into a more superficial layer (Okada et al. 2004).

6 Summary

Based on the above reviewed reports, we conclude that the central respiratory chemoreceptor is located mainly in the most superficial layer of the rostral ventral medulla, especially in the rostral ventrolateral medulla (nucleus retrotrapezoideus), rostral parapyramidal region (nucleus parapyramidalis superficialis) and rostral midline (nucleus raphe pallidus). We also propose a cell-vessel architecture model for the central respiratory chemoreceptor. Primary chemoreceptor cells are mainly located beneath large surface vessels within the marginal glial layer of the medullary surface, and surround fine penetrating vessels that branch from a large surface vessel (Fig. 4). These primary chemoreceptor cells monitor the perivascular PCO_2/pH level (Okada et al. 2006). Some respiratory neurons in the rostral medulla are chemosensitive but their role is secondary. Definitive identification of chemosensitive sites and primary chemoreceptor cells need further studies.

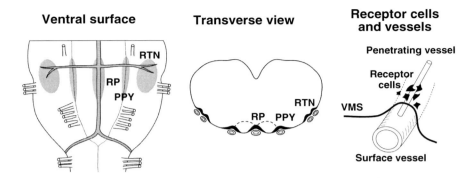

Fig. 4 Localization of central chemoreceptive regions and our cell-vessel architecture model for the central respiratory chemoreceptor. *Left panel*: Sites of chemoreceptive regions on the ventral medullary surface. The chemoreceptive regions are located mainly in the nucleus retrotrapezoideus (RTN), nucleus parapyramidalis superficialis (PPY) and nucleus raphe pallidus (RP). *Middle panel*: Sites of chemoreceptive regions shown on a transverse plane at the rostral medullary level. They have a common anatomical feature in RTN, PPY and RP; the medullary surface portions are covered, compressed and indented by surface vessels, and have the thickened marginal glial layer. *Right panel*: Chemoreceptor cell-vessel architecture model. Primary chemoreceptor cells are mainly located within the marginal glial layer beneath a large surface vessel and the indented ventral medullary surface (VMS), and surround fine penetrating vessels that branch from a large surface vessel

Acknowledgments This work was supported by the Grant-in-Aid for Scientific Research from the Japan Society for Promotion of Science (19200021 and 20590218), the ISM Cooperative Research Program (2006-ISM-CRP-2029, 2007-ISM-CRP-2034), the 2006-2007 ISM Research Projects Grant and the Keio Gijuku Academic Development Funds.

References

Bernard DG, Li A, Nattie EE. Evidence for central chemoreception in the midline raphé. J. Appl. Physiol. 1996; 80: 108–115.
Bradley SR, Pieribone VA, Wang W, Severson CA, Jacobs RA, Richerson GB. Chemosensitive serotonergic neurons are closely associated with large medullary arteries. Nat. Neurosci. 2002; 5: 401–402.
Coates EL, Li A, Nattie EE. Widespread sites of brain stem ventilatory chemoreceptors. J. Appl. Physiol. 1993; 75: 5–14.
Feldman JL, Del Negro CA. Looking for inspiration: new perspectives on respiratory rhythm. Nat. Rev. Neurosci. 2006; 7: 232–242.
Issa FG, Remmers JE. Identification of a subsurface area in the ventral medulla sensitive to local changes in PCO_2. J. Appl. Physiol. 1992; 72: 439–446.
Ito Y, Oyamada Y, Okada Y, Hakuno H, Aoyama R, Yamaguchi K. Optical mapping of pontine chemosensitive regions of neonatal rat. Neurosci. Lett. 2004; 366: 103–106.
Kawai A, Onimaru H, Homma I. Mechanisms of $CO_2/H+$ chemoreception by respiratory rhythm generator neurons in the medulla from newborn rats in vitro. J. Physiol. 2006; 572: 525–537.
Kawai A, Ballantyne D, Mückenhoff K, Scheid P. Chemosensitive medullary neurones in the brainstem-spinal cord preparation of the neonatal rat. J. Physiol. 1996; 492: 277–292.

Kuwana S, Okada Y, Natsui T. Effects of extracellular calcium and magnesium on central respiratory control in the brainstem-spinal cord of neonatal rat. Brain Res. 1998; 786: 194–204.

Li A, Nattie EE. Focal central chemoreceptor sensitivity in the RTN studied with a CO_2 diffusion pipette in vivo. J. Appl. Physiol. 1997; 83: 420–428.

Mitchell RA, Herbert DA. The effect of carbon dioxide on the membrane potential of medullary respiratory neurons. Brain Res. 1974; 75: 345–349.

Nattie EE, Li A. Central chemoreception in the region of the ventral respiratory group in the rat. J. Appl. Physiol. 1996; 81: 1987–1995.

Okada Y, Chen Z, Jiang W, Kuwana S, Eldridge FL. Anatomical arrangement of hypercapnia-activated cells in the superficial ventral medulla of rats. J. Appl. Physiol. 2002; 93: 427–439.

Okada Y, Chen Z, Jiang W, Kuwana S, Eldridge FL. Functional connection from the surface chemosensitive region to the respiratory neuronal network in the rat medulla. Adv. Exp. Med. Biol. 2004; 551: 45–51.

Okada Y, Chen Z, Kuwana S. Cytoarchitecture of central chemoreceptors in the mammalian ventral medulla. Respir. Physiol. 2001a; 129: 13–23.

Okada Y, Chen Z, Yoshida H, Kuwana S, Jiang W, Maruiwa H. Optical recording of the neuronal activity in the brainstem-spinal cord: application of a voltage-sensitive dye. Adv. Exp. Med. Biol. 2001b; 499: 113–118.

Okada Y, Kawai A, Mückenhoff K, Scheid P. Role of the pons in hypoxic respiratory depression in the neonatal rat. Respir. Physiol. 1998; 111: 55–63.

Okada Y, Kuwana S, Masumiya H, Kimura N, Chen Z, Oku Y. Chemosensitive neuronal network organization in the ventral medulla analyzed by dynamic voltage-imaging. Adv. Exp. Med. Biol. 2007a; 605: 353–358.

Okada Y, Kuwana S, Oyamada Y, Chen Z. The cell-vessel architecture model for the central respiratory chemoreceptor. Adv. Exp. Med. Biol. 2006; 580: 233–238.

Okada Y, Masumiya H, Tamura Y, Oku Y. Respiratory and metabolic acidosis differentially affect the respiratory neuronal network in the ventral medulla of neonatal rats. Eur. J. Neurosci. 2007b; 26: 2834–2843.

Okada Y, Mückenhoff K, Scheid P. Hypercapnia and medullary neurons in the isolated brain stem-spinal cord of the rat. Respir. Physiol. 1993; 93: 327–336.

Okada Y, Yokota S, Shinozaki Y, Aoyama R, Yasui Y, Ishiguro M, Oku Y. Anatomical architecture and responses to acidosis of a novel respiratory neuron group in the high cervical spinal cord (HCRG) of the neonatal rat. Adv. Exp. Med. Biol. 2008; 648: 387–394.

Oku Y, Masumiya H, Okada Y. Postnatal developmental changes in activation profiles of the respiratory neuronal network in the rat ventral medulla. J. Physiol. 2007; 585: 175–186.

Onimaru H, Homma I. A novel functional neuron group for respiratory rhythm generation in the ventral medulla. J. Neurosci. 2003; 23: 1478–1486.

Paxinos G., Törk I, Tecott LH, Valentino KL. Atlas of the Developing Rat Brain. Academic Press, San Diego, 1991.

Ribas-Salgueiro JL, Matarredona ER, Ribas J, Pásaro R. Enhanced c-Fos expression in the rostral ventral respiratory complex and rostral parapyramidal region by inhibition of the Na+/H+ exchanger type 3. Auton. Neurosci. 2006; 126–127: 347–354.

Sato M, Severinghaus JW, Basbaum AI. Medullary CO_2 chemoreceptor neuron identification by c-fos immunocytochemistry. J. Appl. Physiol. 1992; 73: 96–100.

Takeda R, Haji A. Synaptic response of bulbar respiratory neurons to hypercapnic stimulation in peripherally chemodenervated cats. Brain Res. 1991; 561: 307–317.

Teppema LJ, Berkenbosch A, Veening JG, Olievier CN. Hypercapnia induces c-fos expression in neurons of retrotrapezoid nucleus in cats. Brain Res. 1994; 635: 353–356.

Anatomical Architecture and Responses to Acidosis of a Novel Respiratory Neuron Group in the High Cervical Spinal Cord (HCRG) of the Neonatal Rat

Y. Okada, S. Yokota, Y. Shinozaki, R. Aoyama, Y. Yasui, M. Ishiguro and Y. Oku

Abstract It has been postulated that there exists a neuronal mechanism that generates respiratory rhythm and modulates respiratory output pattern in the high cervical spinal cord. Recently, we have found a novel respiratory neuron group in the ventral portion of the high cervical spinal cord, and named it the high cervical spinal cord respiratory group (HCRG). In the present study, we analyzed the detailed anatomical architecture of the HCRG region by double immunostaining of the region using a neuron-specific marker (NeuN) and a marker for motoneurons (ChAT) in the neonatal rat. We found a large number of small NeuN-positive cells without ChAT-immunoreactivity, which were considered interneurons. We also found two and three clusters of motoneurons in the ventral portion of the ventral horn at C1 and C2 levels, respectively. Next, we examined responses of HCRG neurons to respiratory and metabolic acidosis in vitro by voltage-imaging together with cross correlation techniques, i.e., by correlation coefficient imaging, in order to understand the functional role of HCRG neurons. Both respiratory and metabolic acidosis caused the same pattern of changes in their spatiotemporal activation profiles, and the respiratory-related area was enlarged in the HCRG region. After acidosis was introduced, preinspiratory phase-dominant activity was recruited in a number of pixels, and more remarkably inspiratory phase-dominant activity was recruited in a large number of pixels. We suggest that the HCRG composes a local respiratory neuronal network consisting of interneurons and motoneurons and plays an important role in respiratory augmentation in response to acidosis.

Keywords Voltage-sensitive dye · Voltage imaging · Optical recording · Correlation coefficient imaging · Cross correlation technique · Isolated brainstem-spinal cord preparation · Upper cervical inspiratory neuron · Respiratory rhythm · Respiratory pattern · Immunohistochemistry

---

Y. Okada (✉)
Department of Medicine, Keio University Tsukigase Rehabilitation Center, Izu, Japan
e-mail: yasumasaokada@1979.jukuin.keio.ac.jp

1 Introduction

It has been known that animals that are spinalized at the C1 level can generate respiratory rhythm but those spinalized at the C3 level cannot (Aoki et al. 1980). Recently, it was reported that the mouse head decapitated at the medullo-spinal cord junction cannot produce gasping but those attached with the high cervical spinal cord can (Miyake et al. 2007). These findings suggest that there exists a respiratory rhythm generating mechanism in the high cervical spinal cord (C1–C2 level). As candidate neurons that compose the respiratory rhythm generating network in the high cervical spinal cord, the upper cervical inspiratory neurons (UCINs) have been identified in the lateral edge of the intermediate grey matter at the C1 and C2 segments of the spinal cord (Lipski and Duffin 1986; Douse et al. 1992; Lipski et al. 1993). Also, it has been reported that the UCINs play an important role in modulation of respiratory pattern of the phrenic nerve (Lu et al. 2004). Recently, we have found a novel respiratory neuron group in the ventral portion of the high cervical spinal cord (from the medullo-spinal cord junction to C2 level) in the neonatal rat using techniques of voltage-imaging and whole cell patch recording, and named it the high cervical spinal cord respiratory group (HCRG) (Oku et al. 2008b). In this previous study, we analyzed the firing pattern and cellular morphology of HCRG neurons, and suggested that HCRG neurons play a significant role in the generation of respiratory rhythm and pattern (Oku et al. 2008b). However, the anatomy of the HCRG region and the functional role of HCRG neurons in respiratory control have been still largely unidentified. In the present study, we intended to analyze the detailed anatomical architecture of the HCRG region and the responses of HCRG neurons to acidosis in order to further understand the anatomy and the functional role of HCRG neurons in respiratory control.

2 Methods

2.1 Animal Preparation

Experimental protocols were approved by the animal research committees of Keio University School of Medicine, Shimane University School of Medicine and Hyogo College of Medicine. We used neonatal rats (n=11, 0–4 days old) as previously described (Okada et al. 2007; Oku et al. 2007, 2008b). In the histological analysis, each animal was deeply anesthetized with diethyl ether and perfused transcardially with 4% paraformaldehyde in 0.1 M phosphate buffer (PB, pH 7.3). After perfusion, the brainstem and the spinal cord were removed, postfixed in the same fixative, and then saturated with 20% sucrose in PB. In the functional analysis by optical recording, each animal was anesthetized with diethyl ether, quickly decerebrated at the intercollicular level, and the brainstem and the spinal cord were isolated.

2.2 Histological Analysis

Detailed anatomical feature of the HCRG region was analyzed by immunohistochemistry using a neuron-specific marker and a marker for motoneurons. For this purpose, tissue samples containing the HCRG region (n=4) were cut serially into transverse sections with 40 μm thickness on a freezing microtome. The technique of immunostaining was as previously described (Yokota et al. 2007, 2008). Briefly, the sections were incubated overnight in primary antibodies; mouse anti-neuronal nuclei (NeuN; Chemicon/Millipore, Billerica, MA, USA) and goat anti-cholineacetyl transferase (ChAT; Chemicon/Millipor, Billerica, MA, USA) diluted 1:500 and 1:100, respectively by phosphate buffered saline (pH 7.3) containing 0.2% Triton X-100, 3% normal horse serum and 3% normal rabbit serum. Subsequently, the sections were incubated in Cy3-conjugated anti-goat IgG (Jackson, West Grove, PA, USA) and biotin-conjugated anti-mouse IgG (Vector, Burlingame, CA, USA), and then incubated in Alexa488-conjugated streptavidin (Invitrogen, Carlsbad, California, USA). After mounting on gelatinized slides, the sections were observed under a confocal laser scanning microscope (FV300, Olympus Optical, Tokyo, Japan), and the number and size of NeuN-positive cells with and without ChAT-immunoreactivity in the squarely set area (177 × 177 μm) that covered the HCRG region were analyzed using image analysis software (Image J, NIH, Bethesda, Maryland, USA). Data were presented as mean ± S.D.

2.3 Optical Recording

Isolated brainstem-spinal cord preparations (n=7) were incubated in an artificial cerebrospinal fluid (aCSF) containing a voltage-sensitive dye, di-2-ANEPEQ (0.1–0.2 mM, Invitrogen, Carlsbad, California, USA) for 40 min with 95% O_2 and 5% CO_2 (Okada et al. 2007; Oku et al. 2007, 2008a, b). After staining, the preparations were transferred to a 2 ml recording chamber, and pinned with the ventral surface up. The recording chamber was continuously superfused with a control aCSF at a rate of 3 ml/min. The temperature of the superfusate was controlled at 27 ± 1°C. The control aCSF contained (in mM): NaCl 124, KCl 5.0, $CaCl_2$ 2.4, $MgSO_4$ 1.3, KH_2PO_4 1.2, $NaHCO_3$ 26, glucose 30; it was equilibrated with 95% O_2 and 5% CO_2. The pH was 7.4 in the experimental condition. The control aCSF was replaced with test superfusates that differed in either CO_2 or HCO_3^- concentration. In 3 preparations the concentration of aCSF CO_2 was elevated to 8% (pH 7.2) to simulate respiratory acidosis, and in 4 preparations the concentration of HCO_3^- was reduced to 10 mEq/L to simulate metabolic acidosis (pH 7.2).

Inspiratory-related respiratory activity was monitored from the C4 ventral root (C4VR) with a suction electrode. The raw nerve signal was amplified, band-pass filtered ($\lambda = 15$ Hz $\sim$3 kHz), and fed into a time-amplitude window discriminator. The window discriminator generated a TTL-level pulse signal when the C4 activity exceeded the preset minimal threshold level, and then the TTL pulse was fed into the optical recording system. In addition, the filtered signal was full rectified and integrated using a leaky integrator with a time constant of 100 ms. Activity of respiratory neurons in the HCRG region was analyzed using a fluorescence macro zoom microscope (MVX-10, Olympus Optical, Tokyo, Japan) and an optical recording system (MiCAM Ultima, BrainVision, Tokyo, Japan) as previously described (Okada et al. 2007; Oku et al. 2007, 2008a, b). Briefly, preparations were illuminated through a band-pass excitation filter ($\lambda = 480 \sim 550$ nm), and epifluorescence through a long-pass barrier filter ($\lambda > 590$ nm) was detected with a CMOS sensor array. Optical signals were sampled at 50 Hz. A total of 256 frames were recorded starting at 64 frames before the onset of inspiratory C4VR activity, and averaged 20–25 times.

2.4 Analysis of Optical Imaging Data

Cross correlation coefficients between the integrated C4VR activity and each pixel were calculated as previously described (Okada et al. 2007; Oku et al. 2007, 2008a, b). Briefly, cross correlation coefficients were estimated for the lag ranging between −1 s and 1 s. Then the maximum of cross correlation coefficients (CCmax) and the lag at which the cross correlation became maximal (LAGmax) were estimated for each pixel. We defined pixels having CCmax > 0.3 as respiratory-related. Activation timing of each pixel was characterized by LAGmax. A negative LAGmax value indicates that the overall membrane depolarization of cells within the pixel precedes the C4VR activity.

3 Results

3.1 Cellular Architecture of HCRG Region

In the most ventral portion of the ventral horn in the high cervical spinal cord, which corresponded to the HCRG region (Oku et al. 2008b), we found NeuN- and ChAT-positive cells, which were large in size and oval or polymodal in shape (NeuN-positive/ChAT-positive cells; 237 ± 10 cells/square, major diameter 31.6 ± 5.3 μm; n=223, 4 rats; 6 sections/rat). At the level of C1 segment, a large cluster of NeuN-positive/ChAT-positive cells was found in the most ventral portion of the ventral horn, and a small cluster of NeuN-positive/ChAT-positive cells was seen in the medial portion of the ventral horn. At the level of C2 segment, three clusters of NeuN-positive/ChAT-positive cells were found in the most ventral, medial and lateral portions of the ventral horn; the most ventral cluster was larger than the medial and lateral clusters. In and around these clusters of NeuN-positive/ChAT-positive cells, there were a large number of NeuN-positive cells without ChAT-immunoreactivity, which were small in size and oval or fusiform in shape (NeuN-positive/ChAT-negative cells; 356 ± 15 cells/square, major diameter; 22.1 ± 3.7 μm, n=257; 4 rats; 6 sections/rat), which are considered to be interneurons (Fig. 1).

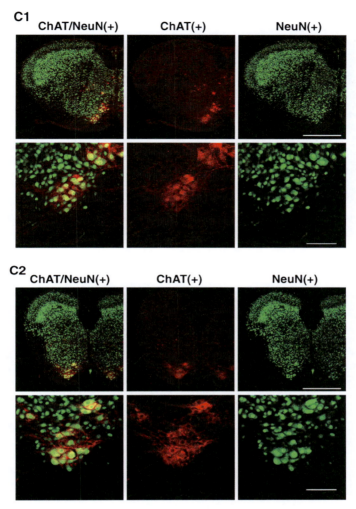

Fig. 1 NeuN/ChAT double immunopositive cells shown as yellow, ChAT positive cells (motoneurons) shown as red and NeuN positive cells (neurons) shown as green at low and high magnifications. *Upper* and *lower panels* correspond to the C1 and C2 levels, respectively. Scale bars, 500 μm in low magnification pictures and 100 μm in high magnification pictures

3.2 Changes in Respiratory-Related Activities in HCRG Region Associated with Acidosis

In optical experiments, we visualized changes in the spatiotemporal profiles associated with acidosis by comparing the CCmax and LAGmax of each pixel of the same preparation between different pH/CO_2 environments, and by mapping the result on a photograph of the preparation as a pseudocolor image. A representative pseudocolor

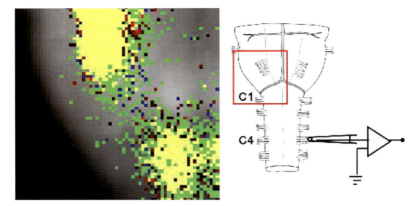

Fig. 2 *Left*: A representative pseudocolor image characterizing a change in the spatiotemporal activation pattern associated with respiratory acidosis. The upper cluster corresponds to the preBötzinger complex/ventral respiratory group, and the lower cluster corresponds to the HCRG. Pixels in blue lost respiratory modulation, pixels in red gained preinspiratory phase-dominant activity, pixels in green gained inspiratory phase-dominant activity after acidosis, and pixels in yellow are respiratory-modulated both before and after acidosis. *Right*: A schema of the isolated brainstem-spinal cord preparation. The square (1.5 × 1.5 mm) indicates the area of the pseudocolor image

image characterizing the changes in the spatiotemporal activation pattern associated with respiratory acidosis is shown in Fig. 2. Both respiratory and metabolic acidosis showed the same pattern of changes in their spatiotemporal activation profiles, and the respiratory-related areas were enlarged in the HCRG region. After acidosis was introduced, preinspiratory phase-dominant activity was newly recruited in a number of pixels, and more remarkably inspiratory phase-dominant activity was recruited in a large number of pixels (Fig. 2).

4 Discussion

We have revealed the detailed neuronal arrangement in the HCRG region of the neonatal rat by immunohistochemically staining NeuN and ChAT. In the HCRG region, i.e., in the most ventral portion of the ventral horn of C1–C2 spinal cord, two types of neurons were found; small NeuN-positive/ChAT-negative cells and large NeuN-positive/ChAT-positive cells. Small NeuN-positive/ChAT-negative cells are considered to be interneurons, because they are negative for ChAT. Large NeuN-positive/ChAT-positive cells are classified as motoneurons based on their size, shape and positivity of ChAT. At least a part of these motoneurons should include those innervating respiratory accessory muscles such as scalene muscles. The respiratory activity of these motoneurons should coincide with those recorded from the C1 ventral root in the isolated brainstem spinal cord preparation (Fig. 1 in Smith et al. 1990). In the HCRG region, motoneurons (ChAT-positive cells) are located in a narrower region than optically identified respiratory-related region, as evident when comparing Figs. 1 and 2. Because both these small interneurons and large motoneurons are located in the most ventral portion of the ventral horn, respiratory-related optical signals in the HCRG region should reflect activities of both interneurons and motoneurons.

We have processed optical imaging data by computing CCmax and LAGmax, and succeeded to characterize the changes in spatiotemporal profiles, i.e., responses to acidosis, of the pixels in the HCRG region. These results indicate that combination of optical imaging and computation of cross correlation coefficients, i.e., correlation coefficient imaging, is a powerful tool to map respiratory-related regions in the brain.

By applying correlation coefficient imaging, we have demonstrated that acidosis, either respiratory or metabolic, enlarges respiratory-related area in the HCRG region. Among newly recruited respiratory-related pixels, a large number of pixels gained inspiratory pattern as shown by green dots in Fig. 2. This finding suggests that the HCRG neurons contribute to increase ventilation by directly driving accessory respiratory muscles as well as by transmitting excitatory signals to the respiratory neuronal network. Although the number of pixels was small, there was preinspiratory phase-dominant activity recruited by acidosis as shown by red dots in Fig. 2. This observation is compatible with our hypothesis that HCRG interneurons are involved in respiratory rhythm generation. We suggest that the HCRG composes a local respiratory neuronal network consisting of interneurons and motoneurons independently of the UCINs and plays an important role in respiratory augmentation in response to acidosis. Further studies are necessary to elucidate the precise role of HCRG interneurons in respiratory rhythm generation and pattern modulation and to strictly identify the involved interneurons.

Acknowledgments This work was supported by the Grant-in-Aid for Scientific Research from the Japan Society for Promotion of Science (19200021 and 20590218), the ISM Cooperative Research Program (2006-ISM-CRP-2029, 2007-ISM-CRP-2034), the 2006-2007 ISM Research Projects Grant, the General Insurance Association of Japan and the Keio Gijuku Academic Development Funds.

References

Aoki M, Mori S, Kawahara K, Watanabe H, Ebata N. Generation of spontaneous respiratory rhythm in high spinal cats. Brain Res. 1980; 202: 51–63.
Douse MA, Duffin J, Brooks D, Fedorko L. Role of upper cervical inspiratory neurons studied by cross-correlation in the cat. Exp. Brain Res. 1992; 90: 153–162.
Lipski J, Duffin J. An electrophysiological investigation of propriospinal inspiratory neurons in the upper cervical cord of the cat. Exp. Brain Res. 1986; 61: 625–637.
Lipski J, Duffin J, Kruszewska B, Zhang X. Upper cervical inspiratory neurons in the rat: an electrophysiological and morphological study. Exp. Brain Res. 1993; 95: 477–487.
Lu F, Qin C, Foreman RD, Farber JP. Chemical activation of C1–C2 spinal neurons modulates intercostal and phrenic nerve activity in rats. Am. J. Physiol. Regul. Integr. Comp. Physiol. 2004; 286: R1069–R1076.

Miyake A, Yamada K, Kosaka T, Miki T, Seino S, Inagaki N. Disruption of Kir6.2-containing ATP-sensitive potassium channels impairs maintenance of hypoxic gasping in mice. Eur. J. Neurosci. 2007; 25: 2349–2363.

Okada Y, Masumiya H, Tamura Y, Oku Y. Respiratory and metabolic acidosis differentially affect the respiratory neuronal network in the ventral medulla of neonatal rats. Eur. J. Neurosci. 2007; 26: 2834–2843.

Oku Y, Kimura N, Masumiya H, Okada Y. Spatiotemporal organization of frog respiratory neurons visualized on the ventral medullary surface. Resp. Physiol. Neurobiol. 2008a; 161: 281–290.

Oku Y, Masumiya H, Okada Y. Postnatal developmental changes in activation profiles of the respiratory neuronal network in the rat ventral medulla. J. Physiol. 2007; 585: 175–186.

Oku Y, Okabe, Hayakawa T, Okada Y. A respiratory neuron group in the high cervical spinal cord discovered by optical imaging. Neuroreport 2008b; 161:281–290.

Smith JC, Greer JJ, Liu GS, Feldman JL. Neural mechanisms generating respiratory pattern in mammalian brain stem-spinal cord in vitro. I. Spatiotemporal patterns of motor and medullary neuron activity. J Neurophysiol. 1990; 64: 1149–1169.

Yokota S, Oka T, Tsumori T, Nakamura S, Yasui Y. Glutamatergic neurons in the Kölliker-Fuse nucleus project to the rostral ventral respiratory group and phrenic nucleus: a combined retrograde tracing and in situ hybridization study in the rat. Neurosci. Res. 2007; 59: 341–346.

Yokota S, Tsumori T, Oka T, Nakamura S, Yasui Y. GABAergic neurons in the ventrolateral subnucleus of the nucleus tractus solitarius are in contact with Kölliker-Fuse nucleus neurons projecting to the rostral ventral respiratory group and phrenic nucleus in the rat. Brain Res. 2008; 1228: 113–126.

Systemic Inhibition of the Na$^+$/H$^+$ Exchanger Type 3 in Intact Rats Activates Brainstem Respiratory Regions

R. Pasaro, J.L. Ribas-Salgueiro, E.R. Matarredona, M. Sarmiento and J. Ribas

Abstract Selective inhibition of the Na$^+$/H$^+$ exchanger type 3 (NHE3) increases the firing rate of brainstem ventrolateral CO$_2$/H$^+$ sensitive neurons, resembling the responses evoked by hypercapnic stimuli. In anesthetized animals, NHE3 inhibition has also been shown to stimulate the central chemosensitive drive. We aimed to analyze the respiratory-related brainstem regions affected by NHE3 inhibition in anaesthetized spontaneously-breathing rats with intact peripheral afferents. For that, c-Fos immunopositive cells were counted along the brainstem in rats intravenously infused with the selective NHE3 inhibitor AVE1599. A rostral extension of the ventral respiratory column which includes the pre-Bötzinger complex was activated by the NHE3 inhibitor. In addition, the number of c-Fos positive cells resulted significantly increased in the most rostral extension of the retrotrapezoid nucleus/parapyramidal region. In the pons, the intravenous infusion of AVE1599 activated the lateral parabrachial and Kölliker-Fuse nuclei. Thus, selective NHE3 inhibition in anaesthetized rats activates the respiratory network and evokes a pattern of c-Fos expressing cells similar to that induced by hypercapnia.

Keywords Central chemoreception · c-Fos immunohistochemistry · Pre-Bötzinger complex · Ventral respiratory column · Retrotrapezoid nucleus · Parapyramidal region · Lateral parabrachial nucleus · Kölliker-Fuse nucleus · Ventral respiratory group

1 Introduction

Peripheral and central respiratory chemoreceptors are important to maintain the level of arterial P$_{CO2}$ and [H$^+$] under different conditions and it is generally accepted that central chemoreceptors are widely distributed in the brain (Nattie, 2001; Feldman et al., 2003). In mammals, respiratory rhythm pattern is generated by

---

R. Pasaro (✉)
Department of Physiology and Zoology, University of Seville, Seville, Spain
e-mail: mrpasaro@us.es

respiratory neuronal network oscillations (Smith et al., 2007), which are originated within a bilateral column of ventromedullary neurons – the ventral respiratory column (VRC) – and is modulated by the pons. The VRC includes several rostro-caudally arranged compartments, the Bötzinger, the pre-Böt complex (pre-Böt) and the rostral and caudal ventral respiratory groups (rVRG and cVRG, respectively) (Smith et al., 2007). Pacemaker respiratory neurons located in the pre-Böt play a key role in respiratory rhythm generation (Wu et al., 2005), while the VRG acts as modulator of the breathing pattern (Hilaire and Pasaro, 2003). In addition, other medullary structures, mainly the retrotrapezoid nucleus/parapyramidal region (RTN/Ppy) (Okada et al., 2002) and the raphe pallidus nucleus, both involved in central chemoreception, intervene to modulate the respiratory network (Feldman et al., 2003; Guyenet et al., 2008; Ribas-Salgueiro et al., 2003).

The stimulus sensed by central chemosensitive cells could be molecular CO_2 as well as a decrease in extracellular or intracellular pH, or the three stimuli altogether (Eldridge et al., 1984; Nattie, 2001; Shams, 1985; Teppema et al., 1983; Wiemann et al., 1998). One system which has been demonstrated to be involved in the control of breathing is the Na^+/H^+ exchanger type 3 (NHE3). In organotypic cultures, NHE3 inhibition evoked a increase in the neuronal firing-rate that mimicked the response to hypercapnia (Wiemann et al., 1999). In addition, a selective inhibitor of the NHE3 augments the central respiratory drive in the working-heart-brainstem preparation (Wiemann et al., 2008). NHE3 appears to be expressed in the brainstem although its precise cellular localization has not been yet defined. Interestingly, there is an inverse correlation between P_{CO2} and brainstem NHE3 expression (Wiemann et al., 2005).

Immunohistochemical detection of the c-Fos protein, which is the expression product of the activated c-fos proto-oncogene, allows the identification of activated cells in the CNS. The distribution of c-Fos has been extensively used to localize cell populations responsive to chemosensitive stimuli, such as hypoxia or hypercapnia (Berquin et al., 2000; Teppema et al., 1997). To test the hypothesis that the central chemoreceptor network and respiratory-related regions may be affected by NHE3 inhibition we have analyzed the c-Fos immunopositive cell distribution along the pontomedullary rostro-caudal extension after systemic application of a selective NHE3 inhibitor.

2 Methods

Experiments were performed in 18 adult Wistar rats following the European Community Council directive (86/609/EEC) for the Care and Use of Laboratory Animals. Animals were anaesthetized with i.p. injection of chloral hydrate 5% (1 ml/100 g) and the left femoral vein was cannulated. After 15 min of recovery, a continuous infusion of the NHE3 inhibitor AVE1599 was applied during 10 min (Aventis Pharma Deutschland GmbH; 0.5 and 2 mg/kg, prepared in 0.1M phosphate buffer (PB), pH=7.4). Control animals were infused with the vehicle.

One hour after the end of the drug or vehicle administration, animals were transcardially perfused with saline followed by 4% formaldehyde in 0.1 M PB. Brains were removed, postfixed overnight, and stored in cryoprotectant until freeze sectioning at 40-μm. Sections from vehicle and AVE1599-infused rat's brains were processed in parallel. First, the sections were incubated with 70% (15 min) and 50% (10 min) ethanol. After washing in PB saline, sections were incubated (72 h, 4°C) with a rabbit anti c-Fos antibody (1:4000; Santa Cruz Biotechnology, Santa Cruz, USA) followed by incubation with a biotinylated donkey anti-rabbit IgG (1:5000; Jackson-ImmunoResearch Labs., West Grove, USA, overnight, 4°C) and the standard avidin-biotinylated peroxidase complex (ABC; Vector Labs., Burlingame, USA, 1 h, room temperature). Peroxidase activity was revealed with a solution containing 0.02% diaminobenzidine, 0.01% H_2O_2 and 0.04% nickel ammonium sulfate. Finally they were counterstained with Neutral Red (1%) for reference purposes.

The sections were analyzed using a light microscope (Olympus BX61). Counting of c-Fos-positive cells was performed at 120 μm intervals (1 in 3 series) from −6.5 to 0.6 mm interaural with the help of the Paxinos and Watson Atlas (1997) as previously described (Ribas-Salgueiro et al., 2006). All values are expressed as mean ± s.e.m. Means were compared using ANOVA with Bonferroni correction (p<0.05) (SigmaStat 3.1). The 3D reconstruction of the labeled cells within the pons and medulla was made with Neurolucida 8.20.

3 Results

The administration of 0.5 and 2 mg/Kg AVE1599 induced a significant increase in the number of c-Fos labeled cells in the rostral extension of the VRC (from −3.9 to −3.1 mm interaural) (Fig. 1A; Fig. 2). This rostral activated region includes the pre-Böt and the rVRG. In addition, a significant increase in c-Fos positive cells was also found in the caudal VRC (−6 to −5.5 mm interaural), only after the 0.5 mg/Kg infusion (Fig. 1A; Fig. 2). A significant increase in the number c-Fos-labeled cells was found in the rostral extension of the RTN/Ppy, ranging from −2.4 to −2 mm interaural (Fig. 1A; Fig. 2). In the pons, the intravenous administration of 0.5 and 2 mg/Kg AVE1599 induced an increase in the number of c-Fos immunoreactive cells in the lateral parabrachial and Kölliker-Fuse nuclei (LPB/KF) (Fig. 1A–E; Fig. 2). The number of c-Fos-positive cells was not significantly altered by the AVE1599 treatment neither in the nucleus of the solitary tract (Fig. 2) nor in the raphe pallidus (Fig. 2) and in the area postrema (not shown).

4 Discussion

Our results clearly show that central respiratory-related regions are activated after selective inhibition of NHE3 in anesthetized spontaneously-breathing rats with intact peripheral afferents. We raise the hypothesis that the effect of NHE3 inhibition on the central respiratory network could be mediated by central chemoreceptor activation.

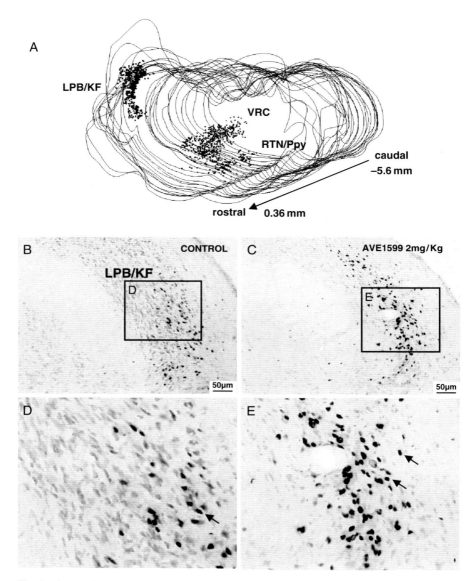

Fig. 1 Distribution of c-Fos-positive cells in the rat brainstem after selective NHE3 inhibition.
A: 3D view of c-Fos immunoreactive cells in the right side of an AVE1599 2 mg/Kg infused rat brainstem showing the distribution of the activated nuclei: retrotrapezoid nucleus/parapyramidal region (RTN/Ppy), ventral respiratory column (VRC), and lateral parabrachial and Kölliker-Fuse nuclei (LPB/KF) (coordinates: mm with respect to the interaural line). **B–C**: c-Fos-positive cells in the LPB/KF. **D** and **E** are higher magnification fields of the inserts in B and C, respectively

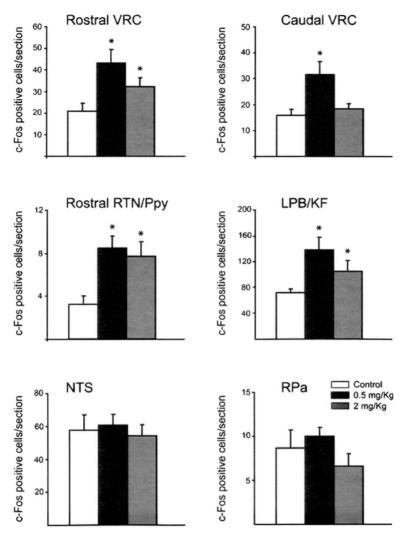

Fig. 2 c-Fos positive cell quantification in respiratory-related brainstem regions after systemic administration of different concentrations of AVE1599. Each graph represents the mean number of c-Fos positive cells per 40 μm section in rats intravenously infused with 0.1M phosphate buffer (control, *white bars*) or with the NHE3 inhibitor AVE1599 (0.5 mg/Kg or 2 mg/Kg, *black* and *grey bars*, respectively). The represented rostro-caudal extensions are the following (in mm with respect to the interaural line): rostral (−3.9 to −3.1) and caudal (−6 to −5.5) ventral respiratory column (VRC), rostral retrotrapezoid nucleus/parapyramidal region (RTN/Ppy) (−2.4 to −2), lateral parabrachial and Kölliker-Fuse nuclei (LPB/KF) (−0.4 to 0.1), nucleus of the solitary tract (NTS) (−6.6 to −3.3) and raphe pallidus nucleus (RPa) (−5.2 to −2). The results are the mean ± s.e.m. of 6 animals. * $p < 0.05$, compared to the control group, ANOVA with Bonferroni correction

The effect of different NHE3 inhibitors on breathing has been widely tested under several experimental conditions. In anaesthetized and vagotomized rabbits, the systemic NHE3 inhibition significantly modified central CO_2-respiratory response (Kiwull-Schöne et al., 2001), while in anaesthetized piglets increased the threshold for a laryngeal reflex that causes cessation of respiration (Abu-Shaweesh et al., 2002).

In our study, several brainstem regions presented an increased number of c-Fos labeled cells as a result of the AVE1599 administration, such as the pre-Böt which has a determinant role in the regulation of the breathing frequency (Wu et al., 2005). The caudal VRC was also activated after NHE3 blocker (0.5 mg/Kg) administration. This region contains both expiratory and inspiratory bulbospinal premotor neurons that constitute an output component of the central pattern generator (Hilaire and Pasaro, 2003). The rostral RTN/Ppy was activated by the two doses, a region where central chemoreceptors are located (Nattie, 2001; Mulkey et al., 2004). The activation of RTN/Ppy cells could lead to synaptic activation of neurons in the rostral and caudal VRC. Indeed, bidirectional connections between the RTN and the cVRG, rVRG and the pre-Böt have been described (Rosin et al., 2006). An increased number of c-Fos-positive cells was also found in the LPB/KF after the systemic infusion of the NHE3 inhibitor. This is a region involved in the modulation of breathing via multiple connections with the spinal cord, brainstem and forebrain structures (Gaytan and Pasaro, 1998). Furthermore, the LPB/KF region is also activated after central chemoreceptor hypercapnic stimulation (Berquin et al., 2000; Teppema et al., 1997). The LPB/KF receives inputs from both the RTN and the pre-Böt (Rosin et al., 2006; Gaytan and Pasaro, 1998), which are also activated after the AVE1599 administration. However, other respiratory-related regions were not activated after the NHE3 inhibitor administration, such as the nucleus of the solitary tract, the raphe pallidus and the area postrema.

Intracellular acidification induced by NHE3 inhibitors might activate the central pattern generating neurons via chemoreceptors located in the RTN/Ppy and/or the rVRG/pre-Böt regions. Alterations in mechanisms regulating intracellular pH could account for pathologies associated with respiratory depression such as neonatal apnea, sleep apnea or sudden infant death syndrome.

Acknowledgments Supported by Junta de Andalucía PAI grant BIO183.

References

Abu-Shaweesh, J.M., Dreshaj, I.A., Martin, R.J., Wirth, K.J., Heinelt, U. & Haxhiu, M.A. 2002, Inhibition of Na(+)/H(+) exchanger type 3 reduces duration of apnea induced by laryngeal stimulation in piglets, *Pediatr Res*, 52: 459–464.

Berquin, P., Bodineau, L., Gros, F. & Larnicol, N. 2000, Brainstem and hypothalamic areas involved in respiratory chemoreflexes: a Fos study in adult rats, *Brain Res*, 857: 30–40.

Eldridge, F.L., Kiley, J.P. & Millhorn, D.E. 1984, Respiratory effects of carbon dioxide-induced changes of medullary extracellular fluid pH in cats, *J Physiol* 355: 177–189.

Feldman, J.R., Mitchell, G.S. & Nattie, E.E. 2003, Breathing: rhythmicity, plasticity, chemosensitivity, *Annu Rev Neurosci*, 26: 239–266.

Gaytan, S.P. & Pasaro, R. 1998, Connections of the rostral ventral respiratory neuronal cell group: an anterograde and retrograde tracing study in the rat, *Brain Res Bull*, 47: 625–642.

Guyenet, P.G., Stornetta, R.L. & Bayliss, D.A. 2008, Retrotrapezoid nucleus and central chemoreception, *J Physiol*, 586: 2043–2048.

Hilaire, G. & Pasaro, R. 2003, Genesis and control of the respiratory rhythm in adult mammals, *News Physiol Sci*, 18: 23–28.

Kiwull-Schöne, H., Wiemann, M., Frede, S., Bingmann, D., Wirth, K.J., Heinelt, U., Lang, H.J. & Kiwull, P. 2001, A novel inhibitor of the Na$^+$/H$^+$ exchanger type 3 activates the central respiratory CO_2 response and lowers the apneic threshold. *Am J Respir Crit Care Med*, 164: 1303–1311.

Mulkey, D., Stornetta, R.L., Weston, M.C., Simmons, J.R., Parker, A., Bayliss, D.A. & Guyenet, P.G. 2004, Respiratory control by ventral surface chemoreceptor neurons in rats, *Nat Neurosci*, 7: 1360–1369.

Nattie, E.E. 2001, Chemoreception and tonic drive in the retrotrapezoid nucleus (RTN) region of the awake rat: bicuculline and muscimol dialysis in the RTN. *Adv Exp Med Biol*, 499: 27–32.

Okada, Y., Chen, Z., Jiang, W., Kuwana, S. & Eldridge, F.L. 2002, Anatomical arrangement of hypercapnia-activated cells in the superficial ventral medulla of rats, *J Appl Physiol*, 93: 427–439.

Paxinos, G. & Watson, C. 1997, *The rat brain in stereotaxic coordinates*, 3nd edn, Academic Press, Sydney.

Ribas-Salgueiro, J.L., Gaytan, S.P., Crego, R., Pasaro, R. & Ribas, J. 2003, Highly H$^+$-sensitive neurons in the caudal ventrolateral medulla of the rat, *J Physiol*, 549: 181–194.

Ribas-Salgueiro, J.L., Matarredona, E.R., Ribas, J. & Pasaro, R. 2006, Enhanced c-Fos expression in the rostral ventral respiratory complex and rostral parapyramidal region by inhibition of the Na$^+$/H$^+$ exchanger type 3, *Auton Neurosci*, 126/127: 347–354.

Rosin, D.L., Chang, D.A. & Guyenet, P.G. 2006, Afferent and efferent connections of the rat retrotrapezoid nucleus, *J Comp Neurol*, 499: 64–89.

Shams, H. 1985, Differential effects of CO_2 and H$^+$ as central stimuli of respiration in the cat, *J Appl Physiol*, 58: 357–364.

Smith, J.C., Abdala, A.P.L., Koizumi, H., Rybak, I.A. & Paton, J.F.R. 2007, Spatial and functional architecture of the mammalian brainstem respiratory network: a hierarchy of three oscillatory mechanisms, *J Neurophysiol*, 98: 3370–3387.

Teppema, L.J., Barts, P.W., Folgering, H.T., Evers, J.A. 1983, Effects of respiratory and (isocapnic) metabolic arterial acid-base disturbances on medullary extracellular fluid pH and ventilation in cats. *Respir Physiol*, 53: 379–95.

Teppema, L.J., Veening, J.G., Kranenburg, A., Dahan, A., Berkenbosch, A. & Olievier, C. 1997, Expression of c-fos in the rat brainstem after exposure to hypoxia and to normoxic and hyperoxic hypercapnia, *J Comp Physiol*, 388: 169–190.

Wiemann, M., Baker, R.E., Bonnet, U. & Bingmann, D. 1998, CO_2-sensitive medullary neurons: activation by intracellular acidification, *Neuroreport*, 9: 167–170.

Wiemann, M., Schwark, J.R., Bonnet, U., Jansen, H.W., Grinstein, S., Baker, R.E., Lang, H.J., Wirth, K. & Bingmann, D. 1999, Selective inhibition of the Na$^+$/H$^+$ exchanger type 3 activates CO_2/H$^+$-sensitive medullary neurons, *Pflugers Arch*, 438: 255–262.

Wiemann, M., Frede, S., Bingmann, D., Kiwull, P. & Kiwull-Schone, H. 2005, Sodium/Proton exchanger 3 in the medulla oblongata and set point of breathing control. *Am J Respir Crit Care Med*, 172: 244–249.

Wiemann, M., Piechatzek, L., Göpelt, K., Kiwull-Schöne, H., Kiwull, P. & Bingmann, D. 2008, The NHE3 inhibitor AVE1599 stimulates phrenic nerve activity in the rat, *J Physiol Pharmacol*, 59: 27–36.

Wu, M., Haxhiu, M.A. & Johnson, S.M. 2005, Hypercapnic and hypoxic responses require intact neural transmission from the pre-Bötzinger complex, *Respir Physiol Neurobiol*, 146: 33–46.

Nitric Oxide in the Solitary Tract Nucleus (STn) Modulates Glucose Homeostasis and FOS-ir Expression After Carotid Chemoreceptor Stimulation

M. Lemus, S. Montero, S. Luquín, J. García and E. Roces De Álvarez-Buylla

Abstract We evaluate in rats the role of NO in the solitary tract nucleus (STn) after an anoxic stimulus to carotid body chemoreceptor cells (CChrc) with cyanide (NaCN), on the hyperglycemic reflex with glucose retention by the brain (BGR) and FOS expression (FOS-ir) in the STn. The results suggest that nitroxidergic pathways in the STn may play an important role in glucose homeostasis. A NO donor such as sodium nitroprusside (NPS) in the STn before CChrc stimulation increased arterial glucose level and significantly decreased BGR. NPS also induced a higher FOS-ir expression in STn neurons when compared to neurons in control rats that only received artificial cerebrospinal fluid (aCSF) before CChrc stimulation. In contrast, a selective NOS inhibitor such as Nω-nitro-L-arginine methyl ester (L-NAME) in the STn before CChrc stimulation resulted in an increase of both, systemic glucose and BGR above control values. In this case, the number of FOS-ir positive neurons in the STn decreased when compared to control or to NPS experiments. FOS-ir expression in brainstem cells suggests that CChrc stimulation activates nitroxidergic pathways in the STn to regulate peripheral and central glucose homeostasis. The study of these functionally defined cells will be important to understand brain glucose homeostasis.

Keywords Nitric oxide solitary tract nucleus · Glucose · Sodium nitroprusside · Fos · Cerebrospinal fluid · Carotid chemoreceptor

1 Introduction

The mammalian brain depends on an efficient supply of oxygen and glucose. Carotid chemoreceptor cells that are strategically located at the initiation of brain circulatory system are of unique importance for CNS glucose homeostasis. CChrc function as peripheral glucose sensors, which in response to hyperglycemia or

E.R. De Álvarez-Buylla (✉)
Universidad de Colima, Av. 25 de Julio S/N, Col. villas de San Sebastián Colima, Col. 28045 Mexico
e-mail: rab@ucol.mx

glucopenia release neurotransmitters to excite the adjacent nerve terminals of the carotid sinus nerves (CSN) (Álvarez-Buylla and Roces de Álvarez-Buylla, 1975; López-Barneo, 2003, Zhang et al., 2007). Afferent pathways reach the STn through CSN to participate in respiratory and cardiovascular functions, as well as in the CNS energy metabolism (Mifflin, 1996). But the precise mechanism of the glucose sensor activity and central projections to achieve central glucose regulation remains unclear.

Nitric oxide (NO) has been identified in several brain areas implicated in a diversity of biological functions (Moncada et al., 1991), such as blood pressure regulation and neuroendocrine responses (Housley and Sinclair, 1988; Sunico et al., 2005). NO synthases and their mRNA are present in the hypothalamic magnocellular neurons in the supraoptic and paraventricular nuclei (Kadowaki et al., 1994) where vasopressin and oxytocin are synthesized. NO induces ATP depletion in the CNS and modulates metabolic pathways during glycolysis (Almeida et al., 2005). Furthermore, in vitro experiments showed that NO functions as a neuromodulator by several mechanisms, which include reduction of the excitability of the CChrc in the hypoxic chemoreception (Prabhakar, 1994). In addition to the well known activity of CChrc in oxygen sensing, we found that NO likely participates in brain glucose homeostasis (Montero et al., 2006). The findings that mitochondria contain an isoform of NO synthase, which produces significant amounts of NO for regulating their own oxygen demands (Brodsky et al., 2002), indicates that NO may be important for glucose metabolism.

Our laboratory has previously observed that NO donors centrally applied into the cisterna magna, increase hyperglycemic response and BGR during normoxia in anesthetized rats. By contrast, during the hypoxic state, as after CChrc stimulation with NaCN, central NO administration inhibits the glucose parameters studied. In the same way, carotid baroreceptor activation induces NO production in the STn and c-fos proto-oncogen expression after CREB phosphorylation through a soluble guanylate cyclase/GMPc/proteinkinase G1 (Chan et al., 2004). In a further attempt to define the significance of centrally produced NO after CChrc stimulation on glucose homeostasis we have manipulated the hyperglycemic reflex with BGR after applying an anoxic stimulus to the CChrc, infusing a NO donor or a NOS inhibitor into the STn. Selective concentrations of the nitroxidergic drugs, capable of altering nitric oxide levels in blood, were subsequently used to study their effects on the activity of STn neurons measured as FOS protein expression (Fos-ir).

2 Methods

2.1 Animals, Surgical Procedures and CChrc Stimulation

Male Wistar rats (280–300 g) were used. They were housed at 22–24°C on a 12:12 h light:dark cycle. Al procedures were in accord to the Guide for the Care and Use of Laboratory Animals from the National Research Council. Rats were anesthetized with a bolus injection of sodium pentobarbital (3 mg/100 g i.p.) and a continuous i.p. infusion of the same anesthetic (0.063 mg/min). The rats were maintained with artificial respiration controlling body temperature.

CChrc stimulation was performed by slowly injecting NaCN (5 μg/100 g in 0.25 ml saline) into the left carotid sinus. To ensure that responses were due to NaCN reaching only the rat's left carotid sinus, it was temporarily isolated from the cephalic circulation, while the right carotid sinus was denervated (Álvarez-Buylla and Roces de Álvarez-Buylla, 1975). Both, the left external carotid artery (beyond the lingual branch) and the internal carotid near the jugular foramen were temporarily occluded (15–20 s); the perfusion liquid that bathed the left carotid body was withdrawn by means of a catheter introduced into the lingual artery (Álvarez-Buylla and Álvarez-Buylla, 1988).

2.2 STn Microinjections

Glass cannulas (Microcaps, Drummond) were placed in the STn according Paxinos and Watson (1986) coordinates: P=12.7 mm, L=1.5 mm, V=7.7 mm At the end of the experiment the whole brain was rapidly extracted and stored at $-70°C$ until cut on a cryostat (Leica VT 1000E) at 40 μm, to verify the injection place.

2.3 Drugs

Sodium cyanide (NaCN); artificial cerebrospinal fluid [aCSF (NaCl, KCl, $MgCl_2$, $CaCl_2$, ascorbate)]; saline (100 nl); sodium nitroprusside (NPS, 5 nmol); N-nitro-L-arginine methyl ester (L-NAME, 10 nmol).

2.4 Blood Collection and Measurements

Blood samples were withdrawn from catheters inserted into the femoral artery and jugular sinus without interrupting the circulation in these vessels (0.12 ml of arterial blood and 0.12 ml of venous blood proceeding from the brain). Blood samples were drawn at t= -10 min and t= -5 min (before nitroxidergic drugs infusion into the STn was made), and t=5 min, t=10 min, t=20 and t=30 min (after CChrc stimulation). In a complete sampling period, 1.2 ml of blood was taken (less than 8% of total blood volume). Fluid loss was replaced by injecting 0.3 ml saline after each pair of samples was taken. Plasma glucose concentration was measured by the glucose-oxidase method in mg/dl, in a Beckman Autoanalyzer. Brain glucose uptake was calculated in mg/dl, between glucose concentration in the abdominal aorta and glucose concentration in the jugular sinus.

The data are expressed in absolute values as mean ± SEM. The statistical comparisons were performed using Student's paired t-test, and ANOVA for repeated measurements when comparisons were made between groups. $*P<0.05$, statistical significant when the result was compared with its own basal; $^+P<0.05$, statistical significant when comparison was made between control group and experimental groups.

2.5 Immunohistochemistry

After the last sample was taken, animals were perfused intracardially (t=40 min after CChrc stimulation) with warm isotonic saline solution followed by ice-cold 4% paraformaldehyde in 0.1 M phophate-buffered saline (PBS). The brain was removed, post-fixed by submersion in PBS overnight. Serial transverse sections of the medulla oblongata were cut on a cryostat (50 μm) and collected in PBS. Free-floating sections that contained the STn were processed for immunohistochemistry using a primary sheep polyclonal antiserum (Jackson Merk) generated against the N-terminal of rat FOS protein to detect FOS-like immunoreactivity (FOS-ir) (Kobelt et al., 2004). As a secondary antiserum, biotinilated BA-1000 antirabbit IgG made in goat (Vector lab) was used. Visualization of immunoreactive product was enhanced with diaminobenzidine (DAB). Positive FOS-ir cells were counted manually.

2.6 Experimental Protocol

Experiment 1: infusion of aCSF (control) in the STn 4 min before CChrc stimulation with NaCN; arterial glucose, BGR and FOS-ir expression in STn neurons in rats were determined. Experiment 2: same as in "experiment 1", after an infusion of a NO donor (sodium nitroprusside- SNP) in the STn 4 min before CChrc stimulation with NaCN. Experiment 3: same as in "experiment 1" after an infusion of a NOS inhibitor (N-nitro-L-arginine methyl ester- L-NAME) in the STn 4 min before CChrc stimulation with NaCN.

3 Results

In the rats in which aCSF (100 nl) was injected into the STn 4 min before CChrc stimulation with NaCN (5 μg/100 g) an increase in arterial glucose levels accompanied by an increase in BGR was observed (Fig. 1). When a SNP (5 nmol/100 nl) injection was made in the STn 4 min before CChrc stimulation, arterial glucose levels also increased, while the increase in BGR was inhibited. In this case BGR decreased down to 13.6 ± 3.8 at t=10 min, and to 13.3 ± 2.4 mg/dl at t=20 min. Comparing these values with those obtained after an aCSF infusion, significant differences were obtained ($P=0.02$ and $P=0.03$ respectively, ANOVA for repeated measures) (Fig. 1). When a NOS inhibitor (L-NAME) was injected in the STn 4 min before CChrc stimulation, the hyperglycemic reflex did not result in significant changes, while BGR increased significantly in comparison with control experiments ($P=0.04$, ANOVA for repeated measures) (Fig. 1).

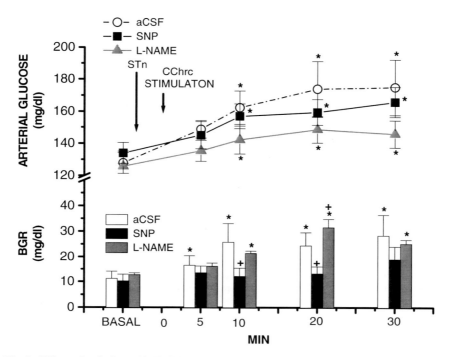

Fig. 1 CChrc stimulation with NaCN (5 μg/100 g 4 min after an aCSF injection (n=5) or a SNP infusion (5 nmol/100 nl) (n=6) or a L-NAME infusion (10 nmol/100 nl) (n = 4) in the STn in anesthetized rats (n=6). aCSF, artificial cerebrospinal fluid; L-NAME, Nω-nitro-L-arginine methyl ester; NaCN, sodium cyanide; SNP, sodium nitroprusside; STn solitary tract nucleus. The values are means ± SE, *$P<0.05$, compared with their own basal, Student t-test; +$P<0.05$ comparing the control group (aCSF), ANOVA for repeated measures

To verify whether the STn participates in the hyperglycemic reflex with BGR, the number of FOS-ir positive neurons in this nucleus was determined 40 min after CChrc stimulation and unilateral injections of aCSF (control), SNP or L-NAME in the STn 4 min before the stimulation. Figures 2 and 3 show the results obtained in FOS-ir expression distribution in these groups. In control rats, the number of positive-marked cells 40 min after CCrhc stimulation was 56±4; while when SNP was injected, the number of positive STn neurons increased up to 71±7 after CChrc stimulation ($P=0.05$, ANOVA one way, SNP vs aCSF); the injection of L-NAME in the STn reduced the number of FOS-ir cells down to 30±2 in the STn neurons after CChrc stimulation ($P=0.001$, ANOVA one way, L-NAME vs aCSF).

4 Discussion

Our findings suggest that nitroxidergic drugs have a unique relationship with CChrc stimulation effects on glucose variables. When aCSF injections were made into the STn 4 min before CChrc local stimulation with NaCN in control rats, a significant

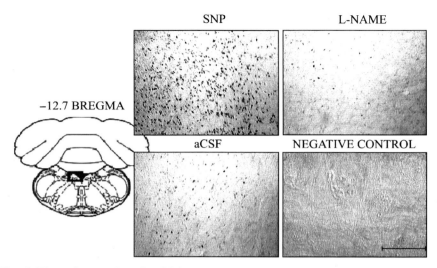

Fig. 2 Photomicrographs of FOS-ir neurons in the STn after: aCSF (100 nl), SNP (5 nmol/100 nl) or L-NAME (10 nmol/100 nl) infusions in the STn rats 4 min before CChrc stimulation of normal anesthetized rats, 100 X. Black dots represent individual nuclei with FOS-ir expression. Large numbers of FOS-ir cells were detected after SNP infusion. The enclosure shows the STn in a brain stem section. Scale bar: 1 mm

increase in arterial glucose concentration and BGR levels were obtained. An infusion of SNP into the STn 4 min before CChrc stimulation elicited an increase in arterial glucose levels with a significant reduction on BGR (when compared with the results obtained in control rats). Chemoreceptor activation also induced a transcriptional activation of c-fos gene in STn of rats that received a NO donor, indicating a

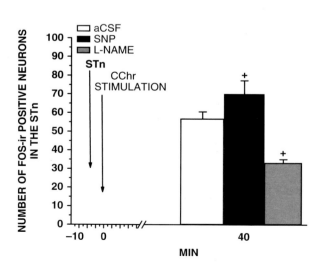

Fig. 3 Number of Fos-ir neurons in the STn 40 min after CChrc stimulation with NaCN and a previous infusion of aCSF (100 nl), or SNP (5 nmol/100 nl), or L-NAME (10 nmol/100 nl) in the STn in normal anesthetized rats. The values are means ± SE, *$P < 0.05$ with aCSF, $^+ P < 0.05$, comparing aCSF group with SNP or L-NAME groups, ANOVA one way

robust Fos-ir neuron expression. We conclude that nitroxidergic neurons in the STn, participate in the glycemic responses obtained here after CChrc stimulation, as was previously observed for baroreceptor stimulation (Talman et al., 2001). Moreover, L- NAME infusion into the STn 4 min before CChrc stimulation induced a significant increase in BGR, decreasing FOS-ir neurons in STn sections. This discrepancy could be due to differential effect of L-NAME in the STn depressing glutamate and adenosine as well as the hyperglycemia induced by 2-deoxyglucose (Sugimoto et al., 1997). NOS inhibitors also reduce baro-cardiopulmonary reflexes (Dias et al., 2005).

As was previously observed in in vitro experiments, NO induces a dual effect on carotid chemosensory discharges depending on the PO_2 level. Low levels of PO_2 inhibit the increase in chemosensory discharges, whereas during normoxia, NO increases the discharges (Iturriaga et al., 2000). SNP also reduces the increase in sensitivity and amplitude in the CSN discharges induced by acetylcholine, but L-NAME enhances it (Alcayaga et al., 1999). We cannot dismiss a possible inhibitory effect of NO on STn blood vessels. The increase in Fos-ir positive cells in the STn after SNP is probably due to NO responding neurons activation after CChrc stimulation with an anoxic stimulus, inhibiting the BGR. By the contrary, L-NAME effects were the opposite. It is also possible that GABAergic activity to inhibit the anoxic stimulus on BGR has been reflected in STn neurons increasing FOS-ir expression.

In conclusion, unilateral microinjections of a NO donor as SNP into the STn before CChrc stimulation with NaCN induced significant changes in BGR, when the results were compared with those obtained in control rats that received aCSF in the STn. In these experiments, FOS-ir expression in STn neurons also changed, indicating a significant increased when compared with STn cells obtained in control experiments. These local changes could be able of modulating the CChrc function by means of nitroxidergic pathways to participate in glucose homeostasis.

Acknowledgments We thank M.Sci. Arturo González González for technical assistance. Project supported by FRABA 330/2005 and CONACYT P49376-Q.

References

Alcayaga, J., Barrios, M., Bustos, F., Miranda, G., Molina, M.J. & Iturriaga, R. 1999, Modulatory effect of nitric oxide on acetylcholine-induced activation of cat petrosal ganglion neurons in vitro, *Brain Res.* 825, 194–198.

Almeida, A., Cidad, P., Delgado-Esteban, M., Fernández, E., García-Nogales, P. & Bolaños J.P. 2005, Inhibition of mitochondrial respiration by nitric oxide: its role in glucose metabolism and neuroprotection, *J Neurosci Res* 79: 166–171

Álvarez-Buylla, R. & Álvarez-Buylla, E. 1988, Carotid sinus receptors participate in glucose homeostasis, *Respir Physiol*, 72: 347–360.

Álvarez-Buylla, R. & Roces de Álvarez-Buylla, E. 1975, Hypoglycemic conditioned reflex in rats: preliminary study of its mechanism, *J Comp Physiol Psychol*, 88: 155–160.

Brodsky, S.V., Gao, S., Li, H. & Goligorsky, M.S. 2002, Hyperglycemic switch from mitochondrial nitric oxide to superoxide production in endothelial cells, *Am. J. Physiol Hear Cir Physiol*, 283: H2130–H2139.

Chan, S.H., Chang, K.F., Ou, C.C. & Chan, J.Y. 2004. Nitric oxide regulates c-fos expression in nucleus tractus solitarii induced by baroreceptor activation via cGMP-dependent protein kinase and cAMP response element-binding protein phosphorylation. *Molec Pharmacol*, 65, 319–25.

Dias, A.C., Vitela, M., Colombari, E. & Mifflin, S.W. 2005, Nitric oxide modulation of glutamatergic baroreflex, and cardiopulmonary transmission in the nucleus of the solitary tract, *Amer J Physiol: Heart Cir Physiol*, 288: H256–H262.

Housley, G.D. & Sinclair, J.D. 1988, Localization by kainic acid lesions of neurons P transmitting the carotid chemoreceptor stimulus for respiration in rat, *J Physiol*, 406: 99–114.

Iturriaga, R., Villanueva, S. & Mosqueira, M. 2000, Dual effects of nitric oxide on cat carotid body chemoreception, *J Appl Physiol*, 89: 1005–1012.

Kadowaki, K., Kishimoto, J., Leng, G. & Emson, P.C. 1994, Up-regulation of nitric oxide synthase (NOS) gene expression together with NOS activity in the rat hypothalamo-hypophyseal system after chronic salt loading: evidence of a neuromodulatory role of nitri oxide in arginine vasopressin and oxytocin secretion, *Endocrinol*, 134: 1011–1017.

Kobelt, P., Tebbe, J.J., Tjandra, I., Bae, H.G., Ruter, J., Klapp, B.F., Wiedenmann, B. & Monnikes, H. 2004, Two immunocytochemical protocols for immunofluorescent detection of c-Fos positive neurons in the rat brain, *Brain Res Prot,* 13: 45–52.

López-Barneo, J. 2003, Oxygen and glucose sensing by carotid body glomus cells, *Curr Op Neurobiol*, 13: 493–9.

Mifflin, S.W. 1996, Convergent carotid sinus nerve and superior laryngeal nerve afferent inputs to neurons in the NTS, *Amer J Physiol*, 271: R870–R880.

Moncada, S., Palmer, R.M. & Higgs, E.A. 1991, Nitric oxide: physiology, pathophysiology and pharmacology, *Pharmacol Rev*, 43: 109–142.

Montero, S., Cadenas, J.L., Lemus, M., Álvarez- Buylla, E. & Álvarez- Buylla, R. 2006, Nitric oxide in brain glucose retention after carotid body receptors stimulation with cyanide in rats, *Adv Exp Med Biol*, 580: 293–300.

Paxinos, G. & Watson, C. 1986, The rat brain in stereotactic coordinates, Academic Press, San. Diego.

Prabhakar, N.R. 1994, Neurotransmitters in the carotid body, *Adv Exp Med Biol*, 360: 57–69.

Sugimoto, Y., Yamada, J., Yoshikawa, T. & Horisaka, K. 1997, Inhibitory effects of nitric oxide synthase inhibitor, N(G)-nitro-L-arginine methyl ester (L-NAME), on 2-deoxy-D-glucose-induced hyperglycemia in rats, *Biol Pharmacol Bull*, 20: 1307–1309.

Sunico, C.R., Portillo, F., González-Forero, D. & Moreno-López, B. 2005, Nitric oxide-directed synaptic remodeling in the adult mammal CNS, *J Neurosci*, 25: 1448–1458.

Talman, W.T., Dragon, D.N., Ohta, H., & Lin, L.H. 2001, Nitroxidergic influences on cardiovascular control by NTS: a link with glutamate, *Ann New York Acad Sci*, 940: 169–78.

Zhang, M., Buttigieg, J. & Nurse, C.A. 2007, Neurotransmitter mechanisms mediating low-glucose signalling in cocultures and fresh tissue slices of rat carotid body, *J Physiol*, 578: 735–750.

Airway Receptors and Their Reflex Function – *Invited Article*

J. Yu

Abstract Sensory information in the lung is generated by airway receptors located throughout the respiratory tract. This information is mainly carried by the vagus nerves and yields multiple reflex responses in disease states (cough, bronchoconstriction and mucus secretion). Airway receptors are also essential for breathing control and lung defense. A single sensory unit contains homogeneous or heterogeneous types of receptors, providing varied and mixed behavior. Thus, the sensory units are not only transducers, but also processors that integrate information in different modes.

Keywords Mechanoreceptors · Chemosensitive receptors · Visceral afferents · Sensory neuron · Lung reflex

1 Introduction

The respiratory centers are under the constant influence of afferent signals from different sources. The lungs carry sympathetic nerves (Yu, 2005) and are richly innervated by the vagus (Coleridge & Coleridge, 1994; Dey et al., 1999). This review focuses on the vagal afferents, which provide important inputs to the central nervous system about the mechanical and chemical status of the lung. These inputs regulate breathing and initiate body defense responses, especially in cardiopulmonary diseases such as heart failure, acute respiratory distress syndrome, chronic obstructive pulmonary disease, and asthma. The mechanical and chemical environment is altered in the diseased lung. Mechanoreceptors may undergo phenotypical changes to express neurokinins (Hunter et al., 2000), and there is increasing evidence that the vagus plays a crucial role in neural immune interaction (Otmishi et al., 2008). Airway sensory activation may cause neurogenic inflammation to intensify local

---

J. Yu (✉)
Department of Pulmonary Medicine, University of Louisville, Louisville, KY 40292, USA
e-mail: j0yu0001@louisville.edu

inflammatory responses, or may initiate an inflammatory reflex through the vagus nerve to suppress pro-inflammatory cytokine production to quench the inflammatory response (Tracey, 2002). Airway sensors, especially chemosensitive ones, serve as biosensors to monitor the lungs' inflammatory status by detecting inflammatory mediators (Lin & Yu, 2007) and cytokines (Yu et al., 2007). To understand lung pathophysiology, a comprehensive analysis of airway sensor behavior and function is clearly needed.

2 Classification

By convention, airway receptors are classified into three types: (1) SARs (Schelegle & Green, 2001), (2) rapidly adapting receptors (RARs) (Sant'ambrogio & Widdicombe, 2001), and (3) C-fiber receptors (CFRs) (Lee & Pisarri, 2001). SARs and RARs are mechanosensitive and connect to myelinated fibers. CFRs are chemosensitive receptors linked to unmyelinated fibers. However, there are other receptor types. Another myelinated afferent was recently identified. It is neither RAR nor SAR in type and has a conduction velocity in the $A\delta$ range (4–15 m/s). This receptor shares many characteristics with CFRs, including chemosensitivity and low discharge frequency even during high pressure lung inflation (Yu, 2002a). It is stimulated by hypertonic saline, hydrogen peroxide, and bradykinin, but also has a high mechanical threshold. The sensor is called the high threshold $A\delta$ receptor (HTAR) and has been identified in vivo and in vitro (Widdicombe, 2003). Another addition to the classification system is myelinated mechanoreceptors that are activated on deflation. They are called deflation activated receptors (DARs) (Yu, 2005). Furthermore, neuroepithelial bodies (NEBs) in the airways are thought to be receptors; these are aggregated, richly innervated neuroendocrine cells (Adriaensen et al., 1998; Cutz & Jackson, 1999). However, nothing is known about their afferent discharge patterns. Lastly, cough receptors have been re-proposed (Canning et al., 2006).

Among the RAR and SAR mechanosensors are intermediates having overlapping characteristics (Yu, 2005). Figure 1 shows a regular resting discharge related to airway pressure, which is a major SAR characteristic. However, its adaptation index at 75% falls into the RAR category. This unit adapted more when the inflation pressure was higher, and became silent at 15 cmH$_2$O (rapidly adapting nature), but discharged at a constant rate at 7 cmH$_2$O (slowly adapting nature). Some receptors are activated only with lung inflation and some only on deflation (Bergren & Peterson, 1993). Others respond to both inflation and deflation. Conventional classification is based only on receptor response to lung inflation. Inflation and deflation are different stimuli, and lung inflation cannot be used to characterize receptor deflation response (Yu, 2005). It is difficult to understand how a receptor can respond to opposing forces. These puzzles fundamentally challenge the current view of sensory receptors.

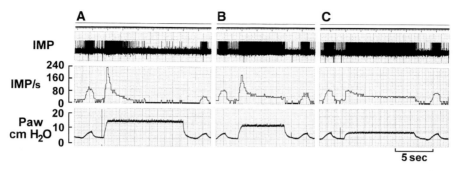

Fig. 1 A sensory unit having SAR-like regular discharge during lung inflation, with an adaptation index of 75% at 15 cmH$_2$O inflation pressure similar to the RAR. *Traces*: IMP, unit activity; IMP/s, unit activity expressed as impulses per second; Paw, airway pressure. The lungs were sequentially inflated to 15 (**A**), 12 (**B**) and 7 (**C**) cmH$_2$O. Note that higher inflation pressure produces a faster adaptation rate. After adaptation, the unit activity almost ceased at 15 cm H$_2$O and reached a plateau at a pressure less than 7 cm H$_2$O, which is indicated by the same activities at 7 and 12 cm H$_2$O. Clearly, this unit has an initial RAR component and a delayed, long lasting, SAR component. Adapted from a previous report (Yu, 2000) with permission

3 Morphology

How sensory units behave depends on their structure. However, receptor morphology is not well described. Even the best known SARs appear as hand-drawn illustrations in the literature, with little agreement as to structure, shape, location, or orientation (Baluk & Gabella, 1991; During et al., 1974; Krauhs, 1984; Yamamoto et al., 1995). Information regarding their sub-cellular structure is scanty. This greatly impedes understanding receptor physiology. The current review provides preliminary information on this issue.

Receptor structures in central and peripheral airways are similar and plant-like (Larsell, 1922). However, SARs in larger airways are bigger and more complex (Fig. 2). A parent axon can give off branches forming several receptor structures with many knob-like or leaf-like endings. These knobs or leaves are buried in the smooth muscle layer. SARs have been reported in parallel with smooth muscle (Baluk & Gabella, 1991; Yamamoto et al., 1995). However, the sensory structure of the SAR has never been defined. In the peripheral airway, the direction of the smooth muscle fibers is irregular. Therefore, it is hard to envision a parallel relationship. Typical SARs have many terminal endings (knobs). Each knob is an encoder (Fig. 2) or basic sensory device that can independently generate action potentials (Yu, 2005). The encoder can be stretched in any direction. No matter how the smooth muscle is oriented in relation to the receptor, muscle contraction will stretch the encoder to activate it.

With advances in immunohistochemistry, neural tracing, and confocal microscopy, airway receptor structures can be examined in detail and evaluated objectively. Na$^+$/K$^+$-ATPase is an excellent marker for airway receptors (Wang & Yu, 2004).

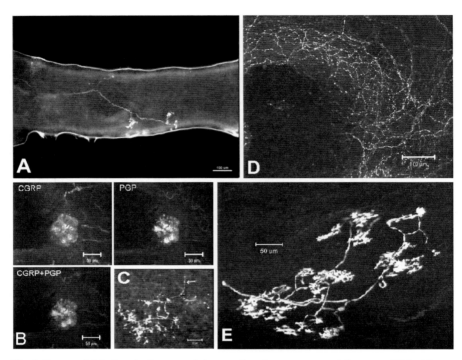

Fig. 2 Immunohistochemical approach to examine airway sensors. A is an unpublished image from our laboratory. C (Wang & Yu, 2004), and B, D, and E (Yu, 2002b) are adapted from recent reports with permission. A shows a SAR unit (stained for Na$^+$/K$^+$-ATPase) in a medium sized airway (400 μm in diameter). The structures in medium sized airway are similar to but simpler than those in the trachea (E). B, a projection image of a neuroepithelial body (NEB) shows double staining for CGRP and PGP. Note that the NEB is located at an airway opening, protrudes into the airway, and is innervated by nerve fibers with different chemical compositions. C is a projection image, showing a sensor in the ventral tracheal epithelium. This structure is 100 × 80 × 12 μm in length, width, and depth, respectively with the axon (indicated by the arrow) of 1.2 μm. D is an image of a lobular bronchus, showing densely distributed varicose nerve fibers that are immunoreactive with SP. These varicose nerves are present in all airways, but their density is much higher centrally than in the periphery. E is a presumptive SAR in the trachea, which is immunoreactive to Na$^+$/K$^+$-ATPase. This structure is embedded in the smooth muscle and is 400 × 150 × 70 μm in length, width and depth respectively

Figure 2E illustrates a receptor structure identified in the trachealis muscle. The parent axon splits into branches, each forming a receptor structure embedded in a muscle band. Vagal nerve fibers are known to penetrate the airway epithelium and connect to receptor structures there (Fig. 2C) and in the lamina propria. In addition, Substance P-immunoreactive (SP-IR) receptor endings have been observed (Fig. 2D). SP-IR fibers are believed to be sensory and connected to the CFRs, although the possibility that they are HTARs needs to be excluded. Besides histochemical techniques, neural tracer can be used to identify airway sensors (Fig. 3).

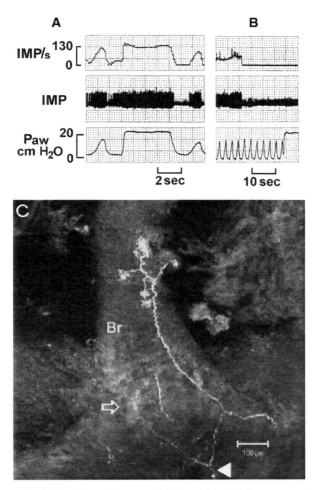

Fig. 3 Structure of a SAR with activity recorded from the rabbit vagus nerve. Neural tracer, DiI (1,1' – dioleyl – 3,3,3', 3' tetramethyl-indo carbocyanine), was injected into the nodose ganglion and transported to the lung periphery (3 weeks). Abbreviations are as in figure 1. The adaptation index shows a typical SAR during constant pressure inflation (**A**). The unit ceased firing and did not respond to lung inflation or touching after the receptive field was removed (**B**). (**C**) A micrograph from the confocal microscope shows a projection image of the receptive field, which contains the SAR identified in **A** and **B**. The SAR is located in an airway about 180 μm in diameter (Br). The axon runs outside the bronchiole wall and penetrates the wall and then branches, forming many twigs. The knob-like endings are buried in the wall. The parent axon (indicated by a *large white arrow head*) gives off multiple branches to receptor structures. Two can be identified in this image. One is clearly shown, and the other structure can be vaguely seen in the wall at the branching point of the bronchioles (indicated by an *arrow*). Adapted from a previous report (Yu, 2005) with permission

Unlike other airway receptors, NEBs are well defined structurally (Cutz & Jackson, 1999). The nerve fibers emanate from local ganglia, the spinal cord and nodose ganglia (Adriaensen et al., 1998; van Lommel et al., 1998). NEBs contain numerous bioactive substances and peptides [calcium gene related peptide (CGRP), SP, and protein gene product 9.5 (PGP 9.5) to name a few], which can be targeted by many antibodies (Fig. 2B). SARs are neither SP-IR nor CGRP-IR. Therefore, SARs and NEBs are distinct. Using this technique, the chemical composition of each airway receptor type can be characterized. Detailed images of vagal sensory receptors and their terminal formations may provide new insights into receptor function.

4 Structure Function Relationships

The importance of identifying a pulmonary receptor's structure is to understand its function. Linking a receptor's morphology to its electrophysiological behavior is possible using immunohistochemical and neuronal tracing techniques that provide detailed receptor structures (Wang & Yu, 2004); and isolating and blocking a single receptor or part of the receptor for physiological and morphological study (Yu & Zhang, 2004). Combining these techniques allows examining a given receptor by characterizing its electrophysiology and then dissecting the receptor for morphological identification (Yu, 2005). Using these techniques, SAR structures have been identified (Fig. 3). Such isolation and characterization techniques can be used to identify other receptors and directly explore their structure-function relationships and sensory unit coding mechanisms.

Unit activity may be recorded from a parent axon that supplies many SARs. In other words, a sensory fiber (or main axon) can receive information from several receptive fields (Yu & Zhang, 2004). Two important issues arise. Where is the action potential generated? How are electrical activities from several receptive fields integrated to give the final activity in a sensory unit? In many cases, eliminating the function of one field by local anesthesia does not decrease peak discharge frequency, but decreases the total discharge of the unit (Fig. 4). This suggests the unit contains multiple encoders (Yu & Zhang, 2004). We observed heterogeneous encoders in a sensory unit, for example, where one encoder senses inflation and the other deflation. In other instances, one encoder may be rapidly adapting and the other slowly adapting (Fig. 1). Exploring such sensory function-structure relationships is required to fully understand the role of these sensors under normal and disease conditions. Combining the techniques of electrophysiology, anatomy, histochemistry, and microscopy (confocal and electron), different structures can be mapped and linked to a specific receptor type.

Based on the above information, a mechanosensory unit should be defined as a functional unit that contains multiple encoders (or receptors) with different characteristics, including RARs, SARs, and DARs (Yu, 2005). Each is capable of sensing one aspect of lung mechanics. A unit's behavior depends on its encoder composition. Airway sensory units are not only transducers but also processors. Significant information is integrated at the intra- and inter-encoder levels, resulting in the final output to the central nervous system.

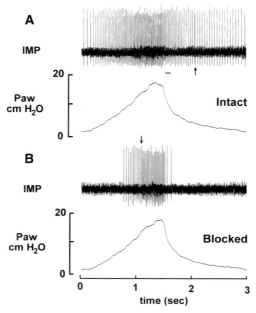

Fig. 4 Activity in an SAR unit with two encoders before (**A**) and after (**B**) blocking one with lidocaine by microinjection. The bar in **A** indicates where unit activity switches from encoder **A** (*phasic encoder*) to encoder B (*tonic encoder*). The maximal frequency of encoder B is somewhere between the frequency at the bar and at the peak, which is the maximal frequency of encoder **A**. The *arrow* in **B** indicates where encoder **A**'s discharge is about to exceed encoder **B**. Drawing two vertical lines at the arrow in **B** and at the beginning of the bar in **A** crossing **A** divides the figure into three sections. The activity during deflation (*first* and *third sections*) is determined by encoder **B**, whereas the activity during inflation (*second section*) is determined by encoder **A**. Thus, the pacemaker switches back-and-forth between the two encoders within a ventilator cycle. Adapted from a previous report (Yu & Zhang, 2004) with permission

5 Reflex Function

SARs operate at birth (Fisher & Sant'ambrogio, 1982) and initiate the Hering-Breuer reflex. Activating SARs inhibits inspiration, stimulates expiration, decreases airway tone, and suppresses sympathetic nerve activity, causing peripheral vasodilatation (Yu, 2002b). SAR input is also responsible for sinus arrhythmia. Activation of SARs is also believed to relieve dyspnea (Paintal, 1995). By definition, RARs are mechanoreceptors. RARs sense the dynamics of the stimulus, whereas SARs sense the magnitude of it. It is known that RARs are activated when breathing is stimulated as lung compliance decreases. However, discussing reflex effects of RARs is difficult because the receptor is not well defined. The assumption there are only three types of airway sensors complicates the issue. Many investigators ascribe reflex effects to RARs by default whenever SARs and CFRs are ruled out. Reflex effects induced by chemical substances may come from HTARs. Deflation responses, such as dyspnea and hyperventilation during acute pneumothorax, previously assigned to

RARs, should mainly be attributed to DARs. DARs are defined as mechanoreceptors activated by lung deflation, which include those connected with RARs and SARs, and pure deflation receptors. Activation of DARs possibly stimulates inspiration, suppresses expiration, and shortens expiratory duration (Yu, 2005).

Activating HTARs may initiate the excitatory lung reflex, which includes hyperpnea and tachypnea, increased inspiratory time and duty cycle, and reduced expiratory time and activity. In lung disease, these effects increase inspiratory muscle load and fatigue (Yu, 2002a). Multiple functions have been proposed for NEBs (Sorokin & Hoyt, 1990), such as oxygen sensors (Cutz & Jackson, 1999) and nociceptors (Sorokin & Hoyt, 1990; Yu, 2002b). NEBs may interrelate with HTARs, because they are connected with myelinated afferents and possess the bioactive substances that can activate nociceptors (Yu, 2002b). According to their blood supply, CFRs can be categorized into pulmonary and bronchial sub-types. Pulmonary CFRs are also called juxta-capillary or J receptors (Lee & Pisarri, 2001). Activation of CFRs causes a triad of responses, consisting of apnea followed by rapid shallow breathing, bradycardia and hypotension. CFR activation reflexively causes vasodilation, secretion, and constriction in the airways.

CFRs and HTARs are chemosensitive and are referred to as nociceptors because they are stimulated by noxious stimuli. Neuro-immune interaction is one of the mechanisms to control inflammation intensity in a variety of diseases (Otmishi et al., 2008). Airway nociceptors play an important role in neuro-immune interaction. These biosensors can be activated by a variety of inflammatory mediators, modulators or cytokines, including tachykinins, calcitonin gene related peptide, prostaglandins, bradykinin, adenosine, 5-HT, histamine, reactive oxygen species, TNF-α and IL-1β. They may serve as biosensors to monitor inflammation in the lung by detecting inflammatory substances to regulate immune responses to trauma, infection or inflammation (Goehler et al., 2000). Indeed, they are activated in different lung inflammatory disease models, such as endotoxemia (Lai et al., 2002), acute respiratory distress syndrome caused by intravenous injection of oleic acid (Lin et al., 2007), lung ischemia-reperfusion injury, and asthma. Stimulation of these sensory receptors may activate the CNS and exert anti-inflammatory effects through neural humoral responses (Otmishi et al., 2008; Tracey, 2002).

6 Concluding Remarks

Vagal afferents send vital information for breathing control and regulating immune responses. Airway sensory units detect such information and are the initiating points for protective or defensive mechanisms. However, our understanding of sensory receptor fundamentals (morphological structure, activation mechanisms, sensory behavior, and reflex function) remains incomplete. A receptor can be identified physiologically by its discharge pattern, morphologically by its structural pattern, and chemically by its chemical composition (or genetic markers). A new system for objective and comprehensive classification of airway receptors is needed to generate new theories on the mechanisms of sensor operation, and thereby provide a foundation for future assessment of their roles in respiratory physiology and pulmonary disease.

Acknowledgments This work was supported by a grant from NIH HL-58727.

References

Adriaensen, D., Timmermans, J. P., Brouns, I., Berthoud, H. R., Neuhuber, W. L. & Scheuermann, D. W. (1998) Pulmonary intraepithelial vagal nodose afferent nerve terminals are confined to neuroepithelial bodies: an anterograde tracing and confocal microscopy study in adult rats. *Cell Tissue Res*, 293, 395–405.

Baluk, P. & Gabella, G. (1991) Afferent nerve endings in the tracheal muscle of guinea-pigs and rats. *Anat Embryol.(Berl.)*, 183, 81–87.

Bergren, D. R. & Peterson, D. F. (1993) Identification of vagal sensory receptors in the rat lung: are there subtypes of slowly adapting receptors? *J Physiol*, 464, 681–698.

Canning, B. J., Mori, N. & Mazzone, S. B. (2006) Vagal afferent nerves regulating the cough reflex. *Respir Physiol Neurobiol*, 152, 223–242.

Coleridge, H. M. & Coleridge, J. C. G. (1994) Pulmonary reflexes: neural mechanisms of pulmonary defense. *Annu Rev Physiol*, 56, 69–91.

Cutz, E. & Jackson, A. (1999) Neuroepithelial bodies as airway oxygen sensors. *Respir Physiol*, 115, 201–214.

Dey, R. D., Satterfield, B. & Altemus, J. B. (1999) Innervation of tracheal epithelium and smooth muscle by neurons in airway ganglia. *Anat Rec*, 254, 166–172.

During, M. v., Andres, K. H. & Iravani, J. (1974) The fine structure of the pulmonary stretch receptor in the rat. *Z Anat Entwicklungsgesch*, 143, 215–222.

Fisher, J. T. & Sant'ambrogio, G. (1982) Location and discharge properties of respiratory vagal afferents in the newborn dog. *Respir Physiol*, 50, 209–220.

Goehler, L. E., Gaykema, R. P., Hansen, M. K., Anderson, K., Maier, S. F. & Watkins, L. R. (2000) Vagal immune-to-brain communication: a visceral chemosensory pathway. *Auton Neurosci*, 85, 49–59.

Hunter, D. D., Myers, A. C. & Undem, B. J. (2000) Nerve growth factor-induced phenotypic switch in guinea pig airway sensory neurons. *Am J Respir Crit Care Med*, 161, 1985–1990.

Krauhs, J. M. (1984) Morphology of presumptive slowly adapting receptors in dog trachea. *Anat Rec*, 210, 73–85.

Lai, C. J., Ho, C. Y. & Kou, Y. R. (2002) Activation of lung vagal sensory receptors by circulatory endotoxin in rats. *Life Sci*, 70, 2125–2138.

Larsell, O. (1922) The ganglia, plexuses, and nerve-terminations of the mammalian lung and pleura pulmonalis. *J Comp Neurol*, 35, 97–132.

Lee, L. Y. & Pisarri, T. E. (2001) Afferent properties and reflex functions of bronchopulmonary C-fibers. *Respir Physiol*, 125, 47–65.

Lin, S., Walker, J., Xu, L., Gozal, D. & Yu, J. (2007) Respiratory: Behaviours of pulmonary sensory receptors during development of acute lung injury in the rabbit. *Exp Physiol*, 92, 749–755.

Lin, S. X. & Yu, J. (2007) Effects of arachidonic acid metabolites on airway sensors. *Acta Physiologica Sinica*, 59, 141–149.

Otmishi, P., Gordon, J., El-Oshar, S., Li, H., Guardiola, J., Saad, M., Proctor, M. & Yu, J. (2008) Neuroimmune interaction in inflammatory diseases. *Clin Med: Circ, Respir and Pulm Med*, 2, 35–44.

Paintal, A. S. (1995) Some recent advances in studies on J receptors. *Adv Exp Med Biol*, 381:15–25.

Sant'ambrogio, G. & Widdicombe, J. (2001) Reflexes from airway rapidly adapting receptors. *Respir Physiol*, 125, 33–45.

Schelegle, E. S. & Green, J. F. (2001) An overview of the anatomy and physiology of slowly adapting pulmonary stretch receptors. *Respir Physiol*, 125, 17–31.

Sorokin, S. P. & Hoyt, R. F. (1990) On the supposed function of neuroepithelial bodies in adult mammalian lungs. *News Physiol. Sci*, 5, 89–95.

Tracey, K. J. (2002) The inflammatory reflex. *Nature*, 420, 853–859.

van Lommel, A., Lauweryns, J. M. & Berthoud, H. R. (1998) Pulmonary neuroepithelial bodies are innervated by vagal afferent nerves: an investigation with in vivo anterograde DiI tracing and confocal microscopy. *Anat Embryol (Berl.)*, 197, 325–330.

Wang, Y. & Yu, J. (2004) Structural survey of airway sensory receptors in the rabbit using confocal microscopy. *Acta Physiol sinica*, 56, 119–129.

Widdicombe, J. G. (2003) Overview of neural pathways in allergy and asthma. *Pulm Pharmacol Ther*, 16, 23–30.

Yamamoto, Y., Atoji, Y. & Suzuki, Y. (1995) Nerve endings in bronchi of the dog that react with antibodies against neurofilament protein. *J Anat*, 187, 59–65.

Yu, J. (2000) Spectrum of myelinated pulmonary afferents. *Am J Physiol Regul Integr Comp Physiol*, 279, R2142–R2148.

Yu, J. (2005) Airway mechanosensors. *Respir Physiol Neurobiol*, 148, 217–243.

Yu, J. (2002b) An overview of vagal airway receptors. *Acta Physiologica Sinica*, 54, 451–459.

Yu, J. (2002a) Pulmonary reflex and ventilatory failure. *Recent Research Developments in Respiratory and Critical Care Medicine* (ed. by S. G. Pandalai), pp. 55–68. Research Signpost.

Yu, J., Lin, S., Zhang, J., Otmishi, P. & Guardiola, J. J. (2007) Airway nociceptors activated by pro-inflammatory cytokines. *Respir Physiol Neurobiol*, 156, 116–119.

Yu, J. & Zhang, J. (2004) A single pulmonary mechano-sensory unit possesses multiple encoders in rabbits. *Neurosci Lett*, 362, 171–175.

Airway Chemosensitive Receptors in Vagus Nerve Perform Neuro-Immune Interaction for Lung-Brain Communication

H.F. Li and J. Yu

Abstract Recently, IL-1β has been found to activate airway C-fiber receptors (CFRs) and high threshold Aδ receptors (HTARs), which may influence innate immune response via activation of the central nervous system. The present study aims to determine whether such a stimulatory effect is restricted to IL-1β or applies to other pro-inflammatory cytokines. In anesthetized, open-chest, and mechanically ventilated rabbits, we recorded single unit activity from vagal nociceptors and examined their response to microinjection of TNF-α (1 µg/ml, 20 µl) directly into the receptive fields. Both CFRs and HTARs had similar responses. Their activity increased from 0.12±0.05 to 0.93±0.16 imp/s (n=15; $P<0.001$). This stimulatory effect of TNF-α was significantly attenuated by mixing with neutralizing antibody (10 µg/ml, 20 µl). The activities were 0.31±0.09 and 0.57±0.16 imp/s for control and injection of the TNF-α mixture (n=9; $P<0.01$), respectively. These nociceptors did not respond to location injection of normal saline. Our results show that TNF-α, like IL-1β, can activate airway nociceptors. It lends support to the hypothesis that airway nociceptors in the lung mediate the innate immune response.

Keywords Pro-inflammatory cytokines · TNF-alpha · Visceral afferents · Sensory neuron · Reflex · Sensory receptors

1 Introduction

There are at least five airway sensor types: slowly adapting receptors (SARs), rapidly adapting receptors (RARs), deflation activated receptors (DARs), high threshold A-delta receptors (HTARs), and C-fiber receptors (CFRs) (Coleridge & Coleridge, 1994; Yu, 2005). SARs, RARs, and DARs monitor mechanical changes, whereas HTARs and CFRs sense chemical status. They are chemosensitive receptors or nociceptors (Yu, 2008). As in other tissues, nociceptors in the lung are activated

J. Yu (✉)
Pulmonary Medicine, University of Louisville, Louisville, KY 40292, USA
e-mail: j0yu0001@louisville.edu

during inflammatory processes (Belvisi, 2003; Coleridge & Coleridge, 1984; Lee & Pisarri, 2001; Undem & Carr, 2001). Intense research into tissue inflammation has increased our understanding of the molecular, cellular and therapeutic effects of cytokines and the receptors through which they operate. However, little is known about the role of the airway sensors in this regard. Tumor necrosis factor-alpha (TNF-α) and interleukin-1β (IL-1β) are important pro-inflammatory cytokines during the acute phase of lung inflammation or injury. It is increasingly clear that the neural and immune systems interact with one another. Cytokines mediate such interaction (Goehler et al., 2000). Recently, a vagal-vagal inflammatory reflex has been proposed to regulate innate immune responses and inflammation (Tracey, 2002). Activation of vagal afferents causes reflex activation of vagal cholinergic efferents to inhibit production of TNF-α and other pro-inflammatory cytokines through alpha7 nicotinic receptor-mediated mechanisms (Pavlov & Tracey, 2006). Such vagal efferent anti-inflammatory pathways can be targeted for the treatment of pathological conditions with cytokine overproduction. The present studies test the hypothesis that the vagal nerves play a key role in immune response through a reflex mechanism to control inflammatory intensity. More specifically, we seek to demonstrate that airway nociceptors may act as biosensors to detect the level of pro-inflammatory cytokines, including TNF-α.

2 Methods

2.1 Animals and Anesthesia

Experimental procedures were approved by the IACUC of the University of Louisville. The rabbit was anesthetized with 20% urethane (1 g/kg, iv). The trachea was cannulated low in the neck and the lungs were mechanically ventilated. Positive-end-expiratory pressure (PEEP) was maintained by placing the expiratory outlet under 3 cm H_2O. Airway pressure was measured with a pressure transducer. The chest was widely opened to allow for location of the receptive field.

2.2 Nerve Recording

Single-unit activities from the vagal afferent were recorded according to conventional methods (Lin et al., 2007). The vagus nerve was separated from the carotid sheath, placed on a dissecting platform, and covered with mineral oil. A small afferent bundle was cut from the vagus nerve and was dissected into thin filaments for recording action potentials. The electrodes were connected to a High Impedance Probe from which the output was fed into an amplifier. Action potentials from a single unit of the vagal sensory receptors were recorded and counted by a rate meter at 0.1 s intervals. The receptive field was located by identifying the most sensitive point on the lung surface with a glass rod having a 0.5 mm round tip. These sensory receptors (HTARs and CFRs) were identified by their discharge pattern (Yu, 2008), and confirmed by verifying their receptive fields in the lung and by their conduction velocities (4 to 20 m/s for HTARs and less than 1.5 m/s for CFRs). Sensory unit responses to TNF-α were examined with and without neutralizing antibody against TNF-α. The agent was injected into the receptive field with a needle (30 GD) in a volume of 20 μl. Airway pressure and afferent activities were recorded by a thermorecorder.

3 Results

To test the hypothesis that airway nociceptors may act as biosensors in monitoring the inflammatory cytokines, we recorded 24 single-unit activity of HTARs (n=16) and CFRs (n=8) in anesthetized, open-chest, and mechanically ventilated rabbits. Since both CFRs and HTARs had similar behavior and responses, we reported their data together. These units had low, sporadic, irregular discharges and were insensitive to changes in lung mechanics, such as lung inflation. Figure 1 illustrates one such unit. Please note that the HTAR is not activated after the PDG injection but a

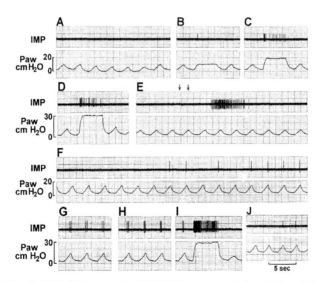

Fig. 1 A high threshold A-delta receptor (HTAR) in response to local micro-injection of TNF-α (1 μg/ml, 20 μl). A, B, C, D are PEEP removal, lung inflations at 10, 20, 30 cm H$_2$O. **E** denotes right atrial injection of PDG (1 mg/ml, 100 μl). Two arrows indicate the start and end of injection time. **F** is the response to topical injection of TNF-α. TNF-α was injected at the second ventilatory cycle. The HTAR became active at 11 s following the injection and discharged at two impulses per ventilatory cycle at 3 min (**G**) and at three impulses at 4 min (**H**) after injection. The HTAR is also more responsive to mechanical stimulation at 5 min (**I**) (compare with Panel D). The effect was long lasting. Receptor activity was still elevated one hour following the injection (**J**). The conduction velocity of the afferent is 10 m/s

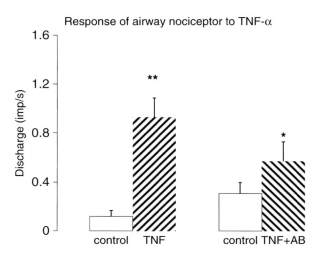

Fig. 2 The nociceptor discharges increased significantly following microinjection of TNF-α (1 μg/ml, 20 μl) into the receptive field. The response was attenuated by mixing the TNF-α (at final concentration of 1 μg/ml) with neutralizing antibody (AB, at final concentration of 10 μg/ml, 20 μl). *P<0.01 and **P<0.001 when compared with control (n=15)

CFR is. The CFR has its unique action potential shape with background electrical noise in the middle of action potential. This is in contrast to the shape of HTAR, which has background noise close to one end of action potentials (Yu, 2005). The nociceptors were stimulated several fold within several seconds following microinjection of (1 μg/ml, 20 μl) exogenous TNF-α into the receptive fields. Their activity increased from 0.12±0.05 to 0.93±0.16 imp/s (about eight fold; n=15 (10 HTARs and 5 CFRs); P<0.001). This stimulatory effect of TNF-α was significantly attenuated by mixing with neutralizing antibody (10 μg/ml, 20 μl) (Fig. 2). The activities were 0.31±0.09 and 0.57±0.16 imp/s for control and injection of the TNF-α mixture (about two fold; n=9 (6 HTARs and 3 CFRs); P<0.01), respectively. Injection of normal saline into the receptive field did not change receptor behavior in this control group.

4 Discussion

TNF-α and IL-1β are cardinal early mediators of the innate host inflammatory response. TNF-α is produced by a variety of inflammatory cells: neutrophils, mononuclear phagocytes, NK cells, endothelial cells and mast cells. During acute inflammation, TNF-α along with IL-1β are essential in neutrophil recruitment and activation. They enhance the immune response by induction of endothelial cells to express chemo-attractants on their cell surfaces that recruit neutrophils to the area of the injury (Martin, 1999; Tracey, 2002). Thus, neutralization of TNF-α has been tried to treat inflammatory diseases, especially sepsis. While it is unclear about the mechanisms that TNF-α activates the airway sensory neurons, many reports indicate that TNF-α can directly activate sensory neurons (dosal root ganglia neurons) by increasing voltage activated Na^+ currents (Czeschik et al., 2008), such as tetrodotoxin-resistant (TTX-R) Na^+ currents (Jin & Gereau, 2006). TNF-α also increases intracellular Ca^{++} via external influx or internal release (Pollock et al., 2002). This activation is mediated through TNF receptor 1 (Jin & Gereau, 2006; Pollock et al., 2002).

Increasing evidence suggests the immune and neural systems interact with one another, and the interaction not only at central level but also at peripheral level. This interaction is propagated by cytokines released during inflammatory processes, such as sepsis and acute tissue injury, and the vagus nerve performs an important role in the interaction (Goehler et al., 2000). The interaction mechanisms operate in inflammatory diseases (Otmishi et al., 2008). The present studies test the hypothesis the vagus nerves link the lungs' immune and neural systems by transmitting information through airway nociceptors. Our results are consistent with the hypothesis and support the notion that the vagus nerves play a role in reflex immune responses, and these effects may be mediated by airway nociceptors, such as HTARs and CFRs.

Recently, it has been reported that IL-1β stimulates airway nociceptors (Yu et al., 2007). In the present studies, we extend the scope of observation to show that this stimulatory effect is not restricted to IL-1β but also applies to other pro-inflammatory cytokines, such as TNF-α. This result further supports that the airway nociceptors are sensitive to pro-inflammatory cytokines. Therefore, they may serve to detect the tissue cytokine levels for communication with the brain. Airway nociceptors were stimulated by a variety of exogenous inflammatory mediators. These mediators include histamine, H_2O_2, 5-HT, bradykinin, and prostaglandins (Lee & Pisarri, 2001). Nociceptors were also stimulated by intravenous injection of oleic acid (Lin et al., 2007) or lipopolysaccharide (LPS) (Lai et al., 2002). Oleic acid (Ishitsuka et al., 2004) and LPS (Morrison & Ryan, 1987) are agents that release endogenous inflammatory mediators, including cytokines and chemokines, leading to lung inflammation and injury. Taken together, these results show that airway nociceptors were stimulated by exogenous inflammatory mediators and by agents that cause endogenous mediator release. These results demonstrate that nociceptors, acting as biosensors, online monitor the concentrations of inflammatory substances (including cytokines), detecting the intensity of inflammation. Thus, airway nociceptors may play a crucial role in reflex immune responses.

In conclusion, vagal nociceptors in the lung may mediate the innate immune response through neuro-immune interactions. They act as biosensors to detect inflammatory information by monitoring levels of pro-inflammatory mediators, including cytokines.

Acknowledgments This work was supported by a grant from NIH HL-58727 and Dr. Li is a recipient of AHA fellowship (0825694D).

References

Belvisi, M.G. (2003) Sensory nerves and airway inflammation: role of A delta and C-fibres 1. *Pulm Pharmacol Ther*, **16**, 1–7.
Coleridge, H.M. & Coleridge, J.C.G. (1994) Pulmonary reflexes: neural mechanisms of pulmonary defense. *Annu Rev Physiol*, **56**, 69–91.

Coleridge, J.C.G. & Coleridge, H.M. (1984) Afferent vagal C fibre innervation of the lungs and airways and its functional significance. *Rev Physiol Biochem Pharmacol*, **99**, 1–110.

Czeschik, J.C., Hagenacker, T., Schafers, M. & Busselberg, D. (2008) TNF-alpha differentially modulates ion channels of nociceptive neurons. *Neurosci Lett*, **434**, 293–298.

Goehler, L.E., Gaykema, R.P., Hansen, M.K., Anderson, K., Maier, S.F. & Watkins, L.R. (2000) Vagal immune-to-brain communication: a visceral chemosensory pathway. *Auton Neurosci*, **85**, 49–59.

Ishitsuka, Y., Moriuchi, H., Hatamoto, K., Yang, C., Takase, J., Golbidi, S., Irikura, M. & Irie, T. (2004) Involvement of thromboxane A2 (TXA2) in the early stages of oleic acid-induced lung injury and the preventive effect of ozagrel, a TXA2 synthase inhibitor, in guinea-pigs. *J Pharm Pharmacol*, **56**, 513–520.

Jin, X. & Gereau, R.W. (2006) Acute p38-mediated modulation of tetrodotoxin-resistant sodium channels in mouse sensory neurons by tumor necrosis factor-alpha. *J Neurosci*, **26**, 246–255.

Lai, C.J., Ho, C.Y. & Kou, Y.R. (2002) Activation of lung vagal sensory receptors by circulatory endotoxin in rats. *Life Sci*, **70**, 2125–2138.

Lee, L.Y. & Pisarri, T.E. (2001) Afferent properties and reflex functions of bronchopulmonary C-fibers. *Respir Physiol*, **125**, 47–65.

Lin, S., Walker, J., Xu, L., Gozal, D. & Yu, J. (2007) Respiratory: Behaviours of pulmonary sensory receptors during development of acute lung injury in the rabbit. *Exp Physiol*, **92**, 749–755.

Martin, T.R. (1999) Lung cytokines and ARDS: Roger S. Mitchell Lecture. *Chest*, **116**, 2S–8S.

Morrison, D.C. & Ryan, J.L. (1987) Endotoxins and disease mechanisms. *Annu Rev Med*, **38**, 417–432.

Otmishi, P., Gordon, J., El-Oshar, S., Li, H., Guardiola, J., Saad, M., Proctor, M. & Yu, J. (2008) Neuroimmune interaction in inflammatory diseases. *Clin Med: Circ, Respir and Pulm Med*, **2**, 35–44.

Pavlov, V.A. & Tracey, K.J. (2006) Controlling inflammation: the cholinergic anti-inflammatory pathway. *Biochem Soc Trans*, **34**, 1037–1040.

Pollock, J., McFarlane, S.M., Connell, M.C., Zehavi, U., Vandenabeele, P., MacEwan, D.J. & Scott, R.H. (2002) TNF-alpha receptors simultaneously activate Ca2+ mobilisation and stress kinases in cultured sensory neurones. *Neuropharmacology*, **42**, 93–106.

Tracey, K.J. (2002) The inflammatory reflex. *Nature*, **420**, 853–859.

Undem, B.J. & Carr, M.J. (2001) Pharmacology of airway afferent nerve activity. *Respir Res*, **2**, 234–244.

Yu, J. Airway receptors and their reflex function. Springer. 2008. *Adv Exp Med Biol*. This volume

Yu, J. (2005) Airway mechanosensors. *Respir Physiol Neurobiol*, **148**, 217–243.

Yu, J., Lin, S., Zhang, J., Otmishi, P. & Guardiola, J.J. (2007) Airway nociceptors activated by pro-inflammatory cytokines. *Respir Physiol Neurobiol*, **156**, 116–119.

The Role of NOX2 and "Novel Oxidases" in Airway Chemoreceptor O_2 Sensing

Ernest Cutz, Jie Pan and Herman Yeger

Abstract In pulmonary neuroepithelial bodies (NEB), presumed airway chemoreceptors, classical NADPH oxidase (gp91 phox, NOX2) is co-expressed with O_2 sensitive K^+ channels (K^+O_2) and functions as an O_2 sensor. Here we examined related NADPH oxidase homologues "novel oxidases "(NOX 1, 3&4) and their possible involvement in O_2 sensing. For immunolocalization we used specific antibodies against various NADPH components and K^+ (O_2) subunits to label NEB in rat /rabbit lung and NEB related H146 tumor cell line. For gene expression profiling of NEB cells microdissected from human lung, and H146 cells, we used custom MultiGene-12TM RT-PCR array that included NADPH oxidase components and homologues /accessory proteins (NOX1-4, phox-p22, p40, p47, p67, Rac1, NOXO1 and NOXA1) and K^+O_2 channels (Kv -1.2, 1.5, 2.1, 3.1, 3.3, 3.4, 4.2, 4.3;TASK1-3). In rat lung, NOX2, NOX4, p22phox, Kv3.3 (and Kv3.4 in rabbit) and TASK1 localized to the apical plasma membrane of NEB cells, and membrane or sub-membrane regions in H146 cells. NEB and H146 cells expressed all NOX proteins except NOX3, as well as all K^+O_2 channels, except Kv1.5 and Kv4.3. Co-immunoprecipitation using Western blot multicolor Quantum dot labeling showed NOX2 molecular complexes with Kv but not with TASK, while NOX4 associated with TASK1 but not with Kv channel proteins. Hypoxia -induced serotonin release was inhibited in H 146 cells by siRNA to NOX2, while siRNA to NOX4 had only a partial effect, implicating NOX 2 as the predominant NEB cell O_2 sensor. Present findings support NEB cell specific plasma membrane model of O_2 sensing, and suggest unique NOX/K^+O_2 channel combinations for diverse physiological NEB functions.

Keywords Pulmonary neuroepithelial bodies · Oxygen sensing mechanism · Oxygen sensitive K^+channels · NADPH oxidase · Airway oxygen sensor · Hypoxia signaling

E. Cutz (✉)
Division of Pathology, Department of Pediatric Laboratory Medicine, The Research Institute, The Hospital for Sick Children and University of Toronto, Toronto, Ontario, Canada
e-mail: ernest.cutz@sickkids.ca

1 Introduction

Pulmonary neuroepithelial bodies (NEB) form innervated clusters of amine (serotonin, 5-HT) and peptide containing cells that are widely distributed within the airway mucosa of human and animal lungs (Cutz 1997, Van Lommel et al. 1999, Linnoila 2006). NEBs are thought to function as hypoxia sensitive airway sensors involved in the control of breathing, particularly during the perinatal period (Cutz and Jackson 1999, Bolle et al. 2000).

NEB cells express a membrane delimited O_2 sensing molecular complex "O_2 sensor" composed of O_2 sensing proteins coupled to an O_2 sensitive K^+ channel (K^+ O_2) (Youngson et al. 1993, Wang et al. 1996). A potential candidate for O_2 sensing protein include the heme-linked nicotinamide adenine dinucleotide phosphate oxidase (NADPH oxidase), similar to the one identified in neutrophils (Quinn and Gauss 2004). Several lines of evidence support NADPH oxidase as the principal O_2 sensor in NEB cells including co-expression of various components of NADPH oxidase (i.e. gp91phox and p22phox) and O_2 sensitive voltage gated K^+ channel subunit $KV_{3.3a}$ (Wang et al. 1996, Cutz et al. 2003). Studies using the small cell lung carcinoma (SCLC) cell line H-146, an immortalized cell model for NEB cells, also support the role of the oxidase as an O_2 sensor (O'Kelly et al. 2000, Hartness et al. 2001, Searle et al. 2002). The definitive proof comes from studies of a mouse model with NADPH oxidase deficiency (OD; gp91^{phox} k/o) (Fu et al. 2000, Kazemian et al. 2002). Taken together, the above studies provide strong evidence for the hypothesis of a membrane delimited mechanism for O_2 sensing by NEB cells which may differ from other O_2 sensing cells and may be also age dependent (Weir et al. 2005).

Other potential candidates for an O_2 sensor proteins in NEB cells include recently identified homologues of NADPH oxidase (also referred to as "low output" oxidases) that are found in a variety of non-phagocytic cells (Lambeth 2004). While the founder protein, *gp91^{phox}* (*Nox2*) is predominantly expressed in phagocytic cells where it plays a critical role in host defense, *Nox 1* is expressed mostly in colonic epithelium, whereas *Nox 3* is expressed mainly in the fetal kidney and inner ear. *Nox 4*, originally described as a renal oxidase (Renox), is widely expressed in various tissues, including lung, placenta, pancreas, bone and blood vessels where it may be involved in various cellular processes such as cell proliferation, apoptosis and receptor signaling (Lambeth 2004). In the present study we have investigated the gene expression, localization and potential role of NOX2 and other NOX proteins as well as O_2 sensitive K^+ channels involved in O_2 sensing using the H146 tumor cell model and native NEB cells of human and animal lungs.

2 Methods

2.1 Cell Lines and Tissues

The classical SCLC cell line H146 and a lung carcinoid cell line H727 were obtained from ATCC (Manassas, VA). Control cell line included a promyelocytic cell line

(HL60, representative of neutrophils). Normal human lung tissue was obtained from a lung resection specimen from a 6 month old male infant and from human fetuses (~20 weeks gestation) obtained at autopsy, with the approval of the Human Subjects Review Committee of The Hospital for Sick Children. Lung tissues from neonatal rats and rabbits (1–5 day old) were embedded in polyethylene glycol (OCT medium)) and then snap frozen on dry ice. All animal procedures were approved by the Animal Care Committee of the Hospital for Sick Children.

2.2 Laser Capture Microdissection (LCM)

Serial frozen sections of lung (8 μm) were post- fixed in 70% ethanol at −4°C for 5 min and air-dried for 15 min at room temperature. After re-hydration in PBS, the sections were incubated in 20% normal donkey serum in 4% BSA to suppress non-specific binding. To identify NEB we used monoclonal anti-synaptic vesicle 2 (SV2) antibody (1:20 dilution; Developmental Studies Hybridoma Bank, Iowa City, IA). After incubation with biotin-conjugated secondary anti-mouse IgG antibody, streptavidin-Texas Red X conjugate was applied (Molecular Probes; Eugene, OR) for 30 min at RT. LCM was performed using the Pix-Cell II™ system with CapSureTM HS LCM caps (Arcturus Engineering, Mountain View, CA) under the fluorescent microscope. Thermoplastic film containing captured cells was peeled off caps and transferred to a 0.5 ml microcentrifuge tube with 20 ml of extraction buffer (PicoPure RNA isolation kit; Arcturus Engineering). Up to 6 tubes of captured NEB and airway epithelial cells were pooled and RNA was extracted with the PicoPureTM RNA isolation kit. To obtain sufficient RNA from these cells for RT-PCR array profiling, two-round amplification procedures were performed using a RiboAmp RNA Amplification kit (Arcturus Engineering, Mountain View, CA).

2.3 Gene Expression Profiling

For cell lines, approximately 1×10^5 cells were placed in RNA lysis buffer, homogenized, and applied to RNA purification columns (SuperArray, Frederick, MD). The bound RNA was treated with DNAse I, washed and eluted. Two hundred ng of total RNA (from cell lines or from LCM/amplification samples) were processed with RT annealing mixture in custom NADPH MultiGene-12 RT PCR profiling array kit (see Table 1) and custom O_2 sensitive K^+ channels MultiGene-12 RT PCR profiling array kit (see Table 2), respectively. Amplified products and DNA molecular weight markers (100 bp DNA Ladder; Invitrogen) were separated by electrophoresis in 2% agarose gels.

2.4 Immunohistochemistry

Cell lines were plated on poly-L-lysine -coated glass coverslips, fixed in Zinc-formalin solution (NewcomSupply, Middleton, WI) for 10 min at room temperature

Table 1 Nucleotide sequences of human NADPH subunits primers used for NOX MultiGene-12 RT-PCR Profiling analyses

GenBank	Symbol	Description	GeneName	Size (bp)	Ref_post (bp)
NM_000101	CYBA	Cytochrome b-245, alpha polypeptide,$p22^{phox}$	CYBA	101	381–399
NM_000397	CYBB	Cytochrome b-245, beta polypeptide,$gp91^{phox}$	CGD/GP91-1	118	1350–1368
NM_000631	NCF4	Neutrophil cytosolic factor 4, $p40^{phox}$	NCF/P40PHOX	191	1165–1185
NM_000265	NCF1	Neutrophil cytosolic factor 1, $p47^{phox}$	NOXO2/P47PHOX	97	754–772
NM_000433	NCF2	Neutrophil cytosolic factor 2, $p67^{phox}$	NOXA2/P67PHOX	101	1557–1577
NM_016931	NOX4	NADPH oxidase 4, NOX4	KOX/KOX-1	88	1643–1665
NM_015718	NOX3	NADPH oxidase 3, NOX3	GP91-3	107	1652–1672
NM_007052	NOX1	NADPH oxidase 1, NOX1	GP91-2/MOX1	88	623–645
NM_006908	RAC1	Ras-related C3 botulinum toxin substrate 1, RAC	MIG5/P21-RAC1	126	2091–2111
NM_172168	NOXO1	NADPH oxidase organizer 1, Noxo1	P41NOX/P41NOXA	82	360–378
NM_006647	NOXA1	NADPH oxidase activator 1, Noxa1	NY-CO-31/P51NOX	105	483–501
NM_002046	GAPDH	Glyceraldehyde-3-phosphate dehydrogenase	G3PD/GAPD	168	389–407

(RT), and permeabilized in 0.02% saponin in PBS with 0.1% bovine serum albumin. Cells were incubated with 4% bovine serum albumin with 10% goat serum at room temperature for 30 min. Cells were then stained at 4C overnight with anti-gp91^{phox}, anti-NOX4, anti-p22^{phox}, anti-p47^{phox} and anti-p67^{phox} polyclonal antibodies (Table 3) followed by 1 hr incubation with Texas Red-conjugated donkey anti-rabbit/goat IgG.

Cryostat sections of lung were cut at 60–80 μm and fixed in zinc formalin solution at RT. Dual immunofluorescence (gp91^{phox} plus SV2 antibodies/p22^{phox} plus SV2) labeling was performed on sections permeablized with 0.5% Triton X-100 in PBS for 10 min and incubated primary antibodies and FITC/Texas Red conjugated 2nd antibodies. For co-localization of NADPH subunits, double immunofluorescence staining was performed using goat anti-gp91^{phox} Nter antibody along with rabbit anti-p22^{phox}, p47^{phox} and p67^{phox} antibodies, respectively. Images were scanned with the multitracking mode on a Leica confocal laser scanning microscope (model TCS 4D) with SCANWARE software. The excitation wavelengths of the

Table 2 Nucleotide sequences of human K+ channel subunits primers used for potassium channel MultiGene-12 RT-PCR Profiling analyses

GenBank	Symbol	Description	GeneName	Size (bp)	Ref_post (bp)
NM_002246	KCNK3	Potassium channel, subfamily K3, TASK1	OAT1/TASK1	95	593–611
NM_003740	KCNK5	Potassium channel, subfamily K5, TASK2	TASK-2/TASK2	103	1102–1121
NM_016601	KCNK9	Potassium channel, subfamily K9. TASK3	TASK-3/TASK3	169	385–404
NM_004978	KCNC4	Potassium voltage-gated channel, shaw-subfamily4, Kv3.4	HKSHIIIC	188	737–756
NM_004980	KCND3	Potassium voltage-gated channel, Shal-subfamily3, Kv4.3	KCND3L	171	2258–2278
NM_004974	KCNA2	Potassium voltage-gated channel, shaker-subfamily2, Kv1.2	HBK5/HK4	191	736–754
NM_002234	KCNA5	Potassium voltage-gated channel, shaker-subfamily5, Kv1.5	HCK1/HK2	168	739–757
NM_004977	KCNC3	Potassium voltage-gated channel, Shaw-subfamily3, Kv3.3	KSHIIID	147	1300–1319
NM_004975	KCNB1	Potassium voltage-gated channel, Shab-subfamily1, Kv2.1	DRK1/h-DRK1	122	867–885
NM_012281	KCND2	Potassium voltage-gated channel, Shal-subfamily2, Kv4.2	RK5	118	2532–2551
NM_004976	KCNC1	Potassium voltage-gated channel, Shaw-subfamily1, Kv3.1	KV4/NGK2	142	570–590
NM_002046	GAPDH	Glyceraldehyde-3-phosphate dehydrogenase	G3PD/GAPD	168	389–407

Table 3 Primary antibodies used

Antibodies	Epitope	Host	Dilut for IF	Dilut for W/B	Sources
Anti-gp91^{phox}	Cter	rabbit	1:400	1:500	Dr Dagher
Anti-NOX4	Whol	rabbit	1:400	1:600	Dr Lambeth
Anti-gp91^{phox}	Nter	goat	1:200	1:400	Santa Cruz
Anti-p22^{phox}	Cter	rabbit	1: 500	1:500	Dr Dagher
Anti-p22^{phox}	Nter	rabiit	1:500	1:500	Dr Dagher
Anti-p47^{phox}	Cter	rabbit	1:1000	1:5000	Dr Dagher
Anti-p67^{phox}	Whol	rabbit	1:500	1:1000	Dr Dagher
Anti-Kv 1.2	Cter	goat	N/A	1:500	Santa Cruz
Anti-Kv 2.1	Cter	rabbit	N/A	1:1000	Alomone
Anti-Kv 3.4	Nter	rabbit	1:200	1:1000	Alomone
Anti-Kv 3.3	Cter	rabbit	1:400	1:1000	Alomone
Anti-Kv 4.2	Cter	goat	N/A	1:500	Santa Cruz
Anti-Kv 4.3	Nter	goat	N/A	1:1000	Santa Cruz
Anti-SV2	N/A	Mouse	1:20	N/A	Hybrydoma B

krypton/argon laser were 488 nm for FITC and 568 nm for Texas Red. Images were processed with Photoshop CS (Adobe).

2.5 Western Blotting and Co-immunoprecipitation

To obtain cell membrane proteins, H146 cells were briefly vortexed in hypotonic HEPES solution with 1X protease inhibitor cocktail and centrifuged at 800 g for 10 min at 4°C. The pellets were extracted in PBS containing 0.001% Tween 20 and 1X protease inhibitor cocktail.

For co-immunoprecipitation studies, 300 μg from each protein sample was co-precipitated with either NOX2/NOX4 or Kv channels/TASK1 antibodies using the Seize@X Protein G Immunoprecipitation kit (Pierce Inc.). For Western blots and triple-Quantum Dot (Q-dot) labeling, samples were loaded in 7% SDS-PAGE (30 ul/lane) and transferred onto a PVDF membrane. After blocking, membranes were processed through co-incubations with goat anti-NOX2, rabbit anti-Kv4.3, (or Kv3.3, Kv2.1 and Kv3.4) and mouse anti-beta-actin antibodies, and co-incubated overnight with rabbit anti-NOX4, goat anti-TASK1 and mouse anti-beta-actin antibodies, followed by incubation in Qdot® Secondary Antibody Conjugates (Qdo705 goat anti-rabbit IgG (red)/Qdot655 horse anti-goat IgG (green) and donkey anti-mouse-AMCA (blue) conjugates diluted 1:1000 in blocking buffer. ImageStation 2000 MM system with Filtered Epi Illumination (KODAK 415/100 bp) was used to capture images by CCD camera.

2.6 Plasmid Construction and Transfection of si RNA Silencing

Silencer® siRNA duplexes specific for the human *NOX2* gene (5'-GGAUACUAAC CAAUAGGA Utt -3' and 5'-AUCCUAUUGGUUAGUAUCCtt -3'), *Silencer*® siRNA duplex specific to human *NOX4* (5'-XCACCACCACCACCACCATT-3' and 5'-AAUGGUGGUGGUGGUGGU GTT-3') and a non-silencing human control duplex (directed against the target sequence, 5'-AATTCTCCGAACGTGTCACGT-3') were purchased from Ambion Inc (Applied Biosystems Canada, Streetsville, Ontario). For stable transfections with siRNA, hairpin siRNA (NOX2 and NOX4) templates were ligated and cloned in p*Silencer* 4.1-CMV neo vectors (Ambion) by a custom service from GenScript Corp (Piscataway, NJ). Propagated NOX2 and NOX4 plasmids were tranfected into $\sim 1 \times 10^5$ suspension cells in 35 mm culture dishes using LyoVec reagent (InvivoGen, San Diego, CA). Transfected cells were selected with G418 and knockdown expression levels confirmed by western blots, immunoflourescence and RT-PCR.

2.7 Hypoxia Induced 5-HT Release

To study the effects of siRNA silencing of NOX 2 and NOX 4 on hypoxia induced 5-HT release, H146 cells transfected with respective si RNAs were exposed to hypoxia (2% O_2) for various time intervals (30 min, 1 hr, 3 hrs and 6 hrs) and the amount of 5-HT (ng/ml) released into the culture media measured using a highly sensitive ELISA assay as previously reported (Pan et al. 2002, 2006). Cells exposed to normoxia served as a negative control (baseline release) whereas cells incubated with diphenylene iodonium (DPI, 10 μM), a blocker of NADPH oxidase, was used as a positive control.

3 Results

3.1 Gene Expression Profiling of mRNA's for NOX and O_2 Sensitive K^+ Channels

In native NEB recovered by LCM, mRNA s for all NOX proteins including accessory proteins, except for NOX3 were abundantly expressed. The gene expression profile in SCLC cell lines was similar to native NEB cells except for the lack of expression of NOXA1 (an accessory protein homolog of p47) in H146 cells, while the better differentiated H727 cells showed an expression profile similar to native NEB cells (Fig. 1a). The neutrophil cell line H60, as expected, expressed mRNAs for all NOX proteins including all accessory units. On the other hand, airway epithelium obtained by LCM from the same lung samples, lacked NOX2 and NOX 3 expression respectively (Fig. 1a).

The gene expression profile of mRNAs for O_2 sensitive K^+ channels in native NEB cells included strong expression of TASK 1–3, and voltage activated K^+ channels (Kv), Kv 1.2, Kv 3.3, Kv3.4 and Kv 4.2. Expression of mRNAs for Kv 2.1 and 3.1 was realively weak and there was no signal for Kv4.3, previously demonstrated in rabbit neonatal NEB (Fu et al. 2007), suggesting possible species variation (Fig. 1b). In H146 and H727 cells, the majority of O_2 sensitive K^+ channel mRNAs were expressed, although the signal intensity varied (Fig. 1b). In control cell lines, the expression profile was variable with only some TASK or Kv mRNA's present.

3.2 Immunohistochemical Localization of NOX and K^+ Channel Epitopes

Immunohistochemical studies using H146 cells confirmed membrane and sub membrane localization of gp 91 phox/NOX2 (Fig. 2a), that co-localized with p22, and mostly cytoplasmic expression of p47 (not shown). Expression of NOX4 epitope was similar to NOX2, with predominantly membrane and sub membrane localization (Fig. 2b). In native NEB s of rat neonatal lung, NOX 2 was prominently localized in the apical cytoplasm facing the airway lumen, as would be expected of an airway sensor monitoring intraluminal O_2 concentration (Fig. 2c). The immunohistochemical localization of K^+ channel epitopes showed distinctive membrane staining of apical membrane in rabbit neonatal NEB, particularly with antibody

Fig. 1 Gene expression profiling for NADPH oxidase and O₂ sensitive K⁺ channel mRNAs in human NEB cells obtained by laser capture micro dissection (NEB-LCM), SCLC tumor cell lines (H146, H727) and control cells (HL-60 promylocytic/neutrophil cell line; Ep-LCM airway epithelium obtained by laser capture microdissection). Note lack of expression of NOX 3 in all but HL-60 cells and no expression of NOXA1 in H146 cells

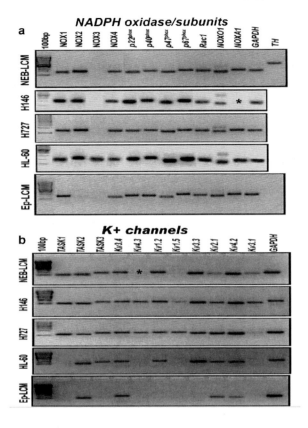

against Kv3.4 (Fig. 2d, e), and a more diffuse membrane staining with antibodies against Kv3.3 (in rat neonate) and TASK 1 in human fetal lung (not shown).

3.3 NOX/K⁺ Channel Protein-Protein Interactions

To study protein-protein interactions between NOX and O₂ sensitive K⁺ channels we used isolated cell membranes of H146 cells and a co-immunoprecipitation method with multicolor Quantum dot technique. In experiments with antibodies against Kv3.3/NOX2 distinct complexes of the two epitopes were obtained (Fig. 3a). In cells exposed to hypoxia (2%O₂, 6 hrs), there was significant up regulation of NOX 2 protein expression without apparent change in signal for Kv3.3 protein. On the other hand, pre-treatment of H146 cells with NOX2 siRNA significantly reduced NOX2 protein expression. Furthermore, Kv3.3/NOX2 complexes were easily dissociated with 0.005 % Triton 100, suggesting "weak" intermolecular binding. In other co-immunoprecipitation experiments we have found that while other Kv channels (i.e. Kv3.4) also formed similar complexes with NOX2, they failed to do so when NOX4 antibody was used (not shown). While TASK 1 did not form complexes with

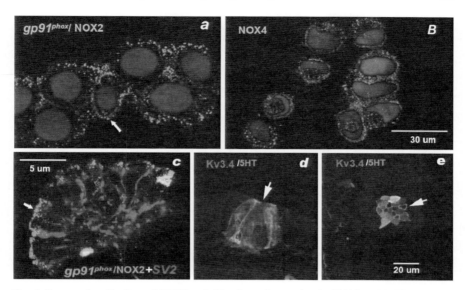

Fig. 2 Immunolocalization of NOX and Kv channel proteins in H146 and NEB cells. (**a**) Membrane and sub-membrane localization of gp91/NOX2 in H-146 cells using C-terminal directed antibody. (**b**) Localization of NOX 4 in H146 cells appears similar to NOX 2 shown in Fig. 1a. (**c**) Immunolocalization of NOX2 in NEB cells of neonatal rat lung is restricted to the apical cytoplasm facing the airway lumen (*arrow*). Double immunostaining with SV2. (**d**) Immunolocalization of Kv 3.4 subunit in NEB of neonatal rabbit lung with positive reaction confined to apical plasma membrane (*arrow*). Double immunostaining with 5-HT localized in NEB cytoplasm. (**e**) En-face view of NEB in the same sample as in Fig. 2d with distinctive linear immunostaning outlining apical cytoplasm (*arrow*)

NOX 2, it readily associated with NOX4, that was also up regulated by exposure to hypoxia (Fig. 3b).

3.4 Effects of siRNA Silencing of NOX Protein Expression on Hypoxia Induced 5-HT Secretion

Control H146 cells, maintained in normoxia showed minimal or no 5-HT release during the duration of experiments (up to 6 hrs) (Fig. 3c). The positive control cells (without siRNA silencing), showed expected time dependent hypoxia-induced 5-HT release as previously reported (Pan et al. 2002, 2006). H-146 cells transfected with NOX2 siRNA showed significant inhibition of 5-HT release after 30 min, 1 and 3 hrs exposure to hypoxia that was identical to normoxia baseline (Fig. 3c). However at 6 hrs there was an increase in 5-HT release ($p < 0.1$) above normoxia control but significantly less then the hypoxia control suggesting partial recovery of NOX2 protein expression (Fig. 3c). Using NOX4 siRNA, 5-HT release was also reduced but was less marked as compared to NOX2. Classical NADPH inhibitor, DPI (positive control) showed the expected inhibitory effect on 5-HT release.

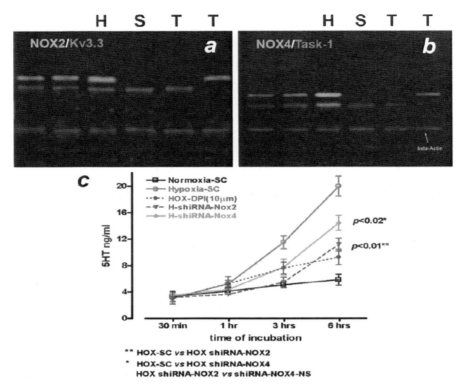

Fig. 3 (**a**) NOX2/Kv3.3 complexes in co-immunoprecipitation experiment of H146 cell membrane protein extracts using multicolor Q-dot method with antibodies against NOX 2 and KV3.3. (H) Up regulation of NOX2 protein expression after exposure to hypoxia. (S) Silencing of NOX 2 protein expression after pre-incubation with NOX2 siRNA; (T) dissociation of NOX2/Kv3.3 complexes after treatment with 0.005 Triton 100. (**b**) NOX4/TASK 1 molecular complexes using NOX4 and TASK 1 antibodies and methods as above. (H) Up regulation of NOX4 protein expression and possibly TASK1 by hypoxia and (T) dissociation of NOX4/TASK1 complexes by 0.005 Triton 100. (**c**) Effects of siRNAs for NOX2 and NOX4 on hypoxia-induced 5-HT release from H146 cells. Under normoxia (control), no significant change (baseline) over duration of experiment (6 hrs). Significant time dependent release of 5-HT in control cells exposed to hypoxia. Significant inhibition of hypoxia-induced 5-HT release after incubation with DPI (10 μM); after pre-treatment with siRNA to NOX2 after 30 min, 1 hr and 3 hrs with a partial recovery at 6 hrs. Response with siRNA for NOX4 is less marked at 3 and 6 hrs compared to NOX2

4 Discussion

The summary and conclusions of our studies reported here include: (a) Gene profiling studies revealed that both native NEB cells and related tumor cell lines express a whole range of NOX and O_2 sensitive K^+ channel proteins that comprise the membrane delimited "O_2 sensor molecular complex" (Wang et al. 1996, Cutz et al. 2003). This gene profile differs from adjacent non-endocrine airway epithelial cells. (b) Immunolocalization studies showed that "O_2 sensor molecular complex"

(NOX2/Kv3.3 etc) is localized to the plasma membrane and sub-membrane regions in representative SCLC tumor cell line (H-146), and in NEB cell apical plasma membrane as expected for an airway based sensor. (c) Co-immunoprecipitation studies showed that in H-146 cells, NOX2 but not NOX 4 forms molecular complexes with O_2 sensitive Kv channel subunits, while TASK 1 complexes with NOX4 but not with NOX2. This is in agreement with recent studies of Lee et al. (2006) showing that in HEK293 cells which endogenously express NOX 4, the activity of transfected TASK-1 channel was inhibited by hypoxia. This hypoxia response was significantly augmented by co-transfection of NOX 4, but not with NOX $2/gp91^{phox}$. The O_2 sensitivity of TASK-1 was abolished by NOX 4 siRNA and NADPH inhibitors, suggesting that NOX 4 may represent the O_2 sensor protein partner for TASK-1. Taken together these observations suggest a possibility of a diversity of "O_2 sensors", even within the same cell type, matching specific NOX proteins with particular O_2 sensitive K^+ channel types (i.e. NOX $2/KV_{3.3a}$; NOX4/TASK-1). Interestingly short term hypoxia (6 hrs) up regulated both NOX2 and NOX4 protein expression while the expression of O_2 sensitive K^+ channel proteins was unaffected. (d) Hypoxia induced 5-HT release was inhibited/reduced after siRNA silencing of NOX2 but less for NOX 4, confirming that NOX 2 is the predominant functional O_2 sensing protein in NEB related tumor cell line, and by inference in native NEB (Wang et al. 1996, Cutz et al. 2003).

Present findings support the organ and NEB specific plasma membrane model of O_2 sensing, and suggest unique NOX/K^+ channel combinations that may serve diverse NEB physiological functions in the lung.

Acknowledgments Supported by grants from Canadian Institute for Health Research, MOP-12742, MPG-15270.

References

Bolle T, Lauweryns JM, VanLommel A. 2000, Postnatal maturation of neuroepithelial bodies and carotid body innervation: a quantitative investigation in rabbit, J Neurocytol, 29:241–248.

Cutz E (Ed) 1997, Cellular and molecular biology of airway chemoreceptors. Landes bioscience, Austin, Texas.

Cutz E, Jackson A. 1999, Neuroepithelial bodies as airway chemoreceptors, Respir Physiol, 115:201–214.

Cutz E, Fu XW, Yeger H, Peers C, Kemp PJ. 2003, Oxygen sensing in Pulmonary Neuroepithelial Bodies and related tumour cell model. In Oxygen sensing: Responses and adaptation to hypoxia, eds. Lahiri S., Prabhakar H., SemenzaG., Marcel Dekker, New York, pp 567–602.

Fu XW, Wang D, Nurse CA, Dinauer MC, Cutz E. 2000, NADPH oxidase in an O_2 sensor in airway chemoreceptors: Evidence from K^+ current modulation in wild type and oxidase deficient mice, Proc Natl Acad Sci (USA), 97:4374–4379.

Fu XW, Nurse C, Cutz E. 2007, Characterization of slowly inactivating Kv alpha current in rabbit pulmonary neuroepithelial bodies:effects of hypoxia and nicotine, Am J Physiol Lung Cell Mol Physiol, 293:L892–L902.

Hartness ME, Lewis A, Searle GJ, O'Kelly I, Peers C, Kemp PJ. 2001, Combined antisense and pharmacological approaches implicate hTASK as an airway O_2 sensing K^+ channel, J Biol Chem, 276:26499–26508.

Kazemian P, Stephenson R, Yeger H, Cutz E. 2002, Respiratory control in neonatal mice with NADPH oxidase deficiency, Respir Physiol, 126:89–101.

Lambeth JD. 2004, Nox enzymes and the biology of reactive oxygen, Nature Rev Immunol 4: 181–189.

Lee YM, Kim BJ, Chun YS, So I, Choi H, Kim MS, Park JW. 2006, NOX4 as an oxygen sensor to regulate TASK-1 activity, Cellular Signalling, 18 :499–507.

Linnoila RI. 2006, Functional facets of the pulmonary neuroendocrine system, Lab Invest, 86: 425–444.

O'Kelly I, Lewis A, Peers C, Kemp PJ. 2000, O_2 sensing in airway chemoreceptor- derived cells: protein kinase C activation reveals functional evidence for involvement of NADPH oxidase, J Biol Chem, 275:7684–7692.

Pan J, Bear C, Farragher S, Cutz E, Yeger H. 2002, CFTR modulates neurosecretory function in pulmonary neuroendocrine cell related tumor cell line, Am J Resp Cell Mol Biol, 27:553–560.

Pan J, Copland I, Post M, Yeger H, Cutz E. 2006, Mechanical stretch-induced serotonin release from pulmonary neuroendocrine cells :implications for lung development, Am J Physiol Lung Cell Mol Physiol, 290:L185–L193.

Quinn MT, Gauss KA. Structure and regulation of neutrophil respiratory burst oxidase: comparison with physiological oxidases. J Leukocyte Biol 2004;76: 760–81.

Searle GJ, Hartness ME, Peers C, Kemp PJ. 2002, Lack of contribution of mitochondrial electron transport to acute O_2 sensing in model airway chemoreceptors, Biochem Biophys Res Commun, 291: 332–337.

Van Lommel A, Bolle T, Fannes W, Lauweryns JM. 1999, The pulmonary neuroendocrine system:the past decade, Arch Histol Cytol, 62:1–16.

Wang D, Youngson C, Wong V, Yeger H, Dinauer MC, Vega-Saenz de Meira E, Rudy B and Cutz E. 1996, NADPH oxidase and hydrogen peroxide-sensitive K^+ channel may function as an oxygen sensor complex in airway chemoreceptors and small cell carcinoma cell lines Proc Natl Acad Sci (USA), 93:13182–87.

Weir EK, Lopez Barneo J, Buckler KJ, Archer SL. 2005, Acute oxygen sensing mechanisms, New Engl J Med 353::2042–2055.

Youngson C, Nurse C, Yeger H, Cutz E. 1993, Oxygen sensing in airway chemoreceptors, Nature (London), 365: 153–156.

Recruitment of GABA$_A$ Receptors in Chemoreceptor Pulmonary Neuroepithelial Bodies by Prenatal Nicotine Exposure in Monkey Lung

X.W. Fu and E.R. Spindel

Abstract Pulmonary neuroepithelial bodies (NEB) act as airway oxygen sensors and produce serotonin, a variety of neuropeptides and are involved in autonomic nervous system control of breathing, especially during the neonatal period. We now report that NEB cells also express a GABAergic signaling loop that is increased by prenatal nicotine exposure. In this study, cultured monkey NEB cells show hypoxia-evoked action potentials and hypoxia-sensitive K$^+$ current. As shown by both immunofluorescence and RT-PCR, monkey NEB cells synthesize and contain serotonin. The monkey NEB cells express the β2 and β3 GABA$_A$ receptor subunits, GAD and also express α7, α4 and β4 nicotinic receptor (nAChR) subunits. The α7 nAChR is co-expressed with GAD in NEB. The numbers of NEB and β3 GABA$_A$ receptor subunits expressed in NEB cells in lungs from control newborn monkeys were compared to lungs from animals that received nicotine during gestation. Prenatal nicotine exposure increased the numbers of NEB by 46% in lung and the numbers of NEB cells expressing GAD and GABA$_A$ β3 receptors increased by 67% and 66%, respectively. This study suggests that prenatal nicotine exposure can modulate NEB function by increasing the numbers of NEB cells and by increasing both GAD expression and β3 GABA$_A$ receptor subunit expression. The interaction of the intrinsic GABAergic system in the lung with nicotinic receptors in PNEC/NEB may provide a mechanism to explain the link between smoking during pregnancy and SIDS.

Keywords Neuroepithelial Bodies · Nicotinic acetylcholine receptors · GABA$_A$ Receptors · Hypoxia-sensitive K$^+$ channel

1 Introduction

Pulmonary neuroepithelial bodies (NEB) form innervated cell clusters in airways and express voltage-activated hypoxia-sensitive K$^+$ channels and act as airway

---

E.R. Spindel (✉)
Division of Neuroscience, Oregon National Primate Research Center, Oregon Health & Science University, Beaverton, OR, USA
e-mail: Spindele@ohsu.edu

chemoreceptors (Youngson et al., 1993). Our previous studies showed that NEB cells in hamster express functional heteromeric α3β2, α4β2 and homomeric α7 nicotinic acetylcholine receptors (nAChR) (Fu et al., 2003). Nicotine has previously been shown to regulate the function of the oxygen-sensitive A type K^+ channels in rabbit NEB cells (Fu et al., 2007). Increased numbers of PNEC/NEB cells have been observed in several smoking-associated pediatric lung disorders such as bronchopulmonary dysplasia, cystic fibrosis, sudden infant death syndrome (SIDS), and asthma (Adgent, 2006; Plummer et al., 2000). Deficiency of NEBs' oxygen-sensor function under nicotine and tobacco-specific toxicants may be the one of factors that causes SIDS. It has recently been reported that both $GABA_A$ receptors and GAD are expressed in human bronchial epithelial cells (Xiang et al, 2007), but little is yet known about GABAergic expression in NEB. Little is also yet known about NEB function in nonhuman primates which provide an animal model closest to humans. In the present study, we report that monkey NEB cells show hypoxia-evoked spikes and hypoxia-sensitive K^+ current and that monkey NEB express a GABAegic signaling loop that is increased by prenatal nicotine exposure.

2 Methods

2.1 Monkey Neuroepithelial Cell and Bronchial Epithelial Cell Culture

Lungs were obtained from rhesus monkeys (fetal to 1 year old) sacrificed as part of ongoing protocols. All protocols were approved by the Oregon Health & Science University Animal Care and Use Committee. Whole lungs were dissected and immersed in ice cold MEM medium. The major bronchi were crudely dissected and then incubated overnight at 4°C in 0.1% Protease type 14 (Sigma) in MEM. The cells were washed from mucosal surface of the airway using MEM with 5% fetal calf serum, centrifuged, resuspended, and plated in bronchial epithelium culture medium containing 2% fetal calf serum (Wu et al., 1990). For patch clamp analyses, the cells were plated on Thermanox plastic cover slips (NUNC, Rochester, NY). For immunofluorescence analyses, cells were plated on glass cover slips.

2.2 Patch Clamp Recording of Whole-Cell Voltage and Current Configurations

Action potential and K^+ currents were recorded from cultured monkey neuroepithelial cells using standard whole cell patch-clamp techniques. Identity of cells as neuroepithelial cells was confirmed by staining for vital dye neutral red (0.02 mg/ml) for 10 min at room temperature. The external Krebs solution bathing the NEB had the following composition (in mM): 130 NaCl, 3 KCl, 2.5 $CaCl_2$, 1 $MgCl_2$, 10 $NaHCO_3$, 10 Hepes, and 10 glucose, pH 7.35 ~7.4. To isolate outward K^+ currents, an internal pipette solution with following composition was used (mM): 30 KCl, 100 potassium gluconate, 1 MgCl$_2$, 10 EGTA, 10 Hepes, 4 Mg-ATP; pH adjusted to 7.2 with CsOH. The pipette resistance was 3 to 5 MΩ. The access resistance was ≤ 15 MΩ. The chamber, which had a volume of 1 ml, was perfused continuously with Krebs solution at a rate of 3–4 ml/min. Drugs were applied to the perfusate, and their delivery to the cells was controlled by separate valves. A multiClamp700B (Axon Instruments, Foster, CA, USA) amplifier was used to record for whole-cell currents (voltage clamp) or membrane potential (current clamp). The data were filtered at 5 KHz. Voltage and current clamp protocols, data acquisition and analysis were performed using pClamp9 software and DigiData 1322A interface (Axon Instruments). All data values are given as mean ± S.E.M.

2.3 Immunohistochemistry and Fluorescence

Lung samples were obtained from timed-pregnant rhesus monkeys administered nicotine (1–1.5 mg/kg/d, subcutaneously) or water using osmotic minipumps from days 26–134 or 26–160 of gestation (full term=165d). Lungs were obtained at necropsy and fixed with 4% paraformaldehyde. Fluorescent staining was performed on monkey lung tissue following cryostat sectioning (5 μM). The primary antibodies used were as follows: polyclonal rabbit anti-glutamate decarboxylase 65/67 (GAD 65/67) (1:100, Chemicon, Billerica, MA); polyclonal rabbit anti-GABA$_A$ receptor β3 (1:100, Chemicon); mouse monoclonal α7 nAChR (306, 1:200, Sigma); monoclonal rat anti-serotonin (1:100, Medicorp, Montreal, Canada). The secondary antibodies used were FITC-conjugated horse anti-mouse IgG, anti-rabbit and rat IgG (all 1:400, Jackson ImmunoResearch Laboratories, West Grove, PA). For negative controls, the primary antisera were omitted. Samples were viewed under an Axioskop 2 and Axiovision 4.2 was used for image acquisition (Zeiss, Germany).

3 Results

3.1 Hypoxia Sensitive Action Potential and Outward K^+ Currents in Cultured Monkey NEB Cells

Neuroepithelial bodies (NEB) in primary cultures of bronchial epithelial cells were visualized by vital dye neutral red (0.2 mg ml^{-1}, Fig. 1A). To test whether cultured NEB cells were sensitive to hypoxia, current clamp was used to record membrane potential. NEB cells did not exhibit action potentials when cells were perfused with normoxic Krebs solution. By contrast, hypoxic solution (bubbling 95% N$_2$, PO_2 = 20 mmHg) evoked action potentials from NEB cells (Fig. 1B). The mean membrane potential of cultured NEB cells was -50 ± 8 mV (n=11). Using whole-cell voltage clamp, depolarizing voltage steps from holding potential of -60 mV to

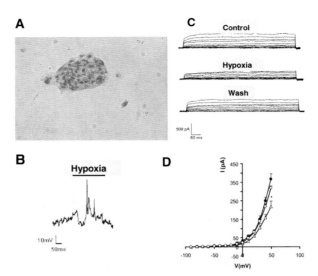

Fig. 1 **Effect of hypoxia on action potential and outward K$^+$ current in cultured NEB cells. A.** Staining of cultured NEB cells using neutral red. **B**. Hypoxic solution induced an action potential in a single NEB cell. **C**. Outward K$^+$ current evoked by depolarization steps from –60 to 50 mV in control Krebs solution and hypoxic solution. Outward K$^+$ current was reduced by hypoxia and a recovery K$^+$ current by washout of the hypoxic solution. **D**. I–V relationship for the current in the control solution (•) and in the hypoxic solution (o) are plotted together with recovery K$^+$ current (∇) against holding voltage. Asterisks indicate significant differences from control ($P < 0.05$)

+ 50 mV activated a delayed rectifier outward K$^+$ current. The activation threshold of this K$^+$ current was around -30 mV. The current amplitudes at test potential of 50 mV were -367 ± 10.5 pA (n=6). Upon exposure to hypoxia, outward K$^+$ current was reversible reduced by 30% (Fig. 1C and D). The *I-V* relationship curve shows that the oxygen sensitive K$^+$ current is voltage dependent (Fig. 1D). These results suggest that monkey NEB cells express oxygen-sensitive K$^+$ channels. Thus hypoxia inhibition of K$^+$ channels plays a role in inducing action potentials in NEB cells.

3.2 Recruitment of GAD and GABA$_A$ β3 Receptors in NEB Cells by Prenatal Nicotine Exposure

The expression of GAD and GABA$_A$ β3 receptors in NEB cells was examined using antisera specific for GAD65/67 and β3, respectively. Serotonin staining was used as a marker for NEB cells. The numbers of NEB was increased 46% in prenatal nicotine exposed-lung compared to control lung (Fig. 2a, b). In NEB cells, both GAD and GABA$_A$ β3 immunoreactivity were expressed on the plasma membrane and submembrane location (Fig. 2a, c). The numbers of NEB cells expressing GABA$_A$ β3 receptor subunits from control newborn monkeys were compared to

Fig. 2 Prenatal nicotine exposure increased GAD and GABA$_A$ β3 receptor expression in NEB cells. Serotonin immunostaining was used as an NEB cell marker. (**a**) Double staining for 5HT and GAD in NEB cells. (**c**) Double staining for 5HT and GABA$_A$ β3 receptor in NEB cells. In lungs from animals which received prenatal nicotine exposure, nicotine increased GAD and GABA β3 receptor expression in NEB cells (panels **b** and **d**). (**e**) nAChR α7 subunit is expressed in NEB cells. (**f**) Double staining of α7 nAChR with GAD in NEB cells

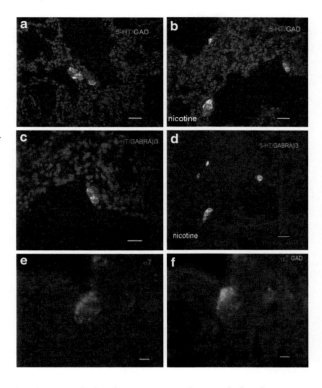

lungs from animals that received prenatal nicotine exposure. Prenatal nicotine exposure increased the numbers of NEB cells expressing GAD and GABA$_A\beta$3 receptors by 67% and 66%, respectively (Fig. 2b, d and Fig. 3). The α7 nAChR subunit (Fig. 2e) is co-expressed with GAD in NEB (Fig. 2f).

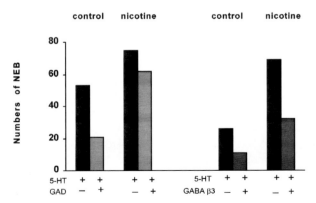

Fig. 3 Summary graph shows the effects of nicotine on the numbers of NEB, and the numbers of NEB cells expressing GAD and GABA$_A$ β3 receptors. Serotonin-positive cells, and cells double stained for 5-HT and GAD or GABA$_A$ b3 receptor cells were counted in fields obtained with a 20X objective. A total of 20 fields were counted per data point. The calibration bar is 30 μm in panels a -d and 50 μm in panels e and f

4 Discussion

Our study shows that NEB cells cultured from monkey lung express oxygen sensors as previously reported for rabbit, hamster and mouse. Because lung development in monkeys is highly similar to lung development in humans this finding provides an important new model to study developmental influences on oxygen sensing by NEB. The oxygen-sensing properties of NEB cells is very important during the neonatal period because NEB cells function as auxiliary chemoreceptors while the function of the carotid bodies is still immature. NEB appear to have a dual function; first through secretion of their amine and peptide products, NEBs may modulate lung growth and differentiation; second, both before and after birth, innervated NEB cells could play an important role as airway chemoreceptors (Linnoila, 2006). Consistent with the first function, NEB cells are associated with stem cell niches in both the proximal and distal airways. One hypothesis to explain the link between maternal smoking during pregnancy and sudden infant death syndrome (SIDS) is that prenatal nicotine exposure alters the oxygen-sensing properties of NEB cells and hence the ability of NEB to respond to hypoxia.

The potential importance of GABAergic signaling in lung has recently been highlighted by Xiang et al (2007) who reported that GAD and $GABA_A$ receptors are expressed in airway epithelial cells of human and mouse and are increased in rodent models of asthma. Thus GABAergic signaling may play a similarly important role in NEB function. In this study we show that NEB cells express a GABAergic signaling loop. Thus our study suggests that nicotine may modulate the function of NEB cells both by increasing the numbers of NEB cells and also by increasing GAD and $\beta 3$ $GABA_A$ receptor subunit expressions in NEB cells. Interestingly, our study also shows that the $\alpha 7$ nAChR subunit is co-expressed with GAD in the same NEB cells. This suggests there may be direct protein-protein interaction between cholinergic and GABAergic systems in NEB cells. In neurons, two proteins, Plic-1 and the GABAA receptor associated protein (GABARAP) have been shown to regulate GABAA receptors. Plic-1, an Ubiquitin-like protein, regulates GABAA receptor a and b subunits (Fiona et al., 2001). GABAA receptor associated protein has been shown to regulate expression of the $\gamma 1$, $\gamma 2s$ and $\gamma 2L$ GABA receptor subunits (Chen and Olsen, 2007). Cultured monkey NEB cells express both Plic-1 and GABAA associated protein. Thus nicotine may modulate function of GABAA receptors in NEB via Plic-1 and the GABAA associated protein. Further studies are needed to clarify this.

Acknowledgments This work was supported by NIH grants RR00163 and HL087710.

References

Adgent MA. 2006, Environmental tobacco smoke and sudden infant death syndrome: A review. *Birth Defects Res (part B)* 77:69–85.
Chen Z and Olsen RW. 2007, GABAA receptor associated proteins: a key factor regulating GABAA receptor function. *J Neurochem* 100: 279–294.

Fiona KB, Kittler JT, Muller E, Thomas P, Uren JM, Merlo D, Wisden W, Triller A, Smart TG and Moss SJ. 2001. GABAA receptor cell surface number and subunits stability are regulated by the ubiquitin-like protein plic-1. *Nat Neurosci* 4: 908–916.

Fu XW, Coline CA, Farraghter SM, Cutz E. 2003, Expression of functional nicotinic acetylcholine receptors in neuroepithelial bodies of neuronatal hamster lung.*Am J Physiol Lung Cell Mol Physiol* 285:L1203–L1212.

Fu XW, Nurse C and Cutz E. 2007, Characterization of slowly inactivating Kv alpha current in rabbit pimonary neuroepithelial bodies: effects of hypoxia and nicotine. *Am J Physiol Lung Cell Mol Physiol* 293:L892–L902.

Linnoila RI. 2006, Functional facets of the pulmonary neuroendocrine system.*Lab Invest* 86: 425–444.

Plummer III HK, Sheppars BJ and Hildegard MS. 2000, Interaction of tobacco-specific toxicans with nicotinic cholinergic regulation of fetal pulmonary neuroendorine cells: Implications for pediatric lung disease. *Exp Lung Res* 26:121–135.

Xiang Y, Wang S, Liu M, Hirota JA, Li J, Ju W, Fan Y, Kelly MM, Ye B, Orser B, O'Byrne PM, Inman MD, Yang X and Lu W. A. 2007, A GABAergic system in airway epithelium is essential for mucus overproduction in asthma. *Nat Med* 13: 862–867.

Wu R, Martin W, Robinson C, St George J, Plopper C, Kurland G, Last J, Cross C, McDonald R and Boucher R. 1990. Expression of mucin synthesis and secretion in human tracheobronchial epithelial cells grown in culture. *Am J Respir Cell Mol Biol* 3:467–478

Youngson C, Nurse CA, Yeger H and Cutz E. 1993, Oxygen sensing in airway chemoreceptors. *Nature* 365:153–155.

Concluding Remarks

Colin A. Nurse

On behalf of the members of ISAC, I wish to begin by extending our sincerest gratitude and appreciation to Professor Constancio Gonzalez and his organizing committee for hosting the XVIIth conference in Valladolid, Spain. The meeting was exceptionally well organized and both the scientific sessions and social events, not to mention the historic setting and ambience, left an indelible mark on all of us. As is customary, the meeting consisted of a series of oral presentations and poster sessions spread over a period of 3 days. This volume consists of over 35 short papers that have captured the essence of many (but not all) of the oral presentations and posters, as well as ~14 authoritative 'Invited Articles' contributed by leading experts in the field. The end result is a voluminous up-to-date resource that is most informative not only to those already familiar with the field of arterial O_2 and CO_2/pH sensing, but also to the uninitiated. Like any other scientific meeting, there were the usual controversies which I see as a reflection that the field is still very much 'alive' and certainly attractive for new investigators.

Most fittingly, the scientific session began with a lecture on "The Discovery of Sensory Nature of the Carotid Bodies", given by the grandson of Fernando De Castro, whose seminal discoveries and remarkable insight in the 1920s laid the cornerstone for many of the presentations featured at this meeting. We were treated to a delightful commentary on the life and works of this brilliant scientist who followed in the footsteps of his preeminent mentor and Founding Father of Neurobiology, Santiago Ramón y Cajal. In one sense then, this meeting has brought us back full circle to where it all began in Madrid some ~80 years ago. As described in great detail in the first Chapter of this volume (by the junior Fernando de Castro), we learnt about the politics of the time, the respect and collaborations between De Castro and Nobel Laureate Corneille Heymans (on the workings of the carotid body), and witnessed the exchange of letters written in their own handwriting. And we saw those wonderful original drawings by De Castro of the carotid body innervation based on histological staining of tissue sections. The beauty and detailed accuracy

---

C.A. Nurse (✉)
Department of Biology, McMaster University, 1280 Main St. West, Hamilton, Ontario, Canada L8S 4K1
e-mail: nursec@univmail.cis.mcmaster.ca

of these drawings can easily stand up to today's images derived from more recent advances in confocal microscopy and multi-label immunofluorescence.

The introductory 'Lecture' by F. de Castro on the first day was followed by several shorter 20 min presentations, and this pattern was followed for the duration of the meeting. Talks on the first day also covered several O_2– and CO_2 -sensing cells/tissues including pulmonary arteries, carotid and aortic bodies, and pulmonary neuroepithelial bodies (NEBs). Among the newer methodologies were attempts at live imaging of the whole carotid body (CB) using two-photon confocal microscopy (presented by H Acker), a technique that will undoubtedly aid our future understanding of organ's function. A new and potentially functional preparation based on dissociated rat aortic bodies and endogenous neurons in culture was introduced (presented by N. Piskuric). Perhaps the most provocative talk came from Adriaensen's and Kemp's laboratories questioning the role of pulmonary NEBs as O_2-sensors (presented by D. Adriaensen). These studies were aided by the use of an attractive mitochondrial styryl dye (4-Di-2-Asp) for vital labeling and visualizing of NEBs in tissue slices. Given the wealth of 'positive' data from Cutz's laboratory showing that NEBs are indeed O_2-sensitive, it is imperative that rigorous controls be carried out to show that the dye does not perturb the O_2-sensing mechanism per se. Cutz's presentation, based on co-immunoprecipitation, siRNA techniques, and 5-HT release measurements, implied that certain NOX isoforms formed functional associations with specific O_2-sensitive K^+ channels in airway NEB chemoreceptors.

The second day began with a lecture by R. Fitzgerald on "Fifty years of Progress in Carotid Body Physiology". This talk provided a historical overview of carotid body responses from the molecular, cellular, tissue, organ and system levels. Many of the issues discussed are highlighted in "Fifty Years of Progress in Carotid Body Physiology" of this volume which covers areas such as ventilatory reflexes, cardiovascular adjustments and 'defense' reactions evoked by CB stimulation. Subsequent talks discussed comparisons between O_2 and acid chemoreception in rabbit CB (presented by C. Gonzalez), neuroimmune interactions in the lung (presented by J. Yu), and optical recordings of calcium transients in petrosal chemoafferent neurons using an *ex vivo* intact preparation (presented by E. Gauda). The laboratories of Monteiro and Gauda presented an interesting talk on the presence of a soluble G-protein-insensitive adenylyl cyclase (sAC) that was selectively activated by CO_2/HCO_3^- in CB tissue. The physiological significance of this sAC remains to be determined. The second session began with a lecture by P. Kumar on "Systemic effects resulting from carotid body stimulation". This lecture (see Invited Article in this volume) discussed natural CB stimuli, integrative aspects between cardiorespiratory and cardiovascular variables, and the role of the CB in cardiorespiratory-related pathologies. Several of the talks presented during the rest of the day (as well as on the final day of the meeting) touched on one of the newer themes that have emerged in CB physiology, i.e. comparison of CB plasticity following intermittent (IH) versus chronic (CH) hypoxia. For example, talks from the laboratories of C. Gonzalez, R. Iturriaga & J. Alcayaga, and K. O'Halloran discussed effects of IH on blood gases and tissue redox status, cardiovascular variables, and respiratory muscle function. The late-afternoon session began with a lecture by F. Powell on "Oxygen sensing in the brain", and the main points are summarized in an Invited Article in this volume. This talk highlighted the power of conditional knock-out models bearing tissue-specific deletion of selected genes (e.g. HIF-1α) on ventilatory function during CH. There was also an interesting talk by M. Gassmann on the role of erythropoietin in enhancing the hypoxic ventilatory response in females (relative to males), via its action at both central and peripheral chemoreceptors. These data might explain why females might cope better during hypoxia-related syndromes.

The final day began with a lecture by Y. Okada on "Central Chemoreceptors: Where and How" and is summarized in an Invited Article in this volume. This talk discussed the use of lesioning, voltage-dye recording, and c-fos immunohistochemical techniques to study the location and properties of CO_2/H^+ excitable neurons in the isolated brainstem/ spinal cord preparation. It was followed by an interesting talk by N. Prabhakar, who extended his pioneering studies on the facilitatory effects of IH on CB chemoreceptor sensitivity (see Invited Article), to neonatal rat adrenal chromaffin cells. This study revealed that whereas CH had a negligible (or suppressive effect) on hypoxia-evoked catecholamine (CA) release from neonatal chromaffin cells, IH dramatically enhanced secretion by augmenting both the quantal event frequency and amount of CA released per event. The rest of the morning session was devoted mainly to studies on the effects of 'gasotransmitters' on O_2-sensitive ion channels. A talk from the laboratory of C. Peers examined the inhibition of the human α_{1C} L-type Ca^{2+} channels by CO in the HEK 293 expression system, and concluded that this inhibition could explain the cardioprotective effects of the CO synthetic enzyme, heme oxygenase-1 (HO-1). Talks from the laboratories of D. Riccardi and P. Kemp identified a critical role of one of a potential four cysteine residues, located at the C-terminus of human large conductance BK_{Ca} channel α-subunit, in the activation of the channel by CO. Also, they reported that CO and H_2S have opposing actions on BK_{Ca} channels heterologously expressed in HEK 293 cells, and raised the interesting possibility of H_2S acting as an endogenous signaling molecule in CB O_2 chemotransduction. It is noteworthy that H_2S synthesizing enzymes were immuno-localized in rat CB type I cells, suggesting the capacity of these cells to make H_2S. A related talk by R.S. Fitzgerald in the afternoon session also highlighted a potential role of H_2S in cat CB function. The laboratories of Evans and Peers discussed further evidence supporting a role of AMP kinase (AMPK) in mediating the hypoxic inhibition of CB O_2-sensitive K^+ channels using recombinant channels expressed in HEK 293 cells. They reported that both BK_{Ca} and TASK-3 (but not TASK-1) channels were modulated by an AMPK activator (AICAR), consistent with a role of AMPK in chemotransduction.

Near the end of the morning session we heard an interesting lecture by J. Ward (joint with M. Evans) on "Mechanisms in Hypoxic Pulmonary Vasoconstriction". This was a succinct and lucid presentation (see Invited Article) that managed to summarize the disparate views and theories on the mechanism(s) of pulmonary vasoconstriction. There was a nice juxtaposition, and critical evaluation, of the three competing theories, i.e. the redox hypothesis, the ROS hypothesis (notwithstanding the controversy as to whether ROS increase or decrease), and the AMPK hypothesis. The afternoon session featured another interesting talk by J Fandrey on "Images of

Cellular Oxygen Sensing" (see Invited Article by Fandrey and Gassmann). This talk featured beautiful high-resolution fluorescence images that tracked the movements of fusion proteins of tagged HIF-1α and HIF-1β to the nucleus. Using FRET analysis the authors analyzed the assembly of the HIF-1 complex by determining protein-protein interactions in living cells. This powerful technique is certain to play a key role in our understanding of HIF-regulated gene expression in O_2 sensitive cells. Some of the closing talks in the afternoon session returned to neuromodulation in the CB. These included a talk from the laboratories of C. Gonzalez, A. Obeso, and E. Monteiro on the important role of adenosine as a CB neuromodulator (presented by S. Conde), and a comprehensive review of this topic appears as an Invited Article in this volume. Additionally, two talks from the laboratories of Shirahata (presented by A. Igarashi) and Nurse addressed the role of GABA and $GABA_A$ receptors in mediating sensory inhibition in the cat and rat CB respectively. Space constraints do not allow me to consider all the talks and posters represented at the meeting, and I apologize for the unavoidable omissions.

The scientific session on the last day was followed by the ISAC business meeting where it was decided that next ISAC conference/symposium will be held in 2011 in Hamilton, Ontario, Canada. As host, I wish to formally extend a warm invitation to all current and future ISAC members to attend this meeting.

Index

A

Acetylcholine (ACh), 19, 22–24, 33, 122, 137, 138, 141, 153, 154, 178, 179, 181, 182, 243, 250, 310, 409, 440
Acidosis, 19, 30, 62, 137, 138, 223, 370, 375, 382, 383, 387–389, 391–393
Acute hypoxia, 35, 40, 45, 83–87, 113, 114, 116, 118, 149, 154, 188, 201, 219, 220, 244, 246, 248, 299, 300, 303–305, 307–309, 329, 330, 332–334, 337, 340, 341, 345, 346, 348, 372, 374
Adenosine, 24, 114, 145–155, 161–167, 178, 250, 251, 409, 418, 450
  receptors, 145–150, 152–154, 166
  transport, 145
Adequate stimuli, 226, 227
Afferent activity, 31, 138, 249, 266, 423
Afferent fibers, 106, 223, 226, 253, 310
Aging, 191–194, 257–262, 265–270
Alveolar hypoxia, 351, 352
AMP-activated kinase, 39, 44
AMP-activated protein kinase, 40, 44, 45, 57–62, 89–91, 93, 351, 356–358, 449
Arterial blood pressure, 227, 229, 230, 265–267, 308, 319, 325, 330, 332
ATP, 19, 23, 24, 33–37, 39–41, 44, 45, 58, 87, 90, 100, 121–123, 137, 138, 141, 145–149, 154, 201, 250, 251, 351, 354, 355, 357, 358, 361, 362, 371, 404, 441
ATP-sensitive potassium channel (mitoK$_{ATP}$), 66, 361–366, 371

B

Benzodiazepine, 169–175
Bicuculline, 169, 171, 173, 174
BK, 35, 36, 49, 52, 65
Blood gases, 226–228, 319–326
Brainstem lesioning, 377

C

Ca^{++} imaging, 187, 188
Ca^{2+} channel, 36, 89–94, 102, 106, 108, 109, 111, 147, 153, 309, 348, 352, 353, 355, 449
Caffeine, 145, 149, 150, 152, 162, 165
Calcium, 22, 30, 32, 43, 44, 49–51, 83, 97–101, 105, 111, 121, 122, 147, 167, 244–246, 299, 302, 304, 351, 358, 371, 381, 416, 448
  channels, 97–101, 103, 105–112, 245
  responses, 190, 243, 245, 299, 302–304
  sensing receptor, 97, 98
  sensitivity, 51, 121
cAMP, 87, 113–119, 145–147, 149, 152, 153, 235–237, 239, 240, 309
Carbon monoxide (CO), 39, 41, 43, 44, 49–55, 65, 66–70, 89–94, 122, 123, 186, 192, 371, 449
Cardiorespiratory reflexes, 223, 224, 227–229, 231, 375
Cardioventilatory acclimatization, 329–334
Carotid body (CB), 1–16, 0, 19–27, 29–37, 40–45, 49, 57, 58, 61, 62, 65–67, 70, 71, 73, 74–81, 83, 84, 86, 87, 89, 90, 97–103, 105–112, 113–119, 121–123, 125–134, 137, 138, 145, 149, 150, 152–154, 161–167, 169–175, 177, 178, 180, 185–187, 191–194, 207–213, 215, 216, 219, 220, 223–231, 235, 236, 238–240, 243–252, 257–262, 265, 266, 269, 270, 273–279, 286, 289, 291, 294–296, 299–315, 320, 329, 330, 342, 346, 348, 354, 369–373, 403, 405, 444, 447, 448
  co-cultures, 29, 33
  slice, 29, 34, 35, 44, 251
Carotid chemoreceptor, 170, 215, 219, 278, 290, 295, 299, 300, 308, 309, 403

451

Carotid sinuses, 1–3, 5–9, 11, 12, 15, 23, 24, 30, 32, 138, 140, 141, 145, 149, 151, 162, 171, 173, 177, 179, 223, 224, 244, 250, 265, 266, 268, 270, 278, 289–291, 293–296, 300, 301, 372, 404, 405
Catecholamine (CA), 34–36, 44, 51, 97–99, 101, 102, 105, 138, 139, 153, 161–163, 165, 166, 226–228, 248–250, 262, 266, 269, 278, 319–326, 345–348, 374, 449
Cell type, 39, 40, 44, 100, 125, 126, 128, 133, 134, 185, 188, 189, 202, 308, 437
Central chemoreception, 377, 379, 382, 383, 395, 396
Cerebrospinal fluid, 389, 403, 405, 407
C-Fos, 370, 377, 382, 395–400, 404, 408, 449
Chemoreception, 9, 22, 40, 42, 45, 61, 70, 86, 97–112, 137–141, 145–155, 162, 177–183, 239, 243, 333, 374, 377, 379, 382, 383, 395, 396, 404, 448
Chemoreceptor cells, 73–81, 105, 106, 111, 125, 145, 149, 150, 152–154, 161–163, 165–167, 190, 215, 219, 220, 257, 377, 378, 383, 384, 403
Chemoreflex, 20, 24, 208, 213, 215, 216, 223, 230, 281–286, 290–293, 295, 296, 329, 372–374
Chemosensitive receptors, 411, 412, 421–425
Chemostimuli, 29, 33, 36, 89
Cholinergic, 22–24, 32, 34, 121, 122, 243, 244, 250, 310, 422, 444
Chromaffin cells, 185, 186, 345–348, 449
Chronic hypoxia, 22, 43, 81, 83, 86, 87, 105, 106, 109, 111, 112, 114, 118, 148, 154, 155, 191, 192, 194, 207–213, 220, 248, 258, 307–311, 369, 372–374, 448, 449
Chronic intermittent hypoxia, 154, 191, 193, 329, 330, 334, 337, 338
Chronic sustained hypoxia, 83, 84, 87, 154, 372
Ciclopirox olamine, 215–220
CO_2/HCO_3^-, 235, 236, 239, 240, 448
Complex II, 361–363
Correlation coefficient imaging, 377, 382, 387, 393
Cross correlation technique, 387
Cystathionine-gamma-lyase, 65–67, 70, 71
Cysteine, 49–55, 66, 372, 449

D
De Castro, 1–16, 20, 26, 447, 448
Denervation, 22, 36, 139, 257–262, 289–295, 370, 373
Development, 30, 32, 34, 37, 226, 230, 243–253, 341, 342, 352, 353, 444

Diazepam, 169, 171–174
Domperidone, 139, 265–270
Dopamine, 19, 22, 23, 26, 107, 114, 122, 137–141, 150, 153, 161–167, 178, 243, 249–252, 265–270, 273–278, 310, 323, 374
Dopamine agonists, 161–164
Dopaminergic, 23, 140, 141, 243, 249, 266, 269, 270, 310
DPPX, 73–81

E
Electron transport chain, 89, 93, 94, 314, 354–357
Environment, 21, 83, 202, 243, 246, 248, 249, 252, 300, 301, 374, 381, 411
Ergoreflex, 281–286
Erythropoietin, 197, 207, 208, 224, 261, 369, 373, 449

F
FIH-1, 197, 201, 202, 204
Fos, 403–409

G
GABA, 33, 103, 169–175, 444, 450
$GABA_A$, 169–175, 439–444, 450
  receptors, 169–175, 434–444, 450
Gene expression, 43, 75, 109, 125–135, 169, 173, 197, 198, 202, 204, 212, 216, 235–240, 252, 258, 262, 369, 372–374, 427–429, 434, 450
Glomus caroticum, 1–5, 7–9, 11, 15, 197
Glomus cells, 19–24, 40, 43–45, 50, 65, 89, 90, 97, 103, 121, 122, 126, 137–141, 162, 169, 172, 177, 178, 180, 182, 183, 185–190, 194, 208, 213, 216, 220, 243–251, 253, 258, 259, 261, 299, 300, 302, 304, 305, 307–311, 313, 314, 354, 371
Glucose, 31, 33, 36, 37, 59, 85, 90, 98–100, 107, 115, 163, 173, 236, 273–276, 279, 302, 354, 389, 403–409, 440

H
HbO_2 saturation, 319, 324
Heart failure, 24, 154, 223, 225, 230, 231, 281, 282, 411
Heart rate, 11, 148, 227–229, 265, 267, 268, 285, 289, 296, 371
  variability, 329–334
HEK, 49, 50, 52, 65–68, 70, 71, 293, 449
Hemeoxygenase (HOs), 37–41, 43, 44, 49, 50, 89, 90, 309, 369, 371, 373, 449
Heymans, J.-F., 1–16

HIF-1α, 37, 148, 149, 198–204, 208, 212, 213, 215, 216, 220, 257–259, 261, 262, 311, 314, 315, 369, 373, 374, 449, 450
High PCO, 121, 122
High-threshold Ca channel, 195
Histamine, 177–183, 418, 425
  receptor, 177, 179, 182
Humans, 3–5, 24, 84, 151, 162, 225, 265, 266, 269, 273, 278, 279, 284, 285, 311, 325, 440, 444
Hydrogen sulfide, 65–71
Hypercapnia, 19, 21, 22, 29–31, 33, 137–139, 174, 223, 225, 226, 228, 236, 247–249, 251, 274, 281–286, 312, 320, 325, 337, 340, 370, 375, 380, 381, 395, 396
Hypercapnic ventilatory response, 337, 341, 342
Hyperoxia, 19, 24, 43, 225, 230, 243, 247, 252, 253, 261, 262, 273–278, 289–296, 299–305, 346, 357
Hypertension, 154, 223, 230, 231, 292, 307, 312, 319, 320, 325, 329–331, 333, 334, 352, 294–296
Hypoglycemia, 278, 279, 403, 409
Hypoxia, 19–23, 29–35, 39–45, 57, 58, 61, 81, 83–87, 89, 90, 93, 94, 97–99, 101–103, 105, 106, 108–112, 113–119, 121–123, 137–139, 145, 148, 149, 151–155, 162–166, 169–171, 173, 174, 178, 186–190, 191–194, 197–204, 207–209, 212, 213, 215–221, 223–226, 228–230, 240, 243–253, 258, 261, 262, 266, 269, 270, 274, 276–279, 299, 300, 302–305, 307–315, 319–326, 329–334, 337–343, 345–348, 351–358, 361, 369–374, 396, 427, 428, 432–437, 439–442
  sensitive K^+ channel, 439
Hypoxia inducible factor-1 (HIF-1), 197–204, 207, 208, 212, 213, 307, 311, 314, 357, 373, 374, 450
Hypoxic pulmonary vasoconstriction, 44, 352, 361
Hypoxic sensitivity, 215, 218, 219, 248, 250, 307, 309, 310
Hypoxic ventilatory response, 33, 174, 213, 215–218, 220, 251, 278, 330, 334, 337, 342, 369, 370, 372–374, 449

I
IBMX, 113, 114, 116–118, 237, 239
Immunohistochemistry, 65, 67, 70, 97, 99, 100, 169–172, 191, 192, 377, 387, 389, 395, 406, 413, 429, 441
Intercarotid gland, 5, 7

Intermittent hypoxia, 154, 191, 193, 207–213, 230, 248, 252, 300, 307, 308, 311–315, 319–326, 329–334, 337–343, 345–348, 448, 449
Intra-acinar arteries, 361–364, 366
Intracellular calcium, 30, 32, 97, 98, 147, 244–246, 299, 302, 304
Ion channels, 21, 39, 57, 62, 65, 98, 147, 226, 246, 247, 249, 307, 309, 358, 369, 371, 449
Iron, 39, 43, 52, 198, 215–220, 371
  chelation, 215–220
Isolated brainstem-spinal cord preparation, 377–383, 387, 389–392, 449
Isolated glomus cells, 138, 246, 299, 302

K
K^+ channel, 34, 37, 39, 40, 42, 43, 45, 57–62, 86, 87, 89, 90, 147, 152, 153, 167, 246, 247, 251, 309, 352, 371, 427–429, 431, 433, 434, 436, 437, 439, 440, 442, 448, 449
KCN, 49, 52, 53, 65, 67, 69, 70
Kölliker-Fuse, 395, 397–399
Kv channel, 351, 355, 427, 432, 434, 435, 437
Kv4 channels, 73–81

L
Last 50 years, 19, 25, 26
Lateral parabrachial, 395, 397–399
Leak K^+ channel, 45, 57, 58
Lectin, 185, 186
L-type, 89–94, 99, 102, 103, 105, 106, 108–111, 153, 352, 353, 355, 371, 449
Lung reflex, 411, 418

M
Marginal glial layer, 377, 381, 383, 384
Marker, 20, 185, 186, 188, 190, 282, 387, 389, 392, 413, 442, 443
Mathematical equation, 125, 126, 128, 133, 134
Maxi K channel, 49, 65, 226
Mechanoreceptors, 411, 412, 417, 418
Mibefradil, 105, 108–112
Microganglia, 1, 4, 7, 8, 20
Mitochondria, 8, 89, 93, 94, 192, 194, 257, 258, 261, 262, 307, 351, 354–358, 361–363, 366, 404
Modification, 15, 111, 125–134, 141, 152, 171, 174, 200, 201, 220, 247, 321, 325
Muscle metaboreflex, 281
Mutagenesis, 49, 50

N

NADPH oxidase, 37, 39–42, 89, 93, 94, 310, 314, 427, 428, 430, 433, 434
Neonatal rat, 84, 87, 252, 299, 313, 345, 346, 377, 378, 381–383, 387, 388, 392, 429, 435, 449
Neonates, 229, 307, 314, 345, 346, 370, 434
Neuroepithelial bodies, 39–42, 44, 412, 414, 418, 426, 427–429, 433–437, 439–444, 448
Neuroglobin, 191–194
Neurotransmitters, 29, 30, 33, 36, 40, 58, 83, 90, 97–99, 101, 103, 105, 106, 122, 137–141, 145, 146, 153–155, 161, 162, 164, 166, 167, 174, 175, 208, 243, 244, 249–252, 261, 307–310, 314, 348, 404
Nicotinic acetylcholine receptors, 439–441
Nitric oxide (NO), 32, 34, 66, 138, 191, 192, 194, 202, 284, 309, 310, 355, 372, 403–409
Nitroprusside, 403, 405–407
NOS-1, 257–259, 261, 262
Nucleotidase, 145–148

O

O_2 sensing, 39–45, 49, 50, 58, 101, 105, 113, 114, 119, 167, 185–190, 197, 198, 201, 202, 244, 300, 307, 308, 355–358, 369–375, 427–437, 448
O_2 sensitive K^+ channels, 37, 42, 45, 57–62, 105, 309, 427–429, 433, 434, 436, 437, 448, 449
O_2 sensor, 37, 40, 42, 44, 167, 200–202, 204, 244, 352, 354, 355, 358, 371, 372, 427, 428, 436, 437, 448
Obstructive sleep apnoea, 209, 230, 319, 320, 329, 330, 337
Optical recording, 377, 382, 387–390, 448

P

Parapyramidal region, 382, 383, 395, 396, 398, 399
Patch clamp, 7, 29, 30, 34, 57, 66, 74, 83, 84, 89, 90, 97, 185, 186, 246, 440
PC-12 cells, 121–123, 185, 186
Peanut agglutinin (PNA), 185–190
Percentage, 79, 85, 93, 118, 125–128, 133, 134, 163, 224, 238
Perinatal, 299–305, 428
Peripheral arterial chemoreceptors, 235, 239, 240, 243, 252, 289, 307, 308
Petrosal ganglion neurons (PGNs), 23, 30, 101, 121–123, 137–139, 150, 177, 250, 251, 308, 309

Petrosal neurons, 29–31, 33, 36, 301, 310
Phosphodiesterases (PDE), 113–118, 146, 147, 237
Physiology, 2, 9, 14, 19–27, 34, 204, 413, 419, 448
Plasma norepinephrine, 319, 325
Plasticity, 194, 225, 227, 231, 252, 307, 312, 337, 338, 341–343, 369, 372, 374, 448
Plethysmography, 215, 216, 331, 337, 339
PO_2, 29–32, 35, 36, 73, 85–87, 115, 137, 162, 163, 171, 178, 183, 200, 224, 225, 227, 243–246, 258, 262, 267, 300, 308, 310, 319, 321, 324, 325, 329–331, 343, 345
Polyamine, 97–103
Postsynaptic, 31, 123, 139, 145, 152, 154, 162, 244, 249, 251, 252
Potassium channel, 49, 50, 65, 66, 70, 83, 361–363, 431
Pre-Bötzinger complex, 369–371, 373, 377–380, 392, 395, 396
Presynaptic, 31, 33, 140, 145, 146, 152, 154, 162, 167, 183, 247, 305
Pro-inflammatory cytokines, 412, 421, 422, 425
Prolyl hydroxylases (PHD), 197, 199, 201, 202, 204
Pulmonary neuroepithelial bodies, 42, 414, 427–437, 439–444, 448
Pyrilamine, 177, 179, 181–183

Q

qRT-PCR, 125, 126, 128, 133, 134, 235, 236

R

Rabbit, 23, 45, 73–81, 106, 111, 137–141, 149, 150, 152, 153, 171, 209
Ratio, 39, 41, 44, 45, 58, 66, 87, 90, 102, 125–135, 187, 200, 258, 261, 330, 333, 334, 351, 355, 357, 358
Reactive oxygen species (ROS), 39, 40, 42, 89, 93, 94, 191, 194, 257, 262, 307, 313, 314, 343, 351, 354–358, 361, 418, 449
Receptor, 21, 29–31, 33, 36, 97, 98, 100, 139, 141, 148–150, 152, 153, 161, 166, 169–175, 177, 179, 182, 183, 198, 207, 208, 212, 213, 224, 228, 229, 251, 252, 310, 314, 373, 412–418, 422–425, 428, 439, 441–444
Recurrent apneas, 307, 311–314, 337, 345, 346
Reflexes, 1–3, 5, 7–9, 11, 14, 15, 19, 25–27, 29, 40, 223–231, 243, 252, 281, 282, 286, 295, 307, 308, 320, 330–332, 338, 341, 362, 372, 374, 375, 400, 403, 404, 407, 409, 411–419, 421, 422, 425, 448

Research, 2, 3, 9, 13–15, 19, 21, 39, 40, 145, 154, 224, 388, 422
Respiratory control, 169, 282, 286, 338–343, 370, 377, 378, 388
Respiratory frequency, 11, 154, 216, 265, 269, 289, 290, 293, 331, 339, 379–381
Respiratory minute volume, 265, 267, 268
Respiratory pattern, 387, 388
Respiratory plasticity, 337, 338, 341–343
Respiratory responses, 24, 169, 174, 248, 281–285, 293, 400
Respiratory rhythm, 321, 372, 373, 377–380, 387, 388, 393, 395, 396
Retrotrapezoid nucleus, 374, 375, 395, 396, 398, 399
Ro 20-1724, 113, 116–118
Rolipram, 113, 116–118
Rostral ventrolateral medulla, 369, 370, 373, 377, 381–383
RT-PCR, 65, 67, 70, 71, 75, 97, 99, 100, 105, 109, 111, 169, 170, 172, 174, 429–432, 439

S
Sarcoplasmic reticulum, 351, 355
Sensory long-term facilitation, 307, 312
Sensory neuron, 8, 20, 24, 248, 411, 421, 424
Sensory receptors, 3, 374, 412, 416, 418, 421, 422
Single unit, 246, 299–305, 421–423
siRNA, 43, 73–75, 79–81, 427, 432, 434–437
Sleep, 25, 154, 169, 207–209, 213, 223, 225, 230, 231, 243, 312, 319–321, 329, 330, 337, 341, 342, 375, 400
Sleep apnoea, 154, 209, 223, 225, 230, 231, 312, 319, 320, 329, 330, 337, 341, 400
Soluble adenylyl cyclase (sAC), 235–240, 448
Spermine, 97–103
Splice variant, 89, 90, 198
Stable transfection, 65, 432
Succinate dehydrogenase (SDH), 37, 361–363, 366, 367

Superoxide anion, 39
Synaptic junctions, 257–259, 262

T
TASK channel, 57, 59
TASK-like, 45, 58, 61, 83–87
Tetraethylammonium (TEA), 73, 81, 85, 90, 100, 245
Tidal volume, 24, 216, 265, 267, 268, 290, 293, 308, 331, 339
TNF-α, 418, 421–425
Transmitters, 19, 22, 23, 84, 97, 137–141, 153, 177, 178, 183, 249, 251, 307, 310, 348
T-type Ca channel, 105–112
Type I cells, 29–36, 40, 42, 45, 57, 58, 61, 62, 83–87, 106, 118, 125, 126, 137, 138, 188, 207, 208, 213, 223, 226, 235, 240, 244, 258, 308, 311, 449
Type II cells, 30, 34, 125, 126, 138
Tyrosine hydroxylase (TH), 34, 35, 78, 99–101, 141, 150, 163, 185–188, 190, 208, 219, 248, 310, 369, 374

U
Upper cervical inspiratory neuron (UCINs), 387, 388, 393

V
Vascular endothelial growth factor (VEGF), 149, 208, 257–259, 261, 262, 307, 311
Ventilation, 23, 24, 29, 30, 84, 145, 150–153, 162, 166, 169, 174, 215–218, 220, 223, 225–230, 251, 262, 265, 266, 269, 270, 274, 281, 282, 284–286, 289, 290, 295, 296, 308, 309, 337, 339–343, 351, 362, 370, 372, 374, 393
Ventilatory chemoreflex, 20, 208, 213, 281–286
Ventral medulla, 377, 379–384
Ventral respiratory column (VRC), 382, 395–400
Ventral respiratory group, 377, 383, 392, 396
Voltage imaging, 377, 382, 387, 388
Voltage-sensitive dye, 377, 382, 387, 389